Energy:

Demand, Conservation, and Institutional Problems

Proceedings of a conference held at
the Massachusetts Institute of Technology, February 12–14, 1973,
by the Massachusetts Institute of Technology Energy Laboratory,
under the National Science Foundation RANN Program Grant GI–36476.

The MIT Press Cambridge, Massachusetts, and London, England

Energy:

Demand, Conservation, and Institutional Problems

edited by Michael S. Macrakis

This book was designed by The MIT Press Design Department. It was set in Linotype Times Roman, printed on Fernwood Opaque, and bound in G.S.B. S/535/13 "Coppertone" by The Colonial Press Inc. in the United States of America

Library of Congress Cataloging in Publication Data
Main entry under title:

Energy: demand, conservation, and institutional problems.

Proceedings of a conference held at the Massachusetts Institute of Technology, Feb. 12–14, 1973.
1. Power resources—United States—Congresses. 2. Interindustry economics—Congresses. 3. Energy conservation—United States—Congresses. I. Macrakis, Michael S., ed. II. Massachusetts Institute of Technology.
HD9545.E58 333.7 74–2257
ISBN 0–262–13091–2

Foreword

In his introductory remarks Dr. Paul Craig of the National Science Foundation briefly described the origins of this conference. During the summer of 1972 the National Science Foundation (NSF) held an energy systems workshop that was attended by grantees and prospective grantees of RANN (Research Applied to National Needs). At that meeting, in addition to the substantive material discussed, two conclusions emerged; first, that NSF was then sponsoring a large fraction of the total ongoing research in energy systems, and second, that there was a great need for mechanisms that could bring together the people working in the energy systems area. This resulted in the MIT Energy Laboratory's acceptance to organize the conference whose proceedings are presented here.

The past year has shown us that energy sources (and natural resources) play a key role in an economic system, and that they can no longer be treated merely as subsidiary inputs to capital and labor in determining production. This transitional period in energy's role is bound to produce a series of erratic shortages of forms of energies (at accustomed prices) as each fuel, in turn, is faced with the problems of resource development, environmental effects, and changing consumption patterns. Further complicating the energy issues are the many intricate, intangible ramifications, such as changing price patterns reflecting costs not previously included, dependence on politically controlled fuel sources, and the wise long-term utilization of nonrenewable resources. The problem that emerges is the handling of our energy feast by developing incentives and market forces that will increase effectiveness of utilization, as well as increase supply, so as to make the least detrimental impact upon the quality of our lives.

Basically, there are three, complementary methods of attacking this problem—effective and farsighted planning methods, energy economics efficiently reflecting the resource and environmental costs, and well-planned technological research on both production and consumption methods. All were stressed at this conference. All will require vigorous support, because, as past energy practices change and at the current rates of consumption, small percentage disruptions will produce large energy shortages; and these are likely to develop quickly, triggered by many interacting and (to a degree) uncontrollable factors.

Recognizing this challenge, I support Dr. Craig's expressed hope that this conference become the first of a continuing series. And like this conference,

I hope each would be directed toward the coordination of efforts and the dissemination of contributions of energy specialists from universities, industries, governmental agencies, and various nations.

David C. White
Director, MIT Energy Laboratory, and Conference Chairman

Cambridge, Massachusetts
January 1974

Introduction

The central issues of energy may be summarized as follows:

How quickly, with what set of energy sources, and at what prices can a supply of energy be secured which satisfies environmental and safety constraints while accommodating prescribed economic growth paths?

In simpler terms one can visualize an economic system such as the one of the United States which, while weakly interacting with the economic systems of the rest of the world, attempts to plan its destiny in terms of projections of graphs such as those found in *Economics* by Paul A. Samuelson (9th ed., McGraw-Hill Book Company, New York, 1973, p. 745). In order to plan, one would have to analyze the classical economic variables such as capital and labor, and more recent refinements such as natural resources, as well as the "intangibles": institutional and educational forms, political climate, and so forth. In recent years there has been a great debate going on here and abroad as to whether there exist adequate, low-cost resources to allow for such planning. Even if the resources exist, their utilization must still be free of serious health hazards, environmental degradation, and, in general, of irreversible effects. As, however, the growth paths of many nations follow the patterns set by the early industrializers—the United States, Britain, and France—a new dimension is added to the proposed analysis: other things are *not* equal, and regional economic systems are strongly coupled with one another.

With this coupling effect comes the realization that not all resources can be imported freely. The resource of greatest concern is energy, both because of its necessity in every possible economic activity and because of the irremediable effects that its use may have. In the United States in particular, the costs of production of energy seem to be high compared with costs in other production areas. The issues therefore center on the development of alternative sources that in some long period will eventually reflect a worldwide equilibrium price which can be comfortably absorbed in the budget of *all* nations. Of course, there is a long list of possible alternative energy sources. But there is uncertainty whether the items in the list will be available on time, at reasonable prices, and with low environmental risks. As with the problems of disarmament, where the dynamics of developing more productive weapons and weapon systems also provides the intellectual basis for their limitation and control, so is it perhaps with energy resources and their conservation. Thus, it is not inconsistent to pursue development while considering other possibilities that will take care of either interim or long-term situations if the supplies for one or the other reason do not materialize.

The traditional path in these uncertain circumstances is to consider conservation as an alternative to increases in supply. The subject matter of this MIT/NSF conference is conservation. To understand the theme of conservation, however, one has to delve into behavioral patterns and demand management in the energy-utilizing society, into issues of growth and externalities, into the technological possibilities for increased productivity of the energy resources, and into the possibilities of substitution of the energy inputs in the economy with, say, capital, labor, or both. Additionally, one must inquire into the institutional structures that will be needed to steer the citizens to modes of conserving energy that may entail temporary or permanent changes in the economic rhythm.

It is with these issues in mind that the papers presented at the MIT/NSF conference of 1973 are now organized in these proceedings. This volume contains thirty-five of the papers presented at the conference in addition to three of the four invited papers. Considering that stringent length restrictions were imposed on this volume and that many authors presented unusually lengthy manuscripts, many excellent papers are represented only as abstracts in the Appendix (A-1–A-28).

The rules that we followed for selection of the papers were the following: papers that appear in other publications are generally omitted, but we have noted in the abstracts where the material might be found. All the papers addressed to solar energy are included in their abstract form. With the help of reviewers the editor has retained those of the remaining papers that contribute to the coherence of the manuscript. For errors of judgment, however, the editor is, of course, solely responsible.

In Part I the proceedings are introduced by the invited paper of Koopmans (Chapter 1), in which he describes the way resources and energy come into economic thinking. First he reviews economic growth models, the methodology for optimization, and the concept of discounting. Then he illustrates these with a capital accumulation model and a model constrained by exhaustible resources entailing a finite survival time. These growth models, however, do not reflect technological change and the phenomenon of substitution of capital and labor for resources. Koopmans argues with some urgency that even though past trends indicate strong substitutability and relative price stability, future possibilities of substitution, particularly in energy, can be perceived only with the help of technologists and can be examined only through continued improvement of estimates and of the methodology for estimation. He then refers to available techniques for estimation under uncertainty (see Chapter 9, for example) and concludes by stating that if all options meet with failure then his earlier abstraction of exhaustible resources model has applicability. Following Koopmans's paper, Nordhaus (Chapter 2) discusses the current system of pricing and allocation of scarce appropriable stocks of natural resources and finds it wanting in view of the lack of active (resource) markets in futures and insurance.

Before proceeding to the next paper we will provide a "back of the envelope" estimate of the effect that energy prices may have on the Gross National Product (dollar GNP) Q. The accounting identity for the economy can be written in the form

$$Q = p_e q_e + \mathbf{p}_c \cdot \mathbf{q}_c, \tag{I-1}$$

where p_e is the price and q_e the quantity of energy in final demand while $\mathbf{p}_c$, $\mathbf{q}_c$ are the vectors of price and quantities of all other consumption goods. The phenomenological connection between GNP and total energy $q_e{}^t$ can be written in the form

$$Q = q_e{}^t/\beta, \; q_e{}^t = q_e + q_e{}^i, \tag{I-2}$$

where $q_e{}^i$ is intermediate energy consumption and where β^{-1} is the productivity of energy in the economy; β is a function of some as yet not fully identified economic activity (perhaps related to the thermodynamic efficiency of energy utilization) and has varied historically between 100,000 Btu per GNP dollar to some 85,000 Btu/GNP of late. We introduce Equation I-2 in Equation I-1, solve for Q, and differentiate the resulting expression with respect to the price of energy. Disregarding the $\mathbf{p}_c$, $\mathbf{q}_c$, and β dependence on p_e, we obtain

$$\frac{\Delta Q}{Q} \approx -\frac{\Delta p_e}{p_e}\frac{p_e\beta}{1 - p_e\beta} \approx -\frac{\Delta p_e}{p_e}(p_e\beta), \qquad (p_e\beta \ll 1). \tag{I-3}$$

In 1973 the price of energy was about 85¢ for 1 million Btu. Assuming that productivity for energy was about 84,000, it follows that $\beta p_e = 0.0714$; as a consequence, a consumer price increase of fuels by, say, 30% would imply a decrease of GNP of about 2.2%, ceteris paribus. Clearly, there is justified obsession with the dissection of β to find its origins as well as an apprehension about possible fuel price increases. We are, of course, referring here to a short-term situation where the impacts of technological, structural, and substitutional changes have had no time to correct and adjust the economic trajectory to a specified path. Traditionally, capital has been substituted for labor; can we now expect substitution of capital and labor for energy in more than just isolated cases? Presumably, any policies that are designed to increase the energy utilization efficiency would have rather important effects, however achieved. On the structural side we expect that by improving the flow of information about technological possibilities we may, through intelligent policies achieve a modicum of energy demand management.

To analyze these problems more naturally, Berndt and Wood (Chapter 3) introduce energy as a separate input in the economic production functions in an attempt to give some interpretation of β. In this paper the authors show that β has an economic interpretation in the context of a complete macroeconomic model of the economy only when capital K and labor L inputs are weakly separable from energy E and other inputs M in the production function $F(K, L, E, M)$. Changes in the ratio will depend upon nonneutral changes in technology and may depend upon changes in the relative prices of all inputs. The implications of these results for developing macroeconomic models that include the demand for energy inputs, and the associated economic accounting structure, cannot be overemphasized. Of primary importance for policy implications is the study of elasticities of substitution between capital and labor as well as between capital on the one hand and labor and energy on the other. If, say, this elasticity is indicative of complementarity rather than substitutability, then important policy implications about conservation and capital formation (the $\dot{K}$ of Chapter 1) may be derived.

Based on a more extensive version of the work described in Chapter 3, Verleger (Chapter 4) attempts to evade the known limitations of input-output and national income account models by developing a flexible input-output-like production function that contains easily identifiable energy and resource sectors; this he then couples to macroeconomic models that provide the dynamics of demand, investment, and so forth. He uses the model to compute demand and consumption projections.

Baughman (Chapter 5) then describes his model, which relies heavily on interenergy substitution possibilities and supply cost calculations. It is a generalized behavioral model with zero price elasticity for demand. With little coupling to the macroeconomic dynamics, it nevertheless describes the past behavior of the energy system well and is believed to be capable of examining medium-term alternative energy policies with some confidence. The model is used here to evaluate the effects of oil import policy and gas regulation on system behavior. In view of the implications of Equation I-2, the changes in these policies that can significantly affect the price, supply, and consumption trends of all forms of energy become critically important, at least for the short term. The potential impact of these policies on the energy outlook for the year 1985 is illustrated through the results of a set of simulations. A discussion of the implications of these policies for development of shale oil resources and coal gasification techniques at current projected costs is included.

Schweizer, Love, and Chiles (Chapter 6) discuss a computer model for examining the supply, demand, and price of primary energy sources and electricity subject to assumptions about future policies and technology changes. Deam and Leather (Chapter 7) describe their efforts to adapt linear programming optimizing techniques used in large international oil companies to the world petroleum modeling while Ward (Chapter 8) gives an overview of the international oil scene.

Part II begins with a paper by Manne (Chapter 9), who deals with electricity-generating investments under uncertainty with respect to the date of commercial availability of breeder nuclear reactors. His planning horizon extends from 1985 through 2025 at five-year intervals while the United States is disaggregated into six regions—each with a distinct load-duration curve and a fossil fuel cost. His computations are based on sequential probabilistic linear programming. There follow four papers that analyze energy issues through the input-output (I/O) tables. The standard I/O, augmented with environmental and resource usage variables, is used by Just (Chapter 10) to establish comparative national economic and environmental impacts of high- and low-Btu coal gasification and the gas turbine topping cycle during the 1980 to 1985 time period. Just projects for 1985 alternative high-, medium-, and low-energy growth futures both with and without these new energy technologies. His results illustrate the high sensitivity of capital investment to the rate of energy use growth and the possible aggravation of the situation by the introduction of high-Btu coal gasification. Additionally he indicates economic mechanisms that will help hold total capital investment within its historical bounds as a percentage of GNP.

In the next paper Istvan (Chapter 11) shows how exponential growth in demand for electric energy has presented the utility industry with severe capacity problems while fundamental technical changes in the industry and new environmental restrictions are affecting the feasible set of investment alternatives. He uses a generalized dynamic I/O model with time-lagged investment to explore interindustry impacts of utility investment strategies including nuclear generation, pumped storage, and pollution control equipment on fossil-fired units. He concludes that increasing the proportion of nuclear generating capacity may have serious macroeconomic implications. Herendeen (Chapter 12) argues that I/O analysis offers a theoretical framework and a large data base well suited to calculating the total energy costs of many classes of products. After describing his methodology for introducing energy into the I/O tables and assessing the limitations of the technique, he presents results on transportation and agriculture which highly reveal the direct and indirect energy intensity of these sectors. Folk and Hannon (Chapter 13) at the Center for Advanced Computation in the University of Illinois describe their efforts to develop a large national linear I/O policy model that will be accessible through the ARPA Network. Their model contains detailed economic activities showing industry demand generated by expenditure categories so that national budgets, individual lifestyles, or scenarios can be evaluated. Energy-output matrices in 367-sector detail contain demand for coal, crude oil, refined petroleum, electricity, and natural gas implied by scenarios while other matrices estimate employment effects and pollution components.

Part III on externalities begins with the invited paper by Dorfman (Chapter 14) who discusses methods for costing pollution and enforcing controls. Dorfman describes the elements of the game theory of pollution control within a less than perfect information environment, when those playing the game are the consumer, the producer, and the administrator. With a knowledge of market discipline and its capability for management of supply and demand one could attempt to duplicate its mechanism in pollution control. Dorfman explores the difficulties of this procedure by examining four methods: pollution effluent limitation, limitation of product output, limitation of product input, and the imposition of taxes on pollution effluent quantities. He compares the implications of each method in the effectiveness of pollution reduction, in efficiency of enforcement, in the production of the least countermeasures on the part of the polluters, and so forth. He concludes that, in view of the difficulties in estimation of externality costs, the best methods still involve compromise, learning by doing, and continous improvement. Following Dorfman's paper, Chapman (Chapter 15) focuses on the interaction between electricity demand growth, likely control costs, and the sulfur emission tax. Using a simple demand function, he concludes that the tax would significantly reduce emissions and damage, have little effect on the electricity demand and nuclear power plant growth, and would result in more benefits than costs, with benefits even greater if the Clean Air Act is not implemented. Spore and Nephew (Chapter 16) then compare the benefits of surface mining defined to be the difference in cost between surface- and deep-mined coal, with the potential benefits from recreational, preferable use patterns in the Big South Fork basin area. While crucial to the analysis is the estimation of the present value of future benefits from the alternative land uses, whereby the influence of technological changes in coal mining are taken into account, they conclude that preservation is the preferred use of land.

Gordon (Chapter 17) then discusses the technological problems that have to be resolved before coal can become an environmentally acceptable fuel at competitive prices and concludes that in the long run there may be some hope for coal. Griffin (Chapter 18) examines the tax on sulfur emissions and concludes that, from a policy perspective, it is preferable to the present situation if the welfare gain from the reduction in sulfur oxide damage exceeds the welfare loss consisting of the pollution control costs, the loss in consumer surplus, and administrative costs. He attempts to measure the welfare gains and losses of the proposed tax as they relate to the electric utility industry, using an extended econometric model describing electricity demand, conversion efficiencies, and fuel input choices.

Habicht (Chapter 19) describes the involvement of the Environmental Defense Fund (EDF) in administrative, legislative, and legal action in Montana aiming to internalize some of the environmental costs of strip mining and coal-fired steam electric generation. In the last paper of this part Buchan (Chapter 20) examines the environmental impact of energy generation and places it in a societal decision-making context. Specific recommendations are made regarding: local referenda on power plant siting, the use of analytical models and establishment of ecosystem quality criteria, and legal reforms. In addition, a new kind of public advocate system is proposed to promote rational long-range planning on power plant construction questions.

In Part IV we have assembled papers that address questions of management of supply and demand of energy. The first two papers (Chapters 21 and 22) discuss the regulation-induced disequilibrium between supply and demand for gas while the third paper attempts to estimate supply costs for oil. Six remaining papers (Chapters 24 through 29) address specifically the questions of demand in electric energy. Stauffer and Jensen (Chapter 21) argue that the anticipated "gas gap" creates a new regulatory need for efficient and equitable curtailment of gas with interim rationing. The economically efficient rationing system would involve the hierarchical ranking of fuel switching costs and the preferential curtailment of the end uses of gas for which such costs are least. Ideally such a system would serve as a market surrogate helping an eventual transition to deregulation. Since a technical-economic survey of all gas usage is unworkably complex, an administratively simple approximation to an "efficient" rationing system is described. (Editor's afterthought: Would it not be useful to apply the same reasoning and techniques to conservation?) Pindyck (Chapter 22) discusses his research in the market structure of the natural gas industry in the United States so as to understand better the effects and problems of regulation of that industry. He shows that the industry is composed of distinct yet interrelated markets— the field market (producers selling gas to pipeline companies) and the wholesale market (pipelines selling gas to consumers). These markets have different spatial characteristics that in turn constrain their interaction. Excess demand for natural gas in the future (particularly on a regional basis) and the impact of alternative regulatory policies on that excess demand can best be understood by looking at the simultaneous interaction of these two markets. Pindyck concludes by giving some preliminary results from the MIT econometric model for gas which also support the gradual deregulation.

Steele (Chapter 23) argues that supply forecasts appear to be demand determined and that despite their availability, industry oil and gas statistics are difficult to analyze. He nevertheless undertakes a limited analysis and projection of future cost trends as they may be affected by governmental considerations.

Mount, Chapman, and Tyrrell (Chapter 24) attribute growth of electricity demand since 1945 to five factors: population, income, and the prices of electricity, substitute fuels (natural gas), and complementary products such as household appliances. The data used are annual observations for the 48 contiguous states from 1946 to 1970. They fit single equation models for three consumer classes (residential, commercial, industrial) with the quantity of electricity as the dependent variable and estimate both constant and variable elasticity models. They conclude that the demand for electricity is generally price elastic for all three customer classes and becomes increasingly so as prices rise; the well-known inelasticity with respect to income is also noted.

Cicchetti and Gillen (Chapter 25) then analyze the pricing of electricity and the cost of service, rate of return and the growth of electric utilities, the elasticity of demand and rate design, and inverted rates versus flat rates to see whether methods can be developed for the control of electric consumption for environmental protection. The paper by Tyrrell (Chapter 26) uses the techniques developed in the earlier paper which he co-authored (see Chapter 24) to conclude that growth in electrical demand may be slowed primarily by price increases.

Khazzoom (Chapter 27) then describes his modeling efforts to understand the Canadian energy system. He observes that in Canada, as elsewhere, the efforts focus on how to obtain more supplies to meet forecast demand while no one seems to worry about whether the public can afford the increased prices necessary to elicit the supplies, and he ruefully ends with the question whether it is really necessary to keep "chasing the demand curve" rather than directing funds and research to conserve energy.

Tansil and Moyers in Chapter 28 examine the growth of residential electricity use for the period 1950 to 1970 from the standpoint of increases in the number of households, appliance saturations, and the average annual electricity consumption per appliance. They derive growth patterns that illustrate the factors accounting for the increase from 1800 kWh per household in 1950 to 7000 kWh per household in 1970. They examine possible monetary and energy savings from the use of additional housing insulation for three climatic regions. In the last paper of this chapter Comerford and Michaelson (Chapter 29) describe the econometric model of an electric utility which will permit the investment decisions to be based on studies of electric demands for various patterns of economic growth, the effects of price elasticities on demand, and so forth.

Papers on transportation of energy and energy in transportation have been brought under Part V of the proceedings for logistic rather than logical reasons. Tankers for oil are elements of the supply, while automobiles, trucks, and airplanes do belong to the consumption end.

Zannetos (Chapter 30) discusses the economics of transportation of fuels and examines the substantial financial requirements for ocean transportation of energy for the coming decade while that of Hale and Deam describes efforts to incorporate transportation issues in the overall world energy models.

Malliaris and Strombotne (Chapter 32) discuss the structure of demand for transportation services and energy both historically and as projected to the year 2020: also the improvements and modifications to existing automobile and truck types that offer an opportunity to reduce relative energy consumption. They argue that advanced engine types are palliatives for subsequent years. For the long term, wayside power and advanced batteries may provide a way to reduce the dependence of surface transportation upon petroleum. The authors also discuss some nontechnological actions that offer substantial energy savings within the transportation sector. Carter and Tierney (Chapter 33) on the other hand discuss the effects of emission controls on oil demand.

In conclusion, Part VI is devoted to papers that explore some issues in the conservation of energy. The word "conservation" is bound to cause serious misunderstandings as it may have different connotations to different people and in different disciplines. Conservation refers to issues of perceived resource scarcity and to issues of intergenerational resource distribution; to aesthetic, environmental, and purely technological considerations; to issues of temporary adjustments to some production miscalculation, shift in technology or resource base; and to issues that run deep into methods for resource allocation, such as pricing versus rationing.

The question of conservation, then, is rather subtle and difficult. Permeating this conference, it has appeared under many guises. In a very generalized way, conservation may be seen through the eyes of growth economics. As such it first appears in Koopmans's exposition of economic growth (Chapter 1): both conservation and capital investment represent abstentions from consumption that affect the balance of benefits and costs between generations. Additionally, conservation can be thought of as the act or policy of preserving resources and the environment for our descendants and is, therefore, an integral part of the reasoning behind Dorfman's discussion of pricing externalities (Chapter 14). As might be expected, all the papers addressing questions of demand for energy and its price elasticity (Chapters 3, 4, 24, 25, 26, 27, and 38) are directed at the understanding of the workings of the energy sector within the economic system so as to implement conservation measures intelligently.

Part VI consists of papers that specifically address the physics and technology of conservation. Keenan, Gyftopoulos, and Hatsopoulos (Chapter 34) note that it is not energy (a quantity always conserved) but entropy that is the scarce resource. They then suggest that to maximize productivity while conserving this scarce resource, thermodynamic irreversibilities should be minimized. As an outstanding example where this is practiced, they cite desalination processes. With current technology, the reduction of irreversibilities in desalination equipment to save on fuel is achieved with large capital investments in equipment. Although an extreme example of substitution of capital and labor for expensive fuel, it is consistent with the basic tenets of economic theory. The question, nevertheless, remains whether the isolated allocation of capital to conserve energy, as exemplified by desalination, can be extended to all processes in the economy that require energy inputs without affecting economic growth patterns: an engineer's design possibility frontiers are always moderated by macroeconomic constraints.

Berg (Chapter 35), following the logic of the previous paper, attempts to identify areas where energy conservation, through effective utilization, could become economically meaningful, while Grot and Socolow (Chapter 36) describe their extensive fieldwork in their attempt to comprehend the patterns of energy utilization in residential communities. The analysis of these patterns along with the transportation system that joins the communities together, will undoubtedly tax the imagination and efforts of future planners. Finally, Berry, Fels, and Makino (Chapter 37) contribute thermodynamic calculations that assess the economics of conservation through recycling of steel used in the manufacture of automobiles. They also comment on the connection between physical thermodynamic valuation and economic valuation.

The connection of economics with thermodynamics has engaged the attention of some scientists (see for example, N. Georgescu-Roegen, *Entropy Law and the Economic Process,* Harvard University Press, Cambridge, Mass., 1971). No doubt, economists feel that the intermediate conceptual stages between thermodynamics of physical and biophysical processes and perception of utility are as yet difficult to trace. But, in view of the heavy burdens placed on analytical work in oxidative or mixing pollution, on resource extraction and processing externalities, on the elasticity of substitution between capital and irreversibilities, and other fields, it would seem that increasingly the economic abstraction of production (perpetual motion machine?) may have to be backed by physical concepts. One might expect to see a production function in the form $F(\mathbf{K}, S, t)$, where $\mathbf{K}$ stands for the generalized vector of capital and labor inputs $\{K_i, L_i\}$, S stands for entropy, and t for time. If so, one could split the output Q into consumption C and investment I, where now $I = \dot{K} - \dot{S}$. The evaporative term $K\delta$ in capital growth theory acquires a physical significance that can describe more than just the depreciation of embodied capital. For a stationary situation with a production function independent of time, the $F(\mathbf{K}, S)$ just proposed has been used in Chapter 3 in the form of $F(\mathbf{K}, E)$. It is a staple in the study of the microeconomics of electric utilities.

It would be interesting to examine the time changes of the production function with energy as an input and to determine the possible effects of reciprocity between energy inputs and technical change, perhaps detecting historical trends in substitution or complementarity of capital and energy. Berndt and Wood, in a recent paper ("Technology, Prices, and the Derived Demand for Energy," Office of Energy Systems, Federal Energy Office, Washington, D.C., February 1974), utilizing an advanced version of their earlier work (Chapter 3 of these proceedings), together with components of the modeling system of Chapter 4, conclude that the "Allen partial elasticity of substitution between energy and labor σ_{EL} is about .65, while that between capital and energy σ_{KE} is -3.5." These results do support the view that capital and energy are historically tied together and that since it is capital that is the investment tool for the economic growth, energy prices and price-elastic demand will, among other things, seriously affect the investment horizon.

The proceedings conclude with a paper by Adams, Foley, and Nielsen (Chapter 38) calling for energy conservation measures that seem to be necessary in view of the perceived international supply situation. They argue for an evolutionary, nondisruptive application of more energy-efficient technology in order to avoid "quick fixes and emergency rationing." Although they do not forecast the price of energy, they describe projected capital needs to ensure supply that is clearing price-inelastic demand. In its pragmatic form, this paper ends the proceedings with a style contrapuntal to the one in the first, more abstract, paper.

Acknowledgments
The conference as well as the proceedings could not have become possible without the support of many individuals. First in the list is Dr. Paul P. Craig of the National Science Foundation, who for the last several years has been imaginatively devoting considerable amounts of time to the issues of energy reflected in the title of the conference.

For chairing different sessions at the conference and for soliciting numerous papers, we thank Professors Morris A. Adelman, Paul W. MacAvoy, Herbert H. Richardson, and David C. White of MIT; Professor Hendrik Houthakker of Harvard University; Drs. Paul P. Craig and Paul F. Donovan of NSF; Professor R. J. Deam of the University of London; Professor Jesse Denton of the University of Pennsylvania; and Mr. David O. Wood of the Federal Energy Office. To those who have patiently reviewed abstracts and papers, as well as to those authors who were kind enough to shorten, revise, or expand their material, go many thanks.

All authors and the editor should consider themselves greatly indebted to the assistant editor, Dr. James Gruhl of MIT's Energy Laboratory, who was responsible for the accuracy, consistency, indexing, and other requirements of the manuscript.

Finally I would like to thank Professor David C. White whose early involvement in the field of energy and deep interest in the societal and environmental implications of the increasing energy utilization have contributed greatly to my education in this area. It goes without saying that his patience, generosity, and ideas added immeasurably to the conference and to the proceedings.

Michael S. Macrakis
Senior Research Associate, MIT Department of Electrical Engineering, and
Conference Co-chairman

Cambridge, Massachusetts
November 1973

I Aggregate Modeling

1 Ways of Looking at Future Economic Growth, Resource and Energy Use*

TJALLING C. KOOPMANS†

It is a great privilege to have been invited to take part in this interdisciplinary discussion. I should perhaps amend an impression that may have been given by the last remark[1] in the kind words of Dr. Macrakis. I am not now a physicist. I was just beginning to be one, and then I changed over to econometrics, and later to economic theory. Perhaps in this company I should apologize for my desertion, but that is what happened. Therefore, I do not and cannot claim expertise on the scientific and engineering aspects of the subject of energy. My comments are more in the nature of responses of an economist who seeks to perceive the outlines of resources and energy problems from the economic point of view. What help can economic thought and techniques give in formulating the problems and in specifying data needs?

Economists use models as much as scientists do. We have our counterparts of the ideal gas model and of the Bohr model of the atom. That is, we also find it helpful to make a simplified picture of reality by approximating assumptions. However, our models have one element that is not present in the work of scientists but does occur in work of engineers. This is the element of valuation, which introduces into the model some expression of a purpose or several purposes to be served. In what we call descriptive (or "positive") economics the purposes would be those of the firms and/or of the consumers. In optimizing (or "normative") economics one would adopt some social optimality criterion. Of course, in all this one applies what I might call the method of deliberate simplification—a method which is, I think, at least as necessary in economics as in the sciences. One leaves out the more diffuse and complicated aspects of what one suspects to be the true face of reality in order to trace the effect of a few more sharply etched traits.

Pertinent to today's subject are the "optimal" economic growth models. Mathematically speaking, the study of optimal economic growth is the economist's counterpart to the engineer's optimal control theory. In economics that topic has a history of its own. As far as I know, it started with a basic and seminal paper by Frank Ramsey in the *Economic Journal* of 1928

* This text is based on a transcript of the invited lecture as given, with some revisions.
† Yale University.

(Ramsey, 1928). This is the same Ramsey who wrote similarly fundamental articles in the foundations of mathematics, in subjective probability, and in optimal taxation. It was a great loss to economic science that a few years later he died at a relatively young age. What Ramsey did in regard to optimal growth could be described as a thought experiment, again a concept quite frequently used in the sciences. I am reminded of Einstein's imaginary experiment of an observer who would travel along with a light wave. As you know, this experiment had revolutionary consequences, some of which occupy us today. It was also an experiment conceived without regard to feasibility. Ramsey's thought has some of each of these traits on a more modest scale. I do think that his experiment has affected and will further affect economic thinking particularly with regard to long-range planning. As to its feasibility, I think that not only the simplification but also the optimization aspect fail to reflect many important realities of our present-day world. True, we have a theorem in economics saying that under certain assumptions optimality can be achieved equally well by perfect markets and by perfect planning. However, neither the markets we have nor the attempts at central planning in other countries, nor even the more universal modern technology, meet the premises of that theorem.

Nevertheless, I believe that it helps to think through under highly simplifying assumptions what would be involved in optimal use of resources over an extended period of time. One starts with an optimality criterion. Then one states constraints that represent just a few important traits of reality. If sense and skill are applied to these tasks, one gains some insight into what optimization by that criterion would require. One also obtains some estimate of the outer limits of what is feasible, a possible check on the performance of market or plan.

There is one other useful by-product. The procedure also yields implicit prices on the commodities that enter into the model, in the following way: assume that you are told that at some future time you will get from outside of the system one small extra unit of some commodity already occurring in the system. What does that add to the result of your optimization? How much does it raise the attainable maximum of the criterion? That number—to be precise, the first derivative of the maximum with regard to the availability of that commodity at that time—is defined as the then shadow price of that commodity. This concept is useful whether or not the shadow price is reflected in observed market prices.

I shall mention later that Ramsey was rather firm as to what was the right optimality criterion in one important aspect, the discounting of future benefits. Later work has branched out in several directions and has taken the form of trying out a variety of alternative optimality criteria, inspecting their results, and then making a judgmental choice of criterion on that basis. I will illustrate these ideas first with an aggregate *capital model* and then with a somewhat stark version of an *exhaustible resource model.* The capital model has been analyzed and exposited a good deal in the literature (see, for instance, Intriligator, 1971 or Koopmans, 1967), so I will be rather brief on that model. After going through these two examples, I will contrast the effects of discounting future benefits and costs in the two cases. Both models assume constancy over time of population, labor force, and technology for the pertinent future. In the capital model this is an indefinite future, in the exhaustible resource model a future limited to the period of survival. The optimality criterion is constructed along similar lines in the two cases. In the capital model it is given by

$$U = \int_0^\infty e^{-\rho t} u(c_t)\, dt, \quad \rho \geqq 0, \tag{1.1}$$

where

c_t = consumption flow at time t,
$u(c)$ = utility flow generated by consumption flow c,
ρ = rate of discounting of future utility flows.

The criterion recognizes the rate of consumption as the ultimate purpose of economic activity and evaluates any consumption flow by a score function (*utility function*) $u(c)$. This function registers a given virtual increment in consumption the more strongly, the lower the consumption flow to which it is applied (see Figure 1.1). The utility flow is then integrated over time, after multiplication by a factor of which Ramsey disapproved. This is the so-called discount factor for future utility. It is an exponential function of time with a negative coefficient in the exponent. This means that as you look further and further into the future you give less and less weight in your criterion to the utility score of consumption then taking place. Ramsey felt that this device "is ethically indefensible and arises merely from the weakness of the imagination." Another British economist, R. F. Harrod (1948), used even stronger language. He spoke of pure time preference—another

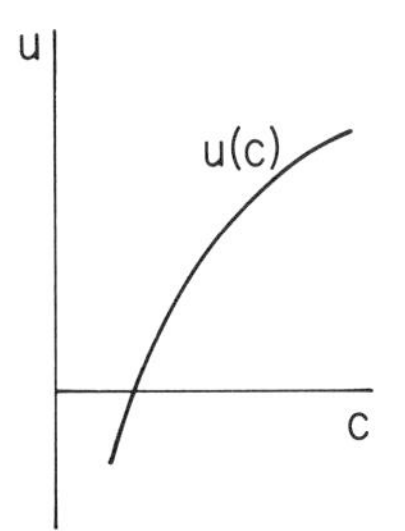

Figure 1.1. The utility of consumption.

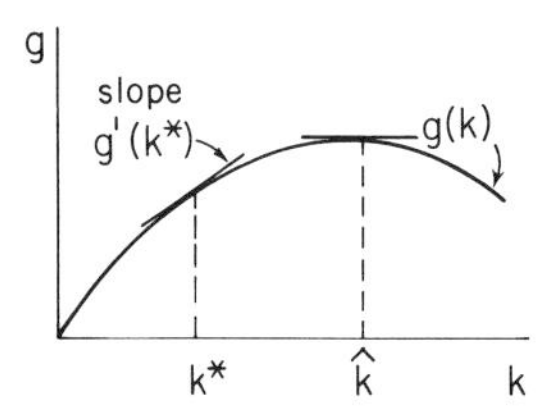

Figure 1.2. Output as a function of capital.

term for discounting of future utility—as "a polite expression for rapacity and the conquest of reason by passion."

The set of *feasible paths* of consumption and capital over time is defined by the constraints

$$c_t + \dot{k}_t = g(k_t), \qquad 0 \leqq c_t, k_t, \qquad 0 < k_0 \text{ (given)}, \tag{1.2}$$

where

$k_t \;\; =$ capital stock at time t, $\dot{k}_t = dk_t/dt$,
$g(k) =$ output flow if capital stock is k.

By Equation 1.2 the output flow $g(k_t)$ resulting at time t from best use of the capital stock k_t is allocated to two purposes, a consumption flow c_t and a net rate of increase $\dot{k}_t$ in the capital stock. The form given to the *production function* $g(k)$ in Figure 1.2 expresses that, due to the constancy of the labor force, successive equal increments in the capital stock entail successively smaller net increments in output. To glimpse the role of this function, we first ignore the given initial capital stock k_0 and look at an arbitrary constant-capital path of the form $k_t = k^*$, say. For any such path the ordinate $g(k^*)$ represents the corresponding constant consumption flow $c_t = c^* = g(k^*)$. The slope $g'(k^*)$ of the tangent to the curve $g(k)$ at $k = k^*$ is

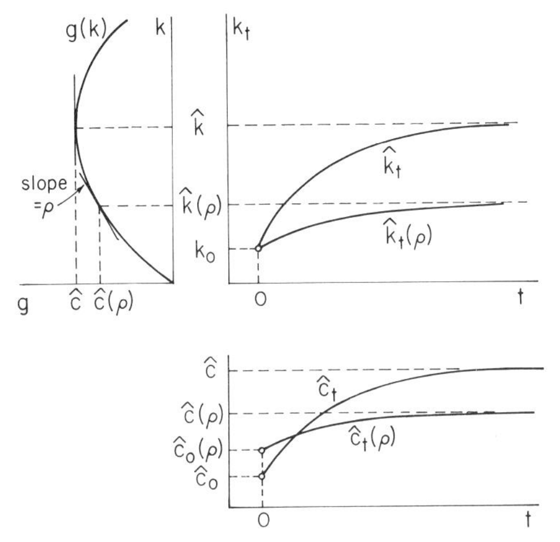

Figure 1.3. Optimal capital and consumption paths in the capital model.

something like a real interest rate. It measures the increase in the indefinitely maintained consumption flow made possible by the investment of a (small) unit increment to the capital stock k^* which thereafter is again held constant. Capital saturation occurs at $k = \hat{k}$, where $g'(\hat{k}) = 0$, and $\hat{c} = g(\hat{k})$ is the highest indefinitely sustainable consumption flow.

Now reimpose the given k_0, allow the capital stock to vary over time, and find those capital and consumption paths $\hat{k}_t(\rho)$, $\hat{c}_t(\rho)$ that maximize the criterion of Equation 1.1 subject to the feasibility constraints of Equation 1.2. Without reproducing the reasoning, let me give the characteristics of the solution shown in Figure 1.3 with time on the horizontal axis. Assume first that you do not discount ($\rho = 0$) and start with an initial capital stock k_0 substantially below $\dot{k}$. Then the optimal capital path $\hat{k}_t$ will initially rise rather steeply and then approach asymptotically the saturation stock $\hat{k}$ that was already identified in Figure 1.2. To show the connection, that figure has now been turned on its side to form the upper left diagram in Figure 1.3. On the other hand, if you discount ($\rho > 0$), then the initial rise is less steep, and your lower asymptote $\hat{k}(\rho)$ is read from the diagram at left by requiring that the discount rate ρ also be the interest rate $g'[\hat{k}(\rho)]$. The corresponding optimal consumption paths are shown in the lower diagram.
If you discount, optimal consumption initially exceeds that without discounting. Therefore, discounting favors the nearby future, but after a time the

two consumption paths will reflect the relative positions of the two capital paths. The reason is that, as all paths flatten out, it is the size rather than the change of the capital stock that determines the output flow, which in turn becomes the consumption flow.

To sum up, discounting favors the present generation. After a sufficient lapse of time it disfavors all subsequent generations about equally.

In the capital model resources do not occur explicitly, only implicitly in the form of the production function. In the exhaustible resource model there is no production, no capital in the sense of produced means of production, only a given stock C of nonperishable food (let us say). Consumption of that stock of food at a rate c_t (that may vary over time) is the only form of economic activity. The population lives at full strength until the stock is exhausted, at which time T life ceases.

The maximand is

$$U = \int_0^T e^{-\rho t} u(c_t)\, dt, \tag{1.3}$$

and the feasibility constraints are

$$\int_0^T c_t\, dt = C, \qquad c_t \geqq c_{\min} \quad \text{for} \quad 0 \leqq t \leqq T, \qquad c_{\min} > 0, \tag{1.4}$$

where

C = total stock of the exhaustible resource,
$c_{\min}$ = minimum consumption flow required for survival,
T = period of survival.

Here, C and $c_{\min}$ are given positive numbers. The survival period T is bounded by

$$0 < T \leqq \bar{T} = C/c_{\min},$$

but T itself is a decision variable.

In contrast with the capital model, population is therefore now also a decision variable, although in a very abrupt way. Once this is the case, one must specify a zero point in the utility scale for the problem to be well defined. (In contrast, in the capital model addition of a constant to $u(c)$ does not change the solution.) We somewhat arbitrarily choose

$$u(c_{\min}) = 0,$$

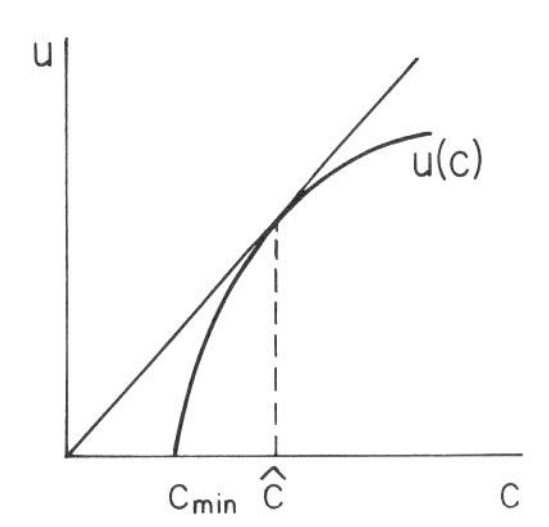

Figure 1.4. Optimal consumption in the resource model (without discounting).

expressing a view that life at the survival minimum does not rate as worth living. The utility function $u(c)$ is defined only for $c \geqq c_{\min}$, and has the shape indicated in Figure 1.4. Since life at the survival minimum does consume the resource, optimality does require a rate of consumption above the minimum. To explore this further, first consider again a constant consumption path at an arbitrary level $c_t = c^* \geqq c_{\min}$. The survival period then is $T^* = C/c^*$, and if $\rho = 0$ the maximand is $U^* = T^*u(c^*) = C[u(c^*)/c^*]$. One sees from Figure 1.4 that, if future utility is *not* discounted, the optimal constant path proceeds at the level $\hat{c}_t = \hat{c}$ corresponding to the point of tangency of $u(c)$ with a ray out of the origin. A simple variational argument[2] shows that this path is then also optimal among all feasible paths, constant or not.

In the case of discounting at a positive rate ρ, the same variational argument still shows that the then optimal path $\hat{c}_t(\rho)$ never dips below the value c that is constantly optimal for $\rho = 0$. However, as shown in the diagram and formula in Figure 1.5, the optimal path $\hat{c}_t(\rho)$ starts at a level $\hat{c}_0(\rho)$ and follows a descending curve such that the level $\hat{c}_T(\rho) = c$ is reached at that time $T = \hat{T}(\rho)$ at which the available stock C is exhausted.

Thus, with discounting the survival period is shorter. Those who live within that period consume no less than without discounting and live better accordingly as they live nearer the present.

This contrast between the effects of discounting on the intergenerational distribution of consumption in the two models can be understood as follows. In the capital model (with low k_0), the undiscounted optimal path favors future generations because (as long as $k_t < \hat{k}$) capital formation in the present permits a greater addition to consumption in the future than the

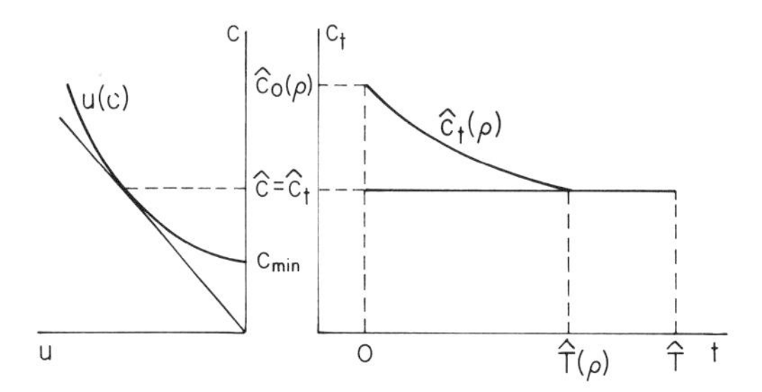

If $\rho > 0$, then $e^{-\rho t}\, u'(\hat{c}_t(\rho)) = u'(\hat{c}_0(\rho))$, and $\hat{c}_{\hat{T}(\rho)}(\rho) = \hat{c}$

Figure 1.5. Optimal consumption path in the resource model (with discounting).

sacrifice of present consumption it requires. This benefit from investment due to the nature of production with capital is diminished by discounting because the sacrifice precedes the benefit, and disappears when k_t reaches $\hat{k}(\rho)$. In the exhaustible resource model there is no net gain over the whole future from postponement of consumption. Therefore, the undiscounted optimal path equalizes consumption over generations. The impatience expressed by discounting then augments early consumption to the detriment of later consumption.

To make one other observation, let us merge the two models into one that has a (consumable) capital good and a (consumable) exhaustible resource. Take as the utility function the sum of a utility flow from direct resource consumption and a flow from consumption of the capital good. Then the ratio of the shadow price of the resource, implicit in an optimal path, to that of the capital good increases over most of the future in proportion to $e^{\rho t}$. This is so regardless of whether life goes on when the resource is exhausted, or whether the resource is essential in the sense that life ceases upon exhaustion—provided the initial stock is a large one. In an ideal competitive market system, this would be reflected in a steady proportional increase in the scarcity price (the market price less cost of recovery) of the resource relative to that of the capital good.

Do the assumptions of this model represent a good first approximation to the role of resources in the modern economy? This is a topic of long-extended debate between conservationists and economists. Note that our model leaves out the crucial phenomenon of technological change. It also leaves out that there are in reality many resources. The prevailing view of economists is that, first, as a resource nears exhaustion, technology advances so that we can deal with lower-grade ores or go to deeper depths to extract. Second, technology is likely to find a substitute that may be more abundant, and can provide an identical or similar service. For instance, the copper in electric wires can be replaced by aluminum or its service taken over by wireless communication.

The economist's view, as I have labeled it, is described persuasively and with ample documentation in the book by Barnett and Morse (1963). They have gone over the record of the last 85-year period and found that the cost of extraction of all resources except those of forestry has consistently decreased at a substantial rate in that period. Now the cost of extraction does not contain what one may call the scarcity price component arising from the fact that the valuable underground stock is being diminished. But that component is included in the market price, and the authors also found an allover decrease in the price of minerals over the entire period.

A more recent study by Nordhaus and Tobin (1972) presented in a symposium of the National Bureau of Economic Research asks a different question of the record of a recent 50-year period: What can we find out about the substitution over time of capital and labor for resources? The simulations that the authors present in an appendix to that paper come out with the result that an elasticity of substitution larger than unity fits the record better. Now the elasticity of substitution is the following concept: raise the price of one input 10% over what it cost previously while keeping the price of another input constant. Assume a cost-minimizing policy for producing the same output before and after the price change. Then what will be the change in the ratio of the quantity of that input to that of the other? If the elasticity of substitution is larger than one, then that ratio will be decreased by more than 10%. The authors found that the record was best represented by a very substantial possibility of substitution.

Of necessity, these are all studies of the past performance of the production system. The question remains: Do these past trends extend into the future? I submit that an answer to that question is not within the capability of the economist. To the extent that an answer is possible, the main contribution should come from scientists and engineers, and especially from those among them who spend a good deal of time and effort looking ahead as to what future technology and resource availability may be.

This observation applies with particular force to the field of energy. I will use the remaining time for a few remarks on modeling of the energy sector. The two models we looked at were theoretical exercises; easy exercises compared with what would be needed in the way of both theoretical and empirical model construction to reflect and express the real complexity of the energy sector.

Most of the available forward projections of energy variables take one variable at a time. What is needed is a mode of projection that considers the principal variables in their interaction. The goal then is to make the projections of total demand for energy and its main components, as related to prices, to the number and income of consumers, and to industrial uses; of the extraction of different fuels; and of the various conversions of fuels into the energy forms consumed, again as related to prices. Finally, all of these projections are to be dependent on a consistent set of assumptions.

At present we are using oil and natural gas as well as coal to generate electricity. It is quite possible that at some future time we will find it economical to use nuclear electric power to generate a liquid fuel. All such conversion possibilities that are actually or potentially important should, in a good model, be represented by the best possible estimates of the ratios of inputs to outputs. To represent technical change, one would distinguish between processes now available (which can be measured reasonably well) and those processes aimed for or hoped for at a later time, again with best present estimates of their ratio of inputs to outputs. These future technological processes are uncertain first with regard to whether they will be available at all; if so, also with regard to time of availability; and finally with regard to the input-output ratios that would characterize them initially and as time goes on. The dependence of all these estimates on research and development efforts for these technologies should in turn be estimated.

I want to make a further distinction between near-future models and long-range models. The emphasis on optimization in the two examples discussed before seems to me somewhat utopian with regard to models for a nearby future. While that emphasis is always desirable for the evaluation of policies, it would be premature in any particular attempt at straight forecasting, say, for the next 5 years. The latter type of forecast would have to take into account actual policies as they stand, whether or not they are efficient or even consistent. One would also wish to estimate how long a given policy will stand, and what would be the effects of alternative policies that may take its place when it is changed.

For longer-range models the Ramsey type of ideas toward optimization are perhaps more appropriate. There is an assumption implicit in the application of these ideas that policies and policy making are improvable. This assumption justifies constructing a long-range model that places more emphasis on what can be achieved if one seeks to satisfy some explicitly stated criterion. Alternative policies may then be tested by comparing their outcomes as estimated by the model.

Another aspect of uncertainty crucial in particular for long-range projections is the future availability and future extraction costs of mineral fuels. There is a need for more systematic estimates than we now have of total availability of important fuel resources in the earth's crust, cross-classified by real cost of recovery (with present and with expected future technologies) and by probability of being found to exist. Many of the availability estimates currently used take their points of departure in known deposits or structures and in costs of recovery with present technology. What is at work here is a preference for what can be estimated with greater accuracy over what most needs estimating. The downward bias inherent in this approach leads to continual upward revision of estimates of total availability of specific minerals as new information accumulates. A recent paper by McKelvey (1972) reviews and suggests concepts and methods that may lead to more inclusive estimates. The case for such estimates is that the best estimate of total ultimate availability, uncertain as it may be, is more relevant for some important purposes than a much more accurate estimate of some more narrowly defined portion of that total. What are needed are methods of estimation that in subsequent revisions are found to have erred on the high and low side with comparable frequencies. Such estimates would need to be supplemented by estimated relationships describing how at any one time exploration and extraction of such resources depend on price (or tax) incentives (or disincentives), and on resources remaining.

The methods and skills of econometrics and of mathematical programming under uncertainty needed for a more systematic model construction for the energy sector of the United States or even the world economy exist and are available. The outstanding characteristic of long-range energy modeling is the importance of uncertainty, and the way in which some of the most important uncertainties pertain to specific conversion processes and to specific resource availabilities. In my limited experience with soliciting expert estimates of the pertinent probabilities, I have found that those best in a position to make best attainable, even if subjective, probability estimates are least willing to do so. The art and technique of eliciting best expert estimates need attention and development. Meanwhile, a wide range of uncertainty will remain. The best line of defense for the model maker in such circumstances is to match each projection based on "best" probability estimates with alternative projections based on equally explicit more "optimistic" and more "pessimistic" assumptions. This may show which estimates are most crucial to needed immediate decisions.

In regard to technological uncertainties, another dimension of the problem is whether the probabilities of success of the several ongoing or potential research and development efforts are approximately independent of one another. As an example, such estimates as I have been able to collect rate at this time the chance that the nuclear breeder development will meet the twin tests of commercial competitiveness and environmental acceptability substantially higher than that for nuclear fusion. But suppose that for some reason or concatenation of reasons the breeder does not meet both tests. Could any of these reasons militate against fusion as well? Or is that altogether a new and independent try? The same question arises, of course, as between nuclear processes and other options, such as geothermal energy, or the various proposals for tapping solar energy on a large scale, and between these other options themselves.

Note that, *if* failure were to meet *all* these options, only *then* would energy resources take the place of the exhaustible resource in the second model we have considered earlier on. The resources of fossil fuels and of U 235 would then, jointly, make up the exhaustible resource essential to life on the scale and at the level to which the world now aspires. *If* the chances of failure of the individual options were to be largely independent, then the probability of the failure of every one of them is much smaller, thanks to compounding.

Notes

1. Editor's note: In introducing Professor Koopmans I alluded to his pioneering contribution in physics; by now a classic, it is known as *Koopmans Theorem* and refers to the determination of the energy requirements for electron transitions in solids (see *Physica, 1,* p. 104, 1933; and Seitz, F., *The Modern Theory of Solids,* McGraw Hill Co., Inc., p. 313, 1940).
2. For a fuller statement of this and subsequent arguments, see Koopmans (1973).

References

Barnett, H. J., and Morse, C. (1963). *Scarcity and Growth, the Economics of Natural Resource Availability,* The Johns Hopkins Press, Baltimore.

Harrod, R. F. (1948). *Towards a Dynamic Economics,* MacMillan, New York, p. 40.

Intriligator, M. D. (1971). *Mathematical Optimization and Economic Theory,* Prentice-Hall, Englewood Cliffs, N.J., Chapter 16.

Koopmans, T. C. (1967). "Intertemporal Distribution and 'Optimal' Aggregate Economic Growth," in W. Fellner et al. (1967). *Ten Economic Studies in the Tradition of Irving Fisher,* Wiley, New York, pp. 95–126.

——— (1973, in press). "Some Observations on 'Optimal' Economic Growth and Exhaustible Resources," Cowles Foundation Discussion Paper No. 356, 29 pp., March 1973, to be published in 1973.

McKelvey, V. E. (1972). "Mineral Resource Estimates and Public Policy," *American Scientist,* pp. 32–40, January–February.

Nordhaus, W., and Tobin, J. (1972). "Is Growth Obsolete?" in *Economic Growth,* Fiftieth Anniversary Colloquium V, National Bureau of Economic Research, New York, pp. 1–80.

Ramsey, F. P. (1928). "A Mathematical Theory of Saving," *Economic Journal, 38* (152), pp. 548–550, December.

2 Markets and Appropriable Resources*

WILLIAM D. NORDHAUS†

In the United States, the prices of appropriable resources have for the most part been determined by market forces. An appropriable resource is one for which all rewards or penalties from services or uses accrue to the owner. Why is it that public policy has accepted a laissez-faire approach to resource pricing?

The basis for allowing market determination of prices is grounded in the theory of general economic equilibrium. General equilibrium theory assumes that there are consumers with initial resources and given preferences and producers with well-defined technical relations. There may be large numbers of time periods and uncertainty about the exact demand or supply conditions. The main requirements are that production sets and preference orderings be convex, and that markets exist for all goods, services, and contingencies. Under the above conditions, a market system will have a general equilibrium of prices and quantities. The equilibrium will be efficient in the sense that there is no way of improving the lot of one consumer without worsening the lot of another. The prices are appropriate indicators of social scarcity given the preferences of the society.

There are several ways that such a system can run into trouble. First, there is nothing in a market system which will guarantee a just distribution of consumption over households or over time. The second proviso is that it is a crucial part of the assumptions of the analysis that all the costs and benefits of a particular process of production be internalized to the decision maker. Finally, the analysis assumes that there are markets for all goods and contingencies. This means that there must be futures markets for, say, petroleum and coal in the year 2000; and there must be insurance markets for different contingencies, such as the possibility that fusion processes never become economically viable. If there are not all the requisite markets, efficiency cannot be guaranteed.

The application of the results to the depletable natural resource sector is straightforward. In considering these we distinguish between *recovery costs* $e(t)$, or the marginal cost per unit output excluding rent or royalties; and *royalties* $y(t)$,[1] which are a reflection of the fact that the resource is judged to be scarce.

* This paper is a summary of work in progress on the topic. I am grateful for the helpful comments of Tjalling C. Koopmans.

† Yale University.

Consider a world of certainty and a time horizon of T years. There are $R(t)$ units of the resource remaining at any point of time t. Let recovery costs be zero up to the limit of resource availability. If alternative assets yield a rate of return $r(t)$, the equilibrium condition for some owners to hold and others to sell the resource is equality of the rate of capital gains on the resource and the interest rate:

$$\dot{y}(t)/y(t) = r(t), \tag{2.1}$$

where $\dot{y}(t) = dy(t)/dt$.

There is a family of solutions to basic Equation 2.1. To get the unique solution, we need to add a terminal condition that all resources are used up at the end of the last period T:

$$[y(0) \text{ is such that } R(T) = 0]. \tag{2.2}$$

Given the interest rate, there will generally be a unique set of prices and royalties satisfying Equations 2.1 and 2.2.

In the case where there are recovery costs, marginal recovery cost $e(t)$ is a decreasing function of total resources remaining $R(t)$. Price is the sum of marginal recovery cost and royalty:

$$p(t) = y(t) + e(t). \tag{2.3}$$

Consider a constant interest rate. The present value of recovery and sale of a resource with recovery costs $\bar{e}$ is $\{p(t) - \bar{e}\}e^{-rt}$, and this is maximized when $\dot{p}(t) = r[p(t) - \bar{e}]$, or

$$\frac{\dot{p}}{p} = r\left(\frac{p - \bar{e}}{p}\right) = \frac{ry}{p}. \tag{2.4}$$

Since $p - \bar{e}$ represents current royalties, the new condition for recovering a resource is that the rate of growth of the price of the resource be equal to the interest rate times the share of royalties in the total resource price. This rate will always be less than the interest rate.

Note that, even though royalty prices satisfy the modified exponential relation, there is no need for *output* prices to rise exponentially. If anticipated improvements in technique lower the cost curve, recovery costs and prices may decline drastically, yet royalties can continue along their modified exponential path. On the other hand, once royalties have reached 90% of price, there is little hope *along an efficient path* for a lower price because of improvements in recovery techniques. (Along inefficient paths all bets are off.)

A further extension of the model considers the functioning of resource markets under uncertainty. In the general equilibrium analysis discussed earlier, the requirement is not only for a complete set of futures markets but also for a complete set of insurance or contingent commodities markets. The insurance markets would span the space of all economically relevant events, such as whether and when breeder reactors become available; what is the future course of population growth; what happens in Mideast politics; whether very large oil reserves in Alaska will be recoverable. In each case, there would be a contingent commodity sold: for example, one barrel of crude in January 1984, if the trans-Alaska pipeline is not built. It can be shown that the system is ex ante efficient as long as there is a complete set of futures markets.

The major problem with actual markets is that a full set of futures and insurance markets is not available. Long-term contracts—rough substitutes for futures markets—are relatively rare. I am unaware of any insurance markets for selling resources contingent on the state of the world.[2]

How serious is the absence of a complete set of futures and insurances markets likely to be? We will consider only two problems that might be serious.

The first possible complication involves the possibility of *myopic decisions*. In the present context, myopia means that the planning horizons of economic agents are relatively short.

Myopic decisions are a possibility because of the absence of futures markets. Recall that expected capital gains play a crucial role in resource decisions. We can conceive of two polar strategies toward investment in resources: either an in-and-out strategy (involving buying on the speculation of short-term gain) or a buy-and-hold strategy (buying with the intention of selling the recovered commodity rather than the land). Up to now we have really assumed that the buy-and-hold strategy was the operative policy. Let us say, however, that investors have an in-and-out strategy, buying titles to resources with an eye to capital gains rather than to selling the resources directly. For simplicity, assume that all investors plan on holding for one period, then selling. This leads to the same equilibrium condition as outlined in Equation 2.1 or 2.4: the market is in equilibrium when investors expect that the (one-period) capital gain on the asset is equal to the (risk-corrected) interest rate. The most important point to note is that there is no unique solution to this condition. *Indeed a path with zero royalties will satisfy the myopic equilibrium—for a while.*

What is missing in this system? The missing element is the "global planner" (or speculator) who calculates out the quantities demanded along a price path to see whether they are consistent with overall availabilities (this is the calculation implicit in Equation 2.2). The reason why pricing of resources might be myopic is that very few planners have the ability, or perhaps even the desire, to perform the operation of checking consistency for a period of several decades.

The second complication is that the nature of the myopic equilibrium condition in Equation 2.1 or 2.4—implying indeterminacy of the level of prices—lends great instability to the price of resources. And this may lead to the operation of destabilizing speculation and to cycles of resource prices. Consider a resource that has zero (or constant) recovery cost, and which is sold only in spot markets. The quantity supplied will then be an increasing function of the difference $(r - \pi)$, where r is the interest rate and π is the anticipated rate of price increase of the resource. Demand, on the other hand, is a decreasing function of price. It can be shown that the system is *stable* only if the elasticity of demand is greater than the response of the rate of growth of production to $(r - \pi)$. Moreover, if the path is stable, it is one where *royalties tend to zero* rather than grow at rate r; the optimal path is clearly inconsistent with the stable paths for such a model.

What can we conclude from this investigation? It can be argued that the current market mechanism is an unreliable means of pricing and allocating scarce appropriable stocks of natural resources. The absence of futures and insurance markets rules out the general theorems usually drawn from general equilibrium theory. This is reinforced by the possibility of instability in such markets. On the other hand, it probably cannot be determined a priori whether current resource usage is too fast or too slow; this can only proceed from carefully constructed econometric and engineering models of the economy.

Notes

1. We use "royalty" to denote the difference between price and marginal cost, the concept being similar to rents on land. Royalties have many other meanings in resource and petroleum economics.

2. Several participants at the conference asked why the sale, option, or lease-hold arrangements currently employed for oil-, gas-, or ore-bearing lands are not good substitutes for futures markets. These contracts are not the equivalent of a full set of futures markets (1) because they represent sale of rights to recover to producers (for example, coal companies) rather than sales to ultimate consumers (for example, utility companies); and (2) because they are spot markets or at best futures markets covering a decade or so. In the cosmic framework of the ultimate exhaustion of fossil fuels or energy sources or phosphorus, these transactions cover a very short span.

3 An Economic Interpretation of the Energy-GNP Ratio*

ERNST R. BERNDT† AND DAVID O. WOOD‡

3.1. Introduction

The ratio of total energy consumed in Btu's to GNP in constant dollars —the energy-GNP ratio—has been used extensively in summarizing the relation between real economic growth and growth in the consumption of energy inputs. Historical studies of movements in this ratio have been performed by Schurr et al. (1960) and Darmstadter et al. (1971), in which they proposed hypotheses to explain its change in 1920 from an increasing to a decreasing trend. Recently Netschert (1972) has suggested that an opposite trend reversal occurred in 1967. His analysis has led him to conclude ". . . that the trend reversal will persist . . . ," although the rate of increase would be significantly less than the first five years would indicate (Netschert, 1972, p. 9).[1] This view is also reflected in a report on energy research needs prepared by Resources for the Future, Inc. (1971) for the Natural Science Foundation. Very little research has been done, however, on the economic interpretation of the energy-GNP ratio.

The purpose of this paper is to extend the Schurr-Netschert analytical framework by presenting an economic interpretation of the energy-GNP ratio that lends itself to rigorous empirical analysis. In the next section we develop a structural model based on the neoclassical theory of the derived demand for factors of production. We show that the energy-GNP ratio has a consistent economic interpretation only under highly restrictive conditions. Even when these conditions are satisfied, we could expect the energy-GNP ratio to vary in response to changes in the relative prices of capital, labor, energy, and intermediate materials. It follows that an increase in the energy-GNP ratio does not necessarily correspond to a structural change in the economy. Rather, the increase in this ratio may

* The research reported in this paper was initiated while the first author was a staff member of the Mathematics and Computation Laboratory. He acknowledges research support from the University of British Columbia. The helpful comments of W. Erwin Diewert, Anthony D. Scott, and Philip K. Verleger are gratefully acknowledged. The views expressed in this paper are those of the authors and do not necessarily correspond to any policy or position of the institutions with which they are affiliated.
† Assistant Professor of Economics, University of British Columbia.
‡ Deputy Chief, Mathematics and Computation Laboratory, Office of Emergency Preparedness, Washington, D.C.

indicate simply that demand is price responsive. For example, as shown in Table 3.1, the energy-GNP ratio for the 1947–1972 period has risen as the relative price of energy fuels has fallen—perhaps firms are simply minimizing costs by adopting more energy-intensive production technologies. If energy demand is price responsive, future growth rates of energy demand could be altered by proper use of policies that affect relative prices. To analyze such policy actions, however, it is necessary to specify an econometric model of the U.S. economy which includes an explicit energy sector. In the final section of the paper we discuss data and accounting problems associated with such an effort, and outline the kinds of econometric investigations that could be undertaken.

3.2. The Demand for Energy in the Context of Cost and Production Functions

We proceed by specifying an aggregate production function F for the economy relating gross output Y to four factor inputs: capital K, labor L, energy E, and other intermediate materials M,[2]

$$Y = F(K, L, E, M). \tag{3.1}$$

The technological structure of the economy, summarized in its production function, relates the maximum amount of output obtainable from any nonnegative combination of the inputs. As such, the production function summarizes basic engineering information. The production function takes on an economic interpretation, however, when assumptions about the behavior of the relevant producing entities are made; namely that, given prices of output and the inputs, producers maximize profits, or alternatively, given the level of output and the prices of the inputs, producers minimize costs.

Corresponding to every production function there exists a cost function that reflects the technology of the economy. We write the cost function corresponding to the production function F as

$$G = G(Y, P_K, P_L, P_E, P_M), \tag{3.2}$$

where P_K, P_L, P_E, and P_M are the input prices of K, L, E, and M, respectively. Structural alterations affecting the production technology F will of course be reflected as changes in its cost function G.

Assuming that the cost function G is homothetic, that is that substitu-

Table 3.1 Historical Statistics on the U.S. Energy-GNP Ratio

Year	Gross Energy Input[a]	GNP (billions of 1958 dollars)	Energy-GNP (1000s of Btu)	Relative Price Index[b]
1947	33.0	309.9	106.4	na
1948	33.9	323.7	104.7	na
1949	31.5	324.1	97.2	na
1950	34.0	355.3	95.7	35.0
1951	36.8	383.4	96.0	32.2
1952	36.5	395.1	92.4	33.3
1953	37.6	412.8	91.1	35.0
1954	36.3	407.0	89.2	35.7
1955	39.7	438.0	90.6	34.2
1956	41.7	446.0	93.5	33.8
1957	41.7	452.5	92.2	35.4
1958	41.7	447.3	93.2	35.2
1959	43.1	475.9	90.6	33.6
1960	44.6	487.7	91.4	33.3
1961	45.3	497.2	91.1	33.6
1962	47.4	529.8	89.5	33.6
1963	49.3	551.0	89.5	33.3
1964	51.2	581.1	88.1	32.8
1965	53.3	617.8	86.3	32.1
1966	56.4	658.1	85.7	31.9
1967	58.3	675.2	86.3	31.9
1968	61.7	706.6	87.3	31.1
1969	65.0	724.7	89.7	31.4
1970	67.4	720.0	93.6	31.6
1971	68.8	741.7	92.7	na
1972[p]	71.3	795.3	89.6	na

[a] Gross energy is the total of inputs into the economy of the primary fuels (petroleum, natural gas, and coal, including imports) or their derivatives, plus the generation of hydro and nuclear power converted to equivalent energy inputs. Source: Division of Fossil Fuels, Bureau of Mines, U.S. Department of the Interior.
[b] The relative price index is the composite fossil fuel price index relative to the industrial wholesale price index. Source: Division of Fossil Fuels, U.S. Bureau of Mines, Department of the Interior.
[p] Denotes preliminary estimate. Energy consumption figure is from the Division of Fuels, Bureau of Mines, U.S. Department of the Interior. The GNP figure is that projected in the October 1972 issue of the *Survey of Current Business*.
na denotes not available.

tion possibilities among the inputs are independent of the level of gross output Y, we may rewrite the cost function Equation 3.2 as

$$G = H(Y)J(P_K, P_L, P_E, P_M). \tag{3.3}$$

If the cost function is twice differentiable and strictly quasi-concave, by Shephard's or Hotelling's lemma[3] we may determine the cost minimizing bundle of input factors by differentiating G with respect to each of the input prices and setting the partial derivatives equal to the quantity of that input.

Thus,

$$K = H(Y)[\partial J/\partial P_K], \tag{3.4}$$

$$L = H(Y)[\partial J/\partial P_L], \tag{3.5}$$

$$E = H(Y)[\partial J/\partial P_E], \tag{3.6}$$

$$M = H(Y)[\partial J/\partial P_M]. \tag{3.7}$$

From Equations 3.4 through 3.7 we can derive input ratios that are a function of the technological structure and prices; for example, we can obtain energy-capital, energy-labor, and energy-other intermediate materials ratios as

$$E/K = \frac{\partial J/\partial P_E}{\partial J/\partial P_K}, \qquad E/L = \frac{\partial J/\partial P_E}{\partial J/\partial P_L}, \quad \text{and } E/M = \frac{\partial J/\partial P_E}{\partial J/\partial P_M}.$$

The economic interpretation of each ratio is due to the behavioral assumption of cost minimization. Note, however, that we have not derived an energy-GNP ratio from the cost minimization assumption. Such a ratio with an economic interpretation can be constructed, but, as we will now show, an additional assumption about the production technology is required. We digress briefly, however, to discuss first the concept of value-added.

The concept of value-added has been employed in national income accounting as a device for allocating the origins of income to the services of the primary inputs, capital and labor. Nominal value-added is therefor the product $P_V V$:

$$P_V V = P_K K + P_L L, \tag{3.8}$$

where P_V is a consistent aggregate price index of capital and labor, and V (real value-added) is a consistent aggregate quantity index of capital and labor.[4]

In the context of a production function, the concept of real value-added (GNP) has an economic interpretation if the production function F can, without any loss, take on the special nested form,

$$Y = F(K, L, E, M) = F_1[B(K, L), E, M] = F_2(V, E, M), \tag{3.9}$$

where B is a function of K and L only.[5] A function satisfying this condition is said to be weakly separable. Heuristically, weak separability permits sequential optimization; we can think of the producer as finding optimal levels of capital and labor with which to produce real value-added, and then finding the optimal levels of real value-added, energy, and other intermediate materials with which to produce gross output Y.

For weak separability to hold, it is necessary and sufficient that technological substitution between K and L be independent of the levels of E and M. It is not surprising, therefore, that the assumption of weak separability places restrictions on substitution possibilities among the inputs. Berndt and Christensen (1973a) have shown that the assumption, Equation 3.8, of weak separability is equivalent to the following restrictions on the Allen partial elasticities of substitution (hereafter denoted by σ):[6]

$$\sigma_{EK} = \sigma_{EL} = \sigma_{EV}, \tag{3.10}$$

$$\sigma_{MK} = \sigma_{ML} = \sigma_{MV}. \tag{3.11}$$

Using the theory of duality, Lau (1972) has shown that, given certain regularity conditions, weak separability of the production function is equivalent to weak separability of its cost function. Hence, weak separability in Equation 3.9 is equivalent to the restriction

$$\begin{aligned} G = H(Y)J(P_K, P_L, P_E, P_M) &= H(Y)J_1[Q(P_K, P_L), P_E, P_M] \\ &= H(Y)J_2(P_V, P_E, P_M), \end{aligned} \tag{3.12}$$

where Q is a function of P_K and P_L only.

Further, since a consistent aggregate price index of real value-added exists if and only if a consistent aggregate quantity index of real value-added exists, by Shephard's or Hotelling's lemma we can now take the derivative of

$$G = H(Y)J_2(P_V, P_E, P_M)$$

with respect to P_V and set it equal to V:

$$\partial G/\partial P_V = H(Y)(\partial J_2/\partial P_V) = V. \tag{3.13}$$

This allows us to derive input ratios:

$$\left.\begin{aligned} \frac{E}{V} &= \frac{\partial J_2/\partial P_E}{\partial J_2/\partial P_V} && (3.14)\\ \frac{M}{V} &= \frac{\partial J_2/\partial P_M}{\partial J_2/\partial P_V} && (3.15)\\ \frac{E}{M} &= \frac{\partial J_2/\partial P_E}{\partial J_2/\partial P_M} && (3.16) \end{aligned}\right\} \text{ where each ratio is a function of } J_2,\ P_K, P_L, P_E, \text{ and } P_M.$$

Equation 3.14 is the familiar energy-GNP ratio; its economic interpretation is owing to the fact that it is derived from the behavioral assumption of cost minimization when the production (or cost) function is weakly separable. The ratio is a function of the prices P_V (which in turn is a function of P_K and P_L), P_E, and P_M, and the technological structure of production as expressed in the cost function J_2.

It is obvious that there is no a priori reason to expect the energy-GNP ratio Equation 3.14 to be constant; the ratio may vary due to changes in the structure of technology as summarized in J_2, and/or changes in relative prices. If technology is constant, the ratio will vary solely in response to price changes. Furthermore, even if technological change is factor augmenting with equal rates of augmentation for all factors (that is, Hicks-neutral), variations in the ratio of energy to GNP will still be due solely to changes in relative prices. However, if relative prices are constant, or if "prices don't matter," that is if the Allen partial elasticities of substitution of Equations 3.10 and 3.11 satisfy the condition $\sigma_{ij} = 0$, for $i, j = K, L, V, E, M$, then changes in the energy-GNP ratio will occur solely in response to nonneutral variations in the structure of technology. This latter position is the one espoused by Netschert (1972), who suggests that recent changes in the energy-GNP ratio represent permanent nonneutral structural changes in the economy.[7]

3.3. Empirical Implementation and Application of the Energy Input Demand Model

In the previous section we have discussed restrictions on production technology which must be satisfied in order for the energy-GNP ratio to have an economic interpretation. We have also noted that these restrictions are equivalent to restrictions on substitution possibilities among the inputs. This suggests at least a twofold motivation for an empirical investigation of input substitution possibilities. First, measures of the elasticity of substitution among factors of production are of interest in their own right and are useful in partial equilibrium studies. Second, if a consistent aggregate index of real value-added exists, then it is justifiable to integrate an energy input demand sector into macroeconomic models whose accounting structure is based upon the concept of value-added. If such a consistent aggregate index does not exist, then the value-added accounting structure employed in typical macroeconomic models will be inappropriate for a model which includes an explicit energy sector. Hypotheses concerning the existence of a consistent index of value-added can of course be tested on the basis of equality restrictions on the estimated elasticities of substitution.

In order to undertake empirical investigations of substitution possibilities among factor inputs, including energy, it is necessary to construct a set of economic accounts giving, in real and value terms, the flows of intermediate products between sectors, the inputs of primary inputs to each sector, and the sales of each sector to final users. These accounts are the interindustry low tables having time and/or regional dimensions. The construction of such accounts appears to be the single greatest difficulty in implementing and investigating the model summarized in Equations 3.4 through 3.7.

A second obstacle to empirical investigation of input substitution possibilities among factor inputs is the choice of a functional specification to characterize production technology. Two attractive candidates include the generalized Leontief form proposed by Diewert (1971) and the transcendental logarithmic form (translog) proposed by Christensen, Jorgenson, and Lau (1972). These specifications are particularly attractive since they place no a priori restrictions on the Allen partial elasticities of substitution and since both may be interpreted as second-order approximations to an arbitrary functional form.

Given the model developed in Section 3.2, the necessary economic accounts, and a suitable functional form to characterize production technology, a variety of significant hypotheses can be tested using appropriate econometric techniques. These include (a) that a consistent aggregate index of real value-added exists, that is, that *K* and *L* are weakly separable from *E*, *M*; (b) the Netschert assumption that all Allen partial elasticities of substitution are zero; and (c) the Netschert assertion that the technological structure changed in 1967. This latter hypothesis could be tested rigorously through the use of a Chow (1960) test.

The estimated elasticities of substitution can be used in analyzing the following kinds of questions: (a) What is the effect on the demand for energy of an increase in its price? (b) By how much has the U.S. federal tax policy of investment tax incentives (the investment tax credit and accelerated depreciation allowances) affected the demand for energy? Note that if energy and capital are complements, introduction of investment tax incentives increases the demand for energy; on the other hand, if energy and capital are substitutes, the introduction of tax incentives decreases the demand for energy. (c) What will be the effect on the demand for labor if the relative price of energy increases? In such an analysis it may be desirable to disaggregate labor into blue- and white-collar workers in order to assess differential impacts. (d) What would happen to the energy-GNP ratio if the relative price of energy rose by a given percentage?

The above questions can be analyzed in a partial equilibrium demand model in which it is assumed that supplies of *K*, *L*, *E*, and *M* are available in unlimited quantities at the present set of prices. Alternatively, the supplies of these factors could be specified exogenously. It might be more desirable, however, to construct a general equilibrium model in which both supplies and demands for the inputs are endogenously determined. Issues and problems associated with such an effort are beyond the scope of this paper. For example, a comprehensive set of energy supply accounts must be constructed. Further, problems of modeling the process of exploration and discovery of new additions to fuel reserves must also be confronted.

Notes

1. Data more recent than that available to Netschert suggests that the trend reversal in the energy-GNP ratio has not persisted through 1971 and 1972. See Table 3.1 for further details.

2. We could of course specify a more disaggregated input classification. For exposition purposes this additional detail is unnecessary. Also note that we do not specify the functional form of the production function *F*. For empirical work the choice of this functional form will of course be an important element in the investigation.

3. For a statement and proof of these results and for further discussion, see Diewert (1971, 1972).

4. The concepts of consistent aggregation, functional separability, and Allen partial elasticities of substitution are discussed in greater detail by Berndt and Christensen (1973a). Empirical implementations of these concepts are found in Berndt and Christensen (1973b, 1973c).

5. This point has been made by Arrow (1972) in the context of a three-input production function—capital, labor, and all intermediate materials. Alternative interpretations of real value-added are discussed in Diewert (1973).

6. Diewert (1972) has extended the Berndt-Christensen theorems.

7. Netschert presents no empirical evidence supporting his position. Related evidence to the contrary has, however, been reported by Chapman, Mount, and Tyrrell (1972) and Anderson (1971, 1972).

References

Anderson, Kent P. (1971). "Toward Econometric Estimation of Industrial Energy Demand: An Experimental Application to the Primary Metals Industry," Rand Corporation Report R–719–NSF, December.

——— (1972). "Residential Demand for Electricity: Econometric Estimates for California and the United States," Rand Corporation Report R–905–NSF, January.

Arrow, Kenneth J. (1972). "The Measurement of Real Value Added," Stanford University Institute for Mathematical Studies in the Social Sciences, Technical Report No. 60, June.

Berndt, Ernst R., and Christensen, Laurits R. (1973a). "The Internal Structure of Functional Relationships: Separability, Substitution, and Aggregation," Washington, D.C., Executive Office of the President, Office of Emergency Preparedness, Assistant Director for Resource Analysis, TR–76, April 1972. Forthcoming *Review of Economic Studies,* October 1973.

——— (1973b). "The Translog Function and the Substitution of Equipment, Structures, and Labor in U.S. Manufacturing 1929–1968," *Journal of Econometrics,* Vol. 1, No. 1, pp. 81–114, March.

——— (1973c). "Testing for the Existence of a Consistent Aggregate Index of Labor Inputs," Washington, D.C., Executive Office of the President, Office of Emergency Preparedness, Assistant Director for Resource Analysis, TR–79, August. Forthcoming, *The American Economic Review,* June 1974.

Chapman, Duane; Mount, Timothy; and Tyrrell, Timothy (1972). "Electricity Demand Growth: Implications for Research and Development," Testimony prepared for the Subcommittee on Science, Research, and Development of the Committee on Science and Astronautics of the United States House of Representatives, June 16.

Chow, G. C. (1960). "Tests of Equality Between Sets of Coefficients in Two Linear Regressions," *Econometrica, 28,* July, pp. 591–605.

Christensen, L. R.; Jorgenson, D. W.; and Lau, L. J. (1971). "Conjugate Duality and the Transcendental Logarithmic Production Function," *Econometrica,* Vol. 39, No. 4, July 1971, pp. 255–256.

Darmstadter, Joel, with Teitelbaum, Perry D., and Polach, Jaroslav G. (1971). *Energy in the World Economy,* Resources for the Future, Inc., The Johns Hopkins Press, Baltimore.

Diewert, W. Erwin (1971). "An Application of the Shephard Duality Theorem: A Generalized Leontief Production Function," *Journal of Political Economy,* Vol. 79, May/June, pp. 481–507.

——— (1972). "Applications of Duality Theory," paper presented to the 1972 Econometric Society meetings, Toronto.

——— (1973). "Hick's Aggregation Theorem and the Existence of a Real Value Added Function," mimeo, University of British Columbia, January.

Lau, Lawrence J. (1972). "Duality and the Structure of Cost Functions," unpublished paper, Stanford University Department of Economics, Stanford, California.

Netschert, Bruce (1972). "Fuels for the Electric Utility Industry, 1971–1985," a report of National Economic Research Associates, Inc., Washington, D.C., to the Edison Electric Institute.

Resources for the Future, Inc. (1968). *U.S. Energy Policies: An Agenda for Research,* The Johns Hopkins Press, Baltimore.

Schurr, Sam H., et al. (1960). *Energy in the American Economy, 1850–1955: An Economic Analysis of its History and Prospects,* The Johns Hopkins Press, Baltimore.

4 The Relationship between Energy Demand and Economic Activity

PHILIP K. VERLEGER, JR.*

Models of aggregate energy systems have generally assumed the price elasticity of demand for energy to be zero.[1] And, although some determine demand from economic forces, feedbacks between these energy models and the economy are assumed to be absent. These assumptions require validation. In this paper some results which indicate that the zero price elasticity assumption is probably not acceptable even in an energy demand model are presented. Further, it is proposed that a better way to examine energy demands is within the framework of a fully simultaneous model of economic activity where long-run substitutions between inputs such as capital, labor, and energy can be evaluated. However, whether or not it is feasible to estimate such a fully simultaneous (or general equilibrium) model, partial equilibrium models with some energy feedbacks into the macroeconomy are offered as an alternative to the zero price elasticity trend extrapolation models.

The paper begins by analyzing the relative strengths of three types of macroeconomic models; input-output, national income account, and production function. Then, although production function models are argued to be the most useful from a general equilibrium point of view, a method for incorporating energy into a national income account type model is developed due to the more general use of the latter type of model. In Section 4.3 the structure of energy demand models is examined to determine whether the zero price elasticity of demand hypothesis can be maintained. Finally, in Section 4.4, forecasts made where this hypothesis is maintained are compared with those made where it is rejected.

4.1. Macromodels of Economic Activity

It is important at the outset to distinguish between three types of macromodels; input-output, national income account (hereafter referred to as NIA models), and production function. Input-output models have been widely discussed elsewhere, as have the NIA type (or, as they are more popularly called, macroeconometric models of the economy). The production function model is relatively new. This form of econometric model is more suitable for the analysis of long-term economic trends than NIA type

* Data Resources, Inc., Lexington, Mass.

models that are designed to capture and project the short-term fluctuations in the economy. As the name implies, the production function model is built around the theory of manufacturing production functions although additional sectors for households, transportation, government, and foreign trade are incorporated. Basically it offers a generalization of input-output type models (which are also designed for long-term forecasting), because it can incorporate a more flexible form of the production function and thus examine substitutions that occur as prices of inputs change.

As originally implemented, models of this type such as those estimated by Jorgenson (1972) incorporated only the flows of capital and labor. But as time-series data on energy flows and prices have become available, the model has been generalized to incorporate energy and resource flows. While this model has not yet been completed, it is possible to discuss the level of detail in it. An input-output framework is used to describe the status of the economy at any moment in time while an econometric model is used to characterize its dynamics. Flows between three groups of suppliers and five types of final demands are distinguished. The suppliers include a final goods and services sector (manufacturing, transportation, commercial activity, and services), an energy production sector (electricity and petroleum refining), and a resource supply sector (coal, natural gas, petroleum, uranium, other minerals, capital, and labor). The final demand sectors cover households, government, imports, exports, and inventories.

The theoretical structure of the model is based on the work of Berndt and Christensen (1973) and Christensen, Jorgenson, and Lau (1973) on the translog production function and its modification to consumer expenditure functions. The resource supply functions are designed along the line specified by Fisher (1964), Erickson (1970), and MacAvoy (1971). The dynamic growth in the model is provided by savings and investment. But, as it is currently being implemented, it is a very aggregative model of the economy. The manufacturing sector is treated as a single entity, as are households. And within the household sector consumption may be divided into as few as three items. Furthermore, regional disaggregation is impossible. This situation results from the detailed information on capital stocks, intersectoral purchases, and prices required to estimate the model. Since these data must generally be constructed from national income account transaction work sheets for every year, greater disaggregation is not feasible. For this reason, as well as to explore the usefulness of simulating NIA models in conjunction with production function type, the energy demand model has also been designed to run within the more disaggregated NIA models.

4.2. Energy Models within National Income Accounts Models

The analysis of energy demand and supply within a NIA type model currently sacrifices the general equilibrium nature of the production function model because the energy sector must be designed and estimated as a recursive block. But the sacrifice is rewarded in several important ways. First, since NIA models already exist, the only effort required is in the estimation of an energy demand-supply sector. Second, because naïve, zero price elasticity of demand energy supply-demand models have already been estimated, the effort becomes one of collecting adequate price data and testing for significant price coefficients. Third, since NIA models tend to be more disaggregated than the production function model, it becomes possible to analyze the supply and demand for energy in a more disaggregated way.

The principal problem in using NIA models such as those developed by Brookings, Wharton, Michigan, or Data Resources, Inc., is in their inherent short-term nature. Although some of the models are simulated for long intervals they are basically designed for short-term forecasting and thus, in estimation, the signals in the data resulting from long-term economic trends are overwhelmed by short-term cyclical information. For this reason one is forced to employ a great deal of judgment in constructing realistic long-term simulations. However, once a satisfactory long-term simulation is established, these models are very useful tools for evaluating the impacts of models because of their higher levels of disaggregation.

The energy model estimated for use with NIA type models is identical to the energy model designed for use with the production function model and similar to it on the demand side. The principal difference between the two is in the prediction of consumption in the industrial sector. This difference results because industrial consumption of energy is projected directly in the production function model but not in a NIA model. This shortcoming is sometimes partially offset because these models can simulate disaggregated production at the two-digit level rather than just a single aggregate and because the output-energy ratio at the two-digit level may be fairly constant or show a regular trend. In the next section the demand side of the energy model adapted to NIA type models is outlined.

4.3. Structure of Energy Demand Models

As they are normally structured, models of energy demands distinguish four classes of energy consumers and one to seven fuels. The major demand sectors are household and commercial, industrial, transportation, and electric utilities. If they are separated, the major fuel demands are coal, natural gas, petroleum, and electricity. In general, these models postulate sector *i*'s consumption of fuel *j*, Q_{ij}, as a function of a single measure of real economic demand X such as population, income, real GNP, or manufacturing production.

$$Q_{ij} = F(X). \tag{4.1}$$

Sometimes real economic demand is even measured by a time trend. While such models may seem terribly naïve, these approaches have worked well during the postwar period as the results of Baughman (1972) indicate.

Historical accuracy, however, is only one criterion for accepting a model. It must also be determined whether the model can adapt to a future where events may differ from historical experience. In this case one wants to determine whether demands that exhibited approximately constant rates of growth during the 1950s and 1960s will continue to do so in the 1970s as prices which formally declined relative to other prices now increase. This is a testable proposition which most analysts, who have constructed aggregate models, have failed to make. However, in their models they would appear to accept the null hypothesis of a zero price elasticity for energy demands in all three major consuming sectors.

The test of the hypothesis of zero price elasticity actually involves two separate parts due to variations in the types of models which others have estimated. First, one must determine whether it is satisfactory to study aggregate energy consumption by sector, and then one must determine whether the price coefficient is significant. Such tests must be made for each sector. Elsewhere, in Verleger (1973), we have presented results for the household sector indicating that an aggregate measure of this sector's consumption does not exist. At the same time it was demonstrated that the null hypothesis of a zero price elasticity was not supported for the household sector's consumption of electricity, petroleum products, or natural gas. These results compare favorably with those of Houthakker and Taylor (1970) who, in a model estimated for a different purpose, also rejected the assumption of zero price elasticities of demand.

In the industrial sector it is more difficult to test the hypothesis that the price elasticity of demand for energy is zero due to the larger array of energy products (especially for petroleum), the dominance of a few industries in total demand, and the lack of adequate price data at the two-digit level. Currently, one is forced to choose between the analysis of consumption at the aggregate level, where prices can be collected, or analysis of consumption at the level of two-digit product demand, where prices are not yet available. In the production function model this problem does not occur because it is not disaggregated to the two-digit level. But in the energy model constructed with NIA models, one can go to the two-digit level. The compromise chosen here is to disaggregate and forsake price information under the assumption that the energy-output ratios in any industry are relatively constant at any given time and are affected indirectly over time as energy prices, along with labor wage rates and capital costs, cause manufacturers to adjust the mix of inputs.

The question of aggregation also arises in the transportation sector when one attempts to test whether or not the price elasticity of demand is significantly different from zero. If one uses a single measure of the sector's total consumption, or even if one disaggregates into the four main products, household consumption of gasoline is still lumped with jet fuel consumption and diesel consumption by trucks and railroads. However, for the moment aggregation is accepted. Further, the zero price elasticity assumption is still accepted, regardless of the level of aggregation.[2]

In the electric utility sector the question of testing for a nonzero price elasticity of demand becomes intertwined with the evaluation of the utilities' abilities to minimize short- and long-run costs by selecting the proper mix of hydro, nuclear, and fossil capacity and by choosing the least cost mix of fossil fuels. Having made the necessary adjustments, it does appear that the utilities do minimize costs by shifting to the lowest cost fuel mix in both the short and long run.[3]

Thus, preliminary results indicate that the demand for energy, when evaluated at the national level, is price sensitive in the household sector but not in the transportation sector. Finally, in the industrial sector the test can not yet be performed due to the absence of adequate data. With these general statements in mind, we will present some of the central equations of a preliminary energy demand model.

The three central equations in the household sector reflect modifications of the work of Houthakker and Taylor (1970). The demand equations are

$$\Delta q_e = -.549P_e + .563\text{PCE},$$
$$(.107) \qquad (.058)$$
$$R^2 = .74,$$
$$\text{DW} = 1.80, \text{ (Durbin-Watson statistic)}$$
$$\text{SEE} = .124 \times 10^4 \text{ Btu per capita (Standard Error of Estimation);} \qquad (4.2)$$

$$\Delta q_g = 1.154 - 6.466P_g,$$
$$(.088) \qquad (3.27)$$
$$R^2 = .11,$$
$$\text{DW} = 2.25,$$
$$\text{SEE} = .430 \times 10^4 \text{ Btu per capita;} \qquad (4.3)$$

$$q_o = 8.829P_{g/p_o} + 3.353\text{PCE} - 5.888P_o + .569q_o(t-1)$$
$$(3.881) \qquad (1.487) \qquad (3.369)$$
$$R^2 = .942,$$
$$\text{DW} = 2.215,$$
$$\text{SEE} = .893 \times 10^4 \text{ Btu per capita.} \qquad (4.4)$$

(Standard errors of coefficients are shown in parentheses.) In Equations 4.2 through 4.4, q_i represents consumption of fuel type i (PCE real per capita consumption expenditures on all goods, and P_i the real deflator for fuel type i). The subscripts for different fuel types are e for electricity, o for oil, and g for natural gas. The consumption data are measured in trillion Btu's while population is measured in millions.[4] The other fuel in the household sector is coal. Due to its diminished role, its consumption is projected by a time trend.

The equations in the industrial sector reflect an attempt to assign specific fuels to specific users. Thus, coal used for coking purposes is directly related to the output of iron and steel, while natural gas consumption is assigned jointly to the iron and steel industry and petroleum refining based on the relative use of this fuel per unit of output. While this approach reflects an improvement over the approach used by National Economic Research Associates, Inc. (1972) and the National Petroleum Council (1971, 1972), who employed either constant dollar GNP or manufacturing production to project the sector's total consumption, it is still not satisfactory because price effects are absent; hence, a reestimation effort is already under way.

The general form of the equations is indicated by the equation for industrial natural gas consumption.[5]

$$\begin{aligned} Q_{ing} = {} & 303.28 + 5340(a_{29}\Delta \mathrm{JFRB29} + a_{331}\Delta \mathrm{JFRB331}), \\ & (36.23) \quad (2798) \\ & R^2 = .14, \\ & \mathrm{DW} = 1.97, \\ & \mathrm{SEE} = .135 \times 10^{12}\ \mathrm{Btu}, \end{aligned} \tag{4.5}$$

where Q_{ing} represents industrial consumption of natural gas, JFRB331 the Federal Reserve Board's index of production by iron and steel producers, JFRB29 the FRB's index of output by petroleum refiners, and Δ represents the quarterly change in these indices. The weights a_{331} and a_{29} represent the share of natural gas in total output by these two producers as taken from the 1963 input-output table. Their values are .0489 and .027.

Transportation demand is analyzed by fuel type with the three minor components simulated off of time trends and the major component, petroleum consumption, related directly to the real consumption of gasoline in the NIA model. This lumps personal and common carrier consumption together, an unsatisfactory situation that must be remedied.

4.4. Long-Term Consumption Projections

Due to the absence of a satisfactory supply sector, price projections must be made exogenously before the model can be simulated. In addition, due to the design of the model, a long-term forecast must be made with the macroeconometric model. For the example presented here, the macroforecast is based on the Data Resources, Inc. (DRI) ten-year forecast CONTROL TEN 6/72. Energy prices are assumed to increase at 6% per year for oil and electricity and 8% per year for natural gas.[6] As an alternative, we consider the case where all energy prices remain constant (this is almost equivalent to the naïve case where the price elasticity is zero).

Before turning to the results, it should be indicated that these are still partial equilibrium solutions to the general equilibrium problem. If energy prices rise at a rate of 6 to 8% per year, certain industries will suffer while others will benefit. Even if these effects should cancel out in terms of measures of aggregate economic performance, the consumption of energy will still be affected. In terms of industries consuming large shares of energy, the critical areas are iron and steel, aluminum, chemicals, paper, petroleum,

Table 4.1 An Example of the Change in Energy Requirements for 1975 and 1980 under Two Alternative Price Assumptions (Trillion Btu)

	1975			1980		
Sector	Case 1[a]	Case 2[b]	Diff. (%)	Case 1[a]	Case 2[b]	Diff. (%)
Household and Commercial						
Oil	6276	5734	−8.6	7452	5619	−24.6
Natural Gas	8637	8146	−5.6	10435	8999	−13.8
Electricity	3828	3762	−1.7	5435	5086	−6.4
Coal	256	256	0.0	160	160	0.0
Total	18997	17898	−5.8	23482	19864	−15.5
Industrial—Total (excludes nonenergy uses)	22488	22488	0.0	26443	26443	0.0
Transportation						
Petroleum[c]	20749	18965	−8.6	25810	18765	−27.5
Other Fuel	1065	1065	0.0	1490	1490	0.0
Total	21814	20030	−8.2	27300	20255	−27.4
Electric Utilities						
Hydro	3115	3115	0.0	3797	3797	0.0
Nuclear	3394	3394	0.0	9377	9377	0.0
Fossil	13550	13347	−1.6	15802	14730	−3.8
Total	20059	19856	−1.1	28976	27904	−3.7
Total Energy Requirements[d]	77043	74021	−3.9	97165	85780	−12.2

[a] In case 1 no change in relative prices is made.
[b] In case 2 the price of petroleum and electricity is increased 6% while the price of natural gas is increased 8%.
[c] Petroleum projection for transportation changes due to the reduced projection of consumption of nondurables as simulated by the DRI model.
[d] The total energy projection reflects the subtraction of electrical consumption in the three final demand sectors to avoid double counting.

and food. The 70% increase of energy prices by 1980, which we assume in our first case, should reduce the demand for their products by some amount, but lacking satisfactory demand functions and cost functions we are hesitant to estimate the total possible reduction.

The results are given in Table 4.1 for each consuming sector with consumption of the household and commercial sector broken down by fuel type. The effect of the price increase is felt in two of the three other sectors, transportation and electric utilities. The impact on the transportation sector results from the reduced real consumption of gasoline projected by the DRI Macro model in response to the 6% per year increase in the price of gasoline.

The projected reductions in consumption of 4% in 1975 and 12% in 1980 are quite high. They result from the assumption that prices will rise 6 to 8% more than the general price level, an assumption which is admittedly extreme. This further emphasizes the need for the incorporation of a full supply-demand model of energy in a general equilibrium model of the economy.

4.5. Conclusion

In this paper the importance of developing a fully simultaneous energy-economic model has been emphasized. Without incorporating energy price effects in the macro- and energy system models, demands have been shown to be biased upward. An initial estimate of the consumption reductions that are obtained was calculated by the incorporation of an energy model within a national income accounts type quarterly model of the economy.

Notes

1. Four well-known models make this assumption; these are by Baughman (1972), Morrison and Readling (1972), The National Petroleum Council (1972), and National Economic Research Associates (1972). In addition, the works of Darmstadter (1972) and others for Resources for the Future, Inc., which have examined the energy-GNP ratio have been forced to make such an assumption.

2. Our preliminary results confirm those of Houthakker and Taylor (1970), who were also unable to reject the zero price elasticity assumption for household consumption of gasoline. This, however, does not mean the hypothesis should be accepted. In the case of gasoline, it may be necessary to disaggregate to state-by-state analysis of demand in order to correct for population shifts from areas with low miles driven per capita to areas with high miles driven per capita.

3. The basic analysis was reported in Verleger (1973). There, using a modification of the abstract, transportation model, preliminary results indicated that the price elasticities for coal, gas, and petroleum were as follows:

Price Change in	Demand Change		
	Coal	Gas	Petroleum
Coal	−.4	.6	.6
Gas	1.2	−3.1	1.2
Petroleum	.1	.1	−.4

These were calculated by changing the price of each fuel, one fuel at a time, by 1%. The analysis was done using national data on consumption and price data taken from *U.S. Minerals Yearbook*. The model is being reestimated using data broken down on a regional basis with the price data taken from the *Edison Electric Yearbook*

4. Due to the linear form of the equations, the coefficients cannot be interpreted as price or income elasticities. The approximate short-run price elasticities are —.08 for electricity, —.6 for oil, and —.7 for gas. The cross elasticity of demand for oil due to a change in the price of gas is .3. The income elasticities for electricity is .152 while it is .3 for oil.

5. The other equations are omitted due to space limitation. They are available, however, upon written request.

6. These oil and electricity assumptions are taken from MacAvoy and Pindyck (1972, p. 61). The natural gas increase is based on their estimate of the price required to maintain a constant reserves-to-production ratio.

References

Baughman, Martin L. (1972). "Dynamic Energy System Modeling—Interfuel Competition," Report 72-1 of the Energy Analysis and Planning Group, School of Engineering, Massachusetts Institute of Technology, Cambridge, Massachusetts, 1972.

Berndt, Ernst W., and Christensen, L. R. (1973). "The Translog Function and the Substitution of Equipment Structures and Labor in U.S. Manufacturing 1929–1968," *Journal of Econometrics, 1*(1), pp. 82–113, March.

Christensen, L. R.; Jorgenson, Dale W.; and Lau, Lawrence J. (1973). "Transcendental Logarithmic Production Frontiers," *Review of Economics and Statistics, 55*(1), pp. 28–45, February.

Darmstadter, Joel (1972). "Energy," Chap. 5 in Ridker, Ronald G., *Population Resources and the Environment,* Vol. 3 in the series of studies published by The Commission on Population Growth and The American Future, Washington, D.C.

Erickson, Edward W. (1970). "Crude Oil Prices, Drilling Incentives and the Supply of New Discoveries," *Natural Resources Journal,* Vol. 10, No. 1, pp. 23–52, January.

Fisher, Franklin M. (1964). *Supply and Costs in the U.S. Petroleum Industry: Two Econometric Studies,* The Johns Hopkins Press, Baltimore.

Houthakker, H. S., and Taylor, Lester D. (1970). *Consumer Demand in the United States: Analyses and Projections,* Harvard University Press, Cambridge, Mass.

Jorgenson, Dale W. (1972). "Long Term Impact of U.S. Tax Policy: Final Report to the U.S. Department of the Treasury," Data Resources, Inc., Lexington, Mass., January.

MacAvoy, Paul W. (1971). "The Regulation Induced Shortage of Natural Gas," *Journal of Law and Economics,* Vol. 14, No. 1, pp. 167–199, April.

———, and Pindyck, Robert S. (1972). "An Econometric Policy Model of Natural Gas," working paper for the Alfred P. Sloan School of Management, Massachusetts Institute of Technology, December.

Morrison, Warren E., and Readling, Charles L. (1968). "An Energy Model for the United States Featuring Energy Balances for the Years 1947 to 1965 and Projections and Forecasts to the Years 1980 and 2000," Bureau of Mines, Information Circular 8384, U.S. Department of the Interior, Washington, D.C.

National Economic Research Associates, Inc. (1972). "Fuels for the Electric Utility Industry 1971–1985," a report of the National Economic Research Associates, Inc. to the Edison Electric Institute, New York.

National Petroleum Council (1971). *U.S. Energy Outlook, An Initial Appraisal 1971–1985,* Vols. 1, 2, and Demand Task Force Report, Washington, D.C.

——— (1972). *U.S. Energy Outlook, A Summary Report of the National Petroleum Council,* Vols. 1 and 2, Washington, D.C., December.

Verleger, Philip K., Jr. (1973). "An Econometric Analysis of the Relationship Between Macro Economic Activity and U.S. Energy Consumption," in M. Searl, ed. (1973), *Energy Modeling, Art, Science, and Practice,* Resources for the Future, Inc., Washington, D.C., pp. 62–102.

5 Energy System Modeling, Regulation, and New Technology*

MARTIN L. BAUGHMAN†

5.1. Introduction

For many uses in our country the competing sources of energy are highly substitutable. The Energy Study Group (Cambel, 1964) wrote as follows:

> While there are some markets for which only one energy form is now economical, as much as 95 percent of total U.S. energy is consumed for purposes in which several or all of the primary energy sources are potential substitutes (directly or through conversion).

The high degrees of substitutability, compounded by the effects of consumer preferences, changing technologies, resource availabilities, regulatory frameworks, and national policy, imply that a comprehensive analysis of the future outlook for energy supply, demand, and fuel prices in the United States is a rather complex undertaking.

This paper reports some initial work done in the development and utilization of an economic model that appears to offer much potential in these analyses. The model is a simulation program combining into a consistent framework a set of economic theories describing energy flows in the aggregated U.S. energy economy. The first sections of the paper describe the overall formulation of the model, its scope, characteristics, and general behavior. Then the results of a series of simulations done with the model are discussed. These discussions help to show how the model can be used, as well as to indicate that analyses of future policy alternatives taken out of a complete energy system context may be misleading.

In summary, the results indicate that the costs of a policy of low foreign dependence in oil supply can be twice what one would expect if only the oil sector itself were analyzed. The potential impact of gas price regulation upon the future energy outlook is discussed, and the role that coal gasification and shale oil might play in mitigating the rise in future energy costs is considered.

* This work has been supported by the Hertz Foundation and the National Science Foundation (Contract GI-32874). The support from these sources is gratefully acknowledged.

† Massachusetts Institute of Technology.

5.2. The Interfuel Competition Model

The interfuel competition model is an engineering-economic simulation program for the medium to long-range (3 to 30 years) interactions of the primary fuels and secondary energy sources (coal, oil, natural gas, nuclear fuels, and electricity) in the U.S. energy markets. Total energy demand is subdivided into residential and commercial, industrial process heating, transportation, and electricity (a consumer as well as a supplier). Structurally, the model consists of an economic framework for matching consuming sector demands with energy supplies in a way that is consistent with consumer preferences and relative fuel prices. This is done via the following features:

1. By combining coupled equations that connect *all* primary fuels and secondary energy sources within a total energy system framework, a novel picture of how the different fuels interact with each other in the marketplace over time is provided.
2. By focusing on the market processes in supply and demand, a method for examining the effects of regulatory policies and environmental constraints on the energy supply and consumption patterns is provided.
3. By considering the time delays involved in capital investments and changes in demand, the inertia of the system is incorporated into the structure.
4. By identification and incorporation of the principal factors influencing marginal and average costs in the models for each energy source, a basis for market pricing is developed.

Basically, the model combines an engineering description of the state of the system (the configuration and magnitude of the physical plant for production and consumption of energy at each point in time) with the economical considerations that affect the changes in state. The model is dynamic in that consumption patterns and cost trends affect the rate of development of new supplies available to the energy markets, which in turn influence the fuel consumption patterns and fuel prices.

The model is applied to the U.S. energy system with no geographical disaggregation and contains the dynamic relationships relevant on a 3- to 30-year time scale. The seasonal fluctuations of demand are not included, and for this reason, the effects of storage and processed goods inventories in smoothing these variations are neglected. The intent is to concentrate on those phenomena that have their effects on prices for periods of years, such

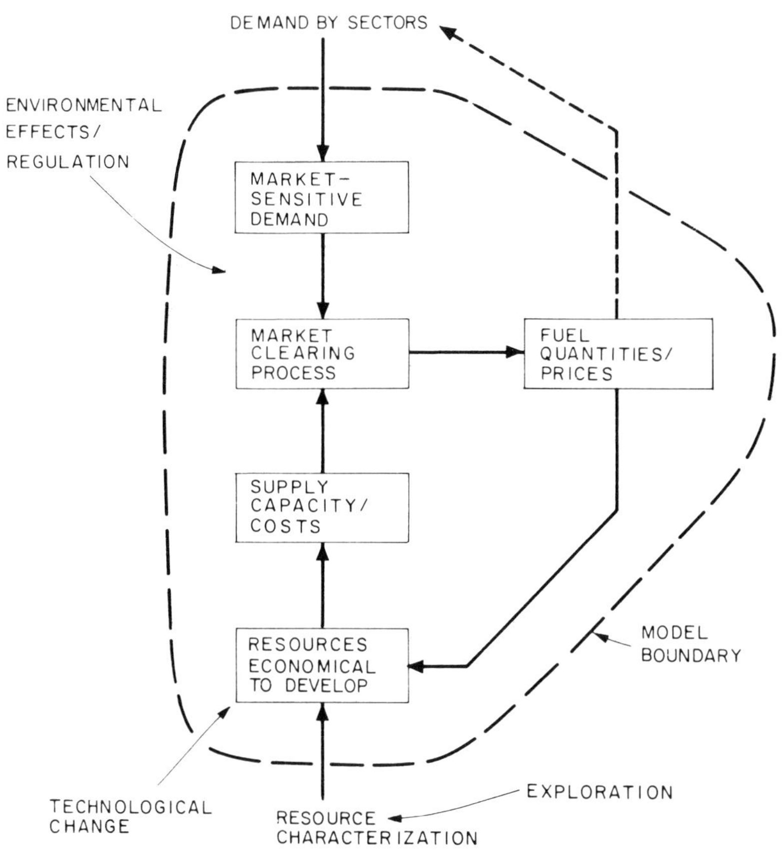

Figure 5.1. Overall model structure—interfuel competition.

as resource depletion, persistent shortages or excesses in production capacity, or lasting exploration successes or failures. In short, what the model does is this: *Given the availability of the fuel resources and the levels of demand for each of the consuming sectors as a function of time, the model simulates the process in which supply production capacity is constructed and resources are depleted, the processes whereby different fuels are chosen to satisfy the demand, and resolves these processes into prices and market shares for each of the forms of supply.*

The overall model structure is depicted in Figure 5.1. The exogenous inputs to the model are shown as the demand by sectors in the upper portion and the resource characterizations in the lower portion. As determined from the rates of growth of demand and the fraction of consumers replacing their energy consumption equipment over each interval of time, some portion of the total demand in each of the consuming sectors will be going to the marketplace to "bargain" for energy. This portion of the total demand is termed the market-sensitive demand in Figure 5.1. The aggregate of those consumers who continue utilizing the same fuels from one time period to the next is termed the base (or "locked in") demand. Depending on the fuel price configurations, the market clearing process matches up supplies of energy to meet the market-sensitive demand.

In general the boundary of the system modeled is given by the dotted line in Figure 5.1. In order for this to be consistent, it is assumed that none of the variables inside the model boundary affects those outside the boundary. The boundary shown in Figure 5.1 therefore has some important implications.

The first of these is that fuel prices computed within the model boundaries do not affect the levels of total sector demand. That is, for each of the primary consuming sectors the total consumption is an input to the model and is assumed to be price inelastic. The cross-elasticities are modeled, that is, the competition among the sources of supply to meet the total level of demand in each sector is the explicit focus within the boundaries.

Second, the model boundaries of Figure 5.1 imply that exploration activities and the resulting additions to reserves therefrom are not dependent on the price variables within the model boundary. It is well known that this is not true. However, the relationship between investment in exploration and the resulting returns is not well understood. For the time being the inputs to the model on the supply side are what are normally considered the results of the exploration process, that is, the rate of additions to reserves and the cost of developing those reserves.

The fact that the total sector demands and exploration process are not price dependent in the model places limitations upon its uses. However, if the user is aware of these limitations, they can be compensated for. Further development to relieve these limitations is under way.

Electricity, as a secondary supplier that utilizes the primary fuels and competes on the marketplace with the primary fuels, is not explicitly shown in Figure 5.1. This is a limitation of the diagram only. In Figure 5.1, think of electricity as simultaneously a supplier and consumer, whose sales to the ultimate consumer are determined in the marketplace, and which simultaneously places a demand on the primary fuels commensurate with those sales. The price of electricity to the consumer is then related to the price that must be paid for the primary fuels, along with the other fixed and variable costs pertinent to that industry.

The environmental, regulatory, and technological effects are shown crossing the boundaries of the model in Figure 5.1. The user must translate these effects into parameters, inputs, and structural changes for the issue of interest. The model can then be used to simulate the implications of these changes on the overall system behavior. In this interactive model, it becomes a tool for aiding in the analysis of future alternatives.

Figure 5.2 summarizes the important assumptions and characteristics of the model. Also included is a broad energy flow diagram to depict the levels of aggregation and interconnections as they exist in the programmed model.

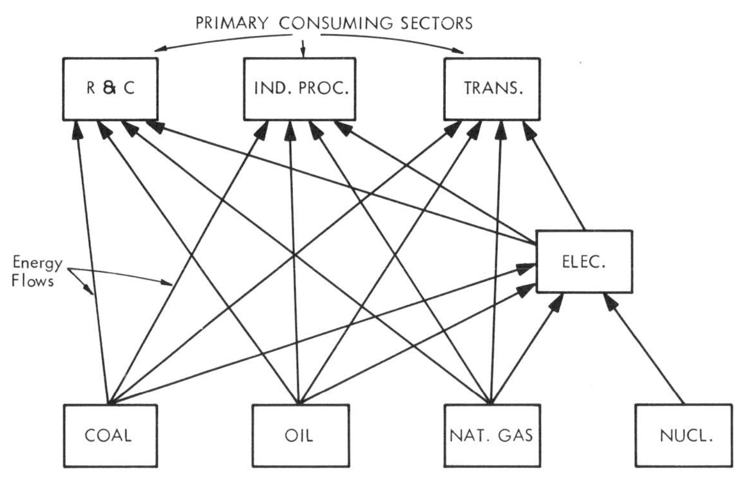

Figure 5.2. Overall model structure—interfuel competition.

Model Characteristics

1. U.S. Aggregated Model.
2. 3- to 30-year Time Scale.
3. Total Primary Sector Demands Inelastic.
4. Only Fuel Prices Used in the Demand Models of the Primary Consuming Sectors.
5. Exploration Not Modeled.
6. Imports and Exports Are Exogenous Time Series.

5.3. Overall Model Behavior—Example

To show the utility of the model, it is expeditious at this point to trace the effects of a hypothesized perturbation through the simulated model behavior. For an example, suppose that we run two case studies. For simplicity, suppose these two cases are identical except for one parameter, the capital costs of fossil-fired generating plants in the electric utility sector. Let us assume for some reason (for instance the cost of pollution abatement requirements) that the capital costs in the second case are higher than in the first. When comparing the model results for the two cases, a series of complex interactions are evident.

In the second case the following phenomena occur:

1. Higher costs of fossil plants relative to nuclear plants.
2. More nuclear generation capacity being constructed.
3. Less demand for fossil fuels by the electricity sector (because a larger fraction of generation is supplied by nuclear).
4. Different average costs of electricity generation (because of different fixed and variable costs of nuclear versus fossil-fired plants).

The interfuel competition exhibits these phenomena, but it also goes beyond these first-order effects to quantify what happens in the rest of the system. In the interfuel competition model the following phenomena also result:

5. Different prices for each of the fossil fuels (the demands for the fossil fuels change in electricity sector, consequently different supply-demand equilibria are reached for all fuels).
6. Different fuel consumption patterns in all the consuming sectors (the fuel prices change because of 5, electricity price changes because of 4, and depending upon all the elasticities and cross-elasticities of demand in each sector the energy consumption patterns change).

This is only one example, but it clearly shows how a change in just one part of the system can affect the entire system behavior. It is the ability to depict this behavior in a complete energy system framework that makes the interfuel competition model useful.

5.4. Model Applications

Before going into the analysis of the topics to be discussed in later sections of this paper, it could be desirable to launch into a discussion of the structural detail and parametric values of each of the component supply-and-

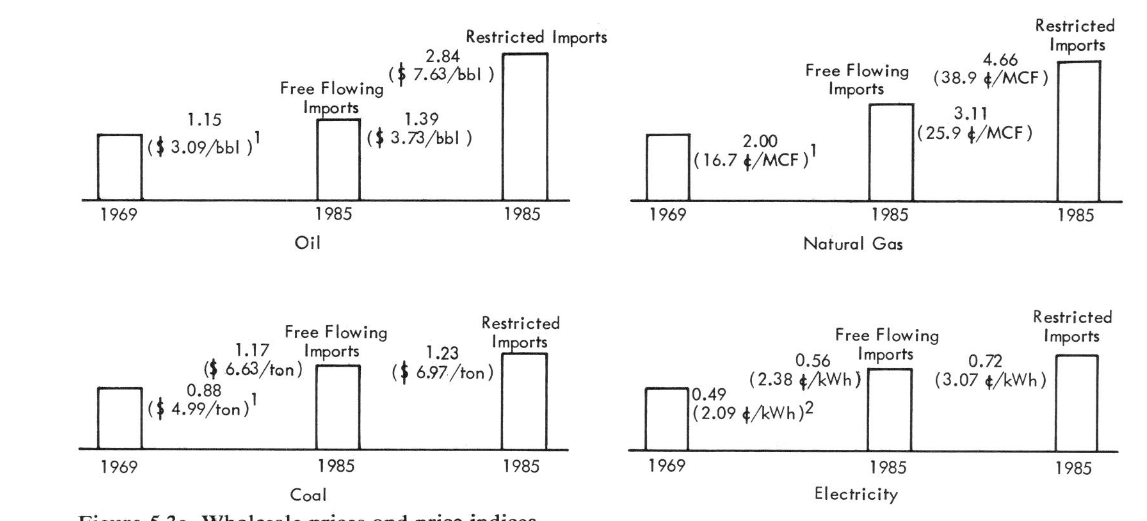

Figure 5.3a. Wholesale prices and price indices.

[1] As reported by the Bureau of Mines
[2] As reported by the Edison Electric Institute
All prices given in 1969 dollars, price indices are relative to 1947 prices in constant dollars.

demand models in the overall formulation. Unfortunately, neither time nor space permits such liberties. However, a complete discussion of the theoretical bases for the model formulation and a comparison of the model performance to historical data from the 1947 to 1969 time period can be found in Baughman (1972). These discussions indicate the model is quantitatively consistent with past behavior of the energy system, and the following sections show how the model can be used to aid future policy studies. The purpose is to quantify what effects a sampling of future alternatives could have on the U.S. energy system.

5.4.1. Oil Imports—The Costs of Low Foreign Dependence

The first set of simulations to be discussed is concerned with oil imports. For illustrative purposes, two scenarios in future oil supply are defined, then the model is used to depict the implications of these scenarios on the future U.S. energy economy.

For the first case let us assume that the future energy system is characterized by relatively free flowing oil imports. This could come about if we assume that prices on the world oil market are fairly stable, remaining at approximately current levels in constant dollar terms. The pressure, under these conditions, would be to hold the line on domestic oil prices and use increased imports to meet the future growth in oil demand.

For the second case, let us assume that the flow of foreign oil into this country is restricted. That is, for some reason (whether balance of payments, national security, or higher prices of foreign oil) the policy is to allow imports to supply only one-third of the U.S. oil requirements (this corresponds to the present fraction supplied by imported oil). In both cases let us assume deregulation of gas prices and growth rates of the primary sector demands corresponding to the recent National Petroleum Council (1972) projections.

Figure 5.3a shows the wholesale prices of each energy commodity that results in 1985. For comparison the 1969 values are also given. The price indices are defined as the price in any given year deflated to 1947 dollars divided by the price in 1947. These are the variables used in the model. For convenience, the price of each fuel in 1969 dollars has been calculated and shown in parentheses.

Note the effect of the restricted import policy on oil prices in 1985. With relatively free flowing imports the price of oil remains fairly constant, increasing by only 20% over the 1969 price. However, with restricted

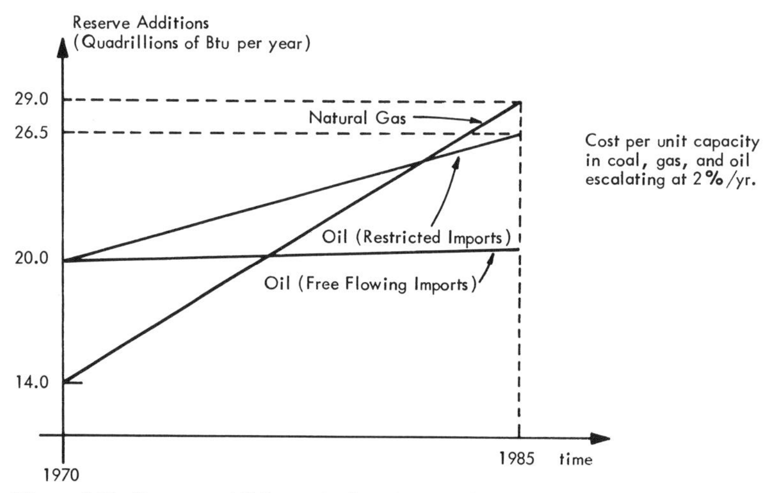

Figure 5.3b. Reserve additions of oil and natural gas.

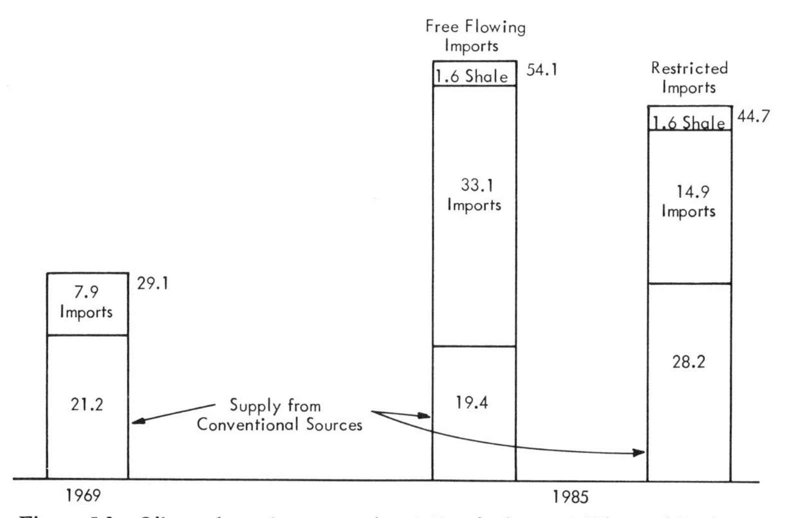

Figure 5.3c. Oil supply and consumption (all units in quadrillions of Btu).

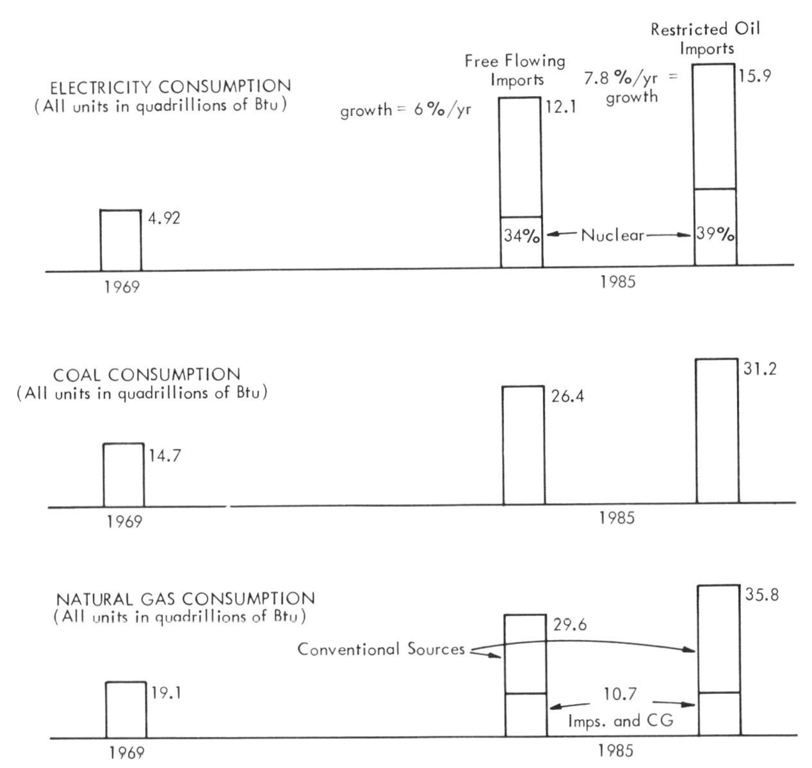

Figure 5.3d. Consumption of other fuels.

imports the price of oil goes to over twice current prices, to about $7.60 per barrel. In Figure 5.3c can be seen the effect this has on oil consumption, decreasing it from 54.1 to 44.7 quadrillion Btu per year in 1985. One might think that this suggests a very low elasticity of demand in oil. This is not necessarily so because of the effect fo the restricted oil import policy has on the prices of the other energy commodities. Their prices in 1985 are also higher with restricted oil imports (see Figure 5.3a). The net effects of all these price interactions are what is displayed in Figure 5.3. The effect on the fuel consumption configuration in each demand sector is shown in Table 5.1.

Now what are the implications of all this? In the first case we assumed relatively stable oil prices and free flowing imports. In order to maintain stable prices domestically (assuming the additions to reserves at historical

Table 5.1 1985 Consumption Summary (all units in quadrillions of Btu's)

	Free Flowing Oil Imports	Restricted Imports	Regulated Gas Prices	Shale Oil @ $5.50
Total Consumption	122.0	129.9	119.6	128.4
Residential and Commercial Consumption				
Coal	0.34	0.31	0.33	0.33
Natural gas	10.8	14.6	14.8[a]	13.3
Oil	11.3	6.2	7.5	7.6
Electricity	5.6	6.9	5.3	6.7
Industrial Process Heating				
Coal	9.9	11.6	8.5	11.5
Natural gas	12.2	13.3	18.8[a]	12.7
Oil	10.2	5.0	5.9	6.2
Electricity	6.5	9.0	5.6	8.4
Transportation Consumption				
Oil	27.8	27.8	27.8	27.8
Other	0.8	0.8	0.8	0.8
Electricity Production	12.1	15.9	11.0	15.2
Electrical Consumption				
Coal	10.9	14.0	9.3	13.4
Natural gas	5.9	7.1	8.0[a]	6.8
Oil	4.8	5.7	3.4	5.4
Nuclear	12.5	18.9	9.8	18.0
Hydro	3.3	3.3	3.3	3.3

[a] As total supply is only 21.7, some sectors must use other fuels, depending upon the rationing scheme.

averages) over 60% of the U.S. oil supply in 1985 must come from foreign sources. At $3.70 per barrel, this indicates a balance of trade deficit in oil of about $20 billion per year. In the second case it was assumed we maintain foreign dependence to 33% of the total oil supply instead of 60%. The result is that domestic prices rise, and at $3.70 per barrel for foreign oil the balance of trade has been reduced to a $9.5 billion deficit. Under the assumption of constant prices for foreign oil, the balance of payments deficit has been cut in half. However, if the prices were to go up on the world oil market to something comparable to U.S. prices, then even at 33% foreign dependence there would still be an oil trade deficit of almost $19 billion in 1985.

The course of action that one should pursue under these conditions is dependent upon a number of factors. The relative attractiveness of the scenarios from the balance of trade point of view depends upon one's expectation for oil prices in the world markets. From the national security point of view, restricting the flow of foreign oil to one-third total supply is obviously better than 60% dependence, but look at the cost. Not only must oil prices rise dramatically to increase the U.S. supply significantly, but also upward pressure on the prices of all sources of energy results. The total price times quantity for each of the sources of energy in Figure 5.3 for the case of relatively free flowing imports comes to about $60 billion per year. However, for the case of restricted imports the same calculation comes to $85 billion. Consequently, to buy the security of decreased foreign dependence and a potential savings of $10 billion per year in the balance of trade, the U.S. consuming public could be paying as much as $25 billion per year more for energy in 1985.

Low foreign dependence has a high price tag. Whether it is worth it or not depends on the value one places on it. What the model illustrates, however, is that it is not enough to look only at oil when evaluating the potential impact of oil import policy. Of the $25 billion increment in costs for the restricted import scenario, only half, or $12.5 billion can be attributed to oil directly. The rest is due to the higher cost of alternative sources. The utility of the model is that the calculation of these second-order system effects is made possible.

5.4.2. Gas Price Regulation

Yet the future outlook for energy may be more pessimistic than that just discussed. The simulations presented in the previous section both assumed deregulation of gas prices. The equilibrium price of gas in 1985 for those simulations was displayed in Figure 5.3a, in one case being almost 56% above current prices, in the other, where oil imports were restricted, it was about 130% above current prices. The rates of additions to reserves and gas supply for each of those cases were shown in Figures 5.3b and 5.3d, respectively. The question to be addressed here is what would be the impact if gas prices were not deregulated.

The price regulation in the gas industry really has three effects. First, it alters the regional allocation of gas in the intrastate versus the interstate markets; second, it affects the intensity of development that takes place on known gas reserves because of the disincentive to invest in more rapid recovery; and third, it decreases the incentive to explore because of the market restrictions in being able to profitably develop and sell newfound gas reserves. The impact of all these factors combined in the gas industry is difficult to make explicit. Certainly the interstate versus intrastate allocation issue cannot be analyzed with a U.S. aggregated model. It can be argued, however, that the price ceilings on jurisdictional gas are effectively price ceilings on nonjurisdictional gas as long as the total nonjurisdictional demand is less than the total supply (Brown, 1972). Let us, therefore, make the naïve assumption that the foregoing condition holds to 1985 to see, at least in the worst case what could be the impact of continued regulation on the total gas supply. For this case it is assumed that gas prices are regulated at their current level, and correspondingly the rate of additions to gas reserves are commensurate with the most recent 5-year average (excluding Alaska).

When this scenario is run on the model, the results shown in Figure 5.4 are yielded for 1985. For comparison, the gas prices and supplies that resulted in the previous simulations with no regulation are also shown. At the low regulated price the total demand for gas in 1985 is 42.5 quadrillion Btu (compared to 29.6 and 35.8 in the previous runs). However, a total supply from conventional domestic sources of only 11.0 quadrillion Btu is available! If we assume this is augmented by about 10.7 quadrillion Btu of gas imports and SNG, as the National Petroleum Council (1972) has projected, an indicated deficit of 20.8 quadrillion Btu still remains.[1] This means that this excess demand will either have to be supplied by other fuels, more imports, or synthetic gas, or go unsupplied. If other fuels are used (as they most likely will be, at least partially), the upward pressure on prices of the other fuels will be even greater than shown in the previous sections.

To be sure, the naïveté of our assumptions forces us to label this a worst case analysis. However, what results in the extreme is clearly the direction that the system moves for the less extreme. The overall effect of the regulation is merely to increase the pressure on the other fuels. In a period when environmentally acceptable domestic alternatives are already under severe pressure, continued gas price regulation could be disastrous.

5.4.3. Are New Technologies the Answer?

The results of Sections 5.4.1 and 5.4.2 may be pessimistic because in some of the simulated results presented the potential that some new sources of supply could have on the future outlook have not been fully exploited. The

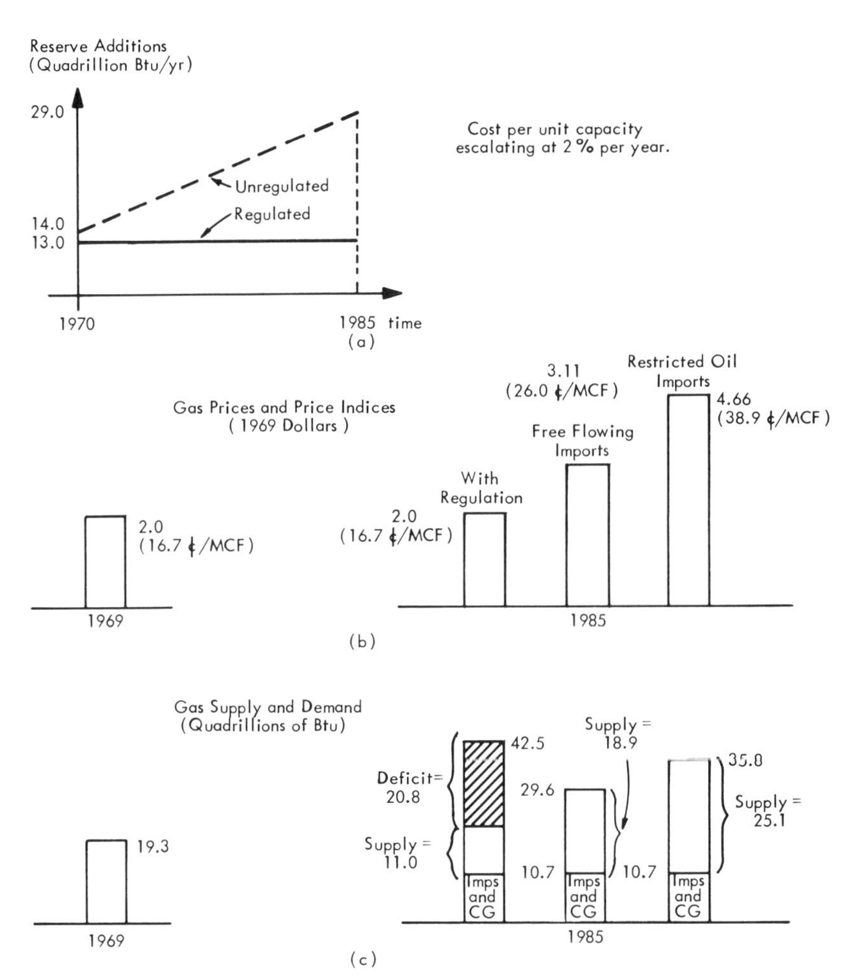

Figure 5.4. Gas price regulation.

purpose of this final section is to discuss how coal gasification and shale oil could change the results presented in the previous sections. These two forms of supply were chosen because they have the attractive feature that plentiful domestic resources exist. To start, let us take some projected costs of supply for these sources and compare them to the prices that resulted in previous simulation.

The author is aware of a whole range of projected costs for each of these supply alternatives, and undoubtedly there exist other projections of which the author is not aware. First let us examine coal gasification.

The projected costs at which it becomes economical to supply synthetic gas from coal usually fall somewhere in the range of 50¢ to over a $1.00 per MCF equivalent wellhead price, but seldom under 50¢ (see, for example, Hottel and Howard, 1971). In the simulated scenarios presented, the average wellhead price of gas never rose above 40¢ per MCF (in 1969 dollars) by 1985. The implication is clear. At these projected costs for coal gasification it will not become a viable large-scale competitor in gas supply by 1985 (at least in the scenarios presented). That is not to say that some demand for 50¢ or $1.00 synthetic gas will not be present in 1985, nor that some synthetic gas from coal will not be consumed in 1985. For isolated markets this could well come to pass. What it does say, however, is that from purely economic considerations one would not expect coal gasification to displace or significantly augment conventional gas supplies, at least for scenarios presented in this paper.

On the other hand, the outlook for shale oil could be quite different. The projections of costs of oil from shale range anywhere from $4.50 to over $6.50 per barrel (see, for example, National Petroleum Council, 1972; and Hottel and Howard, 1971). If we take $5.50 per barrel as a representative number, we see that in the scenario of restricted oil imports presented earlier shale oil could definitely be a competitive source of supply by 1985. In the simulation without oil shale, the price of oil reached about $7.60 per barrel by 1985. Keeping in mind that we are talking about the restricted import scenario, let us discuss the potential impact that shale oil could have on the 1985 supply picture by giving the results of one final simulation. For this run it is assumed that shale oil could be economically produced at $5.50 per barrel and that a completely elastic supply exists at that price. What we want to know is how this alters the supply-demand outlook for in 1985 for the restricted oil import scenario discussed earlier.

The results of this run are summarized in Figure 5.5. There it is seen that oil demand in 1985 is 47.1 quadrillion Btu. At the price of $5.50 per barrel, there would be available from conventional sources a supply of 24.0 quadrillion Btu. Oil imports at one-third total oil consumption would be 15.7 quadrillion Btu. This means that to clear the oil market at $5.50 per barrel there would need to be available 7.4 quadrillion Btu's from shale oil sources. This corresponds to 1.2 billion barrels, or 3.3 million barrels per day of production. This is a large amount, but not unrealistic if the financial resources are committed to have the processing capability available on a time scale commensurate with its need. In addition, having this shale oil available reduces the costs of the one-third dependence calculated previously from $25 billion per year in 1985 to about $15 billion. This is a substantial savings over the case with no shale; however, it still represents a sizable increment compared to the total expenditures of $60 billion resulting in the scenario with 60% dependence.

So the answer to the question posed at the beginning of the section (Are new technologies the answer?) is "no." New technologies may help, but unless some cheaper alternatives become available, the nation is faced with the prospect of either high foreign dependence in oil or higher energy costs.

5.5. Conclusions

When planning future alternatives for the energy system, one could think in simple terms of a national energy policy as one which offers low-cost energy with minimum foreign dependence and minimum environmental impact. Unfortunately, all three of these objectives cannot be met at once. The reason for development of models is so that tools are available for striking a balance among these objectives.

The future outlook in energy supply and demand depends on many factors. The results presented in this paper indicate that substantive changes in the planned or projected outlook in any one sector can have important ramifications for all sectors of the energy economy. The scenarios presented indicate the costs of a policy of low foreign dependence in oil supply could be quite high. They also indicate that the timely development of shale oil resources could do much to mitigate these costs. However, the scenarios presented in this paper are only four possibilities. For a complete analysis of any issue many simulations must be done. There are many alternatives that have not been discussed in this paper. The model presented here,

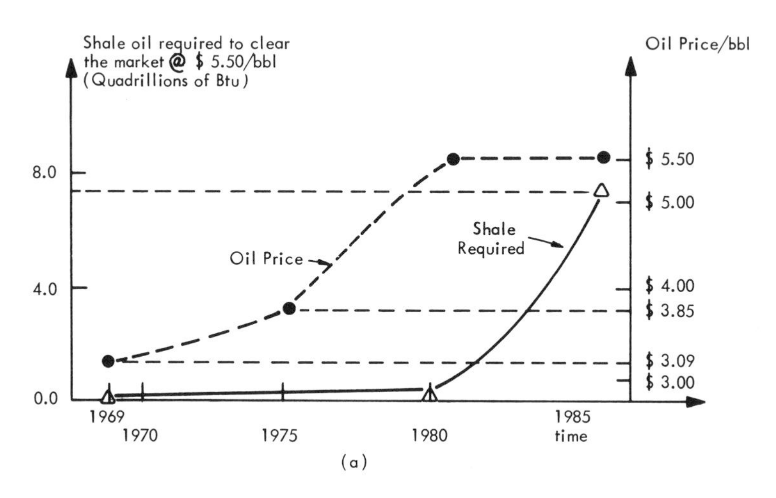

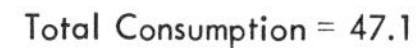

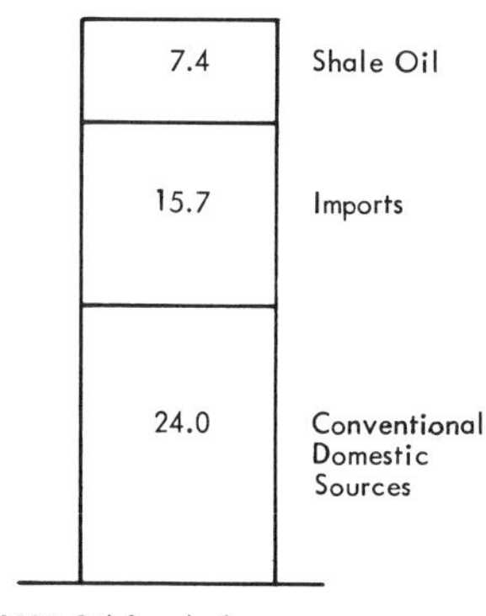

Figure 5.5. Effects of shale oil upon oil price and supply for the restricted oil import scenario.

however, has proved to be a useful tool for many of these analyses, at least on a broadly aggregated level.

Note

1. All runs presented assume the auxiliary supplies of gas are available. Without these additional supplies, the model indicates the deregulated gas prices in 1985 would be significantly higher.

References

Baughman, Martin L. (1972). "Dynamic Energy System Modeling—Interfuel Competition," Ph.D. dissertation, Department of Electrical Engineering, M.I.T., Cambridge, Mass., August.

Brown, Keith C., ed. (1972). *Regulation of Natural Gas Producing Industry,* Resources for the Future, Inc., Washington, D.C.

Cambel, Ali Bulent (1964). *Energy R & D and National Progress,* Energy Study Group of the Executive Office of the President, Lib. Congress Card No. 65-60087, p. XXV, June 5.

Hottel, H. C., and Howard, J. B. (1971). *New Energy Technology—Some Facts and Assessments,* The MIT Press, Cambridge, Mass.

National Petroleum Council (1972). "U.S. Energy Outlook, A Summary Report of the National Petroleum Council," Washington, D.C., December.

6 A Regional Energy Model for Examining New Policy and Technology Changes

P. F. SCHWEIZER,* C. G. LOVE,* AND J. HUNTER CHILES III*

This paper discusses a computer model for examining the supply, demand, and price of primary energy sources and electricity subject to assumptions about future policies and technology changes. The dynamic interactions of supply, demand, and prices are modeled for each primary fuel and electricity for each region and market sector. The energy model is composed of a basic fuel allocation model and a conversion industry model for treating nonprimary fuels such as electricity. Inputs to the model are provided for adjusting parameters and constraints for investigating "what if type" questions on new technology, policy, and environmental restrictions.

6.1. Introduction

Recently, it has become apparent that the United States cannot simultaneously continue its rapid growth in energy usage, maintain low energy prices, protect the environment, and remain relatively independent of foreign suppliers. Because changes in the price and availability of end-use energy forms impact almost every citizen, this situation has provided the incentive for the development of a national energy policy.

In order to evaluate rapidly the consequences of different proposed energy policies, we have developed a computerized energy model. (Baughman, 1972 and Dawson, 1972 describe other modeling efforts.) A computerized model possesses several advantages over proceeding as others have to date (Chase Manhattan Bank, 1972; National Petroleum Council, 1971; and Szego, 1971). The mechanization allows greater detail to be incorporated in the model, and, in fact, provides the capability to implement a model that sequentially simulates in time as opposed to one that makes discrete jumps (of five years for example) that are not strongly interrelated. Thus, the dynamics of changes are captured more accurately. Further, if one intends to answer many "what if" questions, the computerized model constitutes a consistent framework in which to investigate a wide variety of policies rapidly and at low cost. It is pointed out, however, that the existing energy studies serve as sources of data for our model and to a great extent we are using the model as a means of obtaining more information from data that is readily available.

* Westinghouse Electric Corporation, Pittsburgh, Pa.

We plan to use the model to investigate the impact of a wide range of factors on the national energy situation. The following list is indicative of the type of factors that we plan to consider: new technology, environmental restrictions, import quotas, energy conservation measures, rationing, regulations or taxes that force substitutions, and price increases.

In the next section, the model structure is discussed and equations for the model presented. A following section presents a national energy policy case study.

6.2. Model Structure

6.2.1. General Discussion

The general structure of the energy model, as shown in Figure 6.1, is composed of a main allocation model and an auxiliary conversion industry model. The initial data input is composed of information on supply, demand, and price of each fuel and market sector under study. Additional inputs concerning new technology and policy modify parameters within the energy allocation model (EAM).

The EAM uses the mathematical techniques of linear programming to match supply with demand when possible. When shortages occur, specified algorithms are used to determine appropriate substitutes provided they exist and that the unmet demand is of the substitutable type.

The conversion industry model accepts nonprimary end-use demand (that is electricity, low-Btu gas, and so on) and by applying appropriate conversion formulas and efficiencies return this demand to the EAM as a primary fuel demand.

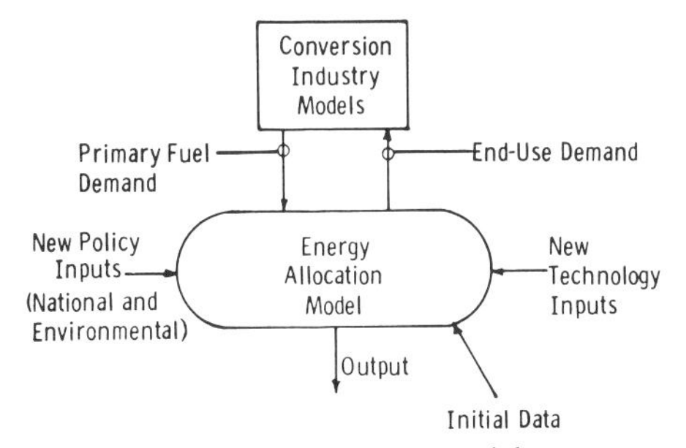

Figure 6.1. Computer energy model.

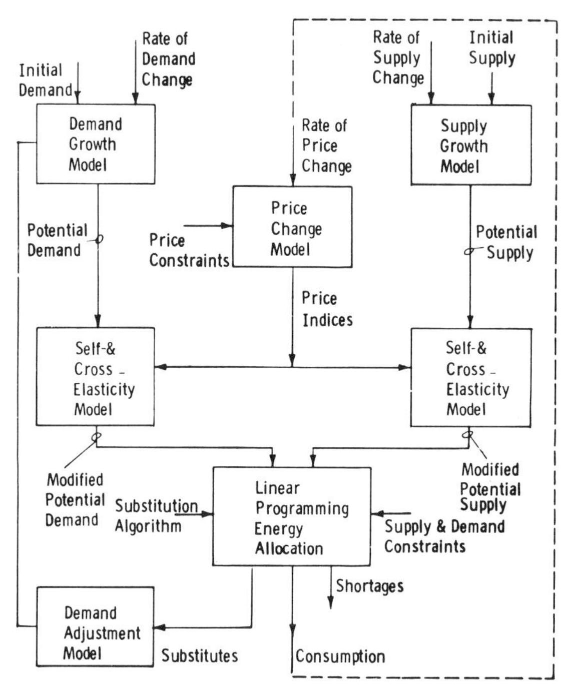

Figure 6.2. Energy Allocation Model.

A functional block diagram of the EAM is shown in Figure 6.2. An initial demand and supply and an initial rate of change of demand and supply for a given starting year are the model inputs. The next years' potential demand and supply are computed using the initial values and rates of change by the demand and supply growth models. To include the effects of prices self- and cross-elasticity coefficients are used to modify the supply and demand. The self-elasticity coefficients provide an increase or decrease in a specific fuel demand or supply based on the price of that fuel while the cross-elasticity coefficients provide a change in demand and supply based on the prices of other fuels or electricity. After the potential supply and demand have been modified by price, they are inputted to the Linear Programming (LP) Allocation part of the model which tries to match supply with demand considering imposed constraints. When potential shortages exist, the LP code uses algorithms inputted by the user to select appropriate substitutes. The final outputs of the LP part of the Energy

Allocation Model (EAM) are the energy consumption, shortages, and substitutes for the various fuels and market sectors. The substitutes are then used to adjust the initial demands to appropriate levels and the consumptions are used (exogenously) in adjusting the various rates of price changes. This model is then used in a repetitive fashion for calculations in successive years.

The purpose of the new technology inputs is to investigate the impact of future major developments (that is, coal gasification, fast breeder reactors, electric vehicles, and so on) with respect to energy supply and demand. From new technologies under investigation, input constraints to the LP on supply and demand are developed to reflect the new technologies' effects on consumption over the appropriate time period. The specification of substitutions and price constraints provide additional means of reflecting the effects of new technologies.

The investigation of market, national, and environmental policies also requires manipulation of EAM parameter inputs. From the new policies under investigation, LP input constraints on supply, demand, and price are developed and substitution policies are modified.

6.2.2. Mathematical Model

The following equations describe the details of the Energy Allocation Model. The equations and order of presentation correspond generally with the model block diagram in Figure 6.2.

Using the initial fuel demand and growth rates, the potential demand at time $t + 1$ is calculated by the following:

$$y(t + 1) = G_d y(t), \tag{6.1}$$

where

$y(t)$ is a 1 by N dimensional column vector representing the potential fuel demands in each market sector and region at time t. The maximum dimension of y, N, equals the product of the total number of fuels, market sectors, and regions;

G_d is an N by N matrix of constant demand growth rates.

In a similar manner the initial supply and growth rates are used to calculate the supply at time $t + 1$:

$$x(t + 1) = G_s x(t), \tag{6.2}$$

where

$x(t)$ is a 1 by M dimensional column vector representing the fuel supplies in each region at time t. The maximum dimension of x, M, equals the product of the number of fuels and regions;

G_s is an M by M matrix of constant supply growth rates.

To include the effects of prices, the potential energy demands and supplies are modified by elasticity coefficients to give quantities that reflect consumer attitudes. The potential demand and supply are modified by

$$y'(t + 1) = (I_d + E_d P_d) y(t + 1), \tag{6.3}$$

$$x'(t + 1) = (I_s + E_s P_s) x(t + 1), \tag{6.4}$$

where

$'$ denotes the modified vectors of supply and demand at time $(t + 1)$;

I_d, I_s are N by N and M by M identity matrices, respectively;

P_d, P_s are F by N and F by M dimensional matrices representing the price indices; elements of P_d and P_s differ by transportation costs;

E_d is an N by F matrix of demand elasticity coefficients;

E_s is an M by F matrix of supply elasticity coefficients.

Linear Programming is used to allocate supply to demand, perform necessary substitution, indicate shortages, and apply constraints to individual and total consumptions. The LP matrices, constraints, and objective function are defined by the following:

$$A\bar{x}(t + 1) = y'(t + 1), \tag{6.5}$$

$$A \stackrel{\Delta}{=} [I_{\text{short}}, A_{\text{sub. policy}}, I_{\text{consumption}}],$$

$$A_{\text{sum}} c_f \leq x'(t + 1), \tag{6.6}$$

$$Z = c'\bar{x}(t + 1), \tag{6.7}$$

where

$\bar{x}(t + 1)$ is a 1 by $(2 + R)N$ dimensional column vector representing the shortages, substitutes, and consumptions for each fuel in each market sector and R regions;

I_{short}, $I_{\text{consumption}}$, are N by N and RN by N identity submatrices, respectively;

R represents the total number of fuel supply regions (may or may not be coincident with market sector regions);

$A_{\text{sub. policy}}$, is an N by N submatrix with coefficients identifying the allowable substitutions of excess fuel supplies for deficient fuels;

c_f is a 1 by $(N \times R)$ dimensional vector representing the consumption of each fuel in each market sector;

Z is the objective function to be minimized;
c' is a vector of weighting coefficients for variables in $\overline{x}(t+1)$.

Equations 6.1 through 6.7 are used in the Energy Allocation Model (Figure 6.2) and in an associated computer code to calculate the energy consumptions, shortages, and substitutes.

6.3. National Energy Policy Example

To demonstrate the application of the model and the associated computer code, an example was selected that might be representative of a future national energy policy. The example problem discussed here will be concerned with energy consumption, substitutes, and shortages beginning in 1970 and projecting to the year 1985. Input data developed from the policy statements below will be used in the Computer Energy Model to compute an Energy Analysis for each of the above-mentioned years. This policy to be investigated is similar to one advocated by Sporn (1972). The policy contains six major statements, and these will be discussed with respect to their implications on input data to the computer energy model.

The first point in this policy advocates that the United States maintain independence of foreign nations by restricting imports of oil and gas. Since present oil imports are nearly 33% of domestic production and rising, it seems realistic for this example to allow imports to rise to 50% in 1980 and then limit them to 50% through 1985. A similar phasing policy was used for liquid natural gas (LNG) imports by allowing them to be 40% of domestic production in 1980 and then to increase to 50% by 1985.

A second point of the policy encourages free competition by allowing natural gas prices to rise to a market level unconstrained by regulation. This is implemented in the model directly through inputs on gas price constraints.

A third point of the policy emphasizes the acceptance of the nuclear energy trend. This is translated for the model purposes by a faster processing of construction and operating permits and the provision of enrichment facilities to allow a rapid growth in both nuclear supply and demand. More specifically, all nuclear plants presently being constructed or planned through 1980 are assumed to meet schedules, and beyond 1980 virtually free substitution of nuclear power plants for fossil-fired plants is assumed.

The fourth element of the policy encourages a conversion to electricity. This is accomplished in the model by the relatively attractive price of electricity, the high supply growth rates, and the substitution algorithms that allow 80% of the potential shortages in industrial and residential/commercial oil and gas to be substituted with electricity.

A fifth element of the policy requires the phasing in of environmental requirements in a manner consistent with the capability of the U.S. economy to absorb these changes. For this example run, this policy element has been implemented by allowing coal (high and low sulfur) to be used by utilities where Environmental Protection Agency (EPA) ambient regulations would not be violated.

The final element of this policy advocates a more efficient conversion, transmission, distribution, and utilization of energy. This is implemented in the model input parameters by requiring transportation demands for oil to be decreased through the use of smaller vehicles, increased mass transit, and increased use of intercity rail transit. The space-heating de-

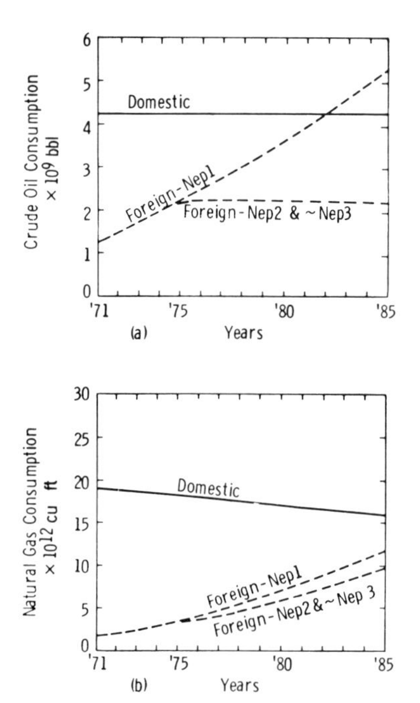

Figure 6.3. Consumption of oil (a) and gas (b) for the projected period of the case study.

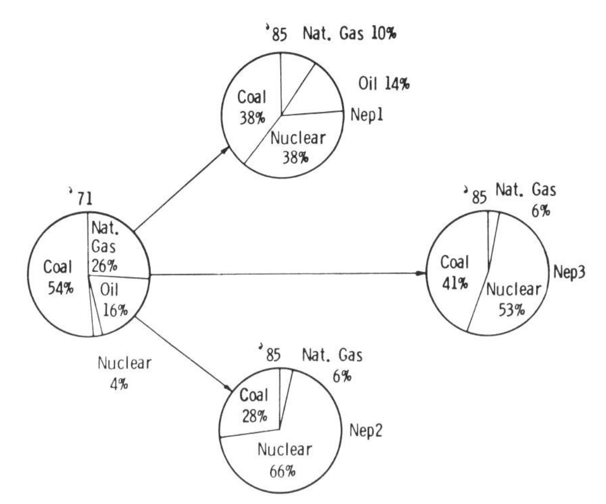

Figure 6.4. Primary sources of energy for the electric utilities for the case studies.

mands for fuel and electricity in the residential commercial sector are adjusted to allow for increased insulation in residential and commercial buildings.

The previous example has been run using the Computer Energy Model for various sets of parameters, and these results are shown in Table 6.1 and Figures 6.3 through 6.5.

Three sets of parameters have been used for three separate runs of the Energy Model. The first set (National Energy Policy, parameter set 1—NEP1), uses unlimited foreign oil and gas imports and allows a moderate coal industry growth of 4% per year. The second set, NEP2, restricts foreign oil and gas imports, allows a moderate coal industry growth (4%), and allows the substitution of most potential shortages (80%) with electricity. The third parameter set, NEP3, is the same as NEP2, with the exception that an accelerated coal industry growth is used (~6% per year).

Table 6.1 shows selected fuel consumption output from the three runs for 1971 and 1985 displayed by market sector and type fuel. The total growth in energy from 1971 (~70 quadrillion Btu) to 1985 for NEP1 (120 quadrillion Btu) is 4.6%. Because of the extensive electrical substitution in NEP2 and NEP3, the consumption in the Electric Utility Sector and total consumption increase significantly.

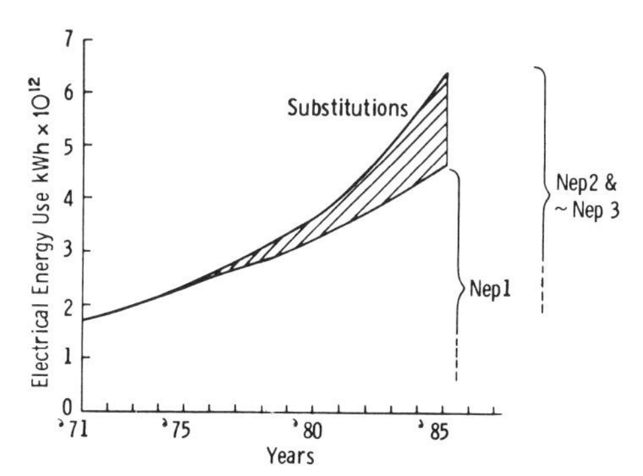

Figure 6.5. End-use electrical energy consumption.

Figure 6.3 shows the computer model estimate of the total consumption of oil and gas for the time period 1971 through 1985. Domestic oil production displays zero growth while foreign imports either rise over 5 billion barrels per year (NEP1) or are constrained to around 2 billion barrels (NEP2 and NEP3). In Figure 6.3b domestic gas production decreases slightly while foreign gas increases under all three parameter sets.

Figure 6.4 shows the primary sources for energy in the Electric Utility Sector for the year 1971 and estimates for 1985. All three runs show declining uses of fossil fuels and increasing use of nuclear to generate electricity. Most likely NEP2 and NEP3 are not attainable in practice because of the limitations of capital, construction capability, and licensing delays. These considerations could be inputted to the model with appropriate data to constrain nuclear growth. This would most likely result in gross shortages since all fossil fuels and nuclear would be constrained.

Figure 6.5 displays the computer estimate of electrical energy consumption from 1971 through 1985. Electrical substitutions for potential shortages of primary fuels are shown added to the base electrical demand.

6.4. Summary

This paper has presented a computer model for examining the supply, demand, and price of primary energy sources and electricity subject to assumptions about future policies and technology changes. The dynamic interactions of supply, demand, and prices have been modeled for each primary fuel and electricity for each region and market sector. The com-

Table 6.1 Energy Analysis from Computer Model Output for NEP1, NEP2, and NEP3[a]

1971 (NEP1, NEP2, and NEP3)	Industrial	Electric Utilities	Transportation	Residential/ Commercial	Total[b]
Coal	4.83	9.33	0	0.12	14.29
Oil	6.73	2.80	16.58	5.81	31.94
Natural gas	10.85	4.74	0	6.81	22.41
Nuclear	0	1.02	0	0	1.02
Total[b]	22.43	17.90	16.58	12.75	69.67
1985 (NEP1)					
Coal	5.40	16.81	0	0.17	22.38
Oil	14.40	6.25	25.20	8.25	54.09
Natural gas	11.40	4.43	0	10.70	26.52
Nuclear	0	16.88	0	0	16.88
Total[b]	31.20	44.37	25.20	19.12	119.89
1985 (NEP2)					
Coal	5.44	17.95	0	1.82	25.21
Oil	14.21	0	22.68	0	36.90
Natural gas	11.40	3.50	0	10.70	25.60
Nuclear	0	44.16	0	0	44.16
Total[b]	31.05	65.62	22.68	12.52	131.88
1985 (NEP3)					
Coal	5.40	26.66	0	1.72	33.78
Oil	14.40	0	22.68	0	36.90
Natural gas	11.40	3.90	0	10.70	26.00
Nuclear	0	35.05	0	0	35.05
Total[b]	31.20	65.62	22.68	12.42	131.74

[a] Units Btu $\times 10^{15}$.
[b] Totals may not add exactly because of roundoff.

puter model has been used to examine the consequences of an example National Energy Policy.

References

Baughman, Martin (1972). "Dynamic Energy System Modeling—Interfuel Competition," Energy Analysis and Planning Group, Massachusetts Institute of Technology, Cambridge, Mass., August.

Chase Manhattan Bank (1972). "Outlook for Energy in the United States to 1985," Energy Economics Division, New York, June.

Dawson, D. O. (1972). "A First Look at Tera," *American Gas Association Monthly*, pp. 23–26, June.

National Petroleum Council (1971). *U.S. Energy Outlook, An Initial Appraisal 1971–1985,* Vols. 1 and 2, July.

Sporn, P. (1972). "The Indispensability of Research," Keynote address, Conference on Research for the Electric Power Industry, December 11.

Szego, G. C. (1971). "The U.S. Energy Problem," National Technical Information Services, PB 207–518, November.

7 World Energy Modeling

R. J. DEAM* AND J. LEATHER*

The Energy Research Unit at Queen Mary College, University of London, was set up six months ago. It has the aim of building mathematical models that would be useful in providing a valid and rational framework for decision making by governments and companies involved in supplying energy. The staff of the Unit have accumulated many years of experience in the construction of linear programming models within the oil industry. Such models, in optimizing some economic criterion, subject to physical, financial, political, and social constraints, have helped to bring about the best use of the resources available and an uninterrupted supply of energy at reasonable cost, albeit at the company level.

It is our belief that the world energy problem is basically the company problem on a larger scale, with more pressure from the political and social constraints, and that the same type of model can be extended to describe the world energy system.

The Energy Research Unit is at present engaged in the construction and validation of such an aggregated model. In the first place, the model will be confined to oil and gas because this section of the energy industry is probably both the most important and the most complex. Important because for many years to come, oil and/or gas will supply the world's marginal energy requirements, and complex because of the variety and multiplicity of raw materials available, refining processes employed, and large-volume products supplied.

The problems of an oil company can be typified by a hypothetical system consisting of, say, two crude oil fields, two refineries, and three product installations adjacent to the market.

As a further simplification, it is assumed that, for the period of time considered, the demand for various refined products at each of the three installations is given. Then the information needed by the management of the company will be

1. How much crude oil is to be shipped from each field to each refinery?
2. How much of each of the products is each refinery to manufacture?
3. How much of each product is to be shipped from each refinery to each installation?

* Queen Mary College, University of London, England.

The constraints on the system are

1. The availability and physical and chemical properties of each crude oil.
2. The selection of refinery processes available at the refineries and the maximum throughputs of each process.
3. The qualities which the products must meet.
4. The mass balance equations.

The data will also include the costs of crude production, shipping, and refining, and it is possible by linear programming (LP) to choose the solution that satisfies the given demands at the least overall cost.

Critics of the linear programming approach, or, indeed, of optimizing techniques in general, frequently base their case on the uncertainties present in the data and the supposed absurdity of attempting to optimize in such a situation. If one were to accept blindly the recommendations of a single solution to the model, such criticisms would be well founded. However, by far the most important aspect of the use of LP models is the analysis that is performed after the first solution is obtained. This can be considered under two closely related headings; postoptimal analysis of a given solution using the marginal values generated on constraints, and cost and quality-ranging techniques; and, where these methods are inapplicable, the further solution of variant cases of the model. Solutions to variant cases can be obtained very much more easily than the first solution since a reasonably good starting point is available.

The economic significance of dual or marginal values generated along with the solution of the LP problem is vitally important. In our very simple model each of the four types of constraints generates such values. Taking them in reverse order, these represent

1. The marginal value of each product at each installation.
2. The marginal value of quality.
3. The marginal value of refining capacity.
4. The marginal value of each crude oil.

Now suppose that the two refineries were owned by different companies. If it were possible to assume perfect and instantaneous knowledge, the formulation and hence the solution of the problem would remain completely unchanged. Under such an assumption the marginal costs of products developed by the LP solution process would represent market prices. The total demand at the three installations would be divided between the two companies in the way that minimized the overall cost, because this is exactly the same solution which a single owner of the two refineries would aim at. If constraints of a political or social nature are imposed on one or other or both of the two companies, these constraints will, insofar as they are restrictive, increase marginal costs, and the balance of market share may be affected. But the equality between marginal costs and prices would still hold.

Taking account of the fact that knowledge, particularly of the future, is imperfect and of the inevitable time lags in the system, one could say that the marginal costs will set the levels toward which market prices will tend. This "perfect competition with time lags" is capable of being upset by the companies' own actions, for example, by unilateral price fixing or by a collusive price or market share agreement. By comparing marginal costs developed by the model with actual market prices, it would be possible to detect areas where the deviation was large and lasting and where intervention by government or by an industrial competitor would be advantageous.[1]

While it is clear that a legal or fiscal constraint imposed on one of the two companies will increase that company's costs, it should be realized that it may well increase the overall costs by a greater amount by also increasing the other company's costs. For example, if the market supplied by one installation is legally prevented from burning high-sulfur fuel oil, this may result in increased demand for one of the crude oils and could force up its price and hence the other company's costs. In this way, the desire for low-sulfur fuel oil in New York may force up fuel prices in Europe.

Thus, when the canvas is broadened from the two-refinery system to the world energy system, it becomes apparent that constraints imposed in one area may have economic repercussions in many or all of the other areas. The ramifications are too numerous for the unaided human brain to comprehend. A model with worldwide coverage is necessary to enable the total impact of existing or proposed legislation to be seen, let alone evaluated.[2]

The initial model will be elaborated in many directions. It will include a description of the transport function, with limitations on the availability of shipping.[3] It will represent the possibility of expanding capacity (in oil fields, shipping, or refining) by means of capital investment. The other primary sources of energy will be included. The financial activities and the corresponding financial constraints may be added, and balance-of-payments questions may be examined. Political constraints may be represented and their impact on the optimal solution measured. And very importantly, the effects on strategy and the marginal costs of changing demand patterns may be followed. The model will eventually be expanded to cover several time periods so that the nature and timing of strategic decisions may be examined (for example, the substitution of one fuel for another, the rate of investment in VLCC's [very large crude carriers] and in deepwater port facilities). A time horizon of at most 15 to 20 years is envisaged, long enough to set the immediate future decisions into their temporal perspective.

Furthermore, given a macromodel of this kind, it becomes possible to feed quantities and values to more detailed micromodels. Consider, for example, a national electricity generating system in which the investment decisions are presumably very dependent both on the availabilities and prices of alternative fuels (gas, coal, oil, nuclear), and also on the need to provide base-load and peak-load generation facilities. The latter calls for a detailed model of existing and proposed generation and distribution equipment, while the supply and cost of various fuels are obtained from the world energy model.

Even before the incorporation of other energy sources, the model will be usable to study many problems, some of which are dealt with in subsequent papers. It is proposed here to discuss product prices and the effects on these of changing demand patterns.

There is no need to describe the details of refinery technology at this juncture since the response of the refinery to a change in the structure of product demand can be described, at least to a first approximation, quite simply. We imagine a refinery as basically a device for splitting crude oil into naphtha ($<200°C$), middle distillate (200 to 350°C), and heavy fuel oil (HFO) ($>350°C$).

Since distillation capacity is being increased, there must presumably be an economic incentive to do so. This requires that

$$\alpha P_N + \beta P_M + (1 - \alpha - \beta)P_H \geq P_C + \text{fully built-up distillation costs.} \quad (7.1)$$

In the perfect-knowledge, perfect-competition situation the equality would hold.

The Western European products market, over the last decade, has been dominated by the fact that the average crude barrel is deficient in middle distillate relative to the demand pattern (see Table 7.1). There have resulted two sorts of measures: the installation of cracking processes to convert HFO to middle distillates and the taking of shorter naphtha cuts to raise the yield of middle distillate by lowering its front-end cut point. The consequence of these measures is, first, that the price of fuel oil is determined by

$$P_F = P_M - \text{fully built-up vacuum distillation unit/cracking costs,} \quad (7.2)$$

and second, that the price of naphtha is very similar to that of gas oil,

$$P_N = P_M. \quad (7.3)$$

These three relationships are sufficient to determine the equilibrium prices of the three products relative to that of crude oil. This model is, of course, a greatly simplified description and ignores the downstream complexities at the refiner's command and the effects of product quality variations on price. That it nevertheless represents the real world to a reasonable degree of approximation is shown by actual market prices at Rotterdam for regular motor spirit (approximately naphtha), gas oil, and HFO over the period up to the late 1960s (see Table 7.2).

Table 7.1 The Middle Distillate Shortage

	Average Demand (Western Europe) (1960–1970) as % of barrel	Yield, as % of barrel on Kuwait	Nigerian Light
Naphtha	25	23	29
Middle Distillate	30	22	35
Heavy Fuel Oil	45	55	36

Table 7.2 Average Bulk Product Spot Prices (Rotterdam) *less* Cost of Freight (Persian Gulf to Rotterdam) 1966–1971[a]

Regular Motor Spirit	$17 per metric ton
Gas Oil	$15 per metric ton
Heavy Fuel Oil	$5 per metric ton

[a] For reference, see this volume's Appendix of Proceedings, Abstract A-1, "Energy Economics," F. M. O'Carroll.

The "price" of crude to which we refer relates to the landed cost in Western Europe and consists, for the marginal crude (which in the period considered was Kuwait), of the marginal cost of production plus taxes and freight. For other crudes, the value relative to Kuwait is determined by a complex of factors including its transport costs and its quality and yield structure relative to Kuwait.

There are several reasons to suppose that this price structure is breaking down because the pattern of demand is changing. There is pressure for fuels to meet more and more exacting quality requirements (for example, sulfur content of HFO) and fuels, such as natural and substitute natural gas that are relatively acceptable environmentally, are in increasing demand.

To some extent the demand for low-sulfur fuel oil (LSFO) can be met by producing relatively more low-sulfur crude oils. Unfortunately, however, there is a tendency for low-sulfur crudes to have rather lower yields of HFO. Furthermore, this difference is exaggerated if two crudes with different yields of HFO are processed to give the same total quantities of naphtha and middle distillates (see Table7.3). Naturally, since fuel oil prices are below crude costs, it is rational to refine for minimum make of HFO. To the extent that the demand for low-sulfur fuel oil cannot be met by use of low-sulfur crudes, it will become necessary to desulfurize the HFO. This in turn will result in some loss of material boiling above 350°C.

Taking these effects together, there would arise a shortage of HFO, and the probable result will be a convergence of prices of low-sulfur fuel oil and middle distillates. In such a situation the installation of a cracking plant would make no sense at all.

The demand for gas is likely to lead to an increased call for naphtha for

Table 7.3 The Residual Fuel Oil Shortage

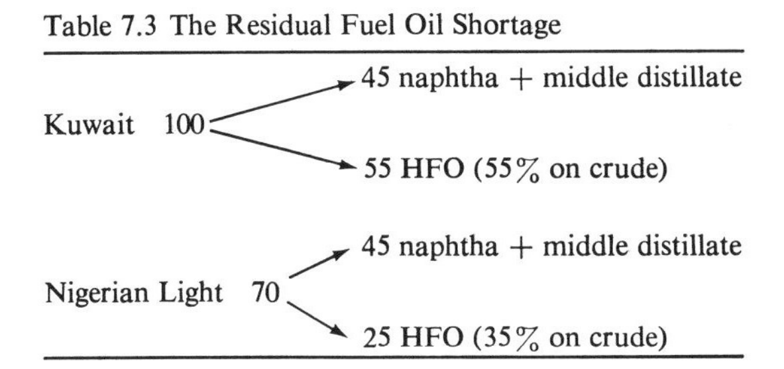

conversion to synthetic natural gas (SNG). The overall result will be that our earlier equations will be replaced by something like the following:

$$P_N = P_G - \text{cost of SNG conversion } (P_G = \text{price of gas, that is, landed price of liquid natural gas, LNG}), \quad (7.4)$$

$$P_M = P_{\text{LSFO}} = P_H + \text{cost of desulfurization.} \quad (7.5)$$

Thus, the former price structure could give way to a quite different one relative to the marginal crude. Furthermore, it is fairly clear that Kuwait will be replaced as the marginal crude, if it has not already been replaced, by Arabian crude.

These conclusions are speculative at present. It is our intention, when solutions of the model become available and when they have been analyzed in depth, either to confirm or deny these speculations. It is, however, an essential part of model building to formulate hypotheses that it is desired to test. When the model confirms the results of prior reasoning, confidence in the model is generated. When there is disagreement, then either the model needs correction or a step has been taken toward a deeper understanding.

Notes

1. See Appendix of Proceedings, Abstract A-1, in this volume.
2. See Appendix of Proceedings, Abstract A-27, and Chapter 8, both in this volume.
3. See Chapter 31, this volume.

8 The Implications of National Policies on World Energy

P. C. WARD*

There is no unified international world energy policy at this point in time, and the prospect of one evolving in the foreseeable future would appear to be very remote.

Individual countries (and companies come to that) pursue policies that they consider will suit themselves best economically or will protect their indigenous industries, commercial interests, or social environment.

National legislation is invariably in the interests of self-betterment in one way or another, those countries that have extensive indigenous energy resources aim to exploit, conserve, or protect them, generally without consideration of the effects their policies have on other, perhaps less fortunate countries. National policies, especially those of major nations such as the United States, often have international ramifications, but ironically, and sometimes fortunately, they do not always achieve their desired intent. There have been in fact several situations, which I will describe later, where legislation has generated quite a different effect from that intended.

The Queen Mary College group, as you have already heard, has constructed a linear programming model representing the world's oil supply refining and distribution system, and one of our projects is to examine in depth the global effects of unilateral political decisions by individual nations or groups of nations. Initially we are concerned with the effects of oil legislation, but the model is being expanded to represent the complete world energy scene, so that in a few months time we will be in a position to examine the worldwide supply and economic effects of national actions with regards to the oil, gas, coal, and power industries, including nuclear power; at the same time we will consider the associated environmental protection and worldwide resources of energy and technology.

The linear programming technique is ideally suited for an investigation of this nature. Our basic oil model has been constructed considering the world as twenty or more discrete refining and demand areas, with a comprehensive representation of the world's crude oil availabilities, transport systems, refining capacities, and area product demands and quality patterns. Area demands will be satisfied on a minimum cost basis, that is

* Queen Mary College, University of London, England.

Table 8.1 Linear Programming Model of Restricted Interchange of Oil Products

Crude Shipping Restriction	Area	Product					
→	Area *A*	Gasoline			=	Total tons	$/ton
		Gas oil	+1	+1	=	Total tons	$/ton
		Fuel oil			=	Total tons	$/ton
→	Area *B*	Gasoline			=	Total tons	$/ton
		Gas oil	−1		=	Total tons	$/ton
		Fuel oil			=	Total tons	$/ton
→	Area *C*	Gasoline			=	Total tons	$/ton
		Gas oil		−1	=	Total tons	$/ton
		Fuel oil			=	Total tons	$/ton
* * *					$\leq$	Availability	$/DW ton
Product Transfers							
Freight Cost			$/ton	$/ton	Σ	Total $	$/ton
Import Tariff Area *A*			$/ton	$/ton	Σ	Total $	$/ton
Import Restriction, Gas Oil to *A*			1	1	$\leq$	Total tons	$/ton
Shipping Restriction			*	*	$\leq$	Availability	$/DW ton

* Shipping Ton Days.

assuming perfect competition, but with known physical, commercial, fiscal, and political restraints superimposed onto the model.

Table 8.1 explains in a very simplified manner, the way in which our linear programming model is being used to assess the national and global, material and economic effects of discriminatory oil legislation.

The table considers only three areas, each with refining capacity but producing only three products. In the real model each of the twenty or so areas includes a sophisticated refinery representation where all conventional types of plant are represented along with new plant and grass-roots refinery building opportunities. Over ten different product types are in fact detailed with comprehensive exchange opportunities between areas.

The estimated production levels of all the known world crude sources are given as availabilities in the model, all major crudes being fully structured for quality and yield. Production expansion opportunities at cost are also included.

If we study Table 8.1, we see that the world's crude oil production capability is allowed to supply any of the three areas *A*, *B*, and *C*. Costs of crude production are included in the model where factors such as pumping, storage, and jetty costs and sheikh's takes can be realistically assessed

The restrictions on the shipping capacity available is represented by the "Crude Shipping Restriction" row, which represents a simultaneous equation, the total tons per day of shipping required for the crude supply operation not being allowed to exceed the figure on the right-hand side, which represents the total shipping availability, also in tons per day. We have, in models considering periods, some years hence, the ability to procure extra shipping capacity at the estimated cost of building, but this is not shown in the diagram.

Product demands, which are represented as equation right-hand sides, must be satisfied either by own area production or imports from other areas. Landing costs and refining costs are included in the model so that the marginal values generated against the product rows will be the equilibrium values or cost of the marginal ton of this particular product ex-refinery in this area.

In a world of perfect competition, unrestricted interchange of products would result, for any one demand pattern, in a least-cost solution based on the relative costs and availabilities of crudes, crude shipping costs, and refining and product freight costs.

Now we are not in a world of uninhibited free competition. Area *A* is quite likely to penalize and restrict imports, say of gas oil (No. 2 Furnace Oil) in order to protect her indigenous industry. This is easily represented in LP terms as can be seen from Table 8.1. The activities "Gas Oil transfers *B* to *A*" and "*C* to *A*" will incur freight costs and import tariffs in the relevant summation rows, and could also be subject to a maximum level of total imports to area *A* as defined by the "import restriction" row. This row will not allow total imports to exceed a prestated level denoted by the right-hand side figure, and at the same time will generate in an optimal solution, a marginal value which represents the unit cost of this restriction at the margin.

Thus, we can produce another solution, but one subject to these additional restraints. From this solution we can assess the total cost to the system of diverging from the previously optimal solution and the cost change within each area. A suitably designed model will also indicate changes in national balance of payment levels as well as indicating, with the marginal or generated values, the unit cost of any individual restriction.

It can be seen then that with a comprehensive linear programming representation of the world's energy operations and requirements, we are able

Table 8.2 Estimated Oil Production and Consumption (million metric tons/annum)

	Production	Consumption	Gap
United States (including Alaska)			
1972	475	750	−275
1977	550	1000	−450
1980	550	1150	−600
Western Europe			
1972	20 (EEC 12)	700 (EEC 550)	−680
1977	120	900 (EEC 700)	−880
1980	150 min 600 max	1000 (EEC 800)	−850 max −400 min

to assess the effect of national policies (or political restraints) on all aspects of the world energy environment. From our solution we can derive the equilibrium relationships between the various energy types and assess the global economic implications of any single or series of actions.

National political actions that influence the world oil scene fall into two camps: those made by the producing countries, with Organization of Petroleum Exporting Countries (O.P.E.C.), United States and U.S.S.R. predominating, and those of the major consumers—United States, Western Europe, and Japan. The United States has always been, and is still the largest single consumer but is being closely followed by Western Europe, as Table 8.2 shows.

U.S. production fell behind consumption to the extent of 275 million tons in 1972 (that is about 5.5 million barrels per day), and even if Alaskan oil is flowing by say 1977 this gap will widen to something in the order of 450 million tons (9 million barrels per day) and is likely to top 600 million by the early 1980s. Without Alaska, the gap could well be in excess of 500 million tons (over 10 million barrels), so in the long run political pressure may be necessary to ensure that Alaskan oil does flow.

I do not intend to dwell on the U.S. energy gap or crisis, call it what you like, but it is interesting to see what political actions preceded it and what political decisions are likely to be taken in the future.

One of the most significant and far-reaching political actions ever to influence the oil world was made in 1959 by the Eisenhower administration. I am talking of course of the U.S. Quota System for oil imports. This legislation was passed in order to protect the U.S. indigenous oil industry, and in essence it achieved its objective. Crude oil prices in the United States were maintained sufficiently high to stimulate production activities, and it has been estimated that between one-third and one-half of domestic production would have been shut down but for the quota restrictions. It also increased coal production in the United States at least for a time and of course put considerably more pressure on natural gas supplies.

However, the significant results of this act really lie outside of the United States, since it has affected both foreign oil producers and consumers to a great degree.

With the imposition of the quota system, foreign producers, notably Venezuela and the Middle East, found themselves with production capabilities in excess of the available demand. Producing countries whose reserves had been established, often by U.S. companies with a view to supplying the North American market, found themselves with dwindling returns through no fault of their own. This stimulated concerted action by the producing countries and, in part at least, resulted in the formation of one of the most powerful political organizations of the oil world, O.P.E.C.

This was not however the only significant result of the U.S. import restrictions. The restricted U.S. import market for crude oil resulted in easier prices for the rest of the world, and the Eastern Hemisphere should certainly regard the U.S. action as a massive act of inadvertent benevolence, as these lower prices gave considerable stimulus to the postwar development of Europe and Japan. It has been stated that "the importance of this event to the economics of Western Europe has been as great as the Marshall Plan in European recovery."

Adjustments to import quotas will affect the current U.S. energy shortage but will also significantly influence the costs of oil to all other nations. These factors we intend to investigate in the near future using our world model.

While we are still with the United States, it might be worthwhile considering what economic (and political) alternatives exist for filling the U.S. energy gap over the next decade.

Political pressures will probably preclude any long-term, large-scale

United States/Soviet gas or oil deal, so United States non-American import requirements are likely to be met by the Middle East, Africa, or Far East sources. Already there have been political/commercial contacts between Saudi Arabia and the United States with a view to Saudi downstream participation in the U.S. refining and marketing operations. While this participation might appear attractive to the Saudis, it is doubtful whether it will be politically acceptable to the United States. Nevertheless, some arrangement between the United States and Middle East countries will soon be arrived at, of that we can be sure.

The current U.S. gas shortage will undoubtedly have to be met by imports, and it is accepted that these alternate supplies will be expensive. The actual method of supplying the United States will probably be decided, not so much on economic grounds but by political and security considerations.

Transporting liquid natural gas (LNG) to North America not only requires extremely expensive bulk carriers but highly capital intensive liquifaction plants located often in politically unstable areas such as the Middle East or North Africa, and this will not appeal greatly to the more security-minded American politicians or international oil companies. Some gas deals with Algeria have already been agreed, but the more acceptable alternative would appear to be either the location of gas plants in areas such as Australia or undertaking a home investment program for synthetic natural gas (SNG) production based on crude or naphtha. The advantages of the latter is in the availability of feedstock from a number of sources and thus lower security risks, and also the promotion of indigenous technologies, a factor that is particularly attractive to certain sections of the U.S. administration. Whatever method the United States adopts to satisfy the home demand for gas will cause balance of payments problems, but, with appreciable gains likely for the participants at the supply end, Australia would stand to benefit considerably from any United States gas deal, whereas Europe, with refining capacity well in excess of that of the United States, would be in a position to supply naphtha feedstock in large quantities to the Eastern Seaboard.

Let us now consider the politics of the next largest demand center—Western Europe. Europe is currently undergoing a transformation, happily a peaceful one, but it remains to be seen whether the European Economic

Community (EEC), now nine nations strong, will ever be in a position to make effective political decisions that will have global implications in the energy scene.

The demand for oil in the nine nations of the EEC is something like double that of Japan and 75% of that of the United States. Estimates of the future indigenous crude production from the North Sea vary considerably. Official estimates are of the order of 150 million tons per year by 1980 (perhaps 15% of demand), but some experts consider the true figure could be well in excess of this. If the lower estimate turns out to be correct, then by 1980 Europe will be looking for imports almost double those required by the United States. Even EEC imports would require about 800 million tons compared with 600 million for the United States. Is there then a danger of a political and economic confrontation between North America and the European Community, both competing for world oil supplies? Now history shows us that official estimates of oil reserves are usually conservative, and, if the North Sea estimates of the optimists turn out to be correct, then European requirements of imported oil in the 1980s is going to be considerably reduced. Europe is unlikely ever to become self-sufficient in oil, but there is a real possibility of EEC production surpassing that of the United States, putting the Community in a very strong bargaining position for the rest of the world's oil production. However, this does hinge not only on North Sea oil but on the European Community evolving a common energy policy, and so far there are no signs that this will happen in the near future. In the opinion of the author, North Sea oil is likely to be a stumbling block to European unity for many years to come. Already, one potential member of EEC (Norway) decided to opt out on hearing certain European politicians state that oil found within the Community should be oil for the Community. (I say "European" there in the strictly non-British context.)

There does not really need to be political pressure to make European oil producers exploit their reserves to the fullest, such that Europe may achieve some of the supply security that has in the past been the unique luxury of the United States. A reduction in the demand for oil imports to Europe from traditional sources is likely to have the effect of stabilizing prices. Added impetus could well be given by community legislation restricting imports of foreign oil so we could, within the next decade, have the reverse situation of the 1960s, this time Europe's imposition of import quotas contributing to the continued economic progress of the United States.

One item of EEC legislation however will result in increased costs to the consumer: this is the mandatory storage regulations. Currently the original six nations are committed to retain reserve storage equivalent to 65 days of the previous year's consumption and it is now proposed to extend this to 90 or even 120 day's stocks. The EEC Commission is not proposing to foot the bill, this is likely to fall on the international oil companies. Eventually the consumer will have to pay, and it has been suggested that the cost of meeting the 90-day requirement could approach 80¢ per ton. EEC countries are, however, proposing that oil produced from indigenous sources should be exempted from the stockpiling obligation so there is an added incentive to exploit North Sea oil to the fullest.

We intend to investigate and quantify the international significance of this legislation, but it is logical to assume that as the major international oil companies will have to bear the brunt of this extra cost then the global cost of products is likely to rise. Furthermore, the traditional suppliers may consider that large stockpiling in Europe could weaken their bargaining power on future price discussions, so they might react by restricting production or increasing charges on the additional crude required.

The six original members of the EEC have not shown any inclination to adopt common policies with regard to oil. In particular, Italy, whose market is dominated by a powerful national oil company, tends very much to "go it alone." Italian purchases of Soviet crude have been much higher than those of other EEC countries, much to the annoyance of her NATO allies. The Italian government also initiated legislation that compelled companies building new refineries to install excess capacity. The original intention of this was to increase home refining and thus reduce the Italian balance-of-payments deficit. However, we consider, and hope to be able to prove with our energy model, that this excess capacity in Italy provided a surplus of products which in turn generated lower product prices in Italy at the expense of higher prices in other countries that had not built sufficient refinery capacity.

The policy of the U.S.S.R. with regard to international oil, despite apprehension in many quarters, has not been of particular embarrassment to the Western World. Figures available of Communist production and demands indicate that, while their surplus is appreciable, it is unlikely to be of global significance relative to the established Western World suppliers. The Soviets are obviously not averse to using their exportable oil to achieve local political ambitions, as for instance in Cuba, but it would appear that their main objective in promoting sales is an economic rather than a political one.

The Middle East has been and is likely to remain, the world's major supplier of oil. Over the past few years two major factors have influenced world supplies of oil from this source; I refer to the formation of O.P.E.C. and the closing of the Suez Canal. Oil companies and consumers have grown to live with and come to terms with both of these happenings, in fact it has been suggested that with the advent of a Red Sea/Mediterranean Arab pipeline there will never be an economic incentive to reopen Suez. While we will be investigating the economics of opening the Canal, the final (if ever) decision on opening it will be made not on economic grounds, but as a result of extensive political (and possibly military) maneuvering.

O.P.E.C. has been successful in negotiating higher returns for developing countries, but what the world really worries about is whether the organization will be able to take concerted action and restrict the production of all members. So far they have been singularly unsuccessful in applying any working quota system, and it is likely that this situation will prevail despite fallacious assurances from various bodies that "oil is worth more in the ground." That O.P.E.C. will aim to get a larger slice of the economic cake is obvious, but what they intend to do with their vast revenues and what global effects the latter will have is something we will investigate. While individual producing nations may have delusions of grandeur and political aspirations to dominate the economies of Western nations, the organization as a whole realizes the importance of the interdependence between producers and consumers, so that while they are likely to squeeze every possible cent out of their customers, they will also have an eye on their long-term future when their supplies of energy have less significance. Participation with or in international majors, or with national governments, or investments in new technologies are but a few of the outlets that may be considered by the nations of O.P.E.C. The result of their actions could be significantly felt by the Western nations over the next decade or so.

Iran, being one of the most progressive of Middle East countries, needs as much money as she can get to finance her ambitious economic development programs. As a result, the shah's latest ultimatum to the oil companies is not out of character. His demands that oil companies should either accept nationalization, with a promise of security of supplies for 25 years or continue operating to the existing agreement until 1979 and then get out altogether, may well have a significant effect on the attitudes of the other O.P.E.C. members. The global implications of his action will have to be carefully considered especially if his condition looks as though it is being infectious.

I have detailed but a few of the items of national legislation that we intend to investigate in the near future with a view to quantifying their short- and long-term national and global effects. Later we will examine the effect of legislation on the different forms of energy: coal industry subsidies, legislation relating to nuclear power, and natural gas, to name but a few.

Only by the use of a multienergy, worldwide integrated model can we begin to appreciate the interactions generated by isolated planning and unilateral actions by the nations of the world. If dangerous and damaging international confrontations are to be avoided and if a logical progression through the advancing technologies of energy production is to be realized, then national policies must at least be compatible on a worldwide basis.

II Disaggregate Modeling

9 Electricity Investments under Uncertainty: Waiting for the Breeder*

ALAN S. MANNE†

9.1. Introduction and Summary

Following the pioneering work of Massé and Gibrat (1957) at Electricité de France (EDF), mathematical programming has come to be widely applied for the optimization of long-term investments in electricity supply. For a survey of the models constructed at EDF and elsewhere, see Anderson (1972). In the United States, the most comprehensive linear programming work seems to have been that undertaken by the Atomic Energy Commission (AEC) for benefit-cost justification of the breeder reactor R&D program. See the AEC documents WASH-1098, WASH-1126, and WASH-1184 (respectively issued in 1970, 1969, and 1972).

Like its predecessors, the present paper is addressed to the selection of an optimal mix of generating plants. The focus is on the problem of uncertainty with respect to the date of commercial availability of breeder nuclear reactors.[1] Sequential probabilistic linear programming is employed. This makes it possible to optimize the mix of fossil, nuclear, and pumped storage plants to be installed during the 1980s—for an assumption that breeder technology will become economical at some point during the 1990s. According to the current round of computations, the initial policy is virtually identical, regardless of whether the breeder availability date is viewed as a random variable or as a point estimate. This conclusion holds not only when future demands are taken as fixed parameters but also when the peak demands are flexible with respect to the price charged.

9.2. Structure of Decision Tree

For major electricity generating units, an investment decision must be taken some 5 to 10 years in advance of the date when a new plant's capacity first becomes available for use. It is for this reason that the first

* Revised version of the paper presented at this conference. This work was supported by National Science Foundation Grant GS-30377 at the Institute for Mathematical Studies in the Social Sciences at Stanford University.

The initial impetus for this research came from conversations with Tjalling Koopmans. Helpful suggestions were received from Juan Eibenschutz, Kenneth Hoffman, William Nordhaus, Milton Searl, and Robert Spencer. The author is much indebted to Roderic Savorgnan and Chiranjib Sen for their assistance with the computations reported here.

† Stanford University.

decision period ($t = 1$) refers to plant capacities that are to become available during the middle 1980s. This is the earliest point to be significantly affected by decisions taken during the middle 1970s. The 1980 capacity mix is already (as of 1973) virtually determined.

Over the future planning horizon, it is convenient to employ 5-year time intervals $t = 1, 2, \ldots$. More exactly then, the initial decisions are those that affect capacities to be brought onstream during the period of 1983 through 1987. For short, we shall generally refer to this 5-year time period ($t = 1$) in terms of the representative midpoint year 1985. Similarly, the second investment decisions ($t = 2$) are those that must be taken during the late 1970s. These affect capacities during 1990, the representative year for the 5-year interval 1988–1992.

It is generally believed that the breeder will eventually become a commercially competitive technology. There is, however, substantial uncertainty as to the date when the breeder will *first* become competitive with conventional LWR (light water reactor) nuclear units. Until this point is reached, breeder capacity will not be installed in significant quantities. For our cost target definition of "commercially competitive," see Table 9.6 and Figure 9.2 later.

In this sequential decision model, the commercial availability date is viewed as a random variable s. Let p_s denote the probability that the breeder will first become available during period s. For illustrative calculations, it has been supposed that this event will occur either in period 2 or 3 or 4. The following numerical values have been adopted for the subjective probabilities: $p_2 = .2$; $p_3 = .4$; $p_4 = .4$. (See Figure 9.1.)

On this decision tree diagram, the vector x_s^t denotes the decisions to be adopted for period t, *given* that the state of world is s. Because it is supposed that the breeder will not be available during the initial time period, the state-of-world subscript is omitted from the vector x^1. That is, the initial decisions must be taken under uncertainty with respect to s. Subsequently, with respective probabilities p_2, p_3, and p_4, the stochastic process will lead to the top, middle, or bottommost branch of the decision tree.

If the breeder becomes competitive during period 2—that is, if $s = 2$ —the uncertainties will be resolved directly at that date. Suppose, however, that $s \neq 2$. Then during period 2 it is not known whether the breeder will first become commercially available during period 3 or 4. Therefore, x_3^2 must be identical with x_4^2. (See Figure 9.1.) According to this tree,

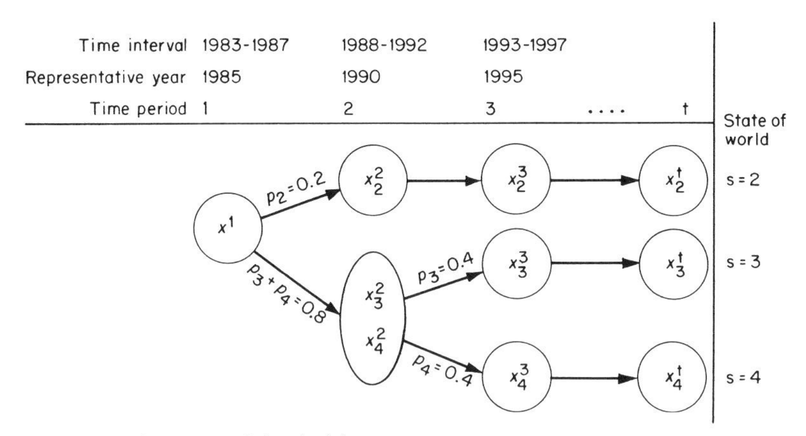

Figure 9.1. Structure of the decision tree.

the uncertainty on the breeder's date has been resolved by the time that decisions must be taken for period 3. Hence, for $t \geq 3$, there is no sequential uncertainty restriction that $x_3{}^t = x_4{}^t$.

To connect this notation with the linear programming variables identified in the next section, note that those variables will be written in a slightly different form: CP(i, t, s), UT(i, k, t, s), WT(l, t, s). For compatibility with computer format requirements, the time period index t and the state-of-world index s will be written on the same line as the rest of the identification for the individual unknowns. Nonetheless, these variables have the same logical structure as the decision vectors $x_s{}^t$. Each denotes a strategy to be adopted during period t, contingent upon knowledge of the state of the world s. The strategy is calculated so as to minimize the expected discounted cost of meeting electricity demands over the entire planning horizon.

9.3. Activities and Cost Coefficients

Table 9.1 defines the indices employed here. First there are the indices t and s, denoting, respectively, the time period and the state of world. There is an index i to distinguish six alternative processes for generating electricity: LWRs, breeders, pumped storage, and three types of fossil units. The index j is employed to approximate the annual load-duration curve by a step function. It is supposed that the peak demands will occur at a con-

Table 9.1 Definition of Indices

Index	Representative Year	Time Interval
$t = 1$	1985	1983–1987
2	1990	1988–1992
3	1995	1993–1997
4	2000	1998–2002
5	2005	2003–2007
6	2010	2008–2012
7	2015	2013–2017
8	2020	2018–2022
9	2025	2023–2027
$s = 2$	1990	State of world: alternative dates for breeder to become less expensive than LWR; p_s is the subjective probability that this cost target is *first* reached in period s; random variable s is unknown in period 1 but is known with certainty in period 3 and thereafter.
3	1995	
4	2000	
$i = 1 \ldots 6$		Plant types
$j = 1 \ldots 3$		Energy blocks: 10, 40, and 50% intervals along load-duration curve
$k = 1 \ldots 6$		Mode of operation of generating plants
$l = 1 \ldots 9$		Level of peak demand (for piecewise linear approximation to shortage cost function); see Appendix 9.A.

stant rate throughout energy block 1, that is, during 10% of the 8760 hours in a year.[2] Similarly, it is supposed that the intermediate demands will occur at a constant rate during 40% of the year (energy block $j = 2$), and that the base-load demands will occur at a constant rate during the remaining hours (block $j = 3$). It will be seen (in Table 9.9) that there may be a ratio of 5:1 or more between the economic value of a kilowatt-hour of energy in the peak-load and the base-load blocks of time.

There is some flexibility in the number of hours per year that each of the plant types may be operated. To distinguish alternative modes of operation, we employ the index k. In Table 9.2, six modes are defined: three for thermal plants and three others for pumped storage. For example, if a thermal plant is utilized in mode 2, this means that it is to be operated at its capacity during block 1 (10% of the year) and also at capacity during block 2 (40% of the year), but that it is to produce no energy during the base-load period (block 3). Since linear programming models allow for convex combinations of activities, other modes of operation are incorporated implicitly.[3] For example, an optimal solution might specify that

Table 9.2 Alternative Modes of Operation
Fraction of year[a] in which plant type i produces energy during demand block j, operating in mode k

	Type $i = 1 \ldots 5$ (thermal plants: fossil, LWR, breeder)			6 (pumped storage)		
	Mode $k = 1$	2	3	4	5	6
Energy Demand Block $j = 1$	.10	.10	.10	.10	.10	.10
2	.40	.40	.00	.23	$-.15$[b]	.00
3	.50	.00	.00	$-.50$[b]	.00	$-.15$[b]

[a] To convert these fractions into hours per year, multiply by 8760. This gives the coefficients $A(j, k)$ employed in the demand constraint rows DM(j, t, s).
[b] Pumped storage units consume energy during these demand blocks, and they produce energy at other times. The pumped storage coefficients here are calculated by supposing that it takes 1.5 kWh of off-peak energy per kWh generated during the peak periods.

7 GW of capacity are to be operated in mode 2 and 3 GW (gigawatts, that is 10^9 or 1 billion watts) in mode 3. This would be interpreted as follows: all 10 GW are to be fully utilized during energy block 1, but are only 70% utilized during block 2.

Table 9.3 defines the activities of the programming model, and Table 9.4 shows how the cost coefficients are computed.[4] First, consider the variables CP(i, t, s), the 5-year capacity increments of plant type i in period t under state of world s. There are initial capital outlays of $C(i)$ per unit of capacity type i.[5] This parameter is multiplied by the probability p_s to obtain the expected future cost in time period t. The expected future capital costs are discounted as of the date that lies 2.5 years before the midpoint of period t. Let β denote a uniform 5-year present worth factor. Then future capital costs in period t are converted to present values by the factor $\beta^{t-.5}$.

For the capacity utilization variables UT(i, k, t, s), the coefficients in the minimand row are calculated by supposing that the operating and maintenance costs OM(i) are incurred whenever the unit is operated at any point during the year. The annual fuel costs are proportional to $H(k)$, the number of hours operated for mode k.[6] The operating, maintenance, and fuel costs are converted into an expected value through multiplying by the probability coefficient p_s. For these annually recurring costs, the midpoint is taken to be representative of the entire 5-year interval. Hence, the annual

Table 9.3 Definition of Activities

The 5-year Capacity Increments, First Available in Period t	CP(i, 1)	3
	CP(i, t, s) [excludes three nonpermissible activities: CP(5, 2, 3), CP(5, 2, 4), and CP(5, 3, 4)]	93
Peak Capacity Utilization, Representative Year t	UT(i, k, 1)	15
	UT(i, k, t, s)	432
Peak Demand Interpolation Weights	WT(l, 1)	9
	WT(l, t, s)	216
Total		768

Table 9.4 Computation of Cost Coefficients

Activity		Present Value of Expected Costs, Coefficient in Minimand Row PV
The 5-year Capacity Increments, First Available in Period t	CP(i, 1)	$[C(i)]\beta^{1-.5}$
	CP(i, t, s); (t = 2 . . . 9)	$p_s[C(i)]\beta^{t-.5}$
Peak Capacity Utilization, Representative Year t	UT(i, k, 1)	$[\mathrm{OM}(i) + H(k)F(i, 1)]5\beta$
	UT(i, k, t, s); t = 2 . . . 9[a])	$p_s[\mathrm{OM}(i) + H(k)F(i, t)]5\beta^t$
Peak Demand Interpolation Weights, Representative Year t	WT(l, 1)	$[u(\bar{q}_{l,1})]5\beta$
	WT(l, t, s); (t = 2 . . . 9[a])	$p_s[u(\bar{q}_{l,t})]5\beta^t$

where

β = $(1/1.10)^5 \approx .62$ = 5-year discount factor at uniform annual discount rate of 10%,
$C(i)$ = initial investment cost per unit of capacity, plant type i (\$/kW); see Table 9.6,
OM(i) = operating and maintenance cost, plant type i (\$/kWyr); see Table 9.6,
$F(i, t)$ = fuel cost per kWh generated, plant type i, period t (\$/kWh); see Table 9.6,
$H(k)$ = hours operated per year, mode k; see Table 9.2,
$u(\bar{q}_{l,t})$ = shortage cost for failure to meet peak demands, level l, period t (\$/yr); see Appendix 9.A.

[a] Exception for $t = 9$: multiply this cost coefficient by $\sum_{\tau=0}^{5} \beta^\tau$. This is an approximation to reduce horizon effects. In effect, we are assuming constant rates of utilization during the 30 years (6 periods) following 2025.

costs are multiplied by the factor of 5 and are discounted by the factor β^t.

In general, this model is formulated as though there were fixed demands to be satisfied at minimum expected discounted costs. Alternatively, some results will be presented for the case of "flexible peak demands"—where prices are raised to the incremental costs of supply during the peak periods, and the level of demands is thereby reduced.[7] To allow for this possibility, there are the peak demand interpolation weight variables $\mathrm{WT}(l, t, s)$. (See Appendix 9.A for details on the interpolation index l and the method of calculating the shortage penalty coefficients $u(\bar{q}_{l,t})$. The shortage costs are annually recurring during period t under state of world s—hence, are multiplied by the probability p_s and by the factor $5\beta^t$.

9.4. Constraints of the Programming Model

Table 9.5 summarizes the complete set of 277 constraint rows employed in the flexible peak demand model. The demand constraints are written in two ways—depending upon the energy block. For the off-peak blocks ($j = 2,3$), the energy demands are fixed and the demand constraints are[8]

$$\mathrm{DM}(j, t, s): \quad \sum_i \sum_k \underbrace{\begin{bmatrix}\text{hours per year available}\\ \text{during block } j,\\ \text{operating mode } k;\\ \text{see Table 9.2}\end{bmatrix}}_{\overline{A(j,k)}} \underbrace{\begin{bmatrix}\text{peak}\\ \text{capacity}\\ \text{utilized,}\\ \text{kW}\end{bmatrix}}_{\mathrm{UT}(i,k,t,s)} \geq \begin{bmatrix}\text{fixed}\\ \text{energy}\\ \text{demands}\end{bmatrix} - \begin{bmatrix}\text{fixed}\\ \text{hydroelectric}\\ \text{energy supplies}\end{bmatrix}. \tag{9.1}$$

For the peak energy block ($j = 1$), the constraints $\mathrm{DM}(1, t, s)$ are written in the identical way except that the right-hand term "fixed energy demands" is replaced by the following:[9]

$$\sum_l \underbrace{\begin{bmatrix}\text{peak}\\ \text{energy demands}\\ \text{at interpolation}\\ \text{level } l;\\ \text{see Appendix 9.A}\end{bmatrix}}_{\bar{q}_{l,t}} \underbrace{\begin{bmatrix}\text{interpolation}\\ \text{weight,}\\ \text{level } l\end{bmatrix}}_{\mathrm{WT}(l,t,s).} \tag{9.2}$$

Table 9.5 Definition of Constraint Rows—Flexible Peak Demands

			Number of Rows
Demand Requirements	DM(j, 1)	(j = 1 . . . 3)	3
	DM(j, t, s)	(j = 1 . . . 3; t = 2 . . . 9; s = 2 . . . 4)	72
Peak Demand Interpolation Weights	WT(1)		1
	WT(t, s)	(t = 2 . . . 9; s = 2 . . . 4)	24
Peak Capacity Utilization	UT(i, 1)	(i = 1 . . . 4, 6)	5
	UT(i, t, s)	(i = 1 . . . 6; t = 2 . . . 9; s = 2 . . . 4)	144
Sequential Uncertainty, Period 2	SU(i)	(i = 3, 4, 6)	3
Requirements, Fossil Fuel	RF(1)		1
	RF(t, s)	(t = 2 . . . 9; s = 2 . . . 4)	24
Total			277

The interpolation weight variables are nonnegative, and must add to unity. Therefore,

$$\text{WT}(t, s): \quad \sum_{l} \text{WT}(l, t, s) = 1. \tag{9.3}$$

The capacity utilization rows UT (i, t, s) represent the link between successive points of time. That is, the amount of capacity available at period t is the sum of the capacities installed during each of the previous periods τ. The available capacity must be sufficient to handle not only the peak requirements but also to provide a reserve for scheduled and unscheduled shutdowns. These constraints are written as weak inequalities to allow for the possibility that it may be optimal to leave some of the older equipment idle:

$$\text{UT}(i, t, s): \quad \sum_{k} \underset{\overline{B(i)}}{\begin{bmatrix}\text{reserve}\\ \text{capacity}\\ \text{factor;}\\ \text{see Table 9.6}\end{bmatrix}} \underset{\text{UT}(i, k, t, s)}{\begin{bmatrix}\text{peak}\\ \text{capacity}\\ \text{utilized,}\\ \text{kW}\end{bmatrix}} \leq \begin{bmatrix}\text{initial}\\ \text{capacities}\end{bmatrix} + \underset{\sum_{\tau=1}^{t} \text{CP}(i, \tau, s).}{\begin{bmatrix}\text{capacity increments}\\ \text{through period } t\end{bmatrix}} \tag{9.4}$$

The sequential uncertainty restrictions SU (i) refer only to period 2. Recall Figure 9.1.) If the breeder is not commercially competitive during period 2, the identical decisions must be taken, regardless of whether it subsequently turns out that the random variable $s = 3$ or $s = 4$. For this, it is sufficient to impose the following restrictions[10] upon the capacity increments during period 2:

$$\text{SU}(i): \qquad \text{CP}(i, 2, 3) = \text{CP}(i, 2, 4) \qquad (i = 3, 4, 6). \tag{9.5}$$

For calculating the minimand coefficients (Table 9.4), it was implicitly assumed that unlimited supplies of fuel would be available to the electricity generating industry at a predetermined price in each future period. In a systems analysis of the entire energy sector, the fuel cost could not be taken as a datum, for it would in turn depend upon the fuel demands of the electricity industry. Here, in a model of the electricity industry alone, we can do no more than estimate the industry's demands at alternative fuel prices. For fossil fuel, this is done through the following definitional rows:

$$\text{RF}(t, s): \quad \sum_{i=1}^{3}\sum_{k=1}^{3}\Big[\overset{\begin{bmatrix}\text{heat rate,}\\ \text{Btu/kWh;}\\ \text{see}\\ \text{Table 9.6}\end{bmatrix}}{\overline{\text{Btu}(i)}}\quad \overset{\begin{bmatrix}\text{hours}\\ \text{operated per}\\ \text{year, mode } k\text{;}\\ \text{see Table 9.2}\end{bmatrix}}{\overline{H(k)}}\Big]\quad \Big[\overset{\begin{bmatrix}\text{peak}\\ \text{capacity}\\ \text{utilized, kW}\end{bmatrix}}{\text{UT}(i, k, t, s)}\Big] = \begin{matrix}\text{requirements}\\ \text{for fossil fuel.}\end{matrix} \tag{9.6}$$

9.5. Demand Forecasts

The "fixed energy demand" estimates are adapted from Federal Power Commission (1971, pp. I-18-23–29). Hereafter, this document will be abbreviated as 1970 NPS (*The 1970 National Power Survey*). This source is out of date but has the advantage of providing a regionally disaggregated as well as a U.S. total demand forecast for 1980 and 1990. Moreover, it provides a consistent set of estimates of installed capacities at those dates.[11]

We have assumed that the growth of electricity demand will slow down after 1990, and that the demand *increments* will grow at the rate of 3% per year in all regions. This implies that the total annual growth rate will decelerate from 6.6% during the 1980s, down to 5.3% during the 1990s, and that it will gradually approach 3%.

The 1970 NPS does not provide a U.S. total load-duration curve but only a forecast of the peak demand and of the total energy requirements. Accordingly, the "fixed energy demand" was extrapolated by supposing that the peak would occur at a constant rate throughout the 10% of the year that corresponds to block 1. The remaining energy requirements were allocated between the two off-peak blocks by supposing that the intermediate power demands (block 2) would lie halfway between the peak and the base load.

9.6 Numerical Values of Cost and Performance Factors

There is no objective way to estimate the cost and performance factors of new technologies. Inherently there is a subjective element in the parameters given in Table 9.6.[12] For example, a critic of the breeder development program will point out that LWRs (light water reactors) are a more conventional technology than breeders. He will then say that LWRs have proved quite unreliable and that the reserve capacity factor $B(4)$ should be much higher than 1.20. A fortiori, it is hopelessly optimistic to set the breeder reserve factor $B(5) = 1.25$. Similar questions can and should be raised about each of the numbers in Table 9.6. These parameters must be viewed as illustrative, and they do not represent an industry-wide consensus.

Given the cost and performance factors for the LWR in Table 9.6, it is fairly straightforward to see what targets must be achieved for the breeder to reach a commercially competitive position. This technology will not be commercially competitive if its capital costs are much more than $50/kW above those of the LWR, or if its operating and maintenance costs are much higher than $8/kWyr, or if the LWR's fuel costs are much lower than $2.50/10^3$ kWh.[13] This is the reasoning that underlies the breeder performance factors of Table 9.6. Clearly these costs will not be achieved by the initial demonstration plants, for breeder technology will require a lengthy period of learning by doing.

Before running these data through a dynamic decision model, it is helpful to make informal calculations such as those in Figure 9.2. For this comparison, it is supposed that each post-1980 type of plant ($i = 3,4,5,6$) will be operated in *one fixed mode*[14] throughout a 30-year service life. The initial capital investment plus the discounted sum of operating, maintenance, and fuel costs will then be a linear function of the fixed operating rate. According to Figure 9.1, breeders will be the least expensive proc-

Table 9.6 Cost and Performance Factors (in 1972 prices)

Plant Type *i*	1 Fossil, before 1970	2 Fossil, 1970–80	3 Fossil, after 1980	4 LWR (light water reactor)	5 Breeder	6 Pumped Storage
Btu(*i*), Heat Rate, Fossil Plants (10^3 Btu/kWh)	10.4	9.5	8.5	—	—	—
F(*i*, *t*), Fuel Cost ($/$10^3$ kWh)	5.20[a]	4.75[a]	4.25[a]	2.50	1.00	0[b]
OM(*i*), Operating and Maintenance ($/kWyr)	6.0	5.0	4.0	6.0	8.0	3.0
C(*i*), Initial Investment ($/kW)	[c]	[c]	250.	350	400.	200.
B(*i*), Reserve Capacity Factor; Reciprocal of Maximum Plant Factor	1.15	1.15	1.15	1.20	1.25	1.05

[a] For fossil plants, the fuel costs in period *t* are obtained by multiplying the heat rate Btu(*i*) by the fuel price in year *t*. In this table, the fuel costs *F*(*i*, *t*) are based upon a 1985 fossil fuel price of $.50/$10^6$ Btu.

[b] For pumped storage, the direct fuel cost is zero, but there is an indirect cost: an input of 1.5 kWh during off-peak hours per kWh generated during the peak hours. These indirect fuel inputs are introduced into the programming matrix through the negative entries in Table 9.2 for energy blocks $j = 2, 3$ and for operating modes $k = 4, 5, 6$. Correspondingly, there are negative output coefficients $\overline{A(j, k)}$ in the demand constraint rows DM(j, t, s).

[c] These two fossil plant types are not candidates for new investment after 1980.

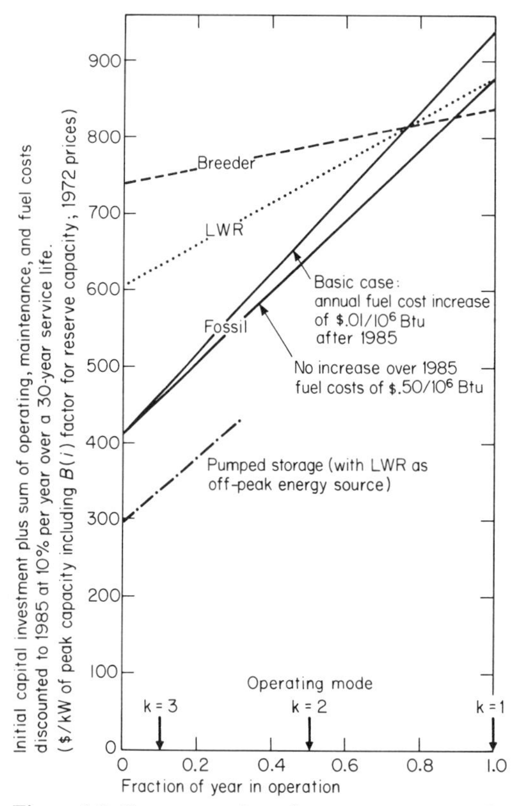

Figure 9.2. Cost comparison for a constant operating rate throughout the service life.

ess for base-load duty (mode 1), fossil plants for intermediate duty (mode 2), and pumped storage for peaking (mode 3).

Before the breeder becomes available, the optimal choice of base-load equipment is quite sensitive to the assumption with respect to fossil fuel prices. If these costs lie much below $\$.50/10^6$ Btu—and all other factors remain the same—fossil plants will be less expensive than LWRs. No elaborate optimization model is then needed to evaluate these two technologies.

Low future fossil fuel costs seem unlikely for the United States—partly because of tightened air pollution standards and partly because of the gradual exhaustion of domestic oil and gas reserves. For these reasons, we have supposed that the 1985 average fossil fuel cost will be $\$.50/10^6$ Btu. This price would be needed to cover the cost of low-sulfur western coal that will be strip-mined in accordance with increasingly stringent land-use regulations and then shipped for distances of 1500 or more miles. Alternatively, this price would be needed to cover the cost of coal gasification or of solvent refining,[15] or to cover the cost of the cleaner fossil fuels: natural gas and low-sulfur residual fuel oil. Given the controversial nature of any price forecast, the programming model has been run on two sets of assumptions: the basic case with an annual fossil cost increase of $.01/10^6 Btu after 1985, and an alternative case based upon no increase over the 1985 fuel costs of $.50/10^6 Btu.

9.7. Numerical Results—Total United States

For the United States as a whole, twelve alternative cases are summarized in Tables 9.7 to 9.9. There are comparisons between fixed versus flexible peak demands, alternative fuel cost assumptions, stochastic versus deterministic decision criteria, and alternative estimates of the breeder's availability date.

First, consider the differences between columns A and B of Table 9.7: the savings from accelerating the breeder's development program so that large-scale commercial use could begin in period 2 (1990). According to Table 9.7, the discounted cost savings would be of the order of $4 to 5 billions, regardless of whether peak demands are fixed or flexible, and regardless of whether or not the annual fossil fuel cost increase is $.01/10^6 Btu. These estimates of the breeder's benefits are consistent with those reported in an as yet unpublished manuscript by T. Cochran but are of course lower than those appearing in Atomic Energy Commission documents based upon a 7% discount rate.

Table 9.7 provides some insights not only into breeder development but also into the potential benefits from new fossil fuel processes. Note that if coal costs could be held to $.50/10^6 Btu—through environmentally acceptable technologies for mining, followed by coal gasification or solvent refining—this would permit a saving of some $5 to 10 billions versus oil and gas with prices rising at the annual rate of $.01/10^6 Btu. Even if there is a low probability of success, the potential benefits might be sufficient to justify a major R&D effort on coal.

Next, consider the cost savings from peak-load pricing policies. According to Table 9.7, these savings could be quite high: $30 to 40 billions. It is possible that the model has understated the extent to which peak-load

Table 9.7 Expected Discounted Costs (unit: 10^9\$; discounted to 1980 at 10% per year; 1972 general price level)

Case	A	B	C	B − A	C − B	
	1990 Availability of Breeder	Basic Case	Point Estimate of Availability Date; Deterministic Criterion for Initial Decision	Cost Savings from Early Availability	Cost Increase from Deterministic Criterion for Initial Decision	Assumed Cost of Fossil Fuel[a]
Fixed Demands	453.5	458.3	458.7	4.8	0.4	Annual increase of \$.01/10^6 Btu after 1985
Flexible Peak Demands	422.4	426.7	426.7	4.3	0.0	
Cost Savings from Flexible Peak Demands	31.1	31.6	32.0			
Fixed Demands	448.5	452.8	453.0	4.3	0.2	No increase over 1985 fuel cost of \$.50/10^6 Btu
Flexible Peak Demands	411.2	415.1	415.1	3.9	0.0	
Cost Savings from Flexible Peak Demands	37.3	37.7	37.9			
Subjective Probabilities:						
p_2	1.0	.2	.2			
p_3	0	.4	.4			
p_4	0	.4	.4			

[a] In all cases, it is supposed that the 1985 fossil fuel cost will be \$.50/10^6 Btu.

and interruptible pricing are already a common practice for large industrial consumers. Nonetheless, the results suggest that it would be worth examining the peak-load question in greater depth.[16] These savings will become increasingly important as we shift away from fossil processes where fuel constitutes a large fraction of total generating costs.

Finally, consider the differences between columns B and C of Table 9.7. Column B denotes the basic case: following an optimal strategy for decisions under uncertainty while waiting for the breeder. From column C, we can observe the effects of a decision criterion that is known to be defective and yet is widely employed for practical calculations: replacing a probability distribution with a point estimate. Recall that 1995 is the

Table 9.8 Capacity Available (gigawatts)

	The 1970 National Power Survey[a]			Results[b]—Basic Case B—1990 (random variable s = date when breeder becomes commercially competitive)			
				Fixed Demands		Flexible Peak Demands	
	Actual	Estimated		$s = 1990$	$s \neq 1990$	$s = 1990$	$s \neq 1990$
Plant type i	Year t 1970	1980	1990	1990	1990	1990	1990
1. Fossil, before 1970	259	259	259	259	259	259	259
2. Fossil, 1970–1980	0	131	131	131	131	131	131
3. Fossil, after 1980	0	0	168	0	0	0	0
4. LWR	6	140	{475	333	650	172	447
5. Breeder	0	0		330	0	287	0
6. Pumped Storage	4	27	70	91	91	27	27
7. Conventional hydro[c]	52	68	82	82	82	82	82
Total Capacity[d]	321	625	1185	1226	1213	958	946
Total Capacity, Annual Growth Rate over Preceding Decade, %		6.9	6.6	7.0	6.9	4.4	4.2

[a] See Federal Power Commission (1971, p. I-18-29).
[b] Assuming that 1985 fossil fuel costs will be \$.50/10^6 Btu and that these costs will increase at the annual rate of \$.01/10^6 Btu thereafter.
[c] It is supposed that negligible amounts of conventional hydro will be added after 1990.
[d] Excludes gas turbines and internal combustion generators. In 1970, their capacity was 6% of the total, but their energy generation was only 1.4%. This total also excludes geothermal plants.

median, the mean (approximately) and the modal [17] estimate of the breeder availability date. The costs in column C were calculated as though the 1985 plant-mix decision ($t = 1$) were based upon the point estimate that the breeder will first be available in 1995 ($t = 3$) but as though subsequent decisions were made in the light of revised information. For example, with probability p_2 the breeder will be available in 1990, and in that event the plant-mix decisions for period 2 would be revised accordingly. Surprisingly enough, the deterministic criterion works well. In the two flexible demand cases, it leads to no increase in expected costs, and in the two fixed demand cases there is an increase of \$.4 billion or less. Caution: from this one example, no general conclusions can be drawn on point estimates versus probability distributions for optimal decisions under uncertainty.

Table 9.9 Incremental Costs—Basic Case B

Demand Block j	Expected Incremental Costs[a] of Supplying Electrical Energy in Year t – 1972 Prices ($/10³ kWh)						Assumed Cost of Fossil Fuel[b]
	Fixed Demands			Flexible Peak Demands			
	Year t 1990	2000	2010	1990	2000	2010	
1	33.5	30.8	32.1	20.0	20.0	20.0	Annual increase of
2	5.7	6.2	5.7	7.8	7.4	9.6	$.01/10⁶ Btu after
3	4.7	3.7	3.8	5.7	4.9	3.1	1985
1	34.0	32.4	32.1	20.0	20.0	20.0	No increase over
2	5.2	5.2	5.7	8.4	7.6	8.9	1985 fuel cost of
3	5.0	4.2	3.8	5.2	4.8	3.7	$.50/10⁶ Btu

[a] Dual variables, rows DM(j, t, s) summed over s and converted from present values into future costs in year t. Pumped storage provides the possibility of converting 1.5 kWh off-peak into 1 kWh of peak energy. Nonetheless, because of capital and operating costs, it is optimal for the expected cost ratios to be much higher than 1.5:1.
[b] In all cases, it is supposed that the 1985 fossil fuel cost will be $.50/10⁶ Btu.

9.8. Numerical Results: Six Regions

Whenever a model covers a geographical area as large as the entire United States, there is a possibility for sizable errors of aggregation. Based upon a countrywide *average* fossil fuel price, Table 9.8 indicates that it is not optimal to install any new fossil fuel plants after 1980. This extreme conclusion might be altered if the model were to allow for the fact that fossil fuel prices differ substantially from one area to another throughout the country.

To quantify this possible bias, Case B has been recalculated separately for each of six geographical areas. It is supposed that regional fossil fuel prices would differ from the countrywide average by the identical amounts in each year throughout the planning horizon. (See the regional differences in Table 9.10.) Adjustments are also made for differences in the load-duration curve and for the heat rates of pre-1980 fossil plants. Except for these differences, the regional programming matrices are the same as that for the United States as a whole. An optimal plant mix is computed separately for each region, as though there were no interregional transmission links.[18]

According to Table 9.10, geographical disaggregation would alter the expected discounted costs by less than $1 billion throughout the planning

Table 9.10 Effects of Regional Disaggregation

Region	Regional Difference in Fossil Fuel Cost from U.S. Average[a] (\$/10^6 Btu)	Expected Discounted Costs, Basic Case B—Fixed Demands[b] (unit: 10^9 \$: discounted to 1980 at 10% per year: 1972 general price level)	
		Annual Increase of \$.01/10^6 Btu after 1985	No Increase over 1985 Fuel Cost of \$.50/10^6 Btu
1. Northeast	+.074	69.6	69.2
2. East Central	−.042	68.0	65.8
3. Southeast	+.026	94.8	93.7
4. West Central	+.013	56.8	56.2
5. South Central	−.021	81.3	79.5
6. West	+.011	88.3	87.6
U.S. Total, Based upon Regional Disaggregation		458.8	452.0
U.S. Total, Based upon Countrywide Average Fossil Fuel Price (from Table 9.7)		458.3	452.8

[a] 1990 regional fuel cost differentials, according to the Federal Power Commission, *The 1970 National Power Survey*, p. I-19-7. Here we have adopted the NPS estimate of the regional differentials but not of the countrywide average. In that document, it is forecast that the 1990 U.S. average fossil fuel cost will be \$.40/10^6 Btu.
[b] In all cases, it is supposed that the 1985 U.S. average fossil fuel cost will be \$.50/10^6 Btu.

horizon. Moreover—provided that fossil fuel prices increase at the annual rate of \$.01/10^6 Btu—it is still optimal to install no fossil capacity after 1980. It is only when these fuel prices remain constant that it becomes optimal to install new fossil capacity in the two low-cost regions.[19]

Table 9.11 indicates that geographical disaggregation could make a difference of up to 30% in estimating the electricity industry's future demands for fossil fuel. The importance of this factor will depend upon the fuel price forecast. With rising fuel prices, no new fossil plants will be installed, and the older units will be retired rapidly. The smaller the fraction of fossil capacity in the U.S. electricity grid, the less significant will be the effects of regional differences in fossil fuel costs.

Table 9.11 Fossil Fuel Demand by Electricity Sector (10^{15} Btu)[a]

Actual	Estimated		Effects of Regional Disaggregation[b]						Assumed Cost of Fossil Fuel[c]
			U.S. total			U.S. Disaggregated into 6 Regions			
1970	1980	1990	1990	2000	2010	1990	2000	2010	
13.6	19.5	25.5							
			8.6	3.1	3.0	11.4	4.0	3.2	Annual increase of \$.01/10⁶ Btu after 1985
			20.2	12.1	14.8	23.5	19.1	19.5	No increase over 1985 fuel cost of \$.50/10⁶ Btu

[a] See Federal Power Commission (1971, p. I-4-2).
[b] All results refer to the basic case B, with the breeder availability date s = 1995, and with demands fixed exogenously.
[c] In all cases, it is supposed that the 1985 U.S. average fossil fuel cost will be \$.50/10^6 Btu.

Appendix 9.A Flexible Peak Demands—Principal Assumptions and Model Formulation

1. For each future time period and region, we already have a "fixed demand" projection for peak electricity. (These estimates are based upon the 1970 NPS, pp. I-18-23–29). After 1990, it is assumed that the *increments* in demand will grow at the rate of 3% per year in all regions. Call the fixed future demand in period t the quantity $\bar{q}_t$—or for short, $\bar{q}$. It is supposed that $\bar{q}$ corresponds to a future reference price $\bar{p}$ = \$12/10^3 kWh. (see 1970 NPS, p. I-19-2.)

2. From 1985 onwards, electricity prices will be raised from \$12/10^3 kWh to the incremental cost of supplying peak demands (energy block 1).

3. Prices will not be lowered from \$12/10^3 kWh in order to stimulate off-peak demands (energy blocks 2 and 3). The off-peak demands will neither be increased nor decreased as a result of the peak pricing policy.

4. The price elasticity of demand for peak electricity is $\eta = -.5$. For simplicity, no distinction is drawn between the short- and the long-run elasticity. Employing the price elasticity $\eta = -.5$, we extrapolate the market demand curve from the reference values $\bar{p}$, $\bar{q}$. This is casual econometrics!

5. Let q denote the future quantity demanded of peak electricity, with an assumption of assuming a flexible price policy. It is supposed that the money value of total benefits is an isoelastic function of q. That is,

$$u(q) = aq^b + c, \quad (9.A.1)$$

where a, b, and c are constants to be estimated from the market demand curve. Setting the incremental benefits equal to the market price p, the demand curve is related to the benefit function as follows:

$$p = abq^{b-1}, \quad (9.A.2)$$

$$\therefore \eta = \frac{dq}{dp} \cdot \frac{p}{q} = \frac{1}{b-1}, \quad (9.A.3)$$

$$\therefore b = 1 + \frac{1}{\eta} = -1. \quad (9.A.4)$$

It can also be seen that

$$a = (\bar{p}/b)\bar{q}^{1-b}. \quad (9.A.5)$$

Finally, the arbitrary constant c is chosen so that zero benefits are associated with the reference demand level $\bar{q}$:

$$u(\bar{q}) = a\bar{q}^b + c = 0. \quad (9.A.6)$$

With this normalization, the expected discounted costs are comparable—both for the fixed and the flexible demand cases in Table 9.7.

6. Having estimated the constants a, b, and c, we may make a piecewise linear approximation to the nonlinear benefit function for period t, $u(q_t)$. The index l identifies the quantity $\bar{q}_{l,t}$ and the associate shortage cost $u(\bar{q}_{l,t})$ for delivering less than the reference quantity $\bar{q}_t$. Here, the index $l = 1 \ldots 9$, and the approximation refers to the following nine points:

$$\bar{q}_{1,t} = \bar{q}_t, \quad (9.A.7)$$

$$\bar{q}_{2,t} = \bar{q}_t[.95], \quad (9.A.8)$$

$$\vdots \qquad \vdots$$

$$\bar{q}_{l,t} = \bar{q}_t[1 - .05(l - 1)], \quad (9.A.9)$$

$$\vdots \qquad \vdots$$

$$\bar{q}_{9,t} = \bar{q}_t[.60]. \quad (9.A.10)$$

This means that the calculations are focused upon the range between 60 and 100% of $\bar{q}_t$. Typically it is optimal for the price to rise so that the peak is cut back to 75% to 80% of the reference demand level $\bar{q}_t$.

Notes

1. According to Bessière (1970, p. B-206), there have been EDF investment programming models incorporating uncertainty with respect to future technologies. These have been based upon minimax regret, not upon the expected discounted value criterion employed here.

2. This numerical approximation to the load-duration curve leads to a bias against one type of generating equipment: gas turbines. Because of their low capital and high fuel costs, it is advantageous to employ gas turbines for reserve peaking capacity but to plan to operate them for less than 10% of a year—only in the event of an unscheduled shutdown of a thermal plant during the peak-load period. To evaluate gas turbines properly, it would require a fourth energy block, for example, a 1 to 3% interval along the load-duration curve. In the interests of simplicity, this fourth block has been omitted, and therefore gas turbines have also been omitted from the model.

3. Table 9.2 rules out some logically possible but uneconomic modes of operation. For example, there is no provision for the possibility that a thermal unit would be operated during blocks 1 and 3 but not during block 2. This mode will be uneconomical whenever the fuel costs are the same in all three blocks and when the incremental value of energy is at least as high in block 1 as in 2 and at least as high in block 2 as in 3. This is the pattern of incremental values observed in all of our solutions to date.

4. Income and property taxes are omitted. These are private but not social costs.

5. With the discounted cash flow model, there is no need to calculate amortization or depreciation. Capital costs are charged directly as of the date at which the plant's capacity first becomes available. Here we have supposed that the physical service life is indefinitely long, but that the useful service life is to be determined endogenously. Typically, it is optimal to employ a thermal plant as a base-load unit during its early years of operation (mode $k = 1$) and then gradually retire it toward peak-load service (mode $k = 3$) in its old age.

6. To calculate $H(k)$ for thermal plants, add up the coefficients in column k of Table 9.2, and multiply by 8760. For example, in mode 2, a thermal plant will be operated for 4380 hours per year.

For pumped storage units, there are no direct fuel costs, and so $H(k)F(6,t) = 0$.

7. For a specific proposal of this type, see Doctor et al. (1972).

8. These constraints are written with the following notational conventions: when an array denotes a set of predetermined coefficients, it is barred above. Unbarred arrays denote the levels of activities or the identification of constraint rows.

For period $t = 1$, it is understood that the state-of-world index s is omitted.

9. Note that the linear programming unknowns are not the demand level $\bar{q}_{l,t}$, but rather the interpolation weight variables WT(l, t, s). For further details on this method of approximation to nonlinear functions, see for example, Westphal (1971, p. 61).

10. These same restrictions could have been imposed in an even simpler form: eliminating the unknown CP(i, 2, 4) and replacing it throughout by CP(i, 2, 3).

11. The 1980 capacities are the "initial capacities" term on the right-hand side of the utilization constraints UT(i, t, s).

12. All cost comparisons are made in terms of the 1972 U.S. price level and are not adjusted for general inflation thereafter.

13. LWR fuel costs might be \$2.00/10^3 kWh with U_3O_8 at \$8.00 per pound but could rise to \$2.50 with the gradual exhaustion of low-cost uranium reserves. See National Petroleum Council (1972, p. 176).

14. By contrast, the dynamic optimization model allows for endogenously determined shifting between base-load, intermediate, and peaking service.

15. For a review of the costs of alternative coal conversion processes, see Hottel and Howard (1971, Ch. 3).

16. No new metering equipment would be needed in order to charge higher electricity prices during peak rather than during off-peak seasons.

17. Since $p_3 = p_4 = .4$, the mode is not unique.

18. An allowance for transmission links might reduce the requirements for peaking capacity, but would probably not have a large impact upon base-load units.

19. If fossil fuel prices remained constant throughout the planning horizon, it would be optimal to install 29 GW of fossil capacity in the East Central region and 18 GW in the South Central during 1983–1987 (period $t = 1$). These capacity increments are to be compared with 132 GW of LWRs and 50 GW of pumped storage for the United States as a whole during this same period.

References

Anderson, D. (1972). "Models for Determining Least-Cost Investments in Electricity Supply," *The Bell Journal of Economics and Management Science, 3,* No. 1, Spring, pp. 267–299.

Atomic Energy Commission (1970). *Potential Nuclear Growth Patterns,* WASH-1098, U.S. Government Printing Office, Washington, D.C., December.

——— (1969). *Cost-Benefit Analysis of the U.S. Breeder Reactor Program,* WASH-1126, U.S. Government Printing Office, Washington, D.C., April.

——— (1972). *Cost-Benefit Analysis of the U.S. Breeder Reactor Program,* WASH-1184, U.S. Government Printing Office, Washington, D.C., January.

Bessière, Francis (1970). "The 'Investment '85' Model of Electricité de France," *Management Science, 17*(4), pp. B192–B211, December.

Doctor, R. D., Anderson, K. P. et al. (1972). "California's Electricity Quandary: III, Slowing the Growth Rate," R-116-NSF/CSA, RAND Corporation, Santa Monica, Calif., September.

Federal Power Commission (1971). *The 1970 National Power Survey,* U.S. Government Printing Office, Washington, D.C.

Hottel, H. C., and Howard, J. B. (1971). *New Energy Technology—Some Facts and Assessments,* MIT Press, Cambridge, Mass.

Massé, P., and Gibrat, R. (1957). "Applications of Linear Programming to Investments in the Electric Power Industry," *Management Science, 3*(2), pp. 149–166, January.

National Petroleum Council (1972). *U.S. Energy Outlook,* Washington, D.C., December.

Westphal, L. (1971). "An Intertemporal Model Featuring Economies of Scale," Ch. 4 in H. B. Chenery, ed., *Studies in Development Planning,* Harvard University Press, Cambridge, Mass.

10 Impacts of New Energy Technology Using Generalized Input-Output Analysis*

JAMES E. JUST†

10.1. Introduction

The research described here examines the economic impacts of three new technologies that could have significant commercial application by 1985:

1. High-Btu coal gasification.
2. Low-Btu coal gasification.
3. Gas turbine topping cycle (combined gas and steam cycle).

It also explores the interaction between energy use growth and total investment in the United States and how this relationship is affected both directly and indirectly by the introduction of these new technologies.

The techniques developed during the research are applicable to any possible new technologies, and provide broad looks at the United States 10 to 15 years hence. The techniques are based on a generalized [1] form of input-output (I/O) analysis and thus can focus on the many interactions between the sectors of a projected future economy. One contribution of the techniques is that microeconomic engineering studies can now be linked to macroeconomic national or regional models. Since engineering studies can handle technological change induced by relative price changes or other variables, this capability facilitates the development of dynamic economic models in which technology depends on relative prices, energy availability, and so forth.

The research utilized a projection of the 1980 economy prepared by the Interagency Growth Project of the Bureau of Labor Statistics (BLS) (U.S. Department of Labor, 1970). These projections were incorporated into a model that contained environmental variables and new technology representations that had been derived from basic engineering studies. The research focused on the economic impacts of investing in these highly capital intensive technologies and of day-to-day operation of such plants. A dynamic model was used to make a series of 1985 projections involving different rates of energy use growth both with and without the new technologies.

The major results document the sensitivity of total capital investment

* This research was supported by the National Science Foundation RANN Program (No. GI-34936), The MITRE Corporation, and M.I.T. Ford Professional Funds.
† Massachusetts Institute of Technology and MITRE Corporation.

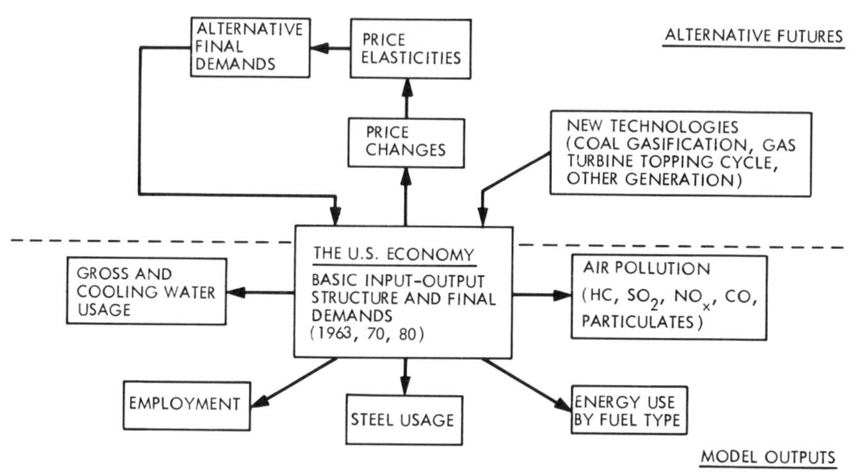

Figure 10.1. Energy-oriented generalized input-output model.

to changes in the energy growth rate and to the adoption of new energy technology. They also illustrate that very small changes in the overall growth rate of personal consumption or government expenditures can restrain total investment to within its historical limits as a percentage of GNP. The significance of these results is that the people of the United States can sustain the huge investment demands created by rapid energy demand growth by reducing the growth rate of personal consumption and government spending by less than 0.1% per year through 1985. Overall GNP growth rate remains unchanged because the sum of the growth rates in investment and noninvestment goods is a constant.

10.2. The Generalized Input-Output Model

The generalized input-output model used in this study is illustrated in Figure 10.1. The core of the model contains the actual and projected input-output structures describing the 1963, 1970, and 1980 economies.

The noneconomic quantities are referred to as accessory variables and are summarized in the bottom half of Figure 10.1. They are assumed to be proportional to the total output of each sector. For example, let S be the total emissions of SO_2 (or any other accessory variable) by the 1980 economy and let $\mathbf{E} = [e_k]$ be the vector of coefficients for SO_2 emissions per dollar of total output for sector. In other words, e_k is the SO_2 emitted per dollar of output of the kth industry. If **X** is the total output vector, then the total SO_2 emission S is the inner product of **X** and **E**, or

$$S = \mathbf{E}^T\mathbf{X} = \mathbf{X}^T\mathbf{E}. \tag{10.1}$$

Similar relationships hold for the other accessory variables.

The boxes in the upper half of Figure 10.1 represent the various means of interacting with the model. These boxes are used to specify the alternative future being investigated. This future can include changes in technology, in size and composition of GNP, and limited effects of price changes. The model is exercised by developing a scenario of possible future events and then constructing a final demand vector to represent the conditions of the scenario and modifying the technological and capital coefficients to include the amount and kind of new technology that are specified. Once these changes are made, the total outputs and investment requirements can be calculated. The values of the accessory variables are then obtained by simple multiplication as already indicated.

When making projections of final demands, it is necessary to calculate the investment required to support that level of final demand. Since final demand itself depends on the investment level, some iterative technique must be used to assure simultaneous solutions for both quantities. A simple two-period model can be used to illustrate this technique. Assume that (1) the same technological coefficient matrix **A** applies for both periods; (2) total final demand **Y** consists of final demand purchases by households and governments $\mathbf{Y}^F$ and capital investment purchases by all sectors of the economy ($\mathbf{Y}^I$), or

$$\mathbf{Y} = \mathbf{Y}^F + \mathbf{Y}^I,$$

and (3) the capital matrix **C** is defined as $\mathbf{C} = [c_{ij}]$, where c_{ij} is the marginal capital purchase from sector i by sector j required to expand the capacity of sector j by \$1 of output. Thus, if $\mathbf{X}_0$ the total output in period T_0 and $\mathbf{X}_1$ were the total output in period T_1, the total new investment required would be $\mathbf{C}(\mathbf{X}_1 - \mathbf{X}_0)$.

The objective is to find for period T_1 the total output $\mathbf{X}_1$ and total final demand $\mathbf{Y}_1$, given the total output in period T_0, $\mathbf{X}_0$, and the noninvestment final demand in period T_1, $\mathbf{Y}_1{}^F$. The model assumes that sectors always operate at 100% capacity so that output can be increased only by capital investment[2] and that capacity can be added in one period. The basic equations for this model are

$$\mathbf{X}_1 = (\mathbf{I} - \mathbf{A})^{-1}\mathbf{Y}_1 = (\mathbf{I} - \mathbf{A})^{-1}(\mathbf{Y}_1^F + \mathbf{Y}_1^I), \tag{10.2}$$

and

$$\mathbf{Y}_1^I = \mathbf{C}(\mathbf{X}_1 - \mathbf{X}_0). \tag{10.3}$$

These equations can be solved for total output $\mathbf{X}_1$ and total final demand $\mathbf{Y}_1$:

$$\mathbf{X}_1 = (\mathbf{I} - \mathbf{A} - \mathbf{C})^{-1}(\mathbf{Y}_1^F - \mathbf{C}\mathbf{X}_0); \tag{10.4}$$

$$\mathbf{Y}_1 = \mathbf{Y}_1^F + \mathbf{C}(\mathbf{X}_1 - \mathbf{X}_0). \tag{10.5}$$

Equations 10.4 and 10.5 can be used to investigate the effect on investment $\mathbf{Y}^1$ and total output $\mathbf{X}$ of changes in the growth rates of individual components of $\mathbf{Y}^F$.

Projections of alternative futures for some year should usually involve the same GNP [3] so that meaningful comparisons can be made. Figure 10.2 describes the projection model used in this study to ensure a constant 1985 GNP of $1.34 trillion[4] (1958 dollars). It is a slight modification of the above two-period analysis. Convergence of the iteration can be assured by modifying the scaling factor. No problems of negative investment were encountered.

10.3. New Technologies

10.3.1. Derivation of Technological Coefficients

The process of deriving technological coefficients for new technologies is best explained using an example. The example we use is taken from a report by the Institute of Gas Technology (IGT) (Tsaros et al., 1968). This report describes a 500-billion Btu/day gasification process that operates via hydrogasification and electrothermal gasification of lignite. This example is intended to illustrate the methodology only. The study itself used extensively modified engineering design studies (Just, 1973) and 110 I/O sectors.

Table 10.1 lists eleven sectors in our hypothetical economy. Figure 10.3 describes the components of the price of pipeline quality gas from such a

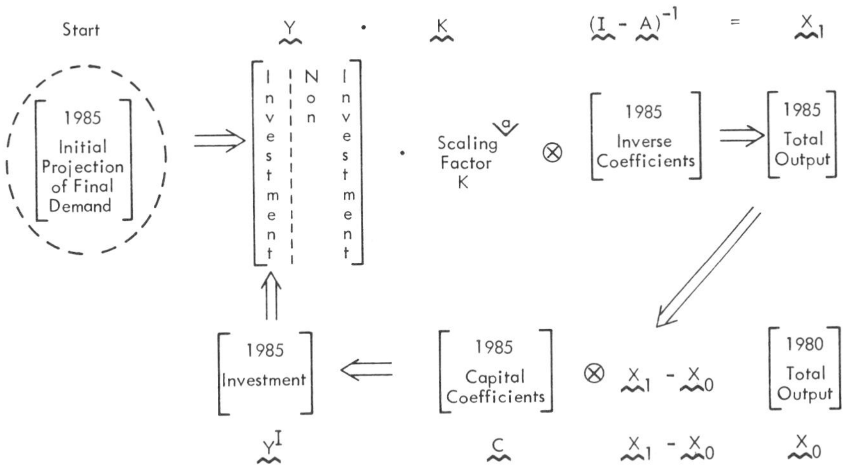

Figure 10.2. Quasi-dynamic projection model.

Symbols
- • Scalar Multiplication
- ⊗ Matrix Multiplication
- ⇒ Equals

[a] Scaling factor is chosen so that GNP $= |\mathbf{Y}| = \sum_i \mathbf{Y}_i =$ \$1.34 trillion.

process. The construction of technical coefficients for coal gasification involves transforming the pie chart of Figure 10.3 into a chart where all purchases are from one of the eleven sectors in the model. A first pass at this process appears in Figure 10.4. Supplies are assumed to be 15% of maintenance, and insurance 10% of local taxes. In this figure, all purchased commodities or services are assigned to the sector that manufactures or supplies them. Retail trade and transportation are ignored in this round. For example, catalysts and chemicals are assumed to be purchased directly from the chemical manufacturing sector even though they may have been purchased from a local distributor.

The convention followed in input-output analysis is that wholesale and retail trade do not purchase any goods for resale. Instead, the purchaser is shown as having bought any particular good directly from the manufacturer at the producer's price (that is, what the manufacturer receives from a wholesale buyer) *and* paying the trade margin or markup directly to the

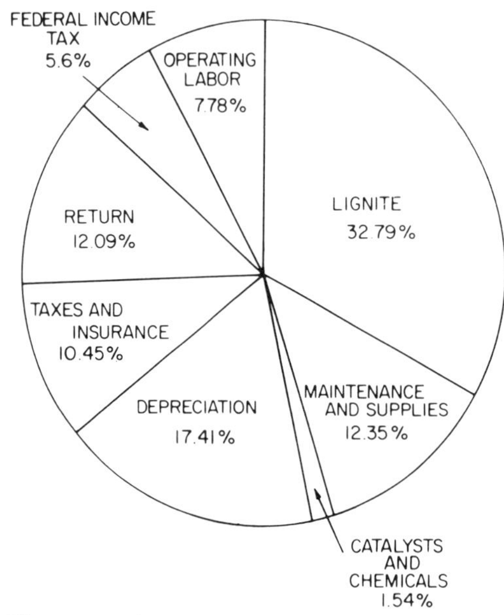

Figure 10.3. Components of pipeline gas price.

Table 10.1 Hypothetical Ten-Sector Economy

Number	Sector Name
1	Agriculture, Forestry, and Fishing
2	Mining
3	Construction
4	Nondurable Manufacturing (Food Processing, Textiles, and so forth)
5	Chemicals, Petroleum Refining
6	Durable Manufacturing
7	Transportation, Communications, Utilities
8	Wholesale and Retail Trade
9	Finance, Insurance, Real Estate
10	Other Services
11	Value-Added
	a. Labor (wages, salaries)
	b. Investors (interest and dividends)
	c. Capital Depreciation
	d. Government (state, local, federal taxes)

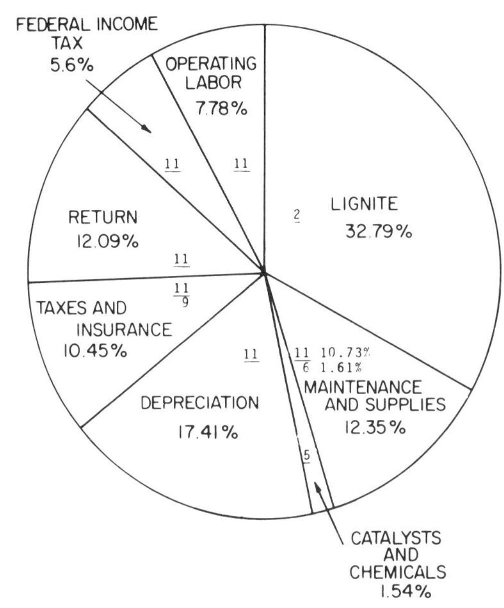

Figure 10.4. Components of pipeline gas price preliminary assignment of sectors.

wholesale and retail trade sector. Thus, any transaction is recorded as two separate entries, one to the manufacturing sector and one to the trade sector. Transportation charges are handled in a manner similar to trade margins. The purchaser is shown as paying the transportation charges directly to the transportation sector.

Figure 10.5 applies these concepts to the IGT example. Here 25% of the price of lignite is assumed to be transportation charges. No trade margin for lignite purchases is included because the company is assumed to buy directly from the mine. Supplies and catalysts and chemicals are assumed to have a 30% trade margin and a 10% transportation margin. All that remains now is to collect and sum all corresponding items.

When dealing with new technologies such as coal gasification, there are many competing processes that perform the same function or yield the same product. In developing the technological coefficients to represent such technology, it is important to make sure that the coefficients are representative of all the processes so that any conclusions from such a study are not sensitive to the exact process chosen. If such representation cannot be done, then sensitivity analysis must be used.

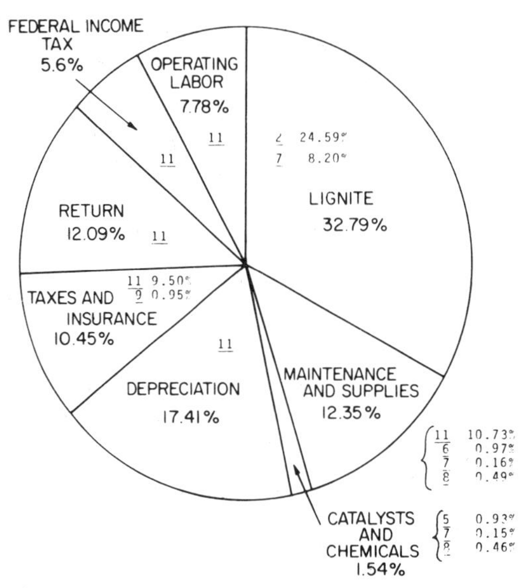

Figure 10.5. Components of pipeline gas price final assignment of sectors (including trade and transportation margins).

10.3.2. Derivation of Capital Coefficients

The process involved in deriving capital coefficients is quite similar to that for the technological coefficients. The basic strategy is to assign each piece of equipment to its producing industry; remove the transportation and trade margins; allocate construction, insurance, engineering, interest, and other charges; and divide the total purchases from each section by the total cost of the plant. The procedure yields a vector whose elements sum to 1.0 and which can be used to allocate each dollar spent on coal gasification plants to the respective industry of origin. From this basic percentage capital distribution vector, the capital coefficient vector can be found by multiplying by the capital/output ratio that describes the dollars of capital investment in plant per dollar of product output from the plant. This is calculated easily by dividing the capital cost of a new plant by the value of its yearly output.

This is the methodology that was followed to derive capital and technological coefficients for the various new energy technologies that are discussed. Detailed derivations, assumptions, and data modifications can be found in Just (1973).

10.3.3. Incorporation of New Technology

The new technologies were incorporated into the I/O framework using the following scheme. Suppose the old technological process for sector i (for example, natural gas production) is represented by the technical coefficient vector $\mathbf{A}_i$ and capital coefficient vector $\mathbf{C}_i$.[5] Next let the new technological process (for example, high-Btu coal gasification) be $\mathbf{A}_N$ and $\mathbf{C}_N$. If the new technology is expected to take over a fraction g of the total production of sector i and a fraction h of total new capacity capital additions by sector i, then the new technical coefficients are

$$\mathbf{A}_i = (1 - g)\mathbf{A}_i + g\mathbf{A}_N, \tag{10.6}$$

where g = fraction of total production supplied by new technology and the new capital coefficients are

$$\mathbf{C}_i = (1 - h)\mathbf{C}_i + h\mathbf{C}_N, \tag{10.7}$$

where h = fraction of total new capacity made up of new technology.

These coefficient column vectors then replace the old ones in the technical and capital coefficient matrices.

10.4. 1985 Projections

The results were obtained by projecting a series of five alternative 1985 futures involving various energy use growth rates, both with and without new technologies. These will be referred to as the Low, Medium, High, High plus Hygas (hydrogasification), and High plus Hygas plus Gas Turbine futures, and are defined now.

All of the projections used the 1980 technical coefficient matrix with some modifications of the energy sectors. The investment component of final demand was recalculated for each projection using the 1975 Battelle capital matrix modified slightly for the new technologies. The initial final demand projection for each alternative differed *only* in the amount of oil, natural gas, and electricity purchases.

The Medium energy use growth rate future assumes a continuation of the 1970–1980 final demand growth patterns and no change in industrial technology from 1980.

The High energy use future reflects a 4% higher final demand (than the Medium future) for oil, natural gas, and electricity and increased industrial consumption of electricity, gas, rubber, and plastics (reflected in slight in-

Table 10.2 New Technologies Investigated

High-Btu Coal Gasification	[1000 Btu/Standard Cubic Feet (SCF)]
Process:	Electrothermal hydrogasification (Hygas)
Data source:	*Electrothermal Hygas Process—Escalated Costs* (Tsaros et al., 1968; Tsaros and Subramanian, 1971)
Originator:	Institute of Gas Technology
Efficiency:	71.7%
Nominal plant size:	500 million SCF/day (90% load factor)
Nominal cost:	Plant: $310–354 million[a] Gas: 54.8–72.4¢/10^6 Btu
Low-Btu Coal Gasification	(173 Btu/SCF)
Process:	1980 Texaco partial oxidation (Hot carbonate scrubbing)
Data source:	*Technological and Economic Feasibility of Advanced Power Cycles* (Robson et al., 1970)
Originator:	United Aircraft
Efficiency:	87%
Nominal plant size:	842 million SCF/day (70% load factor)
Nominal cost:	Plant: $27.5 million[b] Gas: 17.6¢/10^6 Btu
Gas Turbine Topping Cycle	(Combined Gas and Steam Cycle or COGAS)
Process:	1980 high inlet temperature (2800°F) turbine with waste heat boiler steam cycle (using low-Btu gas)
Data source:	*Technological and Economic Feasibility of Advanced Power Cycles* [3] (Robson et al., 1970)
Originator:	United Aircraft
Efficiency:	54.5%[c]
Nominal plant size:	1000 MW (70% load factor)
Nominal cost:	Plant: $94 million Electricity: 5.3 mills/kWh

[a] All dollar figures are in 1970 dollars.
[b] Includes working capital.
[c] Only the efficiency of the COGAS cycle. Overall efficiency is obtained by multiplying the two efficiencies.

Table 10.3 1985 New Technology Modifications

	Capital	Operating
Hygas (coal gasification)[a]	25% of new capacity (gas) additions will be in form of coal gasification.	5% of natural gas demand supplied by coal gasification.
Gas Turbine Topping Cycle (combined with low-Btu coal gasification)[b]	50% of fossil generation (15% of total generation) capacity additions will be added in form of gas turbine topping cycle.	38% of fossil generation (23% of total generation) will be by gas turbine topping cycle.

[a] High + Hygas Future: High Future is modified by the above addition of high-Btu coal gasification (the IGT Hygas process).
[b] High + Hygas + Gas Turbine Future: High Future is modified by the addition of both new technologies indicated above. Note that low-Btu coal gasification is used in conjunction with the gas turbine.

creases in the energy rows of the technical coefficient matrix). These changes assume increased air conditioning and electric heat, worse gasoline mileage, and longer yearly driving distances. All of these projections assume that there will not be a supply limitation on natural gas and that the same domestic to foreign crude and natural gas ratios apply in 1985 that were projected for 1980.

The Low energy use future involved 6% lower final demand than the Medium future) for oil, natural gas, and electricity and better conversion efficiency for electricity conversion and transportation. Two alternative High energy growth futures involving new technologies were also investigated. These technologies are described in Table 10.2. The High plus Hygas future included the introduction of high-Btu coal gasification (Hygas), while the High plus Hygas plus Gas Turbine future included Hygas and the gas turbine topping cycle (supported by low-Btu coal gasification). These technology modifications are described in Table 10.3.

The projection procedure aimed at a GNP of $1.34 trillion (1958 dollars)[6] in 1985 for all five alternative futures. This was accomplished with a balancing procedure using the model of Figure 10.2.

The major assumption in this procedure was that all sectors had the same income elasticity so that a constant scaling factor could be applied to all purchases of final demand. This is a bad assumption for such industries as food and kindred products, but since the conclusions of this study are based on a differential analysis of the various projections and not on the absolute numbers involved, this assumption is not a major problem.

simplify computation, only 1963 and 1980 coefficients were used. Lag operators for the utility partitions were also developed by the author. The maximum lag was taken to be 7 years ($N = 7$). To lessen the effect of growth cessation on the period 1963 to 1980, the terminal year was taken to be 1987.

The historical rate of growth in real output at 7.2% per year was used to extrapolate to terminal year 1987 from observed data to obtain total demands for electricity.

All data were in 1963 dollars.

11.4. Impact of Some Alternative Strategies

One strategy open to utilities faced with exponentially rising demand, increasing costs, and tightening environmental restrictions is improvement of existing technologies. Technological changes work in diverse ways on utility capital and operating costs. To explore the interindustry consequences of such changes, total output paths generated by the observed 1963 utility technological structures and generating mix were compared with total output paths generated from projected 1980 technological structures using the same mix of technologies.

Coal is a typical industry supplying utility current inputs. Figure 11.1

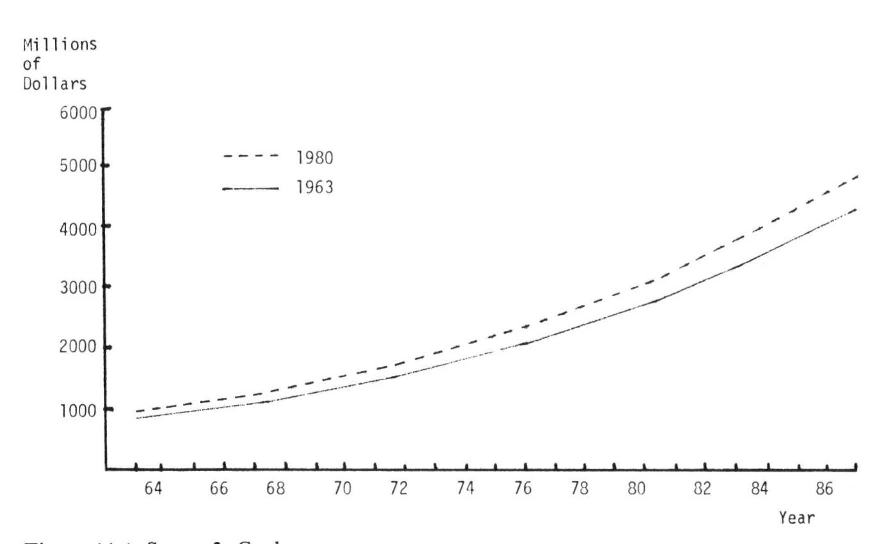

Figure 11.1. Sector 3, Coal.

shows steady growth in total coal output to 1987, the terminal year for the simulation. There is no decline in output although growth, and therefore capital formation, ceases by 1987. This result indicates that direct demand for coal as a utility fuel predominates over any indirect requirement in capital-producing industries. The rate of growth in output with either version of technology is 7.2%, as expected. The increase in coal output with 1980 instead of 1963 technology reflects two underlying trends. One is the decreased efficiency of units due to higher turbine back pressures from artificial cooling systems combined with a shift toward coal relative to natural gas within the model's fossil fuel technology. The other trend is the rise in coal prices due to premiums on low-sulfur coal. Premiums were assumed to be about $1.50 per ton on 1963 prices of about $6.00 per ton, or 25%. Figure 11.1 illustrates that this premium flows through directly to coal industry gross revenues, increasing dollar output of coal.

A steady increase in output indicates industries supplying current inputs. Sectors displaying such output paths are not concerned directly or indirectly with supplying capital needs of the utilities, and therefore are not *directly* impacted by alternative investment strategies. As in the case of coal, investment decisions between technologies (as between coal and other fossil fuels or nuclear power) can affect subsequent current purchases of these outputs. Industries supplying utility current inputs include sectors 3, 12, and 34 to 40. Sector 27 displayed only a very slight decline to 1987.

Intuitively, a major impact of changed technology should occur in the construction industry, sector 5. Figure 11.2 illustrates this impact, showing that 1980 technology will require larger quantities of construction. The fluctuations in construction output from 1963 to 1970 track observed fluctuation in purchases of construction by utilities nicely. A large portion of this increase can be accounted for by the inclusion of artificial cooling methods, either ponds or towers, with costs ranging from $2/kWh to $20/kWh of new capacity. Increased requirements from underground distribution also affect construction output. Construction industry inputs were more narrowly defined than for BEA tables. Exclusion of internal utility design and prime contracting costs reduce the industry's total output and hence the 1980 impact.

The sharp decline in construction output from 1982 to 1986 illustrates the operation of the lag mechanism; less and less construction is required

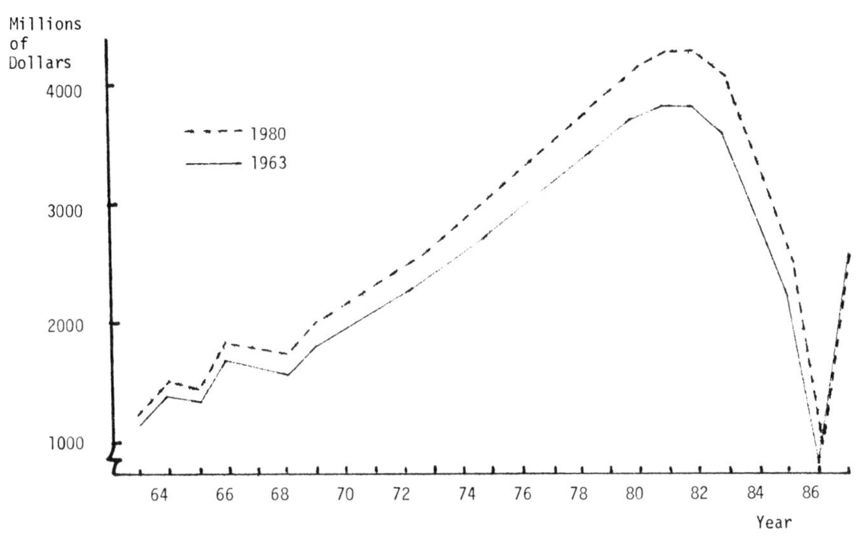

Figure 11.2. Sector 5, Construction.

as the no-growth state approaches. The sharp increase from the 1986 to the 1987 requirement for maintenance and repair construction (given constant demand levels) indicates the level of "disinvestment" in construction, specifically the freeing of construction output previously used indirectly by industries supplying goods to the utilities. Because the model does not distinguish between types of construction output, "disinvestment" occurs which reduces maintenance and repair. This behavior can be used to categorize industries as to investment impact. When output declines monotonically toward zero growth levels in 1987, the industry supplies capital goods and/or current inputs to the utilities but not output to industries which themselves produce capital for the utilities. In some cases the decline is slight, indicating that most goods were sold on current account and direct investment impacts are small. In other industries the decline was large, indicating most goods were sold for capital accounts and large investment impacts. Examples of the former type of industry include sectors 4, 6, 20, 26, and 28, while examples of the latter include sectors 2, 17, 20, 22, and 23. When the decline in output first dips below, then rises to the no growth level, the sector is supplying capital goods to satisfy the direct requirement of the

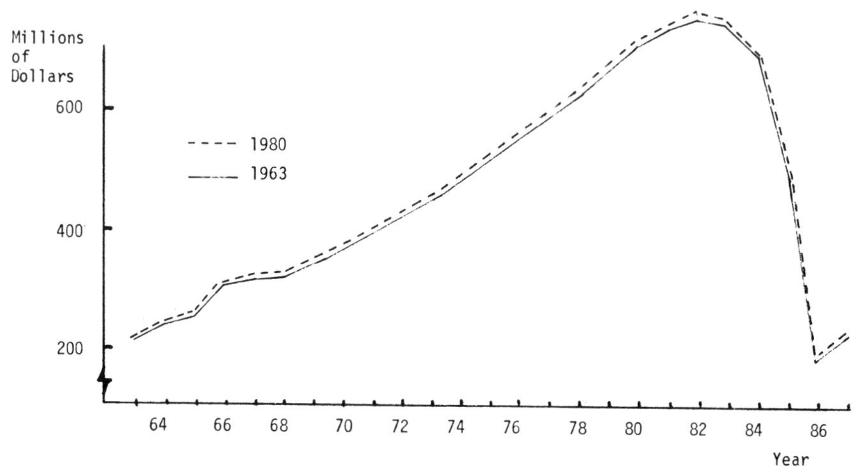

Figure 11.3. Sector 16, Structural metals.

utilities and also the indirect requirements of other industries, indicating great investment impacts. Sectors in this category include 5, 9, 14, 19, 21, and 25.

A result illustrating how the model can be used to predict economic impacts was obtained for structural metals, sector 16, shown in Figure 11.3. Although the output of the industry grows sharply with increasing utility demand, the figure indicates that the sector is relatively unaffected by technological change. The decline of the ending periods indicates that most of the output from the industry goes directly to the capital needs of the utilities, some to indirect capital needs (the dip) and some to replacement (current ending level).

A final illustrative result is the impact of 1980 technologies on industrial electrical apparatus, sector 22, shown in Figure 11.4. The decline almost to zero 1987 output indicates that virtually the entire output of this industry goes directly to utility capital formation. The industry proves, therefore, to be highly sensitive to changes in the level of investment spending by the utilities. More interesting is evidence of increasing efficiency of output from this sector with 1980 technology; the total dollar output of sector 22 required to meet demand growth diminishes with 1980 technology. This can be attributed to developments in extra high voltage (EHV) transmission.

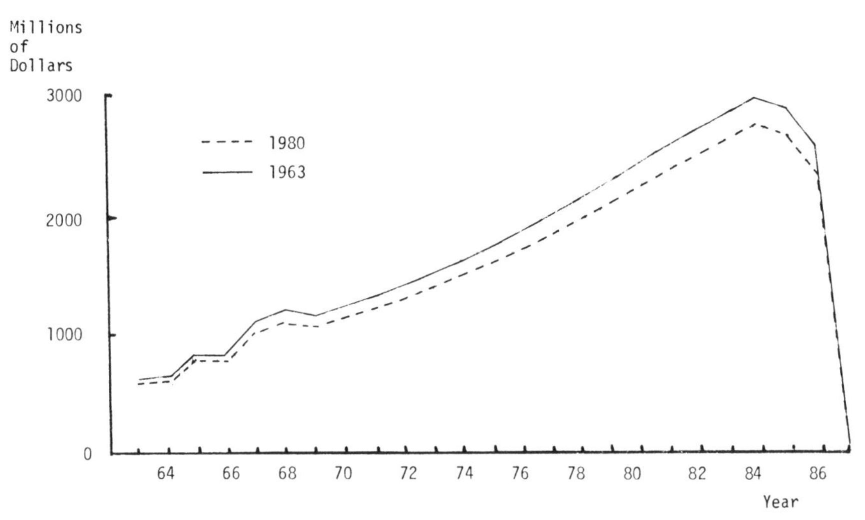

Figure 11.4. Sector 22, Industrial electrical apparatus.

Table 11.1 Generating Technology Mix

Subindustry	Percent of Total Generation Simulation (1)	(2)	(3)	(4)
34 Fossil Steam	80	75	65	60
35 Nuclear	5	10	15	20
36 Hydroelectric	10	10	15	15
37 Other Generation	5	5	5	5
Total	100	100	100	100

A second set of utility strategy decisions involves plant mix—whether to construct nuclear or fossil-fueled generating capacity. To a great extent this decision will be determined by cost of capital, construction period, fuel price, and safety or other environmental considerations. It is unlikely that utilities acting individually will consider the economic impact of these investment decisions unless distortions or bottlenecks become large.

Four different mixes of 1980 generating technologies were simulated for comparison. The percent of total generation for each technology is given in Table 11.1.

The single most important finding of these comparisons is the pervasiveness of economic impacts as the proportion of nuclear generation (and of

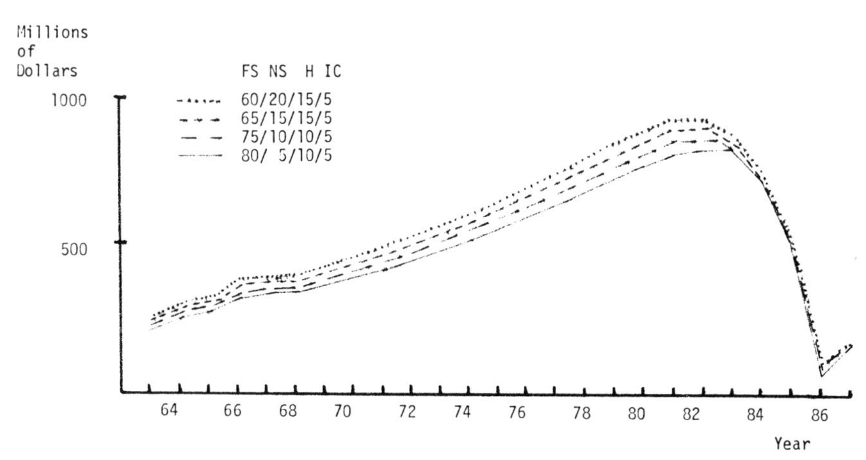

Figure 11.5. Sector 16, Structural metals.

new nuclear plant construction) increases. Of the 33 nonutility sectors, only 4—coal, petroleum, transportation equipment, and transportation—decline as nuclear generation increases. All 4 industries are intimately related to the decline in consumption of fossil fuels as nuclear generation increases. In some cases the increase in output due to increasing nuclear capacity is fairly large, for example in lumber and wood products (construction materials), chemicals (reactor moderator and fuel), stone, glass. and clay (concrete), and business services (insurance and legal services). In most sectors, however, the increase is only a small percent of total output and is the result of indirect demands over time.

Figure 11.5 depicts the change in structural metals output as the proportion of nuclear capacity increases. The changes are regular and gradual. Increases in most other capital sectors follow the same general pattern of gradual increases. The impact on structural metals from increasing nuclear capacity is much greater than from technological changes such as increased pollution controls or fossil-fueled units, the major alternatives to nuclear capacity.

In a sector such as structural metals, the increase in hydroelectric capacity assumed to reflect more pumped storage peaking capacity as nuclear capacity increases does not significantly affect output. The increase in hydroelectric generating capacity does significantly affect some sectors, including construction. The effect is easily seen in engines and turbines, sector

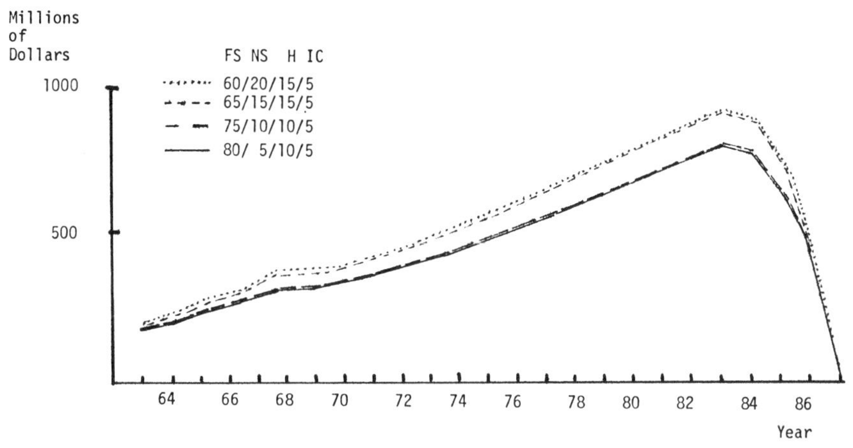

Figure 11.6. Sector 18, Engines and turbines.

18, shown in Figure 11.6. The major difference in output comes not from increased nuclear but from increased hydroelectric capacity. This reflects the fact that the cost of a pumped storage reversible turbine generator unit used for peaking power is, per unit of effective output, more costly than a base-loaded steam unit.

To explore the impact on output patterns of varying generating investment lag structures, the maximum lag was shortened from 7 to 6 and to 5 years with suitably revised lag operators. Only minor shifts in the shape of time path of outputs were observed.

11.5. Conclusions

From this study three principal conclusions can be drawn.

First, the general lagged dynamic input-output model is an effective tool for studying the interindustry effects of industrial growth where investment in plant and equipment is a major factor. Time paths of total outputs indicate major economic impacts. The principal difficulty in such modeling is data collection; as with any input-output analysis, the results are constrained by the linearity of production functions and the absence of substitution.

Second, while the principal changes in utility technologies may involve significant demands on selected industries, in general they do not have a pervasive economic impact.

Finally, increasing the proportion of nuclear generating capacity has pervasive economic effects. Although the fossil fuel/nuclear decision may be made on other bases, the macroeconomic implications should be borne in mind by investment planners.

References

Bureau of Economic Analysis (1969). *Survey of Current Business,* U.S. Department of Commerce, U.S. Government Printing Office, Washington, D.C., November.

Federal Power Commission (1972). *1970 National Power Survey,* U.S. Government Printing Office, Washington, D.C.

Fisher, W. Haldor, and Chilton, Cecil (1971). *An Ex Ante Capital Matrix for the United States, 1970–1975,* Battelle Memorial Institute, Columbus, Ohio, March.

Istvan, Rudyard (1972). *1980 Inputs for Private Electric Utilities,* Interagency Growth Project, Washington, D.C., August.

Leontief, Wassily (1969). "The Dynamic Inverse," in A. P. Carter and A. Brody, eds., *Contributions to Input-Output Analysis,* North Holland Publishing Company, Amsterdam.

12 Use of Input-Output Analysis to Determine the Energy Cost of Goods and Services*

ROBERT A. HERENDEEN†

12.1. Introduction

Two-thirds of U.S. energy goes to uses other than direct personal consumption (that is, residential use or fuel for the automobile). In discussing an individual's total energy requirement, one must therefore pay attention to the indirect energy demand resulting from the demand for goods and services. In particular, one would like to know the energy impact of the production, maintenance, transportation, and marketing of the whole spectrum of consumer products and to predict the energy requirements of hypothetical consumption patterns.

This report discusses use of input-output analysis (I/O) to do this. The methodology is described in some detail, and specific reference is made to potential shortcomings of the approach. Complete results are given elsewhere;[1] here I present several example uses, including calculation of the total efficiency of U.S. energy supply sectors, and the energy impact of agriculture and transportation.

Input-output analysis is used for two reasons. First, there is a large body of data already available from the U.S. Department of Commerce,[2] at a fairly high level of disaggregation (367 sector economy). Second, the conceptual framework of I/O takes into account all steps in the complex manufacturing-sales chain. The latter point is important: results of this work indicate, for example, that automobile manufacturing itself uses only about 6% of the energy needed to produce and market a car.

The task of converting the existing I/O data, which are in terms of dollars, to energy is nontrivial because energy is sold at different prices to different customers. For electricity, for example, there is a factor of 6 in variation.

* This work was begun while the author was a staff member of the Oak Ridge Environmental Program, Oak Ridge National Laboratory, Oak Ridge, Tenn., 37830, and continued at the Center for Advanced Computation, University of Illinois, Urbana, Ill., 61801. At both institutions it was supported by grants from the National Science Foundation RANN program.
† Center for Advanced Computation, University of Illinois, Urbana, Illinois.

12.2. Input-Output Theory: Deriving Energy "Cost"

12.2.1. Dollar Relationships

The data for I/O are dollar sales per year between sectors of the economy. (Of the 367 sectors, 5 are energy producers: coal mining, oil and gas wells, petroleum refining, electrical utilities, and gas utilities.) One then assumes linearity and time invariance to write

$$X_i = \sum_{j=1}^{n} A_{ij}X_j + Y_i, \tag{12.1}$$

where

X_i = the total output (dollars) of sector i,
Y_i = the output (dollars) of i sold to final demand,
A_{ij} = constants, obtained empirically from the data:
A_{ij} = [sales $(i \rightarrow j)$/total sales j] for the study year.

The assumption of constant A_{ij} is that of technological constancy (measured in dollars). It is assumed that Equation 12.1 holds for arbitrary final demands. In matrix form,

$$\mathbf{X} = \mathsf{A}\mathbf{X} + \mathbf{Y} \quad \text{or} \quad \mathbf{X} = (\mathsf{I} - \mathsf{A})^{-1}\mathbf{Y}.$$

The matrix $(\mathsf{I} - \mathsf{A})^{-1}$ for the United States economy has been published (Bureau of Economic Analysis, BEA, 1969) for 1963. The 1967 results are expected this fall.

12.2.2. Introducing Energy[3]

To convert to energy terms, let

$$E_i = \sum_{k=1}^{n} E_{ik} + E_{iy},$$

where

E_i = total energy output (Btu) of energy sector i,
E_{ik} = energy sales (Btu) from i to k,
E_{iy} = energy (Btu) of type i sold to final demand.

Since

$$E_{ik} = \left(\frac{E_{ik}}{X_k}\right) X_k = \frac{E_{ik}}{X_k} \sum_{l=1}^{n} [(\mathsf{I} - \mathsf{A})^{-1}]_{kl} Y_l,$$

$$E_i = \sum_{k=1}^{n} \sum_{l=1}^{n} \frac{E_{ik}}{X_k} [(\mathsf{I} - \mathsf{A})^{-1}]_{kl} Y_l + \left(\frac{E_{iy}}{Y_i}\right) Y_i.$$

The salient points of the various balanced 1985 projections are summarized in Table 10.4. Total investment becomes a larger percentage of the 1985 GNP as energy use increases from low to high. The introduction of high-Btu coal gasification further increases investment, while the introduction of the gas turbine topping cycle (with or without low-Btu coal gasification) decreases it. The output of coal mining is seen to increase dramatically with the introduction of coal gasification. The three illustrated capital-producing industries (plumbing, and so on) respond to different energy use growth rates more than total investment as a whole. Total employment is approximately constant, but there is no indication of how the required skills might change. Certainly more people will be employed in construction and in the capital goods industries for the higher energy growth rate scenarios. Air pollution and steel usage behave as expected. The large decrease in water usage caused by the introduction of the gas turbine topping cycle results from the fact that the gas turbines are air cooled and that the conversion efficiency is higher than the standard generation plant.

The most important fact concerning these balanced projections is not found in Table 10.4. The noninvestment components of the balanced final demand projections were within 0.3% of the initial projections. In other words, only a very slight change in personal consumption and government expenditures was enough to balance the investment demands of the rapidly growing energy sectors. It seems unlikely that most sectors would notice a difference in sales of 0.3% over a 5-year period. If most sectors were growing at the same 4.4% per year, at which GNP is projected to grow, a decrease of 0.3% in sales would decrease the growth rate to 4.35%, hardly a significant change.

10.5. Conclusions and Further Research

The major conclusions of this study are the following:

1. Total investment in general and capital good industries in particular (primarily turbogenerator manufacturers, boilermakers, and construction equipment manufacturers) are quite sensitive to energy use growth rates (especially electricity).

2. The major impacts of introducing the new energy technologies will be on the capital goods industries listed earlier. Operation of the new plants significantly affects only coal mining.

3. Introduction of high-Btu coal gasification will aggravate the demand

Table 10.4 Balanced 1985 Projections (1958 dollars)

	Low	Medium	High	High Plus Hygas	High plus Hygas plus Gas Turbine
GNP					
(billions of dollars)	$1340.8	$1343.0	$1339.0	$1340.9	$1341.0
PCE[a] (% of GNP)[b]	70.2	70.0	69.6	69.3	69.4
Investment (%)[b]	16.6	16.8	17.5	17.7	17.5
Government (%)[b]	13.8	13.8	13.5	13.6	13.6
Total Output					
(billions of dollars)					
Coal mining	5.0	5.1	5.2	6.5	6.6
Plumbing, structural metals	18.2	18.5	19.3	20.0	19.7
Engines and turbines	7.5	7.6	7.9	8.0	8.0
Construction equipment	11.1	11.5	12.5	12.9	12.6
Private Employment					
(millions)	99.2	99.2	99.2	99.2	99.2
Air Pollution					
(million tons)					
Particulates	48.6	49.0	50.0	50.2	50.1
Hydrocarbons	91.7	92.2	92.3	92.3	92.1
SO_2	75.2	76.1	78.2	78.2	78.2
CO	122.7	123.9	124.8	124.8	124.2
NO	30.4	31.8	32.6	32.6	32.5
Steel Usage					
(million tons)	194.0	195.0	198.1	199.6	198.6
Water Usage					
(trillion gallons)					
Gross	278.1	280.6	286.7	291.2	266.5
Cooling	126.0	128.3	134.3	137.8	117.8
Energy Use					
(10^{15} Btu)					
Coal	24.9	25.3	26.0	28.5	28.5
Oil	43.0	43.9	44.5	44.4	44.4
Gas	46.1	46.7	48.5	48.5	48.2
Electricity	33.0	33.8	34.9	34.8	34.8

[a] Personal consumption expenditures.
[b] These percentages total to 100.6% because Net Exports (−0.6%) have been left out

for investment funds, and introduction of the second generation gas turbine topping cycle (with or without low-Btu coal gasification) will decrease the demand.

4. Slight changes in the overall growth rates of total personal consumption expenditures and government spending result in large fluctuations in total investment.

5. If high energy growth continues and if investment is to remain within its historical limits as a percentage of GNP, energy investment will become a larger and larger part of total investment.

6. While interest rates are assumed to be the balancing mechanism between supply of and demand for investment funds, the very act of saving more money (which is induced by higher interest rates) means that less can be spent on consumption goods. This in turn lessens the demand for investment funds because the growth rates of consumption sectors are lower. This indirect effect of interest rates on investment may be quite important.

The policy implications of these types of results can be quite important. How do different sectors of the economy respond to changes in the interest rate? Housing construction seems to be particularly sensitive to interest rates. Will enough skilled construction labor be available to build all of the new required energy facilities? Manpower training programs can be developed if the need for such labor can be predicted long enough in advance. The generalized I/O model can be applied to all of the above questions, either in pointing out the need for policy or in analyzing the effects of new policy.

It is the ability to incorporate engineering studies into the generalized input-output framework that negates many previous objections to input-output analysis. Engineering studies can be used to determine how technology is likely to change if relative price changes or if some fuel becomes unavailable or how technology may improve with time. More work is needed to improve technology forecasting but the potential payoff is great.

Several areas stand out now as both important from the policy decision point of view and as areas where generalized input-output analysis can provide some unique capabilities. Obviously, more techniques than just input-output would be needed to answer the whole question, but input-output will play the central integrating role in these studies. Such areas are

1. Impacts of Alternative Methods of Meeting Oil and Gas Demand. Two extreme cases are possible: (a) the United States can rely on a massive

oil and gas import program to meet its growing energy needs or (b) the United States can stimulate oil and gas development internally. The economy, in terms of employment and sizes of various industries, will be quite different in these two cases. A first approximation to answering these questions could be obtained by ignoring the effects of any price changes in oil or gas products and focusing on the different final demands and industrial structures that might result.

2. Impacts of Multiple Investment Programs (for example, Energy and Pollution Controls). Both the government and industry have goals that entail large investment programs as in the industries' attempts to meet energy demand and the government attempts to control pollution. Generalized input-output analysis is valuable for examining the combined impacts of these various programs on different sectors of the economy.

These are important questions, and the techniques developed in this study can help to answer parts of them. More research is needed to expand the applications of generalized input-output analysis, but I hope that this report has shown that there is a value to such research.

Notes

1. "Generalized" refers to the (a) inclusion of noneconomic variables such as sulfur dioxide emissions or employment within the I/O framework and (b) use of engineering studies to update projections of the technological structure of the economy.

2. Slack variables can be used to modify this assumption, but such considerations are not important at this stage.

3. $\text{GNP} = |\mathbf{Y}| = |\mathbf{Y}^F + \mathbf{Y}^I| = \sum_{i=1}^{N} \mathbf{Y}_i$, where N = number of sectors in model and the magnitude signs indicate arithmetic additions of the vector elements.

4. This GNP represents a 4.4% per year growth rate from the BLS projection of the 1980 GNP. It was calculated by excluding any contribution from Bureau of Economic Analysis sectors 84, 85, and 86 (Government Industry, Rest of the World Industry, and the Household Industry, respectively). These dummy sectors were excluded because they do not interact with other sectors; they only contribute to GNP.

5. Thus the whole technological coefficient matrix could be represented as the partitioned matrix $\mathbf{A} = \mathbf{A}_1 : \mathbf{A}_2 : \ldots : \mathbf{A}_n$. A similar partition holds for the capital coefficient matrix.

6. This GNP represents a 4.4% growth rate from the projected 1980 GNP.

References

Just, James E. (1973). *Impacts of New Energy Technology Using Generalized Input-Output Analysis,* Energy Planning and Analysis Group Report 73-1, M.I.T., Cambridge, Mass., January.

Robson, F. L.; Giramonti, A. J.; Lewis, G. P.; and Gruber, G. (1970). *Technological and Economic Feasibility of Advanced Power Cycles and Methods of Producing Non-Polluting Fuels for Utility Power Stations,* United Aircraft Research Laboratories, East Hartford, Conn., December.

Tsaros, C. L., et al. (1968). *Cost Estimate of a 500 Billion BTU/Day Pipeline Gas Plant Via Hydrogasification and Electrothermal Gasification of Lignite,* Institute of Gas Technology, Chicago, for Office of Coal Research, Department of the Interior, Washington, D.C., August.

———, and Subramanian, T. K. (1971). *Electrothermal Hygas Process, Escalated Costs,* Institute of Gas Technology, Chicago, Ill., for Office of Coal Research, Department of the Interior, Washington, D.C., February.

U.S. Department of Labor (1970). *Patterns of U.S. Economic Growth,* Bureau of Labor Statistics Bulletin 1672, Government Printing Office, Washington, D.C.

11 Interindustry Impacts of Alternative Utility Investment Strategies*

RUDYARD ISTVAN

11.1. Introduction

Brownouts and blackouts in recent years have signaled the energy crisis in electricity. With electricity consumption increasing about 7.2% per year, the utility industry has been hard pressed to construct sufficient capacity to meet demand. The Federal Power Commission estimates that as much as $450 billion of investment by utilities may be required by 1990 (FPC, 1972), and there is some uncertainty about the ability of the industry to fund this requirement. Equally important, however, is uncertainty over the impact that investment spending of this magnitude will have on the economy. It is not clear that the economy can produce the generating, transmission, and distribution facilities needed over the next two decades without major distortions and bottlenecks.

This paper combines a restricted form of a general dynamic input-output model developed in Section 11.2 with specifications and data discussed in Section 11.3 to examine two major sets of issues surrounding alternative utility investment strategies. First, economic impact of projected technological changes to 1980 are explored. Such changes include pollution abatement equipment, extra high voltage transmission, underground distribution, pumped storage peaking plant, and increasing unit sizes. Second, the economic impacts of two major alternative utility investment strategies are explored: fossil fuel generation (with appropriate abatement equipment) and nuclear generation.

The principal finding is that while changing technology will impact some industries large-scale broad economic impacts do not occur. However, construction of nuclear plants tends to place much more diverse strains on the economy. Almost all sectors are affected to some degree.

11.2. Model Formulation

The lagged dynamic input-output model developed for this study is an extension of the Leontief dynamic model (Leontief, 1969). In the Leontief formulation the fundamental input-output accounting identity is modified to separate investment from other final demands with an explicit term:

* Harvard Economic Research Project.

$$\mathbf{X}_t = \mathbf{A}_t\mathbf{X}_t + \mathbf{K}_t + \mathbf{C}_t, \tag{11.1}$$

where $\mathbf{A}_t$ is the technical coefficient matrix for period t, $\mathbf{K}_t$ is a vector of investment for period t, and $\mathbf{C}_t$ and $\mathbf{X}_t$ are, respectively, vectors of consumption and total output for period t. The investment vector $\mathbf{K}_t$ is determined by a simple form of the accelerator model:

$$\mathbf{K}_t = \mathbf{B}_t(\mathbf{X}_{t+1} - \mathbf{X}_t), \tag{11.2}$$

where investment is a function of incremental output and $\mathbf{B}_t$ is a matrix of incremental capital coefficients describing requisite unit capital formation per unit increase in output.

Equations 11.1 and 11.2 assume that investment consumes resources and becomes productive during a single time period. For most major utility plant and equipment expenditures such an assumption is extremely limiting. For example, a fossil-fueled generating station requires about three years to complete. The stream of activities culminating in such a capacity increment impact the economy in more complex ways than are represented by a simple nonlagged formulation. Investment in any period is a subset of the total capacity increments $\mathbf{B}_t$. It is the sum of those fractions of $\mathbf{B}_t$ undertaken during any one period.

This conceptualization can be used to lag capital formation linearly over a number of periods to some horizon N, the maximum construction period. Define a set of lag operator matrices $\mathbf{L}_{0,1,\ldots,N}$ such that the coefficients

$$\sum_{n=0}^{N} (l_{n,i,j}) = 1, \qquad 0 \leq (l_{n,i,j}) \leq 1. \tag{11.3}$$

These lag operators specify the fraction of incremental capital per unit increase in output $(b_{t,i,j})$ that occurs in each of the periods over the construction horizon. If some $(l_{n,i,j})$ coefficient is zero, then no part of investment $(b_{t,i,j})$ occurs in the n lag period; similarly, if $(l_{n,i,j})$ is one, then the total increment to capacity in industry j from i called forth by an increment in output is produced in period n. The $\mathbf{L}$ are a unique function of each incremental capital coefficient $(b_{t,i,j})$ and are empirically observable. Although the lag operators are assumed time independent, such a restriction is easily relaxed.

Total investment for any period t will be the sum of investments coming on line in future periods to horizon N in response to future output increases that are undertaken during t:

$$\mathbf{K}_t = \sum_{n=0}^{N} \mathbf{L}_n \cdot \mathbf{B}_{t+n}(\mathbf{X}_{t+n+1} - \mathbf{X}_{t+n}), \tag{11.4}$$

where $\mathbf{L}_n \cdot \mathbf{B}_{t+n}$ is the dot product.

Combining Equations 11.1, 11.2, and 11.4 yields a general dynamic model with lagged investment:

$$\mathbf{X}_t = \mathbf{A}_t\mathbf{X}_t + \sum_{n=0}^{N} \mathbf{L}_n \cdot \mathbf{B}_{t+n}(\mathbf{X}_{t+n+1} - \mathbf{X}_{t+n}) + \mathbf{C}_t. \tag{11.5}$$

Equation 11.5 is readily solvable by an iterative procedure if the time path of $\mathbf{C}_t$ is known and growth is assumed to cease or become constant beyond some terminal year. The general solution is given by

$$\mathbf{X}_t = \mathbf{G}_t^{-1}\mathbf{C}_t + \sum_{n=0}^{N-1} \mathbf{G}_t^{-1}\mathbf{Z}_{t,n}\mathbf{X}_{t+n} + \mathbf{G}_t^{-1}\mathbf{L}_N \cdot \mathbf{B}_{t+N}\mathbf{X}_{t+N+1}, \tag{11.6}$$

where

$$\mathbf{G}_t = \mathbf{I} - \mathbf{A}_t + \mathbf{L}_0\mathbf{B}_t \tag{11.6a}$$

and

$$\mathbf{Z}_{t,n} = \mathbf{L}_n \cdot \mathbf{B}_{t+n} - \mathbf{L}_{n+1} \cdot \mathbf{B}_{t+n+1}. \tag{11.6b}$$

Equation 11.6 reduces to the simple Leontief dynamic inverse when there is no lagged investment ($N = 0$ and $\mathbf{L}_0 = \mathbf{I}$).

11.3. Specification and Data

To facilitate computation and analysis, a 33-sector aggregation of published 83-order input-output tables was developed. The aggregation was designed to leave maximum detail in capital-producing industries while eliminating detail in consumer-oriented industries. The correspondence between study sectors and 83-order BEA tables (Bureau of Economic Analysis, 1969) is given in Appendix 1 of that publication.

Modeling of the utility industry itself requires more detail than is available from published tables. To obtain additional detail, sector 68.01 (private electric utilities) and sectors 78.02 and 79.02 (public electric utilities) were expunged from published tables, then reincorporated as a set of seven homogeneous technology subindustries 34 through 40 (see BEA Appendix 1, 1969). The resultant 40-order input-output tables are partitioned between a 33-order description of the economy (partition **A**) and a 7-order description of the utilities (partition **U**).

Inputs from the rest of the economy to these technologies are given in the (**A** − **U**) partition; economic interchanges within the utility industry are given in the (**U** − **U**) partition. The matrix is specified so that all electricity is produced in sectors 34 to 37, "sold" to sector 38 (transmission), then "sold" to sector 39 (distribution). The distribution subindustry "sells" to the rest of the economy. By directly varying these coefficients of the (**U** − **U**) partition any mix of the seven major utility technologies can be simulated, and this capability facilitates study of alternative utility investment strategies.

The general model imposes severe data problems. Several simplifying restrictions were placed on the model. The properties of linearity were used to assume that all electricity is sold directly to final demand; other outputs are unaffected by this assumption. Because the (**U** − **A**) partition is therefore zero, utility data collection problems are approximately halved and a single projection of total demand by the whole economy for electricity from sector 39 can be used. By zeroing other final demands, the set of total outputs is produced solely by the interindustry requirements, direct and indirect, of electricity production. Hence, impacts on the economy are discernible separate from impacts caused by growth or change in other final demands.

Two principal sets of data were developed for this model. One set was for the 33-sector aggregation of the economy. Current flows for 1963 come from the published BEA tables (1969). Incremental capital coefficients for 1958 and 1970–1976 for the same partition were obtained from Battelle (Fisher and Chilton, 1971). Capital coefficients for 1963 were obtained by linear interpolation. Because of a lack of 1980 capital tables, only 1963 current and capital coefficients were used for the results reported here. In effect, only technology within the utilities is assumed to change. All investment for this **A** partition of the model is assumed to occur with no lags; that is, $\mathbf{L}_0 = \mathbf{I}$ for this partition. It is recognized that the use of nonlagged 1963 data for the model vitiates conclusions about 1980. However, sufficiently representative results can be obtained to draw qualitative conclusions.

Current and capital input coefficients for 1963 and 1980 for the (**A** − **U**) partition were developed and reported in Istvan (1972). To

Table 12.1 Sales for a Year of a 3-sector Economy[a]

	Crude Oil	Refined Petroleum	Cars	Final Demand	Total Output
Crude Oil	0	10 (40)	0	0	10
Refined Petroleum	5 (5)	5 (5)	5 (5)	25 (25)	40 (40)
Cars	0	0	0	20	20

[a] Figures in parentheses are energy sales; the other are dollar sales.

Define:

(a) $R_{ik} = E_{ik}/X_k$

(b) $S_{ik} = \begin{cases} E_{iy}/Y_i, \; i = k = \text{energy sector}, \\ 0, \text{ otherwise.} \end{cases}$

Then

$$\mathbf{E} = [\mathsf{R}(\mathsf{I} - \mathsf{A})^{-1} + \mathsf{S}]\mathbf{Y} = \boldsymbol{\epsilon}\mathbf{Y}, \tag{12.2}$$

where $\boldsymbol{\epsilon}$ is the total energy matrix. It is a 5 by 362 matrix[4] and ε_{ij} gives the total output (Btu) of energy sector i required for the economy to deliver a dollar's worth of projcct j to final demand, \$1 FD (j). It should be noted that in constructing R, one is constrained to choose the energy flows appropriate to the period and sector definition used in constructing A. In a real economic system R and A are not independent.

As an example, consider the 3-sector economy, whose sales for a year are given in Table 12.1.

$$\mathsf{A} = \begin{bmatrix} 0 & 1/4 & 0 \\ 1/2 & 1/8 & 1/4 \\ 0 & 0 & 0 \end{bmatrix}; \quad (\mathsf{I} - \mathsf{A})^{-1} = \begin{bmatrix} 7/6 & 1/3 & 1/12 \\ 2/3 & 4/3 & 1/3 \\ 0 & 0 & 1 \end{bmatrix},$$

$$\mathsf{R} = \begin{bmatrix} 0 & 1 & 0 \\ 1/2 & 1/8 & 1/4 \end{bmatrix}; \quad \boldsymbol{\epsilon} = \begin{bmatrix} 2/3 & 4/3 & 1/3 \\ 2/3 & 1/3 & 1/3 \end{bmatrix} + \begin{bmatrix} 0 & 0 & 0 \\ 0 & 1 & 0 \end{bmatrix}.$$

Additional information can be obtained from this approach. For example, how much energy is used to make the (say) steel contained in a car? This can be obtained from individual terms in the matrix product $\mathsf{R}(\mathsf{I} - \mathsf{A})^{-1}$. $R_{ik}\,[(\mathsf{I} - \mathsf{A})^{-1}]_{kj}$ is the energy type i supplied to k in order for k to supply enough of its output for the economy to deliver \$1 FD (j).

For the example, writing Equation 12.2 in terms of uncollapsed sums yields

$$\epsilon = \begin{bmatrix} 2/3 & 4/3 & 1/3 \\ (7/12 + 1/12) & (1/6 + 1/6) & (1/24 + 1/24 + 1/4) \end{bmatrix} + \begin{bmatrix} 0 & 0 & 0 \\ 0 & 1 & 0 \end{bmatrix}.$$

Thus, of the total of 1/3 Btu of refined petroleum required to deliver \$1 FD (car), 1/24 Btu is used by crude, 1/24 Btu by the refined petroleum sector itself, and 1/4 Btu by cars. Table 12.2 shows an application of this for the automobile.

12.2.3. A Problem with Primary Energy Sectors

The approach outlined so far does not assure that the amount of refined petroleum required per \$1 FD (j) will not exceed the amount of crude. In fact, it is possible to construct an economy that does this, and yet is monetarily and energetically realistic. The problem is one of allocation—the present method does not use the refined petroleum pricing data (there is a price implied in R_{ij}) to allocate crude; it merely uses the dollar flows. Explicit treatment is needed.

The method I suggest transfers the crude sold to refineries to the refineries (or in general, from primary to the consuming secondary industry), "letting" the refined sector distribute it, and then allocates this use back to crude. In the example, this requires letting $R_{\text{crude}\rightarrow\text{ref}} = 0$ to produce a new $\mathsf{R}(r)$ (reduced) and a new $\epsilon(r)$:

$$\epsilon(r) = [\mathsf{R}(r)(\mathsf{I} - \mathsf{A})^{-1} + \mathsf{S}].$$

The *new* total energy coefficient (called T) has a new crude row:

$$T_{\text{crude}\rightarrow j} = \epsilon_{\text{crude}\rightarrow j}(r) + (E_{\text{crude}\rightarrow\text{ref}}/E_{\text{ref}})\epsilon_{\text{ref}\rightarrow j}(r). \tag{12.3}$$

To demonstrate, change the refined petroleum Btu sales in the example to 5, 5, 10, and 20 Btu. (Note that this implies new prices: a new economy.) Without the correction,

$$\epsilon = \begin{bmatrix} 2/3 & 4/3 & 1/3 \\ 2/3 & 1/3 & 7/12 \end{bmatrix} + \begin{bmatrix} 0 & 0 & 0 \\ 0 & 4/5 & 0 \end{bmatrix}$$

and shows that cars require more refined than crude!

But

$$\epsilon(r) = \begin{bmatrix} 0 & 0 & 0 \\ 2/3 & 1/3 & 7/12 \end{bmatrix} + \begin{bmatrix} 0 & 0 & 0 \\ 0 & 4/5 & 0 \end{bmatrix},$$

Table 12.2 Energy Breakdown for Motor Vehicles (59.03), 1963

		PCINVB[b]					TOTPCP[c]
Row	Description[a]	Coal	Crude Oil & Gas Extraction	Refined Petroleum	Electricity	Gas	Total Primary
1	Food	.18	.47	.81	.28	.24	.36
2	Construction	.03	.80	1.82	.11	.08	.53
3	Textiles	.61	.56	.57	1.32	.58	.50
4	Paper and Lumber	1.61	1.77	1.64	1.31	1.95	1.60
5	Furniture	.01	.01	.01	.02	.01	.01
6	Chemicals	6.25	12.24	10.34	10.44	12.68	9.53
7	Rubber	1.35	1.00	.96	1.91	1.07	1.00
8	Leather	.01	.01	.02	.01	.01	.01
9	Stone, Clay, Glass	.86	3.12	.72	1.61	5.03	2.24
10	Primary Metals	65.11	34.29	24.61	34.78	43.02	41.24
11	Fabricated Metal	.62	1.28	1.32	1.70	1.31	.96
12	Machinery	.87	1.39	1.59	2.32	1.31	1.09
13	Instruments	.69	.84	.65	1.86	1.02	.69
14	Transportation Equipment	15.12	11.01	8.14	25.23	13.64	10.92
15	Transportation Services	.48	11.33	24.78	.69	1.92	7.52
16	Mining (Metal, Stone, Fertilizer)	1.45	1.70	1.46	2.34	1.96	1.47
17	Coal Mining	.89	.24	.35	.80	.16	.40
18	Crude Gas	.07	2.02	.16	.27	.42	1.33
19	Petroleum Refining, and so on	.18	5.39	9.64	.50	2.50	3.55
20	Electric Utilities	1.52	.76	.34	6.01	1.11	.74
21	Gas Utilities	.04	2.31	.24	.03	3.94	1.52
22	Water	.05	.26	.43	.07	.14	.18
23	Wholesale and Retail Trade	.38	2.49	4.43	1.52	1.18	1.69
24	Finance Insurance and Business	.36	1.17	1.16	1.41	1.23	.82
25	Medical	.00	.00	.00	.00	.00	.00
26	Education	.04	.09	.10	.16	.08	.06
27	Advertising	.01	.07	.08	.02	.06	.04
28	Radio, TV, Communications	.11	.32	.44	.39	.24	.23

[a] These sectors represent an aggregation of 367 into 28. Some sectors have been omitted, hence, column sums are less than unity.
[b] $(PCINVB)_{ji}$ is the percentage of the total energy of type i needed to make a motor vehicle that was used in sector j. Thus, 65.11% of the coal needed to make a motor vehicle was used by the primary metals sectors.
[c] $(TOTPCP)_j$ is the corresponding percentage of the total primary energy. Of the primary energy needed to make a motor vehicle, 41.24% was used by primary metals.

and, since $E_{crude \rightarrow ref}/E_{ref} = 1$,

$$\mathsf{T} = \begin{bmatrix} 2/3 & 1/3 & 7/12 \\ 2/3 & 1/3 & 7/12 \end{bmatrix} + \begin{bmatrix} 0 & 4/5 & 0 \\ 0 & 4/5 & 0 \end{bmatrix}.$$

Energy is now "conserved," while some crude is allocated to final demand.

In the Department of Commerce I/O matrix, there are five energy sectors:

Index	I/O Sector	Title
1	7.00	Coal mining
2	8.00	Oil and gas wells
3	31.01	Refined petroleum
4	68.01	Electric utilities
5	68.02	Gas utilities

The $\varepsilon(r)$ is obtained with the following energy flows set equal to 0:

$1 \rightarrow 4$, coal $\rightarrow$ electrical utilities,
$2 \rightarrow 3$, crude oil and gas $\rightarrow$ refined petroleum,
$2 \rightarrow 5$, crude oil and gas $\rightarrow$ gas utilities,
$3 \rightarrow 4$, refined petroleum $\rightarrow$ electric utilities,
$5 \rightarrow 4$, gas utilities $\rightarrow$ electric utilities.

By the same reasoning as earlier[5]

$$T_{ij} = \sum_k C_{ik}\epsilon_{kj}(r), \quad \text{or} \quad \mathsf{T} = \mathsf{C}[\mathsf{R}(r)(\mathsf{I} - \mathsf{A})^{-1} + \mathsf{S}], \tag{12.4}$$

where

$$C_{ii} = 1, \qquad i = 1, 2 \ldots 5,$$
$$C_{14} = E_{14}/E_4,$$
$$C_{23} = E_{23}/E_3,$$
$$C_{24} = (E_{23}/E_3)(E_{34}/E_4) + (E_{25}/E_5)(E_{54}E_4),$$
$$C_{25} = E_{25}/E_5,$$
$$C_{34} = E_{34}/E_4,$$
$$C_{54} = E_{54}/E_4.$$

12.3. Limitations of the I/O Approach

Some of the limitations result from data-handling problems. Several, however, are conceptual and derive mostly from economic conventions used in constructing the published I/O tables:

1. Input-output data are subject to inaccuracies from (1) lack of complete coverage of an industry, (2) restriction of information for proprietary reasons, (3) use of different time periods for different kinds of data. Errors in A may generate disproportionate errors in $(I - A)^{-1}$. We are actively pursuing this last point. Preliminary analysis shows the published $(I - A)^{-1}$ to have a condition number of 5; this is a rough index of the multiplication of error.

2. Thc usc of dollars rather than physical units to express physical dependencies is risky. For example, aggregation can combine two processes whose energy intensities differ widely in the same sector. And dollar economies of scale may be implicit in the A_{ij}, whereas there would be no (or little) corresponding effect in physical terms. We are also concerned with this question.

3. There is the problem of transfers. BEA's I/O sector definition is based on the establishment, rather than activity. For example, if those establishments that produce primary aluminum also produce aluminum castings (amounting to less than 50% of total sales), the primary aluminum sector is credited with their complete output. The secondary output is transferred to the aluminum castings sector, that is, treated as a sale. The corresponding inputs are not transferred. This means that the dollar output corresponding to production of these aluminum castings has been counted twice, but the energy only once. The fraction transferred varies from sector to sector, so that R_{ij} require a correction, details are in Herendeen (in press).

4. A problem arises in capital goods; these are not considered part of the interindustry transactions but are listed as sales to final demand. Conceptually, I could consider the energy to make a steel forming press owned by an auto manufacturer to be as valid an energy contribution to my car as that used to make the steel, but this is not the convention used in I/O. One must learn in detail how this is handled, since capital is defined differently from industry to industry and sways with the winds of the tax laws. Given reliable figures, one can separate the capital for depreciation from that for expansion, and allocate it (as its energy input) to the actual consuming sector. We are actively studying this question.

5. Final demand is measured in producer's, not purchaser's, prices. Since the I/O sectors include wholesale and retail trade, it is possible to make the conversion, including the energy requirements implied in the

markup (as has been done for the automobile in Section 12.3). But for direct use by consumers, it is desirable to convert beforehand to producer's prices. Data on markups are available from the Department of Commerce, and we are now using them.

6. Input-output coefficients change with time, yet we hope to use the results to predict the consequences of hypothetical consumption patterns. Can one quantify their loss or reliability with time? This is a major point, for which much work is needed. We shall be converting the 1967 I/O table to energy terms for comparison with 1963, and we are starting sensitivity analyses to determine which coefficients are most critical. Our feeling is that energy use coefficients, on the whole, change faster than others.

7. Input-output sectors are often too broad. For example, do buses require less energy than cars for manufacture (per dollar)? To answer this, we are currently working on two tacks. One is to gather additional data to allow disaggregation of the "motor vehicles and parts" sector into two. This requires construction of both a row and a column, and an inversion of the new $n + 1$ degree matrix. A second is to use the identity

$$\begin{aligned}(I - A)^{-1} &= I + (I - A)^{-1}A \\ &= I + A + (I - A)^{-1}A^2 \\ &\quad \text{and so on}\end{aligned} \tag{12.5}$$

as the basis for an approximation. For example, for the capital energy cost of a mass transit system, we could adapt Equation 12.5 thus:

$$\Delta E = \Delta E' + CR(I - A)^{-1}\,\Delta Y,$$

where

ΔE = the total energy cost,
$\Delta E'$ = direct energy used,
and ΔY = the bill of the input goods,

stated compatibly with I/O categories. Intuitively, this is just a first-order vertical analysis (that is, an intuitive step by step tracing of inputs) *except* that the energy content associated with the inputs is given by the total energy coefficient. This is called a "hybrid" approach.

12.4. What Is the Energy Needed to Make a Product or Provide a Service?

The total energy coefficient T_{ij} allows an answer when the question is modified to read "to supply a product to final demand." One can allocate all of the U.S. energy budget this way (that is, $\mathbf{E} = \mathbf{TY}$). But this scheme gives absurdly low weight to sectors that sell little of their output to final demand (for example, primary metals). Trying to assign an energy to this kind of sector is difficult. An additional ton of aluminum can be required as a result of many different changes in final demand (for example, for more milk!), and each will have a different energy requirement; there is no unambiguous energy impact of the additional aluminum production. The argument below helps to resolve this.[6]

Let a sector receive (1) direct energy from energy sectors or the "earth" (if an energy sector), (2) embodied energy by virtue of purchases of processed inputs, and let it distribute (1) actual energy (if an energy sector) and (2) embodied energy in its sales.

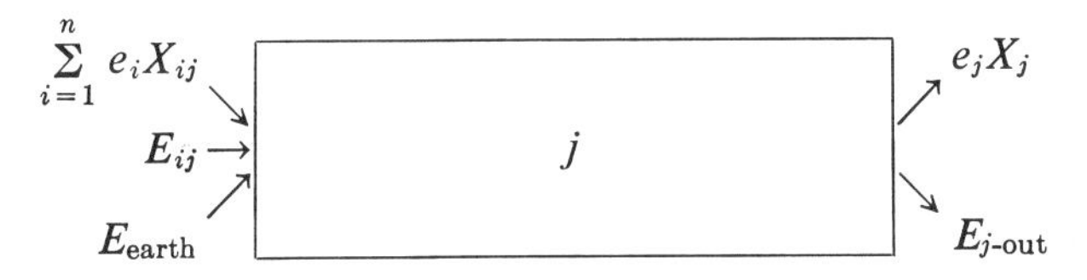

Here we will assume only one energy sector, but generalization is easy. For the energy sector m, $E_{\text{earth-}m} = E_{m\text{-out}}$; these quantities are zero for the other sectors; $E_{ij} = 0$ unless $i = m$. The model assumes that embodied energy content per dollar e_j is the same for all sales by j and does not differentiate sales to final demand. Energy balance requires

$$e_j X_j - \sum_{i=1}^{n} e_i X_{ij} = E_{ij} + E_{\text{earth}} - E_{j\text{-out}}. \tag{12.6}$$

The right-hand side is the energy dissipated in j; call it η_j

$$e_j - \sum_{i=1}^{n} e_i \frac{X_{ij}}{X_j} = e_j - \sum_{i=1}^{n} e_i A_{ij} = \frac{\eta_i}{X_j} = R_j, \tag{12.7}$$

$$\mathbf{e} = \mathbf{R}(\mathsf{I} - \mathsf{A})^{-1}. \tag{12.8}$$

The embodied energy coefficient is thus identical to the total energy coefficient obtained before, but now it is associated with all sales of j, not just those to final demand. If we used total outputs to allocate energy, we would obtain a total energy several times greater than the U.S. energy use, since there is much double counting. One obtains this picture of the economy:

energy enters through the primary sectors, a greater quantity of embodied energy circulates, and an amount equal to the input leaves via final demand.

It is possible to allocate on the basis of total output, but then a disclaimer is needed: "The energy necessary to supply the production of widgets last year is given by the total output times the total energy coefficient. Many of these widgets would not have been produced were there not also production of gizmos which require widgets as inputs. The energy needed to produce gizmos thus includes that to produce the required widgets, and allocating energy this way double counts."

12.5. Selected Applications

The total energy coefficients for the 1963 I/O table are listed in Heerenden (in press). (The main task of that work was to obtain the energies used by the various sectors.) Here I present several uses for the results. All these are subject to as yet unquantified errors, and the capital problem referred to. On the average, my feeling is that the numbers for 1963 are good to ±15%.

12.5.1. Breakdown of the Energy Needed to Make a Motor Vehicle, 1963 (Table 12.2)

This uses the method of Section 12.2.2, which illustrates how the energy to make a product is built up from the energy used by all the suppliers of components of that product. For the auto, over 40% of the total is used by the primary metals sector, and only about 11% is used by the transportation equipment sectors.

12.5.2. Energy Efficiency of the Energy Supply Sectors

The total primary energy coefficient T_j (prim) is defined as the sum of the coefficients for coal, crude oil, and gas extraction and a portion of electricity produced by hydro, converted at the going heat rate. Thus, for energy type i, the fraction (total primary energy)/(1 Btu of type i to final demand) is

$$\left(\sum_k{}' T_{ki}\right)/S_{ii} = T_i(\text{prim})/S_{ii}, \tag{12.9}$$

where $\sum_k{}'$ is the sum referred to. Table 12.3 lists the results for 1963.

12.5.3. Agriculture and Food

1. Agriculture. Ten sectors (1.01–1.03, 2.01–2.07) of the 367 are identified as agricultural (Bureau of Economic Analysis, 1969), and the

Table 12.3 Inverse Energy Efficiency (I.E.E.) of the Energy-Producing Sectors (The total primary energy required to deliver 1 Btu of energy of various types to final demand, 1963. See Section 12.5.2.)

I/O Sector	Title	I.E.E. (Btu/Btu)	I.E.E.[a] Corrected for Imports	E.E.[b] (reciprocal)
7.00	Coal mining	1.024	1.024	0.977
31.01	Petroleum refining	1.082	1.208	0.828
68.01	Electric utilities	3.870	3.870	0.258
68.02	Gas utilities	1.134	1.169	0.855

[a] 10.4% of refined petroleum and 3.0% of gas were imported in 1963; the energy to produce them was expended outside the U.S. economy.
[b] Since these results are based on producers' prices, additional energy would be expended in delivering coal and refined petroleum to most consumers. For electric and gas utilities, this is not a problem.

energy to support their production can be obtained by summing the total primary energy needed to produce each sector's output minus that sold to other agricultural sectors; that is,

$$\sum_{i \in G} T_i(\text{prim})\left(X_i - \sum_{j \in G} X_{ij}\right), \tag{12.10}$$

where G is the set of agricultural sectors. This is the energy needed to allow the aggregated agricultural sector to ship to everyone else (mostly the food processing industry). For 1963, this was 2.2×10^{15} Btu, which is ~4.4% of the U.S. energy budget for that year. The direct energy use for these sectors (with electricity converted to primary by a factor of 3.9 (Table 12.3) was 1.2×10^{15} Btu (this direct figure agrees well with recent ones from Perelman, 1972). Agriculture thus required about as much energy to produce its nonenergy inputs such as fertilizer, equipment, and so on, as it did for direct use.

2. Food and kindred products. The same analysis is applied to the food sectors (I/O 14.01 to 14.32) to yield 3.6×10^{15} Btu, which is ~7.2% of the U.S. total. Much of the agricultural energy has been counted again in this figure. The total figure can be compared with the calorie content of the food, approximately 0.8×10^{15} Btu (Perelman, 1972). The embodied energy is 4½ times the food energy.

3. The energy cost of protein. Several I/O sectors are for high-protein foods, and the energy cost to produce the final processed, packaged product has been calculated as a function of the protein contained. Table 12.4 lists

Table 12.4 Energy Cost of Supply Protein, 1963

I/O Sector	Title	T (primary)[a] (kcal/$)	Producers Price[b] ($/lb)	% Protein[c]	Protein Energy[d] (kcal/lb)	Energy/Calorie[e] Content
14.01	Meat products	14,200	0.50	22	32,600	6.3
14.03	Cheese, natural and processed	15,600	0.30	25	18,800	2.6
14.06	Fluid milk	15,000	0.12	3.5	51,200	6.1
14.12	Fresh or frozen packaged fish	10,100	0.35	20	17,700	6.5

[a] 1 kcal = 4.0 Btu.
[b] Bureau of the Census (1971). Figures are approximate.
[c] (Bowes and Church, 1970).
[d] Marketing energy not included; on the average it would increase figures by 10 to 15%.
[e] Total calories including nonprotein materials.

results for four sectors: meat products, cheese, milk, and fish. It seems that in 1963 fish was a low-energy protein source and milk a high one. For comparison with the work of others, recall that this is the total direct and indirect energy.

12.5.4. Transportation

The direct fuel use for all transportation today is about a quarter of all energy, but much more is required to supply the support: transportation equipment, roads, and so on.

1. Energy impact of the automobile, 1963.

Table 12.5 is an application of the method described in Section 12.4. It involves multiplying the appropriate total primary energy coefficients times the corresponding final demand expenditures associated with the private automobile. In 1963, the average new domestic automobile was worth \$1890 in producer's prices, which implies a total energy cost of 132×10^6 Btu/car (the equivalent of 5 tons of coal). This figure agrees very well with that obtained by Berry and Fels (1972).

The final demand expenditures associated with the automobile were 12.4% of the GNP, but accounted for 20.7% of the nation's energy budget. Only 57% of the automobile energy was for direct use as fuel. These statements can be made without disclaimer since all pertinent expenditures are part of final demand.

2. The energy for all transportation.

This is partly a definitional problem. Twenty-three sectors would be called transportation sectors: highway construction, motor vehicles plus equipment, aircraft plus parts, other transportation equipment, transportation plus warehousing, and auto repair plus services. As described by Hirst and Herendeen (1973), other categories apply some of their final demand expenses to support the automobile (for example, insurance). In addition, the refined petroleum sector fuels almost all transportation. Then an estimate for transportation's total energy is

$$\sum_{i \in H} T_i'(\text{prim})(X_i - \sum_{j \in H} X_{ij}) + \sum_{j \in H'} T_j'(\text{prim}) Y_j' + E_F, \qquad (12.11)$$

where $T'(\text{prim})$ is the total primary energy coefficient diminished by the refined petroleum used by industrial transportation, H is the set of 23 transportation sectors, and H' is the set of sectors which partly support the private car. Here, E_F = direct fuel used by transportation and that burned

Table 12.5 Energy Impact of the Automobile, 1963[a]

	Dollar Flow (10^9 $)	I/O Sector	I/O Coefficient[b] (Btu/$)	Energy 10^{12} Btu	% of Total
Gasoline					
Production	5.86[c]	31.01		5860	57.2
Refining	—	31.01	(0.208 Btu/Btu)[d]	1220	11.9
Retail markup[e]	4.05	69.02	32,700	130	1.3
Oil					
Production	0.83[c]	31.01	—	50[f]	0.5
Retail markup	0.55[g]	69.02	32,700	20	0.2
Auto					
Manufacture	14.43[c]	59.03	70,000	1010	9.9
Retail markup	10.67[c]	69.02	32,700	350	3.4
Repairs, Maintenance, Parts	10.0[c]	75.00	33,700	340	3.3
Parking, Garaging	11.7[c]	75.00	33,700	390	3.8
Tires					
Manufacture	0.83[c]	32.01	99,100	80	0.8
Retail markup	0.55[g]	69.02	32,700	20	0.2
Insurance	8.96[c]	70.04	31,400	280	2.7
Taxes (highway con-) struction)	4.9[e]	11.04	98,500	490	4.8
Total	73.3 (12.4% of GNP)			10,240 (20.5% of total)	100.0

[a] The analysis is described by Hirst and Herendeen (1973). The numbers here differ somewhat since the calculation there was for 1970.
[b] From Herendeen (in press). These are expressed in total primary energy, including a thermal equivalent of hydropower. See Section 12.5.2.
[c] Figures obtained from American Petroleum Institute (1971 ed.), pp. 306, 307, 322 and Bureau of the Census (1971, pp. 536–537). There were 69×10^6 autos registered in 1963, and 7.64×10^6 produced domestically at an average producer's price of $1890; 0.41×10^6 were imported.
[d] See Table 12.3.
[e] Retail gasoline markup and taxes from American Petroleum Institute (1971 ed., pp. 458–459).
[f] Oil/gasoline ratio = 128 on a Btu basis, from Bureau of the Census (1971, p. 537).
[g] Markup of oil and tires assumed to be 40% of purchaser's price.

Table 12.6 Total Energy Impact of Transportation, 1963

Use	Energy (10^{15} Btu)	Percentage of Total U.S. Energy Use[h]
Direct fuel[a]	11.5	23.1
Refining, and so on[b]	2.3	4.6
Highway construction[c,f]	0.57	1.1
Transportation equipment and maintenance[d,f]	2.33	4.7
Transportation services[e,f]	3.52	4.8
Other expenses associated with private automobile:[g]		
Insurance	0.19	0.4
Accessories	0.06	0.1
Retail markup energy:		
Cars	0.26	0.5
Gasoline	0.11	0.2
		39.5[i]

[a] Bureau of Mines figure. See American Petroleum Institute (1971 ed., p. 443).
[b] 20% of direct fuel. See Table 12.3.
[c] I/O Sector 11.04.
[d] Includes tires, maintenance, motor vehicles and parts, aircraft and parts, and other transportation equipment (I/O sectors 32.01, 75.00, 59.01–59.03, 60.01–60.04, 61.01–61.07).
[e] Includes services such as air, water, rail, urban passenger and freight, pipeline transportation (I/O sectors 65.01–65.07).
[f] Energy is obtained by multiplying total output (minus that sold to other transportation sectors), in dollars, times the appropriate energy coefficient from Herendeen (in press) (diminished by the energy used by industry and commerce for transportation).
[g] Various figures in Bureau of the Census (1971). There were 8.05×10^6 cars sold in 1963 at an average markup of $1140; there were 69×10^6 cars on the road paying an average yearly insurance premium of $129; the average car owner bought $14 worth of accessories; 46×10^9 gallons of fuel were bought at a markup of approximately 9 cents per gallon. All these numbers were converted to energy using appropriate (diminished) I/O coefficients.
[h] U.S. energy use = 49.8×10^{15} Btu in 1963 (American Petroleum Institute, 1971 ed., p. 443).
[i] Since undoubtedly some additional transportation expenses have been omitted, this is a lower bound for transportation's total energy impact in 1963.

Table 12.7 Energy Breakdown for Air Transportation (65.05), 1963.

		PCINVB[b]					TOTPCP[c]
Row	Description[a]	Coal	Crude Oil & Gas Extraction	Refined Petroleum	Electricity	Gas	Total Primary
1	Food	3.20	.30	.22	2.28	1.26	1.36
2	Construction	.28	.23	.24	.47	.16	.23
3	Textiles	.31	.01	.00	.28	.01	.12
4	Paper and Lumber	7.33	.23	.10	2.54	1.69	2.94
5	Furniture	.01	.00	.00	.01	.00	.00
6	Chemicals	15.05	.89	.29	11.03	6.68	6.07
7	Rubber	1.11	.02	.01	.61	.18	.43
8	Leather	.02	.00	.00	.00	.00	.01
9	Stone, Clay, Glass	.79	.07	.01	.51	.73	.34
10	Primary Metals	21.57	.47	.14	8.62	4.15	8.48
11	Fabricated Metal	.26	.02	.01	.33	.12	.10
12	Machinery	.75	.04	.02	.85	.25	.29
13	Instruments	1.14	.05	.02	1.29	.32	.43
14	Transportation Equipment	4.47	.16	.11	5.60	.80	1.67
15	Transportation Services[d]	11.95	85.40	92.88	18.66	32.51	53.00
16	Mining (Metal, Stone, Fertilizer)	.89	.06	.02	.80	.48	.36
17	Coal Mining	.82	.01	.00	.31	.03	.32
18	Crude Gas	3.15	2.96	.11	5.27	4.07	2.76
19	Petroleum Refining, and so on	8.65	6.89	4.98	11.19	29.26	6.98
20	Electric Utilities	5.02	.08	.02	8.40	.78	1.74
21	Gas Utilities	.19	.35	.02	.06	3.95	.27
22	Water	.30	.04	.03	.16	.17	.14
23	Wholesale and Retail Trade	2.09	.39	.31	3.50	1.36	.93
24	Finance Insurance and Business	3.65	.38	.17	6.11	2.69	1.45
25	Medical	.02	.00	.00	.03	.01	.01
26	Education	.39	.03	.01	.65	.16	.15
27	Advertising	.04	.02	.01	.07	.09	.02
28	Radio, TV, Communications	.96	.11	.07	1.54	.54	.39

[a], [b], and [c] See notes for Table 12.2.
[d] Almost all of use by row sector 15 is direct fuel use.

in the society to produce that fuel (from Table 12.3, this is roughly an additional 20%).

Table 12.6 lists the results. In 1963, transportation's direct fuel use was 23% of the U.S. use, but the total energy impact amounted to 40%. This is probably a lower bound, as undoubtedly many transportation-associated activities have been neglected (such as industrial construction for parking, which is a capital expense). I would estimate it as high as 45%.

3. Energy breakdown for air travel, I/O sector 65.05.

See Table 12.7. Unfortunately, this sums passenger and freight; we are currently involved in disaggregating it. About 53% of the total energy is fuel. This sort of analysis will have to be applied to transportation systems before the present hierarchy of transportation energy intensities (Hirst, 1972) (which is based on direct fuel use only) can be trusted completely. My opinion so far is that the ordering will not change, and that those of us who flew here did indeed use the most energy intensive mode available.

Notes

1. See Herendeen (in press). The results, with less amplifying material, are contained in Herendeen (1973).

2. See Bureau of Economic Analysis (1969). The BEA has published several amplifying articles, Bureau of Economic Analysis (1969 November, 1971 January, 1971 August, and 1972). The complete tables are also available on tape from BEA.

3. Reardon (1972) has done similar work for a 35-sector breakdown.

4. Five of the sectors use no energy and are deleted.

5. This is still an approximation, as I have not accounted for electricity used by refineries, and so forth. These energy flows are small compared with those included in Equation 12.4.

6. I am indebted to Clark Bullard of the Center for Advanced Computation, University of Illinois, for this approach.

References

American Petroleum Institute (1971 ed.). *Petroleum Facts and Figures,* 1971 edition, New York.

Berry, R., and Fels, M. (1972). "The Production and Consumption of Automobiles," Report of the Illinois Institute for Environmental Quality, Springfield, Ill., July.

Bowes, A., and Church, C. (1970). *Food Values of Portions Commonly Used,* 11th ed., Lippincott, Philadelphia.

Bureau of the Census (1971). *Statistical Abstract of the U.S.*, Department of Commerce, U.S. Government Printing Office, Washington, D.C.

Bureau of Economic Analysis (1969). *Input-Output Structure of the U.S. Economy: 1963,* Vols. 1 to 3, U.S. Department of Commerce, U.S. Government Printing Office, Washington, D.C.

——— (1969, November). "Input-Output Structure of the U.S. Economy: 1963," *Survey of Current Business,* U.S. Deparunent of Commerce, Washington, D.C.

——— (1971, January). "Personal Consumption Expenditures in the 1963 Input-Output Study," *Survey of Current Business,* U.S. Department of Commerce, Washington, D.C.

——— (1971, August). "Interindustry Transactions in New Structures and Equipment, 1963," *Survey of Current Business,* U.S. Department of Commerce, Washington, D.C.

——— (1972). "Definitions and Conventions of the 1963 Input-Output Study," U.S. Department of Commerce, Washington, D.C.

Herendeen, R. (1973). "An Energy Input-Output Matrix for the United States, 1963: User's Guide," Document No. 69, Center for Advanced Computation, University of Illinois, Urbana, Ill., March.

——— (in press). "The Energy Cost of Goods and Services," Environmental Report, ORNL-NSF Environmental Program, Oak Ridge National Laboratory, Oak Ridge, Tenn.

Hirst, E. A. (1972). "Energy Consumption for Transportation in the U.S.," ORNL-NSF-Environmental Report No. 15, Oak Ridge National Laboratory, Oak Ridge, Tenn., March.

———, and Herendeen, R. (1973). "Total Energy Demand for Automobiles," Society of Automotive Engineers, presented at the International Automotive Engineering Congress, Detroit, Mich., January.

Perelman, M. (1972). "Farming with Petroleum," *Environment, 14,* No. 8.

Reardon, W. R. (1972). "An Input/Output Analysis of Energy Use Changes from 1947 to 1958 and 1958 to 1963," Battelle Memorial Institute, Richland, Wash., June.

13 An Energy, Pollution, and Employment Policy Model *

HUGH FOLK AND BRUCE HANNON†

A large national linear input-output policy model is being developed at the Center for Advanced Computation in the University of Illinois at Urbana-Champaign. The model contains detailed economic activities showing industry demand generated by expenditure categories so that national budgets or individual "life-styles" or scenarios can be evaluated. The 367 industry input-output model produces estimates of total demand by industry through manipulation of an activities vector. For example, energy-output matrices in 367 sector detail show demand for coal, crude oil, refined petroleum, electricity, natural gas implied by scenarios. Other matrices estimate employment effects and ten specific pollution components. Thus, employment, energy, and pollution consequences of a shift in expenditures, either on a national or individual basis, can be estimated. The model is easily edited to adapt it to specific applications by nonprogrammers, is remotely accessible through the ARPA Network, and uses innovative and efficient computational facilities.

Energy policy is public policy relating to the production and consumption of energy. Because energy is widely used in production, energy policy cannot be divorced from economic policy in general. The model and modeling system presented in this paper are intended to permit examination of the energy consequences of alternative economic policies and the economic consequences of alternative energy policies.

The interactions of energy, employment, capital, pollution, and other natural resource requirements are extremely complex. Natural gas shortages have already led to employment cutbacks. Emission standards imposed for pollution control have caused interfuel substitution and employment effects and require additional capital expenditures. The task of modeling such a complex system is staggering, and especially so, given the fact that it is not simply a matter of combining models for energy, employ-

* The research was supported in part by NSF grant, GI–35179X.

† Hugh Folk is Professor of Economics and of Labor and Industrial Relations and Bruce Hannon is Assistant Professor of General Engineering, University of Illinois, Urbana-Champaign. Both are members of the Center for Advanced Computation where this research was done. We gratefully acknowledge the assistance of Roger Bezdek, Clark Bullard, David Healy, Robert Herendeen, Al Meyers, Sue Nakagama, Toni Prevedell, and Janet Spoonamore in preparing this report.

ment, pollution, and capital sectors that have already been developed and tested for the separate sectors.

13.1. The Model

The energy policy model (EPM) proposed here grows out of earlier work at the Center for Advanced Computation (CAC) of the University of Illinois in which Hannon attempted to measure the energy consequences of shifts from nonreturnable to returnable beverage containers (Hannon, 1972). Folk attempted to measure the employment consequences of the same shift (Folk, 1972) and Herendeen (1973) examined the different energy prices of various industries.

The problem of determining the impact of a change in spending on a program or commodity is ubiquitous, but relatively little is written about methods. Converting expenditures in program (or object of expenditure) terms to final demand by industry is time-consuming and often arbitrary. The use of input-output theory to convert the final demand pattern by industry to total demand by industry is straightforward but not trivial. Conversion of total demand by industry to the derivation of employment, energy, capital, pollution, or natural resource requirements from total demand by industry is usually quite difficult.

Bezdek and his associates at CAC developed in ERGWORKS a system that accepted a specified vector in final demand by 218 expenditure types, converts it through use of a matrix to a vector of final demands by industry (80 industries), then generates employment for about 200 occupations (Bezdek et al., 1971). This model has been used for occupational forecasts for 1975 and 1980 and to test the feasibility of the Urban Coalition's Counterbudget (Bezdek, 1972).

Babcock (1972) developed an Illinois employment model that extends Bezdek's model and can be used for any regional extension of the mode.

Parallel to the manpower demand models there are population labor force, migration, enrollment, and occupational mobility models under development which together constitute the STEP I model, a complete supply-and-demand model for occupational forecasting.

The energy, capital, pollution, and other natural resource sectors are formally parallel to the employment models, and the general schema for the model is shown in Figure 13.1.

The major parts of the model are the "scenario," which is the set of as-

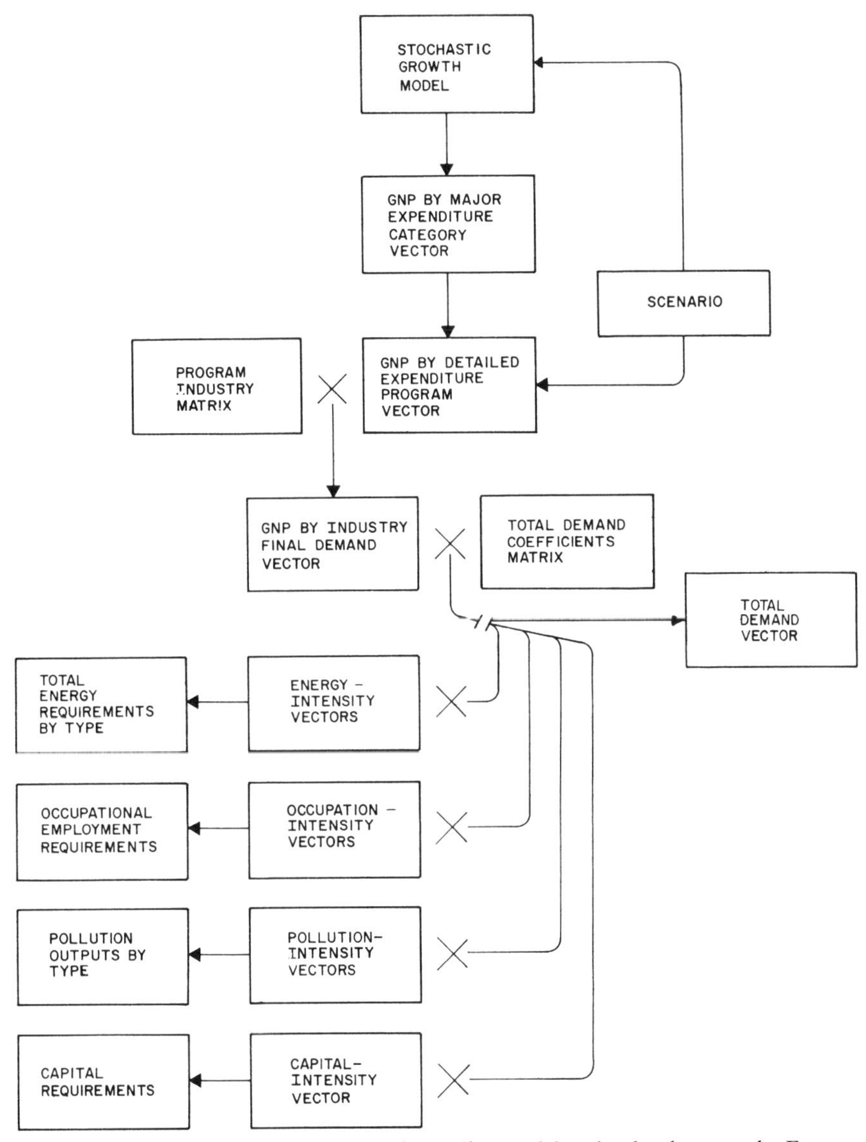

Figure 13.1. Energy, employment, pollution policy model under development by Energy Research Group at Center for Advanced Computation, University of Illinois.

sumptions specifying an economic and legal environment for the model, and the stochastic growth model, which provides feasible consumption, investment, and government expenditure totals for each year to be forecast.

The scenario specifies a vector of expenditures by detailed program for each year $\mathbf{q}_t$. The vector $\mathbf{q}_t$ has m elements representating m different expenditure programs. This number can be indefinitely large. The number used in the 80-industry version of the model is currently 218 activities.

The $\mathbf{q}_t$ vector is converted to expenditures for final demand by industry by a program-industry matrix $\mathbf{P}_t$ for each year, with as many rows as industries and as many columns as there are programs. A program column of $\mathbf{P}_t$ represents the distribution of expenditures of final demand by industry on that particular program. The programs need not include only empirical estimates of new programs, but program vectors for new, proposed, or hypothetical programs can be included. The $\mathbf{P}_t$ matrix is one place in the model in which technological change must be forecast.

Thus, if $\mathbf{Y}_t$ is final demand by industry, then

$$\mathbf{Y}_t = \mathbf{P}_t\mathbf{q}_t. \tag{13.1}$$

Total direct and indirect output by industry is given by the vector $\mathbf{x}_t$, and

$$\mathbf{x}_t = \mathbf{A}_t\mathbf{x}_t + \mathbf{y}_t, \tag{13.2}$$

in which $\mathbf{A}_t$ is the input-output direct coefficients matrix. Matrix $\mathbf{A}_t$ is another part of the model in which technological change must be forecast. It will be possible to consider alternative input vectors, or disaggregate one or more industries to finer detail, or formulate this part of the model as a programming model. Solving Equation 13.2 for $\mathbf{x}_t$, we obtain

$$\mathbf{x}_t = (\mathbf{I} - \mathbf{A})^{-1}\mathbf{y}_t. \tag{13.3}$$

Total energy, employment, capital, and pollution requirements can be obtained by premultiplying $\mathbf{x}_t$ by row vectors that estimate intensity (requirements per unit of total output). Thus, for energy, total energy requirements for energy type i are

$$\mathbf{E}_{it} = \mathbf{e}_{it}(\mathbf{I} - \mathbf{A}_t)^{-1}\mathbf{P}_t\mathbf{q}_t, \tag{13.4}$$

in which $\mathbf{e}_{it}$ is the energy intensity vector for the ith energy type.

Similar estimates can be developed analogously for pollution types, occupational employment, capital, and natural resource types.

If $\mathbf{e}_{it}$ is used as a diagonal matrix instead of as a row vector, then energy requirements are obtained as a column vector that presents the energy required in each industry as a consequence of the specified final demand.

The intensity vectors represent a third part of the model in which technological change must be forecast. It is known that labor productivity changes considerably more rapidly than input-output coefficients and that energy input-output coefficients have tended to change more than others (Carter, 1967).

So far in our work, only energy and total employment intensity vectors corresponding to the 1963 input-output tables have been completed. Work continues on 1967 energy and employment vectors, and these will be updated to current years. Our research concentrates on updating the energy and employment intensity vectors and not on the input-output matrices. We expect some of the pollution vectors will be derived from work now under way at other research centers.

Ultimately, forecasts for each of the components of the model will be made.

13.2. The Modeling System

For a policy model to be useful, it must be generally accessible and inexpensive. For this reason we have designed MEASURE (*M*athematical and *E*conomic *A*nalysis *S*ystem for *U*se in *R*esearch and *E*valuation) as a computer system on the ARPANET. A user with a few hundred dollars for computer time, a computer terminal, and some time should be able to adapt the EPM (or one of the other models in MEASURE, such as STEP I) to his specific needs (such as simulation of another forecasting model).

Today, of course, the use of large data bases of diverse provenance for impact studies or forecasting presents the researcher with serious problems. When government authorities examine the consequences of a proposal (such as the SST, or oil import restrictions, or gas price changes) time is of the essence. There is usually too little time to do any serious data collection, and the use of other people's models and data bases is prevented by problems of accessibility, cost, and lack of documentation. As a result, even if government authorities want reliable, state-of-the-art forecasts, they are unable to get them. Very often the existing models are not conformable with the problem under investigation. Levels of aggregation and assumptions differ, and it is difficult to make appropriate changes in short order.

MEASURE has been designed to meet these problems. It uses virtual memory machines for editing and reporting generation and very large and fast computers such as ILLIAC IV and a 360/91 for numerical processing, superior mathematical algorithms, and general modeling and command languages to reduce computational costs by an order of magnitude or more over comparable conventional systems. The ARPANET provides cheap communications and will soon be publicly accessible. There will be extensive on-line documentation of the system, data, and models.

We can foresee individual researchers preparing complete forecasts in a few days or weeks which are comparable or superior in every way to those that today cost tens of thousands of dollars and require months of work. We hope that this capability will assist in the development of a more open and more even debate on questions of policy than is possible now.

We do not expect there to be a revolution in which politicians hand over the decision-making responsibilities to a cabal of quantitative modelers, but we do expect that modeling of this sort, which is now limited to government agencies, large corporations, and adventitious or farsighted academic researchers, can soon be carried out by a much larger community of users.

13.3. Forecasting

Estimating past coefficients of the EPM and updating them to the present is difficult enough, but forecasting the model will be even harder. In addition to forecasts of the coefficients in the component matrices of the models, feasible and internally consistent scenarios must be developed.

Forecasting the model and developing scenarios will require the use of a large number of persons with highly specialized expertise. Developing a set of consistent final demand by program vectors for each year over a forecast period, for instance, will require participation of political, economic, marketing, and technical experts and the use of a substantial battery of subsidiary and specialized models to make sure that the patterns of final demand specified are feasible (could be produced) and are consistent (for instance, provide enough energy consumption for the energy using goods to be produced).

It is not now easy to incorporate a large number of expert judgments (or

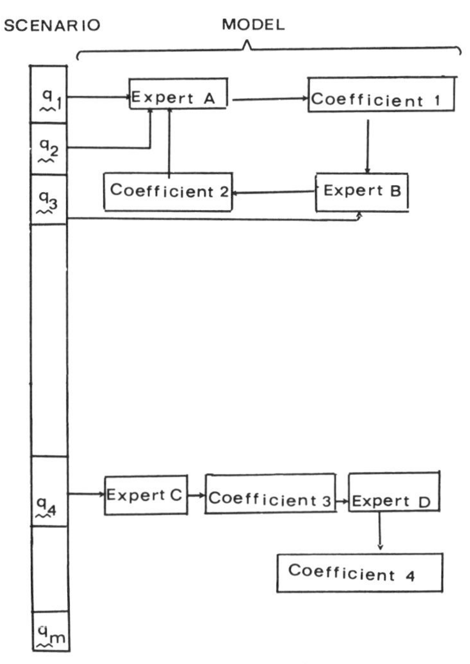

Figure 13.2. Causal ordering of experts.

"implicit models") into formal models because of the difficulties of updating expert opinions and resolving differences between experts by conferencing. We hope to begin such work in one or more industries by identifying a small panel of experts and having them identify the information inputs upon which their predictions (of a matrix row or column, for instance) are contingent. The causal ordering of experts can then be identified as is shown in Figure 13.2.

The scenario is shown as a program vector of m components $\mathbf{q}$; and expert A's prediction of coefficient 1 is influenced by the values taken by components 1 and 2 of the program vector and by the estimate of coefficient 2. Expert B's estimate of coefficient 2 is influenced directly by component 3 of the scenario but is also influenced by the prevailing estimate of coefficient 1. Thus, expert A is influenced directly or indirectly by three of the components. In contrast, expert C's prediction of coefficient 3 is influenced only by component 4 of the scenario, and expert D's prediction of coefficient 4 is influenced only by the prediction of coefficient 3.

In practice, the causal ordering of experts can be expected to become complex. Each component of the model should be covered by more than one expert, and they can be expected to disagree and must conference. Experts who are early in the causal ordering (such as C) must react before those down the tree are called upon. A computer network system such as MEASURE appears to provide a convenient way of mobilizing expert opinion and keeping a forecasting model updated as expectations change.

There are many problems in using live experts in forecasting, but forecasts of the future are necessarily subjective and depend on expert judgment even if they use econometrics, trend projection, bird flight, the inspection of the entrails of sacrificial victims, or other quantitative methods. The identification and incorporation of experts familiar with their obligations into the forecasting system seems to us the best way to approach the problem of forecasting in highly disaggregated systems and, at the same time, identifying and documenting the components of the forecast in a way that good forecasters can be identified and retained and the forecasting community can learn from its past mistakes.

13.4. Preliminary Applications of the Model

While the system is not complete, it has been possible to use the model for some simulations of 1963.

The energy and employment intensities of the 360 industries are presented in Figure 13.3. While a large proportion of the industries are centrally clustered, there are some very energy intensive industries (asphalt coatings and asphalt paving, cement, primary aluminum, building paper, and chemicals) and some very labor intensive industries (hospitals, hotels, credit agencies).

The pattern shown in Figure 13.3 represents the energy and labor requirements of an additional dollar delivered to final demand. It represents, for a consumer, the direct and indirect effect on energy and employment of the expenditure of $1. It does not include any multiplier effects of the expenditure. It is, therefore, inappropriate for use in an impact analysis.

The effects of an increase of a $1 billion expenditure delivered to final demand on each industry offset by a $1 billion decrease in final demand distributed among all other industries in proportion to their share of total GNP (less the industry in question) is shown in Figure 13.4. This represents a possible reallocation, but not a terribly interesting one.

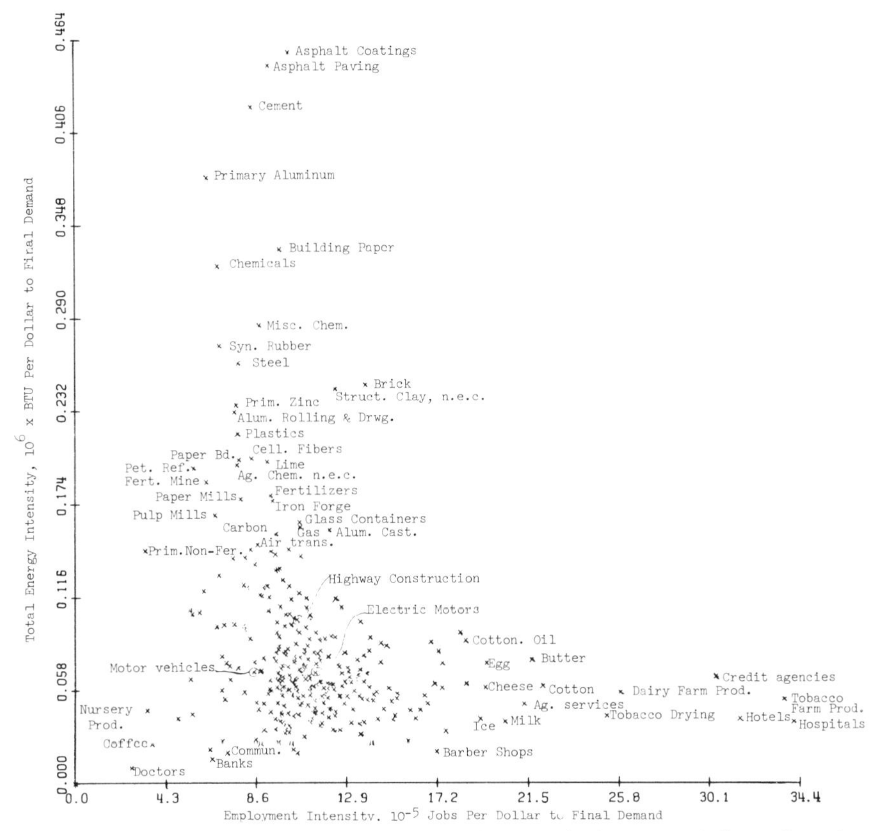

Figure 13.3. Total energy versus employment (direct and indirect) per dollar delivered to final demand, 1963.

Source: Energy Policy Model, Center for Advanced Computation (CAC), University of Illinois, Urbana, Ill., February, 1973. (n.e.c. means not elsewhere classified.)

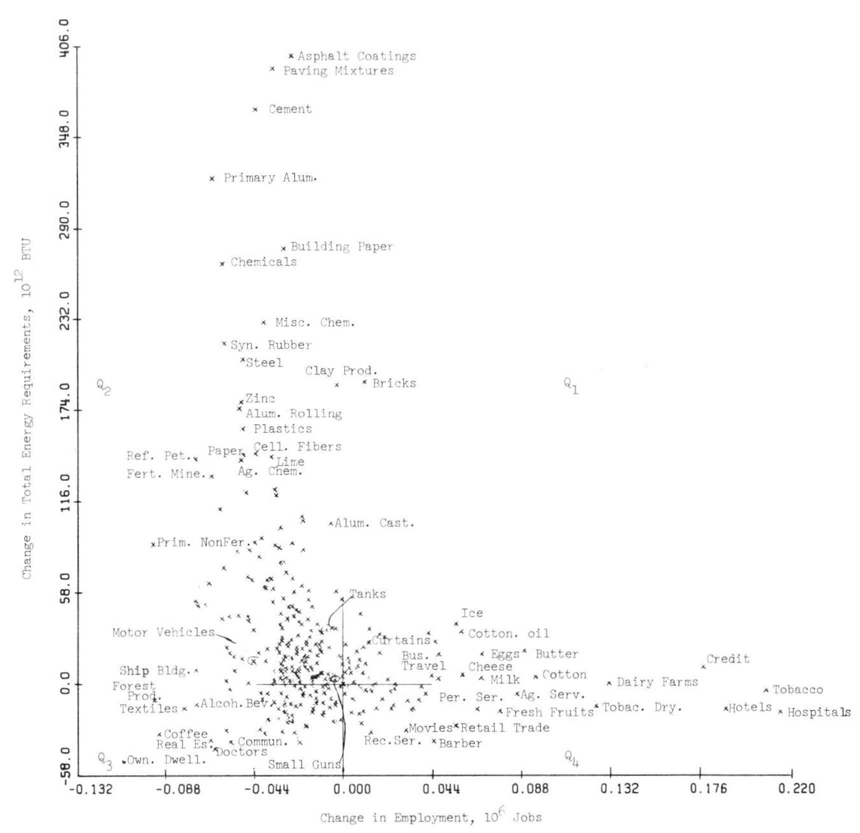

Figure 13.4. Changes in total energy and employment requirements for a $1-billion increase in final demand from the noted industry, proportionately absorbed from all other industries, 1963.

Source: Energy-Employment Policy Model, CAC, University of Illinois, Urbana, Ill., February, 1973.

Yet another way of considering the problem is by examining the effects of 10% proportionate growth in each industry, with an offsetting decrease prorated among the other industries in proportion to their share of GNP (Figures 13.5 and 13.6).

First-quadrant industries are primarily agricultural; second-quadrant industries are basic material production, construction, and fabrication oriented; third-quadrant industries are service oriented with a high degree of technology and high wages; fourth-quadrant industries are service oriented without a great degree of special labor-saving technology and with low wages.

Thus, Figures 13.5 and 13.6 are addressed more to the policymaker concerned about the question of growth. In general, Figures 13.5 and 13.6 are similar to Figure 13.4, that is, most industries remain in the same quadrants but their relative positions have changed. The magnitudes in Figures 13.5 and 13.6 reflect the relative dependence of the U.S. society in 1963 on each of its industries. For example, a 10% increase in delivery to final demand by motor vehicles would have required a direct and indirect energy increase of 34×10^{12} Btu and a decrease in employment of 104,000 jobs (direct and indirect). A 10% increase in deliveries of postal services to final demand would have reduced energy consumption by about 4×10^{12} Btu and increased employment about 36,000 in 1963.

Some intermediate products deliver little to final demand such as steel and primary aluminum. A proportional increase in final demand tends to deflate the artificial importance of the intermediate products seen in Figure 13.4.

The problem with the approach used in Figures 13.4, 13.5, and 13.6 is that the gain in delivery to final demand is absorbed proportionately from all other industries. Quite to the contrary, the product of an industry competes with only a few other products, for example, aluminum with steel and wood as structural members, steel with glass and plastic as food containers. If one industry's gain were at the expense of a few competitors, the complexion of Figures 13.4, 13.5, and 13.6 would change. Suppose, for instance, that a $1-billion gain in primary aluminum deliveries is obtained at the expense of an identical loss in steel deliveries. Then from Figure 13.2, energy use would increase about 116×10^{12} Btu and employment would decrease by 15,000 jobs. A similar increase in primary aluminum deliveries at the proportional expense of all other industries would produce an in-

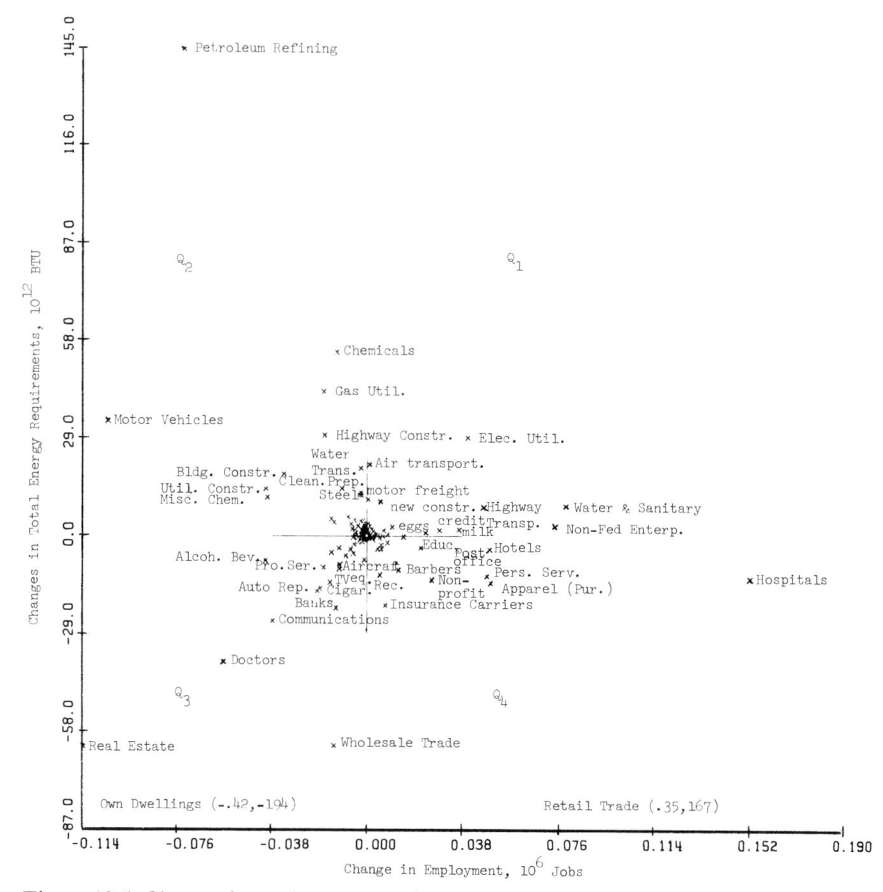

Figure 13.5. Changes in total energy employment for a 10% increase in final demand from the noted industry, proportionately absorbed from all other industries, 1963.

Source: Energy-Employment Policy Model, CAC, University of Illinois, Urbana, Ill., February, 1973.

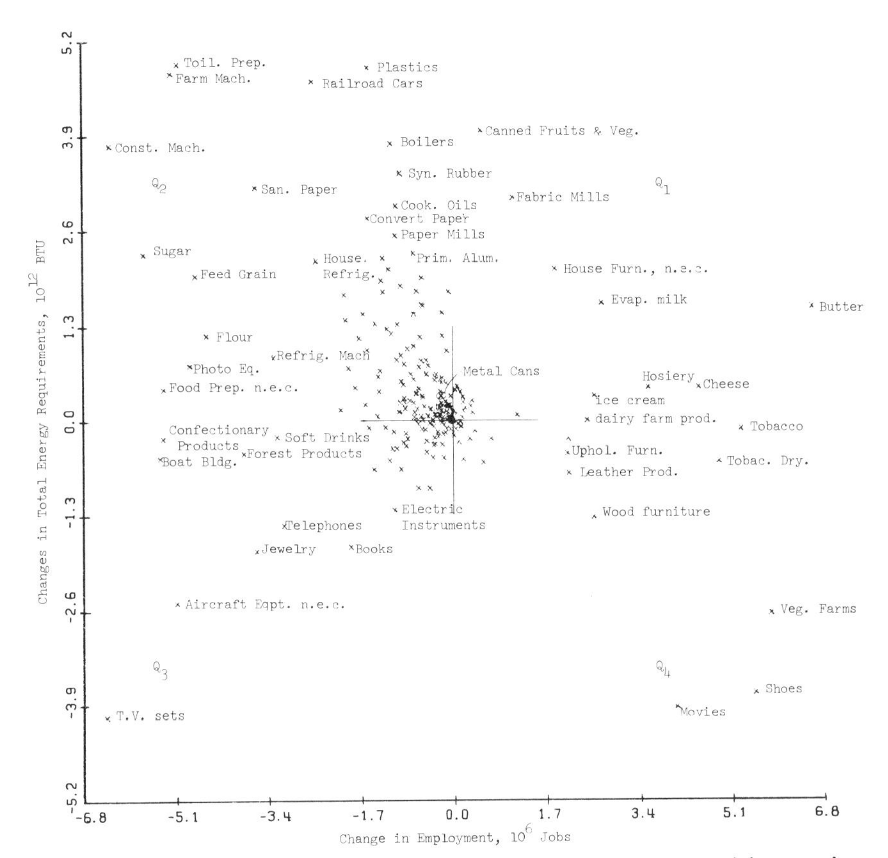

Figure 13.6. Changes in total energy employment requirements for a 10% increase in final demand from the noted industry, proportionately absorbed from all other industries, 1963. An enlargement of the center portion of Figure 13.5.

Table 13.1 Energy and Employment (Direct and Indirect) for the Private Automobile in 1963

Category	Final Demand, 10^9 \$[a]	Energy, 10^{12} Btu[a] (% of total)	Employment × 10^3 Jobs[b] (% of total)
Fuel,			
Produce	5.86	5860 (57.9)	328.3 (4.1)
Refining	—	1220 (11.9)	—
Retail	4.05	130 (1.3)	713.1 (8.9)
Oil,			
Produce	0.83	50 (0.5)	46.5 (0.6)
Retail	0.55	20 (0.2)	96.8 (1.2)
Auto,			
Produce	14.43	1010 (9.8)	1227.6 (15.2)
Retail	10.67	350 (3.4)	1878.7 (23.3)
Tires,			
Produce	0.83	80 (0.8)	58.9 (0.7)
Retail	0.55	20 (0.2)	96.8 (1.2)
Maintenance and Parts	21.7	340 (3.3)	2172.3 (27.0)
Parking		390 (3.8)	—
Highway Construction (fuel taxes)	4.96	580 (5.6)	510.8 (6.3)
Insurance	8.96	230 (2.2)	925.1 (11.5)
Total	73.4 (12.4% of total GNP)	10280 (100.0) (20.6% of total U.S. energy used)	8054.9 (100.0) (12.0% of total employment)[c]

[a] Herendeen (1973).
[b] Nakagama (1973).
[c] Excludes household employment.

creased use of energy of 322 × 10^{12} Btu and a loss of 65,000 jobs (Figure 13.4).

In reality each substitution must be viewed in careful detail. At the Center we have several such competitive product groups under investigation. The most detailed study is that of the bus-car substitution in the intracity region. Preliminary results of the energy and labor cost of the automobile in 1963 are presented in Table 13.1. The automobile consumed 12.4% of the GNP; required 12.0% of the total employment and consumed about 20.6% of the total U.S. energy. This amounts to about 7900 Btu per passenger-mile and 6.2 jobs per 100,000 passenger miles. Preliminary estimates indicate the bus to be about one-third as energy intensive as the automobile.

References

Babcock, Michael W. (1972). "Employment Implications of Alternative Federal Spending Priorities for the Illinois Economy," unpublished Ph.D. dissertation, University of Illinois, Urbana-Champaign, Ill.

Bezdek, Roger H. (1972). "Alternate Manpower Forecasts to 1975 and 1980: Second Guessing the U.S. Department of Labor," Center for Advanced Computation, University of Illinois, Urbana, Ill., paper presented at Annual Meeting of the American Statistical Association, Montreal, Canada, 1972.

———, Lefler, R. Michael; Meyers, Albert L.; and Spoonamore, Janet H. (1971). *The CAC Economic and Manpower Forecasting Model: Documentation and User's Guide,* Economic Research Group Working Paper No. 5, Center for Advanced Computation Document No. 15, University of Illinois at Urbana-Champaign, Urbana, Ill., 58 pp., September.

Carter, A. P. (1967). "Changes in the Structure of the American Economy, 1947 to 1958 and 1972," *Review of Economics and Statistics, 49,* pp. 209–224.

Folk, Hugh (1972). *Employment Effects of the Mandatory Deposit Regulation,* Illinois Institute for Environmental Quality, Chicago and Springfield, Ill.

Hannon, Bruce (1972). *System Energy and Recycling: A Study of the Beverage Industry,* Center for Advanced Computation Document No. 23, University of Illinois, Urbana, Ill., January 5.

Herendeen, Robert A. (1973). *Use of Input-Output Analysis to Determine the Energy Cost of Goods and Services.* Center for Advanced Computation, University of Illinois, Urbana, Ill.

Nakagama, S. (1973). Center for Advanced Computation Document No. 63, Center for Advanced Computation, University of Illinois, Urbana, Ill., February 5.

III Externalities

14 Theory and Practice of Effluent Control*

ROBERT DORFMAN†

The problems of conserving energy and of protecting the quality of the environment are closely related in practice and closely akin in theory. It is therefore not surprising that some nine of the papers at this conference—including this one—lay major stress on the environmental quality range of problems.

In fact, both of these are members of a larger class of problems that have been receiving increasing attention in economics, called the problems of externalities. An externality occurs whenever it is advantageous to a Mr. A to take some action that directly affects the welfare of Mr. B, and when Mr. A is entitled to take it without consulting Mr. B. The typical situation of this sort in the environmental field arises when Mr. A is entitled to discharge waste or heat or noise into land or water or air that Mr. B uses and when it would be inconvenient or expensive for Mr. A to do otherwise. The difficulty in all such cases is that, because Mr. B need not be consulted, the market processes through which economic harmony is normally attained are evaded. Some substitute for economic market adjustments is therefore needed, and I should like to discuss some of the substitutes that are used and proposed, and some of the difficulties that they encounter.

The "theory" part of my discussion will be concerned with the question of how we should like to see the environment used. The "practice" part will deal with the limitations of the social instruments that are available for inducing this ideal usage. I shall discuss these questions by means of a sequence of three examples. The first will be ridiculously simple but will enable us to perceive the basic criterion of ideal usage. On the other hand, it will be too simple to exhibit most of the difficulties of social implementation. The second example will be moderately simple and will serve to bring out some of the practical difficulties of environmental management. The third example will be too hard to formulate or solve mathematically though still falling far short of the complexity of actual situations. In it still further practical difficulties will arise.

In the first example we consider a region with a single source of atmospheric pollution: a factory that emits smoke in strict and unalterable proportion to the amount of its product. The people who live downwind

* Invited lecture.

† David A. Wells Professor of Political Economy, Harvard University.

object to the smoke. The people who buy the product want it, and there is no reason to believe that if the factory were closed down or moved elsewhere it would produce any less smoke or annoy any fewer people. The question is how much smoke and how much product ought to be produced. We leave aside for the moment how this desirable outcome ought to be brought about, if indeed it can be. The first task surely is to perceive what we ought to be aiming for.

Since it is impossible to please everybody, it appears that we should strive for the best resolution on balance, and this can be defined in the following way: the right amount of smoke and product are being produced if a small diminution would harm the users of the product more than it helped the sufferers from the smoke and also if a small increase would benefit the purchasers less than it harmed the sufferers. This sounds like a simple problem in the calculus and it is, provided that we have the requisite data on harms and benefits. To that proviso we now turn.

The product is sold on ordinary markets where its price is determined by the forces of supply and demand. There is a standard argument in economics, which I cannot review, that establishes that a commodity's market price is an adequate measure of its value to the purchaser. It is not, however, a correct measure of the value to society of producing that product. The reason is that if that product were not produced the resources used in making it would be released to produce some other commodity that also has social value. The same economic argument shows that the social value of these alternative products is measured by the cost of the resources used in making the commodity in question, called technically its marginal cost. So the social benefit of making one more unit of the product of the smoky factory is its price minus its marginal cost, and this is also the social cost of forgoing one unit. This calculation, of course, ignores the cost and benefits of the smoke.

Ordinary economic analysis has enabled us to evaluate one member of the equation, the member concerning the social worth of the product itself. It did this because the product itself and the labor, raw materials, and other resources used to make it are traded on economic markets where their social values are registered in the form of prices. Now comes a really tough difficulty. Neither smoke nor exemption from it is a marketed commodity so there is no price that can be used to measure its social value. Presumably the people and businesses who are exposed to the smoke would be willing to pay something to be relieved of it. If we knew how much they would be willing to pay, that would be a possible measure of the social cost of inflicting the smoke on them or the social benefit of reducing that infliction. But we do not know that, and every method that has been suggested for finding it out is beset with difficulties. I want to dwell on that a moment because it is an important and pervasive aspect of the pollution problem, and because it drives home what wonderful and useful things markets are and how difficult it is to manage society without them.

The obvious way to find out how much people would be willing to pay to be relieved of some smoke would appear to be simply to ask them. But that will not serve for at least two reasons. The first is that the people involved do not know themselves how much they would be willing to pay; it is a very "iffy" question, and as a general rule people do not know how much they would be willing to pay for anything until they stand at a counter where it is sold. Second, even if people did know, they would not be likely to tell the truth. The people who are anxious to get rid of the factory and its smoke would be likely to respond with an irresponsibly large figure, while people who fear that they may be asked to contribute in proportion to their expressed eagerness will name unduly small figures in the hope that other sufferers will pick up the tab. Asking people, therefore, will not work. There are other expedients for trying to infer how much people would be willing to pay to be relieved of some pollution, but, without going into an erudite discussion, let me announce the conclusion that all of them are crude and most of them are untrustworthy.

So we can measure only one of the members of the equation that has to be solved to find the right amount of pollution. Nevertheless, the other member is there, and though it is exceedingly hard to observe, in practical instances we may be able to form a sensible judgment about it. Just to give it a name, we call it the marginal externality cost.

To pull the argument together: the social benefit from one more unit of output is its price minus its marginal cost. The social harm from the accompanying smoke is the marginal externality cost. The proper social balance is achieved when these two are equal, or

$$p = \mathrm{MC} + \mathrm{MEC}, \qquad (14.1)$$

because small variations from this level will either reduce benefits more than they reduce harm or increase benefits less than harm. If, as is generally the case, the price of the commodity falls as the quantity increases while the marginal cost tends to rise and the marginal externality cost also tends to rise as the smoke concentration increases, then this formula determines a unique optimum for the quantities of product and smoke produced.

This is a very simple formula and it solves, in principle, the simple problem that I have posed. Indeed, simple as it is, it is the fundamental formula of the economics of pollution and externalities generally.

We have already encountered the first practical difficulty—the virtual impossibility of determining the social cost of pollution or other externalities with anything like the precision that such a formula seems to imply. But there are other difficulties too, and even this ridiculously simple example exhibits some of them. The formula tells us how much pollution "should" be produced. The next question is how to bring this result about. As a preface, notice that it will not happen by itself. If nothing is done, the owners of the factory will find it profitable to continue increasing their output until the price is no longer greater than the marginal cost since they make money on every unit that they sell so long as it sells for more than its marginal cost. In effect the owners of the factory ignore the marginal externality cost, since it is not they who pay it, and the social problem is to induce them somehow to take it into account.

There are several ways in which this can be done, and I shall mention and comment on a few of them. The first method, and a common one in practice, is the method of effluent limitation. Some appropriate government agency simply imposes an upper limit on the amount of smoke the factory is permitted to discharge. Three difficulties are encountered in applying this method. The first concerns information. To establish the correct effluent limit, the government agency must know the data on both sides of the equation. We have already dwelt on the difficulties of estimating the marginal externality cost at different levels of pollutant emission. It is not a trivial matter for a government agency to learn a firm's marginal costs—generally even the firm discovers them only by trial, error, and experience. Second, there is the problem of monitoring. If a limit is imposed, it is necessary to see that it is obeyed, and it happens that measuring the amount of effluent discharged from any source is a difficult and expensive task. Often we are concerned with very low concentrations of pollutants, which are difficult to detect and measure in the laboratory and very nearly impossible in the field. Third is the problem of enforcement. Experience shows that the detection of emissions in excess of the permissible level is likely to trigger a protracted sequence of negotiations followed by litigation during which the infractions are likely to continue at least sporadically. People who have not experienced it can scarcely imagine how long and expensively a corporation can keep the government at bay when an infraction of this sort is alleged.

A second method would be to place a ceiling on the amount of product that the factory can produce. This is equivalent to the first method in the present simplified instance where product and smoke are produced in fixed proportions. This method has the advantage of avoiding the monitoring problem. It has the drawback of direct government regulation of industrial operation at its most sensitive point, and for this reason probably I do not know of any instance in this country where it has been used.

The third method is input limitation, and this is sometimes applied. It avoids the monitoring problem by placing the limit of some material used in the manufacturing process, and it succeeds when the limited material is used in proportion to the effluents emitted. Not only does it avoid the monitoring problem, but it may alleviate enforcement problems because there seem to be fewer respectable legal evasions when this point of control is applied.

And fourth, there is the method, now gaining in popularity, of effluent charges or taxation. In this method the government imposes a tax or charge that is proportioned to the amount of effluent released. One significant advantage of this method is that the government can set the correct charge without having to know anything about the firm's costs. The charge should be set simply so that the tax on each cubic foot of smoke is equal to the marginal externality cost of that release. The firm will respond to that charge by releasing the amount of smoke called for by our basic formula. In the more complicated cases to which we shall come, the charge scheme will be seen to have still greater advantages on the information side. It also has advantages on the enforcement side since, although tax evasion is not unknown in this country, it has proved much easier to collect taxes than to enforce environmental regulations. But the charge scheme does require effluent monitoring, and that is a serious problem.

I think now that we have learned all we are likely to from this simple example. Simple though it was, it served to bring out the fundamental principle and some substantial practical difficulties. But it was too simple to illustrate some significant aspects of the problem of pollution control.

Therefore I should like to substitute a second example that enriches the first in two respects. In this second example we shall allow there to be several sources of pollution and shall admit that each of the sources has abatement measures available which, at some expense, can reduce the amount of effluent per unit of product. We shall proceed as in the first case by first deducing the ideal policy of environmental management and then considering the problem of implementing it.

At this stage I think I can be spared repeating the detailed narrative justification that I gave for the first example. It would follow precisely the same lines. Instead, I proceed directly to the formal statement of this more interesting problem. Consider the following expression:

$$S = \sum_i (p_i x_i - C_i(x_i) - A_i(x_i, a_i)) - D(e_1, e_2, \ldots, e_n), \tag{14.2}$$

$$e_i = f_i(x_i, a_i). \tag{14.3}$$

The summation indicated is taken over all firms that emit the pollutant being controlled. In the parentheses for each firm p is the price of its product and x the quantity produced, so the first term is the gross value of its output. From it are subtracted C, which is the cost of producing x units, and A, which is the cost incurred to reduce pollutant discharges. The quantity A has two arguments, the amount of product x and a, which indicates the pollution abatement measures undertaken. This latter variable is a peculiar one that will cause us some trouble, and I shall have more to say about it. The whole parenthesis then is the social and commercial value of the firm's output allowing for cost of production and pollution abatement but not for the externalities or pollution damages imposed. These last, finally, are subtracted by the function D, which is an externality or pollution damage function whose arguments are the amounts of pollutants discharged by all firms. Finally, we are reminded that the amount of pollution emitted by the ith firm is a function of both its volume of output and the pollution abatement measures that it adopts.

The expression S is therefore the net social value of producing the output x_i in conjunction with the abatement measures a_i. The object of pollution control policy is to choose outputs and abatement measures so as to make this expression as large as possible. One is tempted then to take partial derivatives and see what happens, and that is essentially what I am going to do. But we cannot take partial derivatives straightforwardly with respect to the abatement measures variable because it is not the sort of concept for which differentiation has any meaning. Let me give an example. In the case of a power plant, different values of a would distinguish among such different measures as using low-sulfur-content fuels to reduce the emission of sulfur oxides or using wet scrubbing or some other end-of-pipe treatment. So we have to take a slight detour to avoid illegitimate operations. Let us then denote the pollution costs of any vector of outputs by the function

$$\mathrm{PC}(x_1, \ldots, x_n) = \min_{a_1, \ldots, a_n} \left[\sum_i A_i(x_i, a_i) + D(e_1, \ldots, e_n)\right]. \tag{14.4}$$

That is, the pollution cost of any set of outputs is the smallest attainable sum of abatement costs and pollution damage costs attainable with those outputs. Then the greatest net social value attainable with any given set of outputs is given by this expression:

$$S'(x_1, \ldots, x_n) = \sum_i (p_i x_i - C_i(x_i)) - \mathrm{PC}(x_1, \ldots, x_n). \tag{14.5}$$

The bothersome variable has now been shoved under the rug, and we can differentiate to find

$$\frac{\partial S'}{\partial x_i} = p_i - \mathrm{MC}_i(x_i) - \frac{\partial \mathrm{PC}}{\partial x_i} = 0, \tag{14.6}$$

$$\frac{\partial \mathrm{PC}}{\partial x_i} = \frac{\partial}{\partial x_i} A_i(x_i, a_i) + \frac{\partial f_i}{\partial x_i} \frac{\partial D}{\partial e_i}. \tag{14.7}$$

Now notice that we have obtained a simple generalization of the formula that applied to the elementary case. The social optimum requires that for each product the price be equal to the sum of the marginal cost of production and the marginal pollution cost. In turn the marginal pollution cost is the sum of the marginal cost of abatement and the marginal externality cost. In essence we have found the same social criterion as before, and indeed that criterion runs through all the variations and applications of the pollution abatement problem.

So much for the ideal; we next consider what is involved in attaining it. The methods that we discussed in the simple case are applicable again, though with some complications that we have to take into account, and a new field of options is added by the possibility of abatement measures. Let us now go through some conceivable alternatives.

In the first place it is now more evident than before that no centralized authority can take it on itself to prescribe the amounts of output and the precise pollution abatement measures for all the enterprises that emit a given pollutant in a region. The problem of assembling the requisite data, and the presumptuousness of assuming so much authority completely preclude that direct approach. There is inherently a two-stage decision problem. First, the government or agency selects the regulations or other instruments available to it, and then the individual polluting firms react to those constraints in the usual profit-maximizing way. When the government or agency makes its choice, it has partial, but only partial, information about the data to which the individual polluters are going to react. With this background let us proceed to some more practical methods of control.

It will be most informative this time to start with a tax on polluting discharges, the method that looked so promising in the elementary example. If the control agency takes this route, the typical individual polluter will react by trying to maximize his profit after taxes, or

$$\pi_i = p_i x_i - C_i(x_i) - A_i(x_i, a_i) - T_i(e_i). \tag{14.8}$$

Here the notation is just the same as before except that the individual polluter pays no attention to pollution externalities and does pay attention to T, the tax that he has to pay if he discharges pollutants in the amount e. He will in the first instance choose his abatement measures so as to minimize the pollution costs that he has to bear. Therefore the pollution costs corresponding to any level of output are

$$\min_{a_i} \mathrm{PC}_i(x_i) = A_i(x_i, a_i) + T_i(e_i). \tag{14.9}$$

And the profit corresponding to any level of output is given by

$$\pi_i = p_i x_i - C_i(x_i) - \mathrm{PC}_i(x_i). \tag{14.10}$$

And the most profitable level of output is given by

$$\partial\pi_i/\partial x_i = p_i - \mathrm{MC}_i(x_i) - (d\mathrm{PC}_i/dx_i) = 0, \tag{14.11}$$

$$d\mathrm{PC}_i/dx_i = \partial A_i/\partial x_i + (\partial f_i/\partial x_i)(dT_i/de_i). \tag{14.12}$$

There is a strong family resemblance to earlier formulas, but they are not identical twins. To make them identical twins—that is to say, to induce socially optimal behavior on the part of the firms—the tax schedule would have to be constructed so as to satisfy

$$dT_i/de_i = \partial D/\partial e_i, \qquad i = 1, \ldots, n. \tag{14.13}$$

That is the condition for an efficient set of effluent charges. Now let us consider what that means in practice. This approach partakes of both the advantages and the drawbacks of the tax scheme in the simple case. On the drawback side it does involve effluent monitoring with all its cost and technical difficulties. On the advantage side it is a tax and therefore comparatively easy to enforce, and it does not require the controlling authority to have any information about the internal costs and operations of the firms that it is controlling. This is even more of an advantage in this case than it was before because the internal decisions are more complicated. Allowing the individual firm to choose the proper mixture of level of output and pollution abatement measures under the guidance of artfully selected inducements is surely the effective way to have those choices made efficiently and, really, the only effective way. But on the other hand two new and delicate issues have arisen. In the first place, our ideal tax formula calls for having a different schedule of effluent charges for each firm if the damages resulting from each firm's discharges are different, and this will almost certainly be the case. It is pretty clear that a taxing authority cannot promulgate a separate tax schedule for each taxpayer, so this ideal set of charges is, in fact, not practicable. Instead, there would have to be either a uniform schedule for all effluent emitters or, at best, a few schedules for broad groups or zones. I shall not pursue the mathematics here. It is pretty obvious that with inappropriate tax incentives the individual firm will behave inappropriately and that unnecessary pollution costs will be incurred.

The second new difficulty also results from the numerousness of polluters. Notice that the tax schedule for each polluter requires that his marginal tax for each level of pollution should be equal to the marginal externality's cost at that level of pollution. But that marginal externality cost, in principle and often in practice, depends on the levels of emission of the other polluters. Thus, each polluter's tax schedule depends upon the pollution levels of the other polluters, which cannot be known in advance of setting tax schedules. This simultaneity cannot be evaded unless the externalities inflicted by the several polluters are strictly additive and do not interact in any more subtle way. So, as a general rule the socially efficient tax schedules cannot be found, and if found could not be promulgated. If we choose a pollution charge route, we cannot hope to impose the socially optimal set of charges.

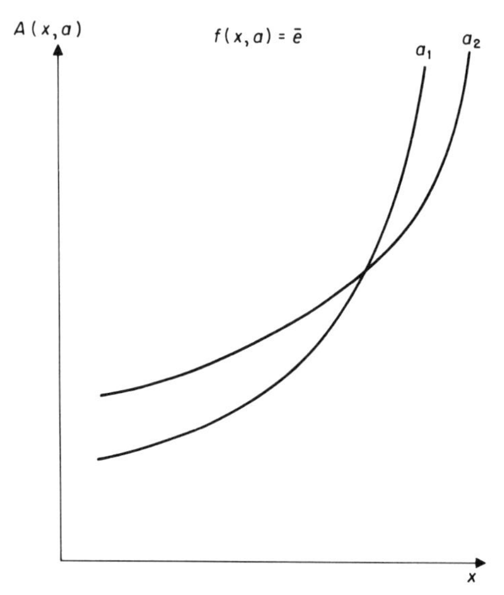

Figure 14.1. Abatement cost as a function of level of output for two abatement techniques.

I have deferred the much more prevalent method of using effluent limits to impose pollution abatement because it is rather untidy technically. The trouble of course is caused by the discontinuous abatement variable a. I shall merely sketch the analysis, trying to avoid the technicalities. If an effluent limit is imposed on a firm and it obeys, the profit that it tries to maximize is its gross revenues minus its production costs minus its abatement costs:

$$\pi = px - C(x) - A(x, a)$$
$$\text{subject to} \quad f(x, a) \leq \bar{e}. \tag{14.14}$$

The output and abatement measures must be chosen so as not to exceed the effluent quota. The relation between the level of output and the abatement costs of living within any particular limit is shown schematically in Figure 14.1. The curve labeled a_1 shows how the costs of any particular strategy, such as using low-sulfur fuels to control sulfur oxide emission, increase as the level of output increases. The costs go up, and rapidly, because the allowable sulfur content of the fuel diminishes as the rate of consumption grows. The curve labeled a_2 is a similar abatement cost curve for some different strategy, perhaps scrubbing or otherwise cleansing the stack gases. For low levels of output, method a_1 is cheaper, but after some point method a_2 becomes the more economical. The point is that in spite of the bad behavior of this variable the abatement cost curve remains continuous though at some places it has only one-sided derivatives. This is regular enough to permit analysis, and the analysis leads to the conclusion that the effluent limit is set properly when the following condition is met:

Additional cost of respecting limit	=	Additional damage from not increasing abatement measures.

The left-hand side is the additional abatement cost that the firm would sustain if it increased its output slightly and strengthened its abatement measures so as to stay within the limit. The right-hand side is the marginal externality cost that would result from the same increase in output without any change in abatement measures. The equality requires that both of these responses to a slight increase in output have the same social cost. It would be quite a task for any environmental control agency to ascertain the effluent limit that met this condition, but if it were permitted some amount of trial and error it might conceivably come fairly close. More information is required than using the effluent tax method, and enforcement problems are likely to be more serious, but on the other hand more flexibility can be permitted in setting effluent limits than in assessing taxes or charges. The contrast between the practicabilities of the two methods does not appear to be as great as in the ultrasimplified example.

There are many other control methods that ought to be considered, for example requiring specific abatement measures, but the analysis is becoming extended, and I have to rush ahead to my final approximation to reality. In this third approximation we introduce explicitly the cost of information gathering, enforcement, and administration that were introduced only for critical purposes in the preceding analyses. Then the social object is to maximize the following expression:

Value of products less value of ordinary private production costs
– costs of pollution externalities
– pollution abatement costs
– costs of administration and enforcement and information.

Now the problem has passed beyond any mathematics that I know. The trouble is all in the last term, and I introduce it to emphasize that administration is socially costly, like everything else, and that in the field of pollution the costs can be substantial. So substantial that an inferior method of control that is cheap and easy to administer may well be preferable to a method that is greatly superior on all counts except the last term. Let me reinforce this by exhibiting the corresponding more sophisticated formula by which the firms respond to measures taken by a control authority. The individual polluters try to maximize

Value of product less ordinary private production costs
− his own pollution abatement costs
− pollution taxes and penalties
− pollution license fees
+ subsidies for pollution abatement
− negotiation and litigation expenses.

This is a very complicated formula. The first two terms are familiar, but the others suggest types of response that we have not encountered up to this point. In particular, this formula recognizes that a firm need not obey an effluent limit, for example, and many firms in fact cheerfully and routinely incur fines of a hundred dollars a day for the privilege of exceeding official effluent limits. When you take such potentialities into account—and they are there—mathematical niceties become rather trivial. I believe that this battery of potential responses is susceptible to fruitful formal analysis, but I do not know of any serious attempt.

Just listing these additional complexities—and there are still others I could bring to your attention if I had time—reinforces the impression left by our second approximation. It is not reasonable to aspire to a control of pollution that at all approximates the theoretical ideals that we illustrated earlier. The data are too elusive and expensive, and the problems of controlling the ultimate decision makers are too slippery. Instead we can strive for changes that are undoubted improvements over the current state of affairs and that do not cost very much to impose and enforce. I think that other economists also are beginning to concede the impracticability of the ideal levels of pollution that they were originally so proud to deduce. There is a general tendency now to abandon the notion that the marginal costs of pollution avoidance should be equated to the marginal externality costs. It is simply too difficult to estimate externality costs, the function D in our equations. In place of that, economists are subscribing to the already conventional practice of establishing environmental quality standards that represent an improvement in environmental conditions but are not too stringent to be attained and enforced. We are going to have to make such compromises with this truly obdurate problem: compromises about our aspirations for environmental purity and compromises about our control strategies. The task I see before economists and other scientists is to understand better the art and implications of compromise in this context.

15 Internalizing an Externality: A Sulfur Emission Tax and the Electric Utility Industry*

DUANE CHAPMAN†

This paper develops a preliminary examination of the effect of a sulfur emission tax on costs, prices, demand growth, type of generation, emissions, and benefits from damage reduction. An engineering-economic model is used to examine consequences of various tax levels with respect to the electric utility industry. With qualification, generalization can be made to the other major sulfur emission sources.

We may draw upon some of the theoretical work (especially that of Baumol, 1972, and Ruff, 1972) to state hypotheses for empirical examination. We shall hypothesize these significant effects of a sulfur emission tax on the electric utility industry: (1) emissions and damage will be reduced; (2) demand growth will decline; (3) nuclear generation (a substitute for coal and oil) will be accelerated; (4) social benefits will exceed social costs; (5) if the tax rate differs from the damage rate, social optimality will differ from market optimality; (6) if the Clean Air Act is not implemented by 1976, the tax will have greater social value. These assertions are not hypotheses in the strict sense of statistical inference. They are tentative assumptions intended to draw out the empirical consequences of the analysis. As will become evident later, the analysis seems to disprove generally the second and third propositions, support the first, fourth, and sixth, and give weak acceptance to the fifth.

15.1. Proposed Emission Taxes

It must be acknowledged at the outset that there is no clear picture of the damage caused by sulfur emissions. Barrett and Waddell in their widely cited study (1970) estimated national air pollution damage at $16 billion in 1968 with the major categories being $8.3 billion sulfur damage ($2.8 billion to residences, $2.2 billion to materials, $3.3 billion to health) and $5.9 billion particulate damage (particularly $2.4 billion to residences,

* This work was supported by the National Science Foundation RANN Program through the Environmental Program at the Oak Ridge National Laboratory and by the Cornell Agricultural Experiment Station. J. M. Ostro, Joseph K. Baldwin, and Anne E. Johnson have been most helpful in its preparation, and Timothy Mount, Timothy Tyrrell, and Martha Czerwinski collaborated in developing the demand analysis section of the paper.
† Cornell University.

$2.8 billion to health). Monetary damage to vegetation and natural ecosystems was thought to be minimal, as were the effects of carbon monoxide, hydrocarbons, and nitrogen oxides. As a national average, the 33.2 million tons of emitted sulfur oxide define damage of $250 per ton of sulfur oxide. Assuming that almost all sulfur oxide is sulfur dioxide, this is $500 per ton of sulfur in 1968 dollars, or $600 in 1972 dollars.[1]

The various proposals for a sulfur emission tax can be compared according to their provisions for time phasing, geographic differentials, and average rates. The Administration's current proposal calls for initiation of a tax in 1976 with three levels of tax rates corresponding to three ambient air quality standards.[2] The Administration's proposal suggests a tax rate of 15 cents per pound of emitted sulfur in regions above the primary standard, 10 cents per pound in regions between the primary and secondary standards, and nothing in regions meeting the secondary standard.

A second proposal by Senator William Proxmire (1972) would have a single rate applicable throughout the country and would have been phased to grow at 5 cents per year per pound emitted over a 4-year period (that is, 5 cents in 1972, 10 cents in 1973, 15 cents in 1974, and 20 cents in 1975 and thereafter).

Each proposal would offer different incentives. The Administration proposal imposes increasing marginal costs by steps on emitters in a region as emissions rise, while the Proxmire version has a constant marginal cost to emitters. The Administration proposal would offer incentive for emitting industries to locate in clean regions, while the Proxmire proposal would not. The Administration proposal would seem to offer some kind of exemption to the Clean Air Act. The Act requires all regions to be out of the primary category by 1976; the tax on emissions in regions not meeting the primary standards would seem to lessen the strength of the requirement to meet this standard. On the other hand, the Administration proposal offers an extra incentive for a region to achieve clean status (that is, the tax rate falls from 10 cents per pound to nothing).

There are arguments on each side of the phasing question. As we shall see later, it is unlikely that significant sulfur removal is technologically possible before 1976. As a practical matter it is unlikely that a tax would be imposed before 1976.

In the subsequent analysis we shall consider three tax rates: (1) $200 per ton sulfur emitted, the median rate in the Administration proposal; (2)

Table 15.1 Sources of Sulfur Emissions, 1967[a]

Source	Nationwide Emissions (million tons of sulfur oxides
Solid Waste Disposal	0.17
Steam–Electric Power Plants	15.40
Petroleum Refineries	2.31
Copper Smelting[b]	2.58
Lead Smelting[b]	0.19
Zinc Smelting[b]	0.45
Sulfuric Acid Production	0.60
Other Industrial Processes	0.15
Residential, Commercial, and Industrial Heating Plants	8.48
Total	30.33

[a] Source: *The Economics of Clean Air*, Administrator of the EPA (1972).
[b] Primary metallurgical processes.

$400 per ton, the ultimate level of the Proxmire proposal; and (3) $600 per ton, the average damage in the Barrett-Waddell study.

15.2. Sulfur Removal and the Electric Utility Industry

As Table 15.1 indicates, the major source of sulfur emissions and presumably damage is the electric utility industry, and the impact of a tax would be greatest here. The remainder of the analysis will focus upon this industry.

A review of current engineering literature indicates that by 1976 processes with 90% removal efficiency are expected to cost between .95 and 1.9 mills per kWh of electricity (or between $2 and $4 per ton of coal).[3] These estimates of control costs may be compared to 1971 U.S. averages of electricity price of 17.8 mills per kWh and coal for utilities costing $8 per ton. Thus, sulfur removal might raise coal-use costs between 25 and 50% and total electricity costs in coal plants from 5 to 10%.

15.3. New York State: A Case Study

A major methodological problem in this study is the definition of the appropriate geographic area. Possible types of areas include (1) the contiguous United States, (2) a multistate region, (3) a single state, (4) one of the

247 air quality regions, (5) a collection of air quality regions, (6) a single electric utility, and (7) an urban area.

The decision here has been to examine the state of New York. Primary reasons for this selection are (1) the previous engineering-economic study by Hausgaard (1971); (2) the states are basic decision-making units with respect to air quality standards and electric utilities; (3) problems of aggregating the New York City area with upstate New York provide qualitative insights into larger aggregation problems; (4) plans for air quality implementation and for electric utilities are relatively well defined and available.

New York would seem to be reasonably representative on the basis of the proportion of sulfur emissions from electric utilities: the percentage is approximately 50% for New York and for the nation.[4] As does most of the country, New York plans to meet primary air quality standards by the use of low-sulfur fuels—an important point we will discuss further.

The major problem in using a specific area rather than the whole nation is the difficulty in generalizing results onto a national basis.

15.3.1. Demand Growth

The question of future growth in electricity demand is a subject of lively controversy. Most observers in industry and government believe that electricity growth can be projected into the future at compound or exponential growth rates. The author and his colleagues believe that demand growth will vary according to changes in electricity prices, population, income, and competing fuel prices.[5]

It is useful to define a single logical framework to compare both types of assumptions. A simple form is sufficient:

$$Q_t = AK(t)Q_{t-1}^{\lambda}(\mathrm{PE}_t/\mathrm{PE}_{70})^{B_1}(N_t/N_{70})^{B_2}(Y_t/Y_{70})^{B_3}(\mathrm{PG}_t/\mathrm{PG}_{70})^{B_4}, \tag{15.1}$$

where t is the year, Q is demand, A is a constant, $K(t)$ is autonomous growth, λ is a lag coefficient, PE is price of electricity, N is population, Y is per capita income, and PG is price of gas. The subscript for consumer class is omitted, and all monetary values are in constant dollars. The B coefficients are elasticities.

The National Power Survey (NPS), (U.S. Federal Power Commission, 1971) projection for New York can be characterized by assuming $A = Q_{70}$, $\lambda = 0$, $B_1 = 0$, $B_2 = 1$, $B_3 = 1$, and $B_4 = 0$. Here $K(t)$ is the autonomous growth not caused by population or income growth. Using the recent U.S.

Table 15.2 Long-Run Elasticity Assumptions for New York

	Residential		Commercial		Industrial	
	NPS Demand (1)	Variable Demand (2)	NPS Demand (3)	Variable Demand (4)	NPS Demand (5)	Variable Demand (6)
Electricity Price	0.0	−1.23	0.0	−1.49	0.0	−1.68
Population	+1.0	+0.94	+1.0	+0.97	+1.0	+1.07
Per Capita Income	+1.0	+0.28	+1.0	+0.84	+1.0	+0.53
Gas Price	0.0	+0.15	0.0	+0.15	0.0	+0.15
Autonomous Growth						
1970–1980	1.84%/yr	0.0	1.60%/yr	0.0	0.11%/yr	0.0
1980–1990	2.30%/yr	0.0	2.06%/yr	0.0	0.56%/yr	0.0
Time Lag Coefficient	0.0	+0.89	0.0	+0.86	0.0	+0.88

Note: The *B* coefficients in Equation 15.1 are equal to (1 − lag coefficient) × (long-run elasticity).

Bureau of Economic Analysis predictions for population and income in New York, the values in the odd-numbered columns in Table 15.2 follow.

By using the Bureau of Economic Analysis population and income predictions by Graham, Degraff, and Trott (1972) for New York and the NPS values in Table 15.2, we derive demand growth at the NPS rates: 6.13% per year for residences, 5.88% for commercial use, and 4.32% for industrial use. Our estimates in the even-numbered columns are quite different. These elasticities differ primarily according to (1) the presence of a price effect, and (2) the presence or absence of autonomous growth. It is of some interest to compare detailed predictions with actual sales in 1971. The variable demand model has less error for two of the three classes as well as for total error. Of course this is just for one year. In the analysis of the tax, both demand models are considered.

15.3.2. The Supply-Demand Model and Growth in Generation

Table 15.3 is used to explain the model employed in examining the effects of a tax. One of 32 cases is reported in Table 15.3: the variable demand model, medium control costs, high tax rate, and assumed nonimplementation of the Clean Air Act as described later. Columns 1 and 2 show projected per capita income and population that are the same in all 32 cases. Columns 3 through 7 show the variable demand and total generation esti-

Table 15.3 Sulfur Tax Analysis with Variable Demand, Medium Control Costs, High Tax Rate, and without Clean Air Act Implementation

	Demand (billion kWh)							Generation (billion kWh)				
Year	($) Per Capita Personal Income (1)	Population (thousands) (2)	Resi-dential (3)	Com-mercial (4)	Indus-trial (5)	Total, Includ-ing Other (6)	Total (7)	Gas (8)	Hydro (9)	Internal Com-bustion (10)	Nuclear Power (11)	Year
1970	5171	18,191	25.2	24.9	27.3	87.4	95.5	8.8	24.8	0.2	4.3	1970
1971	5314	18,447	27.4	26.2	27.3	91.4	99.8	8.8	24.9	0.2	6.4	1971
1972	5461	18,708	29.5	27.6	27.3	95.3	104.1	8.8	25.0	0.2	8.5	1972
1973	5611	18,971	31.6	28.9	27.4	99.2	108.4	8.8	25.1	0.2	10.6	1973
1974	5766	19,239	33.6	30.1	27.5	103.1	112.6	8.8	25.1	0.3	12.7	1974
1975	5925	19,510	35.6	31.4	27.7	107.0	116.9	8.8	25.2	0.3	14.7	1975
1976	6089	19,785	37.5	32.7	27.9	110.8	121.0	8.8	25.3	0.3	17.2	1976
1977	6257	20,064	39.0	33.4	27.7	113.1	123.5	8.8	25.4	0.3	19.3	1977
1978	6430	20,347	40.4	34.2	27.7	115.6	126.3	8.8	25.4	0.3	22.0	1978
1979	6607	20,634	41.9	35.1	27.8	118.5	129.4	8.8	25.5	0.3	25.3	1979
1980	6789	20,918	43.4	36.2	28.1	121.6	132.8	8.8	25.5	0.3	29.1	1980
1981	6952	21,194	44.8	37.2	28.3	124.7	136.2	8.8	25.6	0.3	34.0	1981
1982	7118	21,474	46.2	38.3	28.6	127.8	139.6	8.8	25.7	0.3	38.8	1982
1983	7288	21,757	47.5	39.4	28.9	130.9	143.0	8.8	25.7	0.3	43.5	1983
1984	7462	22,044	48.8	40.5	29.3	134.1	146.4	8.8	25.8	0.4	48.2	1984
1985	7640	22,335	50.1	41.6	29.7	137.2	149.8	8.8	25.9	0.4	52.9	1985
1986	7823	22,630	51.3	42.8	30.1	140.3	153.2	8.8	26.0	0.4	57.5	1986
1987	8010	22,929	52.5	43.9	30.5	143.5	156.7	8.8	26.0	0.4	62.1	1987
1988	8201	23,232	53.7	45.1	30.9	146.6	160.1	8.8	26.1	0.4	66.7	1988
1989	8397	23,538	54.9	46.3	31.4	149.8	163.6	8.8	26.2	0.4	71.2	1989
1990	8598	23,849	56.0	47.5	31.9	153.0	167.0	8.8	26.2	0.4	75.8	1990

Table 15.3 (continued)

	Generation (billion kWh)			Tax, Control Costs (million dollars)			Average Costs and Prices (cents/kWh)					
Year	Coal (12)	Oil (13)	Emission (10,000 tons sulfur) (14)	Tax (15)	Control Cost (16)	Control Cost + Tax (17)	Average Cost, No Tax (18)	Residential Average Price (19)	Commercial Average Price (20)	Industrial Average Price (21)	Damage (million dollars) (22)	Year
1970	26.5	30.9	35.3	0.0	0.0	0.0	2.13	3.09	2.82	1.21	211.5	1970
1971	26.5	33.0	36.3	0.0	0.0	0.0	2.14	3.12	2.84	1.22	217.7	1971
1972	26.5	35.1	37.3	0.0	0.0	0.0	2.16	3.14	2.86	1.23	223.9	1972
1973	26.5	37.2	38.3	0.0	0.0	0.0	2.18	3.17	2.89	1.24	230.1	1973
1974	26.5	39.3	39.4	0.0	0.0	0.0	2.20	3.19	2.91	1.25	236.2	1974
1975	26.5	41.3	40.4	0.0	0.0	0.0	2.22	3.22	2.94	1.26	242.3	1975
1976	26.5	43.0	32.0	192.2	24.7	216.9	2.24	3.25	2.96	1.27	192.2	1976
1977	26.3	43.4	22.9	137.5	49.7	187.2	2.25	3.54	3.23	1.38	137.5	1977
1978	26.1	43.6	13.7	82.4	74.5	157.0	2.27	3.52	3.21	1.38	82.4	1978
1979	25.9	43.6	4.6	27.4	99.0	126.4	2.29	3.51	3.20	1.37	27.4	1979
1980	25.7	43.2	4.5	27.2	98.3	125.4	2.31	3.50	3.19	1.37	27.2	1980
1981	24.6	42.8	4.4	26.5	96.1	122.6	2.33	3.52	3.21	1.38	26.5	1981
1982	23.6	42.4	4.3	25.8	94.1	119.9	2.35	3.54	3.23	1.38	25.8	1982
1983	22.6	42.0	4.2	25.2	92.1	117.2	2.37	3.57	3.25	1.39	25.2	1983
1984	21.6	41.6	4.1	24.5	90.1	114.6	2.39	3.59	3.27	1.40	24.5	1984
1985	20.7	41.2	4.0	23.9	88.2	112.1	2.41	3.61	3.29	1.41	23.9	1985
1986	19.8	40.8	3.9	23.4	86.4	109.7	2.43	3.64	3.32	1.42	23.4	1986
1987	19.0	40.4	3.8	22.8	84.6	107.4	2.45	3.66	3.34	1.43	22.8	1987
1988	18.1	40.0	3.7	22.3	82.9	105.1	2.47	3.69	3.36	1.44	22.3	1988
1989	17.4	39.6	3.6	21.7	81.2	102.9	2.49	3.71	3.38	1.45	21.7	1989
1990	16.6	39.2	3.5	21.2	79.5	100.8	2.51	3.74	3.41	1.46	21.2	1990

mates for the particular cost-price assumptions of this case which are discussed later. Other sales are 11.5% of total sales and transmission losses are 8.4% of total generation.

Predictions of growth in generation by type of process involve too many uncertainties about environmental protection, fuel availability, and cost for any accuracy in assumptions. Herein we shall merely note some educated guesses and offer our own.

The National Power Survey predicted actual declines in the Northeast in generation for coal and oil throughout the coming two decades. They believed nuclear power to be the only major growth process, with modest increases expected for gas turbines, pumped storage, and internal combustion for peaking power. Generation from natural gas and conventional hydropower were thought to be nearly constant.

Hausgaard (1971), focusing on New York, predicted that nuclear plants would account for 50% of new capacity in this decade. Oil generation increases rapidly in the 1970s and peaks in the late 1970s. Coal generation declines slowly to a plateau in the late 1970s. Natural gas generation remains constant. Gas and diesel turbine capacity trebles but remains the smallest source.

The assumptions used here generally follow Hausgaard (1971). Natural gas (column 8) stays at its 1970 level of 8.8 BkWh (billion kilowatt hours). Hydrogeneration (both conventional and pumped storage) accounts for 2% of generation growth and is shown in column 9. Internal combustion generation (column 10) provides .3% of new generation.

Coal, oil, and nuclear generation are defined as providing "base-load growth"—the remaining 97.7% of new generation not made through essentially peak-load processes. Through 1976 generation by these processes is unaffected by the cost of a sulfur emission tax. Coal generation (column 12) is unchanged, and oil (column 13) and nuclear power (column 11) each account for one-half of base-load growth.

Beginning in 1977 (the first year after imposition of a tax), each of these three processes changes its generation because of two factors. First is an autonomous component reflecting other economic conditions, and second is a tax-induced effect. Autonomous nuclear growth increases from 60% of base-load growth in 1976 to 100% in 1980. Tax-induced nuclear growth results from an increased market share of nuclear power. This increased market share is caused by the increased tax and control costs for coal and oil generation relative to average generating costs without the tax. From 1980 to 1990, autonomous nuclear growth is all of the base-load growth and a replacement of 3⅓% of the preceding year's coal generation. These relationships and similar ones for oil and coal are shown in Table 15.4.

15.3.3. The Clean Air Act, Emissions, Taxes, and Control Costs

Sulfur emissions will depend upon the amounts of coal and oil burned, the sulfur content of those fuels, and the proportion of sulfur that is emitted. Following Hausgaard (1971), coal use is .4214 ton per 1000 kWh and oil use is .2725 ton per 1000 kWh. Future sulfur content will depend on municipal, state, and federal regulations. The impact of a sulfur emission tax must be considered in this uncertain context. Hausgaard's base case calculated emissions in the context of a 1% statewide sulfur content average, presumably a mixture of low-sulfur New York City fuels and high-sulfur upstate fuels. Support for this figure is given by the *Economics of Clean Air* Administrator of the Environmental Protection Agency (1972, p. A-2) in noting new emission standards are equivalent to the use of 1% sulfur coal and 1.4% sulfur oil. Taking into account the stricter New York City standards and the relative concentration there of oil generation, one set of analyses will assume that by 1976 the average sulfur content of both coal and oil is 1%.

However, we must take note of the probability of the Clean Air Act *not* being implemented by 1976. The U.S. Environmental Protection Agency stated (1972, p. 10843) ". . . there is strong evidence that the complete implementation of the [state] plans as submitted may not be obtainable in the time prescribed." They believed low-sulfur fuel was insufficient in quantity, and that the area most affected would be New York, the Midwest, and the South Central region. Therefore, another set of analyses is conducted with the assumption that the Clean Air Act is not implemented by 1976, and the average statewide sulfur content is 2%.

If the fuel is untreated, about 90% of the sulfur is passed into the atmosphere. The control costs discussed earlier are based upon 90% removal. If control is implemented in one-fourth of the coal and oil plants in each year from 1976 to 1979, the proportion of sulfur emitted is 90% until 1976; 70%, 50%, and 30% in 1976, 1977, and 1978; and 10% thereafter. If control is not implemented, emissions remain at 90% throughout the period. Emissions appear in column 14.

As discussed previously, three tax rates are analyzed here: (1) $200,

Table 15.4 Generation Changes in Coal, Oil, and Nuclear Power, New York State

I. Nuclear Power

A. 1976–1980

$\Delta GN_t = ANG_t + TNG_t$

$ANG_t = .1 \times J \times .977 \times \Delta G_t$

$TNG_t = ANG_t \times TNMS_t$

$TNMS_t = [TCC_{t-1}(GC_{t-1} + GO_{t-1})]/(.5 \times ACWT_{t-1})$

B. 1981–1990

$ANG_t = .977 \times \Delta G_t + .033 \times GC_{t-1}$

II. Oil Generation

A. 1976–1980

$\Delta GO_t = AOG_t + TOG_t$

$AOG_t = .1 \times (10 - J) \times .977 \times \Delta G_t$

$TOG_t = -TNG_t \times [GO_{t-1}/(GC_{t-1} + GO_{t-1})]$

B. 1981–1990

$AOG_t = 0$

III. Coal Generation

A. 1976–1980

$\Delta GC_t = ACG_t + TCG_t$

$ACG_t = 0$

$TCG_t = -TNG_t \times [GC_{t-1}/(GC_{t-1} + GO_{t-1})]$

B. 1981–1990

$ACG_t = -.033 \times GC_{t-1}$

where ΔGN = nuclear generation growth, t = year, J = years from 1970, ANG = autonomous nuclear generation growth, TNG = tax-induced nuclear generation growth, ΔG = growth in total generation, TNMS = tax-induced nuclear market share increase, TCC = tax plus control cost, GC = coal generation, GO = oil generation, ACWT = average cost without tax, AOG = autonomous oil generation growth, TOG = tax-induced oil generation change, and TCG = tax-induced coal generation change.

Table 15.5 Maximum and Minimum Generation Levels in 1990
(billion kilowatt-hours and percentages)

	Total Generation	Nuclear	Coal	Oil
I. Variable Demand				
A. No tax, No control	173.9	74.0 (42.6%)	18.9 (10.9%)	45.4 (26.1%)
B. High tax, High control cost No Clean Air Act	165.5	75.9 (45.9%)	16.2 (9.8%)	38.1 (23.0%)
II. NPS Demand				
A. No tax, No control	277.0	168.1 (60.7%)	18.9 (6.8%)	52.0 (18.8%)
B. High tax, High control cost No Clean Air Act	277.0	195.2 (70.5%)	11.6 (4.2%)	32.3 (11.7%)
III. 1970 Levels	95.5	4.3 (4.5%)	26.5 (27.7%)	30.9 (32.4%)

$400, and $600 per ton. The high rate is used in the Table 15.5 case, and the total tax is in column 15 of Table 15.3.

From the foregoing discussion of control costs, three levels are examined: 0.95, 1.425, and 1.9 mills per kWh. The illustrative analysis uses the medium level; total control costs are in column 16 of Table 15.3.

15.3.4. Social Benefits and Market Decisions

The basic assumption here is that tax-induced increases in taxes and control costs are passed along to consumers in the next year. This follows from the rate-of-return pricing policies used by the Public Service Commission in New York as well as most other states. Costs in the model consist of two components: an increasing average cost exogenous to the model and tax-related costs. The NPS (FPC, 1971, p. I-19-10) calculates that average power costs in the Northeast will increase 19.9% from 1968 to 1990. This is used on a simple linear growth basis for exogenous average cost increases in the model (column 18 of Table 15.3).

Residential, commercial, and industrial electricity prices (columns 19 to 21) in the model increase at the same rate, and they further increase by the fraction of tax plus control costs to total generating costs in the preceding year. Gas prices are assumed to increase 50% over the period.[6]

Illustrative damages in the absence of a tax depend upon the demand model and the Clean Air Act effectiveness. The National Power Survey demand model has slightly higher damage because its high growth rates call forth more oil and coal generation. Within each demand model implementation of the Clean Air Act standards reduces damage by exactly 50% because of the sulfur content definitions: 1% with implementation and 2% without.

If the tax is imposed and further control undertaken, damage reduction follows emission reduction as in column 22 of Table 15.3.

Social benefit is narrowly defined in each case with a tax. The only claimed benefit is from direct damage reduction. Social cost is simply the control cost.

Both benefits and costs are discounted to 1972 present values. The question of appropriate values for social discount rates remains the subject of much disputation. The value used here is 10%, representative of conclusions reached in recent studies in this area.[7]

The market decision to implement or not provide control is independent of net social benefits. Control will be implemented if taxes plus control costs with removal are less than taxes without removal. In the model it is assumed that developing control equipment for coal alone would be infeasible, so it is required that total control costs (and tax on remaining emissions) be less than the tax without control before control is instituted.

Social benefits and market decisions in the model share control costs. That is, control cost is defined both as social cost and market cost to the utilities. However, on the benefit side, social benefits are equivalent to damage reduction while market benefits are equal to reduced tax liability. As we shall see, whenever the tax rate differs from the damage rate, the social optimum may depart from the market optimum.

15.4. Results and Generalizations to National Policy

In the 26 of the 36 tax cases examined, the market decision is to install the additional control equipment and remove the sulfur. In these 26 cases emissions and damage rise through 1975 and decline therafter, as in columns 14 and 22 in Table 15.3. If we assume each of the demand models, tax rates, control cost levels, and sulfur tax levels are equally likely, then our first hypothesis given earlier is accepted: emissions and damage will be significantly reduced by a sulfur tax.

By definition the National Power Survey demand model shows no response to price changes: all cases project 254 billion kWh in 1990. Within the demand model's range of cases, the high demand is the no-tax case (159.2 billion kWh) and the low demand is the high tax, high control cost case in the absence of Clean Air Act implementation (151.6 billion kWh). This is a small difference—5%—and the second hypothesis is probably false. It appears likely that a sulfur emission tax would not significantly affect demand growth. There are three reasons why there is no important demand reduction from the tax. First, regardless of a tax, the proportion of total generation by coal and oil is expected to decline (see Table 15.5). Second, although $4 per ton high treatment cost for coal is about 50% of coal cost, it is only 1.9 mills of 21 to 25 mills total average cost per kilowatt hour, or 8 to 9% (see column 18, Table 15.3). Third, although electricity prices seem to be the most important factor influencing demand, the lag in demand response further reduces the effect on demand (see Table 15.2).

The third hypothesis asserted that nuclear power generation growth would be substantially increased by a tax. Table 15.5 and Figure 15.1

Table 15.6 Net Benefits and Benefit-Cost Ratios from Sulfur Emission Tax on Electric Utilities in New York (million dollars present value in 1972; ratios in parentheses)

I. Clean Air Act Implemented						
Control Cost	Low		Medium		High	
Demand Model	NPS	Var.	NPS	Var.	NPS	Var.
Tax						
	[a]	[a]	[a]	[a]	[a]	[a]
Low	21.0	10.6	21.0	10.6	21.0	10.6
	(1.7)	(1.7)	[a]	[a]	[a]	[a]
Medium	237.7	213.2	41.9	20.3	41.9	20.3
	(1.8)	(1.7)	(1.2)	(1.2)	[a]	[a]
High	241.2	215.2	95.9	69.7	62.8	28.9
II. No Implementation of Clean Air Act						
Control Cost	Low		Medium		High	
Demand Model	NPS	Var.	NPS	Var.	NPS	Var.
Tax						
	(3.5)	(3.4)	[a]	[a]	[a]	[a]
Low	796.0	729.1	83.8	40.5	83.8	40.5
	(3.5)	(3.4)	(2.4)	(2.3)	(1.9)	(1.8)
Medium	804.7	733.8	662.4	589.5	527.9	447.6
	(3.6)	(3.5)	(2.5)	(2.3)	(1.9)	(1.8)
High	813.3	737.8	673.8	594.6	542.1	453.7

[a] Control is not instituted, and benefit-cost ratio is inapplicable; see text.

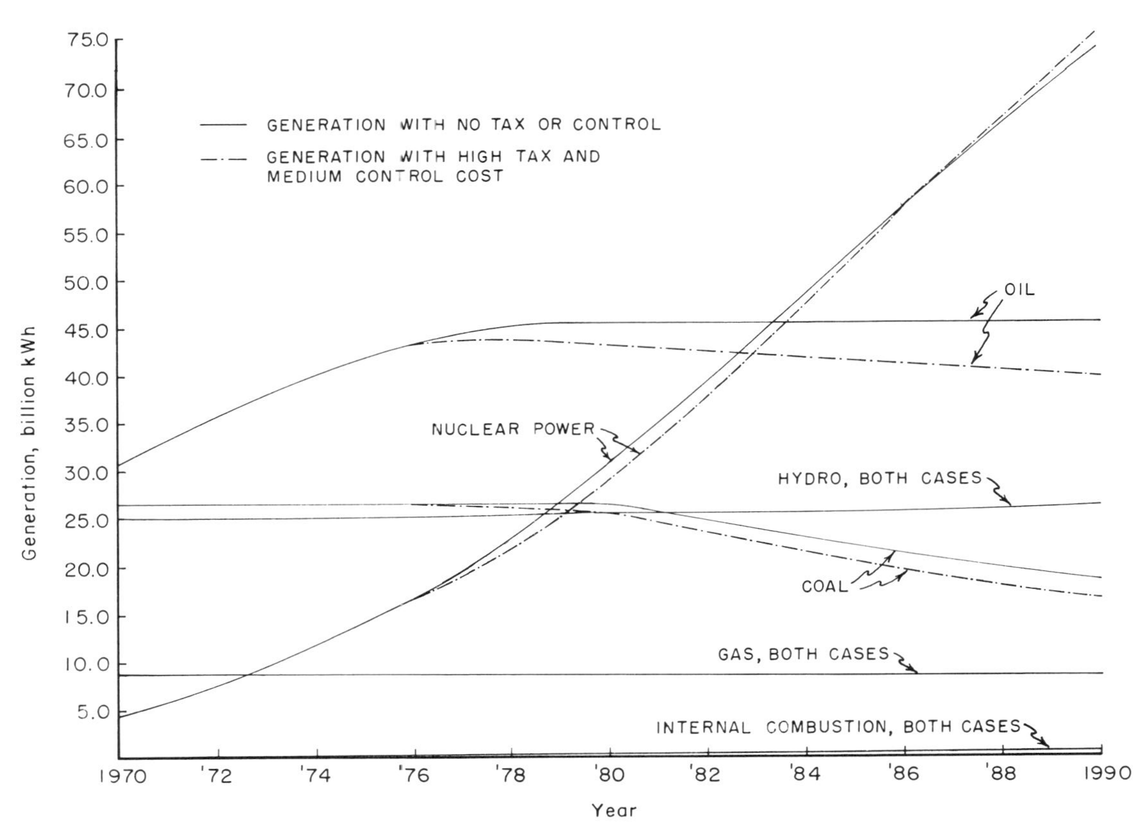

Figure 15.1. Effect of tax on generation levels, variable demand model without Clean Air Act implementation.

display the cases with maximum and minimum generation levels for nuclear power, oil, coal, and total generation for each demand model and for 1970. While there is noticeable difference in 1990 levels of nuclear generation between demand models, there is little difference within each demand model. Suppose this generation growth is expressed in terms of 1000 MWe units. Then in the variable demand model 10 new units (or their equivalent) are needed in New York by 1990.[8] In the NPS demand model, 23 new units are needed without a tax, while in the highest case only 4 more are required. The hypothesis is rejected. Hausgaard (1971, p. 33) anticipated this result.

The fourth hypothesis stated that the social benefits of a tax would exceed its social cost. In Table 15.6 all cases have positive net benefits. Their present value ranges from a low of $11 million to a high of $813 million. In the cases where the tax rate is high enough for the market decision to favor control, net benefits range from $70 to $813 million and benefit-cost ratios from 1.2 to 3.6. In the cases in which the market decision is to pay the tax on full emissions rather than institute control, net benefits range from $11 to $84 million. In these latter cases there is no social cost since control is not effected and a benefit-cost ratio is not defined; benefit arises from demand reduction in the variable demand model and from the slight tax-induced shift from coal and oil to nuclear power in both demand models. While the decisions made within the model are independent of the assumed damage rate, the benefit-cost ratios are directly proportional to the damage rate. Thus, if actual damage is $400 per ton of sulfur rather than $600, all ratios would be reduced by one-third. If actual damage is equal to or greater than the Barrett-Waddell estimate, the conclusion here is that the social benefits of a tax would exceed its social costs.

The fifth hypothesis suggests that if the tax rate differs from the damage rate, social optimality will differ in an important way from market decisions. Table 15.6 indicates that this is only partially correct. Within any triad of assumptions about the Clean Air Act, demand model, and control costs, the maximum net benefits are always with the case with the highest tax rate that equals the damage rate. If the tax rate is so low that control is not effected, social net benefits depart significantly from the maximum possible. However, once the tax rate exceeds the necessary level to promote implementation, there is little further increase in net benefits.

Finally, the presumption that a sulfur emission tax has higher social value if the Clean Air Act has not been implemented seems confirmed by Table 15.6.

To what extent can these conclusions be generalized to other industries and to the nation? The economic choices facing the electric utility industry in New York are representative of utility economics throughout the country with respect to a sulfur emission tax, and utilities account for 50% of sulfur emissions (see Table 15.1). Sulfur is emitted in copper, lead, and zinc primary smelting through the separation of the metal from sulfur in the ores. Similarly, sulfur is emitted in petroleum refining when sulfur is removed from oil, and sulfur emissions are a by-product of sulfuric acid production. Control costs in these industries may be the same or a little lower than for the electric utility industry.[9] Control costs for steam-heated boilers and hot-air furnaces may be much higher than for electric utilities. As a possible lower limit of national benefit, the analysis here is taken as representative of the electric utility industry. As an upper limit, the analysis is representative of all sulfur emitting processes. In the lower limit, all net benefits in Table 15.6 would be multiplied by 23.6. In the upper limit, all net benefits in Table 15.6 would be multiplied by 51.9.[10]

In the case without implementation of the Clean Air Act, medium control costs, medium tax rate, and variable demand (see Table 15.6), the possible high national values are net benefits of $31 billion (present value 1972), social cost of control implementation of $23 billion, and social benefits of damage reduction of $54 billion. The lower limit values would be $14 billion net benefit, $10.5 billion social cost, and $23.5 billion social benefit.

As an ex post note, we may consider the likely consequences of electricity and gas prices much higher than those postulated here. First, since the NPS demand model is insensitive to price changes, there would be no change in demand growth for it. In the variable demand model, it is likely that the impact of electricity price increases will exceed that of gas prices, and demand growth would be reduced in all cases. For both demand models, the assumption of increased costs for all generation processes means the proportional impact of a sulfur tax and its induced control costs have to be lower. Probably none of the six conclusions stated here would be substantively affected by an analysis with higher electricity and gas prices.

In summary, the tentative conclusions which follow from this analysis are that a sulfur emission tax would cause significant reductions in emissions and damage, have little effect on electricity demand growth and nuclear power generation growth, result in greater social benefits than costs, and cause greater social benefits if the Clean Air Act is not implemented than with implementation.

If the Clean Air Act standards should be met in New York by 1976 with emissions being equivalent to the use of 1% sulfur coal and oil, and if control costs should be in the medium to high range, the tax levels currently being discussed—$200 to $400 per ton emitted sulfur—appear to be too low to motivate additional sulfur emission reduction. However, if those standards are not met, a tax might provide incentive to reach and go below the standards.

Notes

1. Adjusting the 1968 figure by the implicit Gross National Product price deflator.

2. See Secretary Connally's transmittal letter (1972) and the U.S. Department of the Treasury (1972) background statement.

3. Recent discussions of engineering economics of sulfur removal include Ketchum (1972), Spaite (1972), Slack and Falkenberry (1971), National Economic Research Associates (1972), and *The Economic Impact of Pollution Control* (EIPC) (U.S. Council on Environmental Quality, 1972).

4. See New York State Department of Environmental Conservation and EIPC.

5. See Mount, Chapman, and Tyrrell, Chapter 24 in these proceedings, and also Chapman, Tyrrell, and Mount (1972).

6. This may be compared to National Petroleum Council (1972) suggestions for wellhead price increases of 80 to 250% by 1985.

7. See Baumol (1968), U.S. Water Resources Council (1971), and U.S. Congress Joint Economic Committee (1968).

8. A 1000 MWe plant operated at 80% load factor would produce 7.013 billion kWh in an average 365.25-day year.

9. The discussion here is derived from *The Economics of Clean Air* (Administrator of the Environmental Protection Agency, 1972), and, to a lesser extent, from *The Economic Impact of Pollution Control* (U.S. Council on Environmental Quality, 1972).

10. For the lower limit ratio, divide national power plant sulfur emissions (17.2 mil-

lion tons) by New York power plant emissions (.65 million ton). For the upper limit ratio, divide total national sulfur emissions (33.9 million tons) by New York power plant emissions. Data for 1968 from EIPC and the New York State Department of Environmental Conservation.

References

Administrator of the Environmental Protection Agency (1972). Annual Report, *The Economics of Clean Air,* March.

Barrett, L. B., and Waddell, T. E. (1970). "The Cost of Air Pollution Damages: A Status Report," mimeo, National Air Pollution Control Administration, Raleigh, N.C., July.

Baumol, W. J. (1968). "On the Social Rate of Discount," *American Economic Revue, 58*(4), pp. 788–802, September.

———, (1972). "On Taxation and the Control of Externalities," *American Economic Revue, 62*(3), pp. 307–322, June.

Chapman, D.; Tyrrell, T.; and Mount, T. (1972). "Electricity Demand Growth and the Energy Crisis," *Science,* 1972, *173*(4062), pp. 703–708, November 17.

Connally, J. B. (1972). "Texts of (1) Letter from Treasury Secretary John Connally to the Senate and House Transmitting Proposed Legislation for a Tax on Emissions of Sulfur into the Atmosphere, and (2) the Administration's Proposed Legislation," Bureau of National Affairs, Inc., Washington, D.C., 1972.

Edison Electric Institute (annual). *Statistical Year Book of the Electric Utility Industry,* New York.

Graham, R. E., Jr.; Degraff, H. L.; and Trott, E. A., Jr. (1972). "State Projections of Income, Employment, and Population," *Survey of Current Business, 52*(4), pp. 22–48, April.

Hausgaard, O. (1971). "Proposed Tax on Sulfur Content of Fossil Fuels," *Public Utilities Fortnightly, 88*(6), pp. 27–33, September 16.

Ketchum, Malcolm R. (1972). "Sulfur Pollution Abatement in the New York Electric Power Industry," New York Public Service Commission Office of Economic Research Report No. 13, Albany, November 20.

Mount, T. D.; Chapman, D.; and Tyrrell, T., "Electricity Demand in the United States," Chapter 24 in these proceedings.

National Economic Research Associates (1972). *Fuels for the Electric Utility Industry 1971–1985: A Report to the Edison Electric Institute,* August.

National Petroleum Council (1972). *U.S. Energy Outlook,* Washington, D.C., December.

New York State Department of Environmental Conservation (1972). *Upstate New York Air Quality Implementation Plan and New York City Metropolitan Area Implementation Plan,* January.

Proxmire, Sen. William (1972). Statement, *Congressional Record,* 92d Cong., 2d Sess., *118*(5), 5276–9, January 24.

Ruff, L. E. (1972). "A Note on Pollution Prices in a General Equilibrium Model," *American Economic Revue, 62*(1), pp. 186–192, March.

Slack, A. V., and Falkenberry, H. L. (1971). "SO_2: More Questions than Answers," *Electrical World,* 1971, pp. 50–54, December 15.

Spaite, P. W. (1972). "SO_2 Control: Status, Cost, and Outlook," *Power Engineering,* pp. 34–37, October.

U.S. Congress, Joint Economic Committee (1968). Subcommittee on Economy in Government, *Economic Analysis of Public Investment Decisions: Interest Rate Policy and Discounting Analysis,* report, and *Interest Rate Guidelines for Federal Decision Making,* hearings, 90th Cong. 2d Sess.

U.S. Council on Environmental Quality (1972). *The Economic Impact of Pollution Control,* Department of Commerce, and Environmental Protection Agency, U.S. Government Printing Office, Washington, D.C., March.

U.S. Department of the Treasury (1972). "Pure Air Tax Act of 1972: Background and Detailed Explanation," Washington, D.C., February 8.

U.S. Environmental Protection Agency (1972). "Air Programs: Approval and Promulgation of Implementation Plans," *Federal Register, 37*(105), 10841–10906, May 31, 1972.

U.S. Federal Power Commission (1971). *The 1970 National Power Survey,* Washington, D.C., December.

U.S. Water Resources Council (1971). "Proposed Priniciples and Standards for Planning Water and Related Land Resources," *Federal Register, 36*(245), 24144–24194, December 21.

16 Opportunity Costs of Land Use: The Case of Coal Surface Mining*

R. L. SPORE† AND E. A. NEPHEW†

16.1. Land-Use Problems: An Overview

Of the many decisions constantly being made both by private market mechanisms and public authorities as regards the allocation of scarce resources, those concerning land use are often the most controversial. Reasons why such should be the case come easily to mind. First, there can be unintended and incidental by-products or impacts associated with many otherwise legitimate uses of land which affect the value of the consumption and production functions of persons or firms located nearby. These impacts, presumably measurable as costs, are external from the point of view of the perpetrators of them in that no incentive exists for taking them into account when the land-use decision is being made. In the interest of both the efficient allocation of resources and equity, remedial action by governmental intervention is called for. The common occurrence of such problems and this manner of their solution is evidenced by the almost universal acceptance of zoning regulations as a means of minimizing the external costs of land use in urban areas (Crecine, Davis, and Jackson, 1967).

Another reason for the controversial nature of land-use decisions lies with the fact that alternative land uses are often incompatible such that the development of land resources for certain purposes results in the irreversible loss of that land for other purposes both now and in the future. The devotion of land to irreversible development can, of course, be legitimate if such decisions are justified following an explicit recognition of the opportunity costs of the land. However, just as in the case of external costs, private market accountings of (here, opportunity) costs can deviate from social accountings. Such a discrepancy can result, for example, when the benefits of certain land use, such as its preservation in a natural state, are of the public good type, for example, wilderness recreation or the option of preserving rare scientific research materials.[1] Since benefits of this nature are not appropriable by private owners of the land, the value of such land will then fail to approximate its true scarcity value or opportunity costs in terms

* Research reported here is sponsored by the National Science Foundation RANN Program under Union Carbide Corporation contract with the U.S. Atomic Energy Commission.

† Economist and Research Staff Member, respectively, ORNL-NSF Environmental Program, Oak Ridge National Laboratory.

of wants. As a result, private market allocations are likely to preserve less than the socially optimal amount of natural environments.

There is an additional dimension to the previous problem that makes it even more serious and a source of controversy. When addressing the question of the allocation of land between development and preservation alternatives, the differential effects of technological progress over time must be considered.[2] The devotion of natural areas to development usually proceeds in order to provide intermediate goods and services (raw materials). For such goods, increases in demand can be met by increasing their supply and, in addition, a wide range of substitutes typically exists. If, under the influence of technology, increased output of the substitute goods can be obtained at supply prices that change over time at rates different from those of the goods obtained from development, then the value of the goods and services flowing from that development will change over time. On the other hand, the preservation of natural habitats yields a flow of final consumption services for which close substitutes often do not exist. Since such services are nonproduced, technological advance is not relevant and the supply cannot be increased. Thus, an increase in demand can increase the value of the service flow and, hence, the value of the preserved land. As a result, the devotion of natural areas to development can have a dynamic opportunity cost that grows over time. Not only do private market allocations preserve less than the socially optimal amount of natural environments, but this optimal amount is likely to be increasing over time. If such development is irreversible, the possibilities for the misallocation of land resources become an ever more serious problem. Controversies over such things as the increased logging in national forests or the development of wild rivers for hydroelectric power reflect just this concern. Of concern in this paper is whether either of the above-mentioned allocational problems of land use are associated with coal surface mining.

16.2. Land-Use Problems Associated with Coal Surface Mining

Clearly, coal surface mining can result in external costs, and it is this aspect that is the source of much current public concern. The adverse environmental impacts of surface mining—water pollution from acid drainage and sedimentation, landslides, destruction of scenic views—are now well documented.[3] Well established also are the theoretical arguments in support of governmental intervention to improve resource allocation (Brooks, 1966; Spore, forthcoming 1973).

What has not as yet been addressed is whether or not the social opportunity costs of devoting land to coal surface mining are in fact being taken into account. The various aspects of this problem can be posed as a series of questions which are presumably amenable to research.

1. To what extent is or will surface mining take place in natural habitats possessing high preservation benefits? Will these benefits be irreversibly lost if surface mining occurs? Precise information regarding the present and future location of coal surface mining vis-à-vis the location of natural environments is not readily available. Isolated instances have occurred where surface mining has been at least temporarily prohibited in favor of the preservation of natural habitats.[4] Of even greater concern is the potential for future conflict. For example, most of the large yet uncommitted blocks of low-sulfur coal reserves occur in the West, of which the federal government is the largest single owner (National Economic Research Associates, 1972, p. 95). Much of this land has remained in the public domain and has not been claimed under the Homestead Act and other land disposal programs because it is not amenable to private development. As such, "it represents the bulk of the remaining roadless areas within the country with much of the nation's scenic and wildlife attractions preserved on . . . [them]." (Krutilla, ed., 1972, p. 2.) These lands are, thus, *de facto* wilderness areas and remain largely unprotected from future development.

Whether or not the act of surface mining will result in the irreversible loss of preservation benefits is fundamental to the problem as it has been stated. As Fisher, Krutilla, and Cicchetti have noted (1972, p. 31), two types of reversal are possible: the restoration of the mined area by a program of direct investment, and natural reversion to the wild. Whether either or both of these processes would be sufficient to regain the benefits of preservation can be answered only by an in-depth study. For example, whether or not the characteristics of restored lands would be acceptable to those who desire wilderness recreation can be determined only by a careful analysis of those environmental features upon which the wilderness experience depends (that is, by estimating the demand for wilderness recreation). The results of such an analysis can be only partially foreseen a priori;

however, it can be noted, for example, that to the extent that man and nature are unable to return a mountainside to its original contour and vegetal cover, and to the extent that recreational values depend upon the absence of aesthetic insults such as a surface-mining highwall, then the loss of preservation benefits will not be reversible.[5]

2. Are substitutes to the intermediate good (coal) produced by surface mining available? For most of its uses, surface-mined coal is an intermediate good or raw material for which several close substitutes exist. In 1970, for example, 62% of all coal mined in the United States was converted to electrical energy in steam electric plants (U.S. Bureau of Mines, 1972, Table 34, p. 48). There is, in general, a high degree of substitutability among the fuels (fossil and nuclear) used in producing electrical energy in a steam plant. In addition, there is a substitutability both in terms of the means of extracting the coal itself (surface and deep mining) and the geographical location of the coal mined. It is reasonable to expect, therefore, that increases in the demand for surface-mined coal can be met by increases in production and that the conditions under which substitutes can be produced are constantly influenced by technology. The value of coal obtained from the surface mining of a specific site can, therefore, change over time.

Thus, the conditions as described for the general theoretical situation presented in Section 16.1 appear to apply to the case of coal surface mining. What must be addressed at this point is whether an analytical framework can be developed whereby the full opportunity costs of alternative land uses are taken into account and a more preferred pattern of land use thereby indicated.

16.3. A Simple Model for the Evaluation of Land-Use Alternatives

Recently, Krutilla and Cicchetti (1972, pp. 1–13) have presented a simple analytical framework for evaluating the benefits of environmental resources. This framework, appropriately modified for the case of coal surface mining, is outlined now.

Consider a tract of land underlain by strippable coal reserves which is also an undisturbed natural environment with unique features of high value for wilderness recreation and for which there are no adequate substitutes of like quality. For simplicity of presentation, it is assumed that these alternative uses are incompatible, that is, that devotion of the land to surface mining would result in the irreversible loss of the recreation benefits. Assuming also that no other alternative uses of the land exist, then a determination of the socially preferred use of the land can be made by comparing the discounted value of the net benefit stream from each alternative.

The net benefit from development by surface mining can be represented as follows:

$$b_d = (B_d - C_d) - (B_a - C_a), \tag{16.1}$$

where

b_d = net benefit from surface mining the tract for coal,
B_d = gross benefit from surface mining the tract for coal,
C_d = cost of surface mining the tract for coal,
B_a = gross benefit of fuel obtained from an alternative source or by an alternative means,
C_a = cost of obtaining alternative fuel.

Since both B_d and B_a are defined to provide an equivalent amount of fuel for, say, conversion to electrical energy, they are equal. Thus,

$$b_d = C_a - C_d, \tag{16.2}$$

or b_d is equal to the resource savings resulting from obtaining fuel by surface mining the tract for coal rather than obtaining it from some alternative source or means.

The net benefit from preservation of the tract for wilderness recreation, in turn, can be given as follows:

$$b_p = (B_p - C_p) - (B_a' - C_a'), \tag{16.3}$$

where

b_p = net benefit from preservation,
B_p = gross benefit from preservation,
C_p = cost of providing the services from preservation,
B_a' = gross benefit from alternative source of recreation,
C_a' = cost of providing recreation services from alternative source.

Since the undeveloped tract is a gift of nature, the costs C_p (other than the opportunity costs represented by b_d in Equation 16.1) are zero. In addition, it is assumed that the demand for wilderness recreation, which would remain unsatisfied if the tract were not preserved, would be distributed across the margin in other sectors of the recreational services industry characterized by free entry and feasibility of augmenting supplies; that is, $B_a' = C_a'$. Thus,

$$b_p = B_p. \tag{16.4}$$

In Section 16.1, it was argued (1) that the value of the goods and services flowing from development can change over time if, under the differential influence of technology, increased output of substitute goods can occur at supply prices that change differently than those of the goods obtained from the development, and (2) the value of the services flowing from preservation can increase over time if they are nonproducible, have no close substitutes, and are subject to increased demand. The effect of these considerations can be examined by introducing them explicitly into a conventional formula for asset valuation. In general, if the value of the annual flow of goods or services derived from an asset changes at a constant annual rate as a result, say, of technological advance or demand growth, then

$$b_t = b_0(1 + \alpha)^t, \tag{16.5}$$

where

b_t = benefit in any year t,
b_0 = benefit of base year,
α = annual rate of change of b.

If, under the influence of technology, the resource savings in year t attributable to development by surface mining, $b_{dt} = C_{at} - C_{dt}$, decreases over time, then by representing $(1 + \alpha)$ by $1/(1 + r)$, the discounted present value of this assumed depreciating stream of annual benefits can be written as

$$b_d = \sum_{t=1}^{T} (C_{a0} - C_{d0})/(1 + i)^t(1 + r)^t = C_a - C_d, \tag{16.6}$$

where

b_d = net benefit from surface mining,
C_{a0}, C_{d0} = base year value of C_a and C_d as defined in Equation 16.1,
T = terminal year to which the discounted annual benefits are summed,
i = rate of discount,
r = differential annual rate of technological change.

If, on the other hand, the annual value of the preservation benefit in year t, B_{pt}, increases over time as a result of increased demand, then the discounted present value of this appreciating stream of annual benefits is given by

$$b_p = \sum_{t=1}^{T} B_{p0}(1 + s)^t/(1 + i)^t = B_p, \tag{16.7}$$

where

b_p = net benefit from preservation,
B_{p0} = gross benefit from preservation in base year,
s = annual rate of growth in preservation benefits.

The implications of introducing the differential effects of technological change and demand growth are of particular significance. In Equation 16.7, for example, introducing the rate of annual appreciation of preservation benefits has the effect of reducing the influence of the rate of time discount and, thus, lengthening the time period over which the annual benefits are summed. Obtaining such a result in this manner is clearly superior to an arbitrary adjusting of the discount rate in order to prevent an otherwise warranted conversion of natural habitats.[6]

16.4. Application of the Analysis: Opportunity Costs of Coal Surface Mining in the Basin of the Big South Fork of the Cumberland River

In this section, results of an application of the foregoing analytical framework are presented. It must be emphasized that the results obtained are based upon a number of simplifying assumptions and are preliminary to a fuller report on the subject currently in progress (Spore and Nephew, in progress). Thus, the procedures and data employed are to be taken as only illustrative of the general approach.

16.4.1. Description of the Study Area

According to a recent interagency report on alternative programs for the development of the Big South Fork of the Cumberland River, it was found that

> The Big South Fork, its gorge, and tributaries constitute an outstanding recreation resource. The river is truly wild and scenic in character, flowing through a narrow, almost uninhabited valley lined on both sides by stately multicolored sandstone cliffs, shrouded with forests. Surprisingly here in the eastern half of the United States, the setting retains an unusual feeling

of naturalness and, in places, the quality of wilderness. There are numerous interesting geologic formations throughout the area, including small caves, natural bridges and arches, waterfalls, scenic side canyons, and palisades. Complementing these features are a wide variety of flora and fauna, and a diverse range of calm and turbulent waters. . . . As a free-flowing system, the Big South Fork possesses an outstanding combination of favorable attributes which make it nationally significant and, therefore, offers the public an opportunity to participate in various forms of high-quality outdoor recreation that are not commonly found elsewhere (Interagency Report, 1970, pp. 68–69).

The study then concluded that

. . . the Big South Fork system, including the main stem, its Clear Fork tributary, and its New River tributary below the town of New River, Tennessee, met all the criteria for inclusion in the national wild and scenic rivers system as a national scenic river (Interagency Report, 1970, p. 68).

On the other hand, in 1971 the nine counties that together contain the Big South Fork basin possessed 52,839,000 tons on strippable reserves, of which about 60%, or 31,703,000 tons, occur in the basin itself. (U.S. Bureau of Mines, 1971, pp. 86, 108; Interagency Report, 1970, p. 32.) In addition, total recoverable reserves, obtainable primarily by deep mining, are estimated to be 245,000,000 tons (Interagency Report, 1970, p. 32). In 1970, approximately 3,062,000 tons of coal were mined in the basin, of which about 55%, or 1,687,000 tons, were obtained by surface mining, primarily on the New River tributary above the town of New River, Tennessee (U.S. Bureau of Mines, 1972, Table 17, pp. 31, 35).

In order to preserve the flow of preservation benefits—primarily wilderness recreation—that the Big South Fork could provide, it must be protected from incompatible uses, not only within the boundary of the area where the recreation would take place (essentially rim to rim of the gorge) but also in adjacent areas (the remainder of the basin). For the purposes of this study, it is assumed that the environmental impacts of surface mining—primarily aesthetic insults—are incompatible with preservation of the Big South Fork, that is, that the benefits associated with the recreational use of a scenic river would be irreversibly lost if the basin were developed for coal surface mining. However, it is assumed that such is not the case with deep mining, if the environmental effects (principally acid mine drainage) of this method of mining are prevented. Other land uses in the basin (forestry, agriculture) are assumed generally to be compatible with either preservation of the gorge, or coal surface mining and will not be considered further.

Thus, in conformance with the discussion in Section 16.3 and Equation 16.2, the net benefit of development of the Big South Fork basin for coal surface mining is the difference in the costs of production between deep and surface mining for that quantity of coal which would be surface-mined in lieu of preservation. What must be demonstrated is whether the opportunity costs of surface mining—the preservation benefits forgone—are greater or less than this resource savings.

16.4.2. Net Benefits of Development by Coal Surface Mining

The first step in the estimation of the net benefits of development is to construct a forecast of the annual surface-mining production that will otherwise occur in the basin. The forecast of annual production employed in this study is presented in Table 16.1 and is based on the following assumptions: (1) all coal mined in the basin will be burned in steam electric plants, (2) the growth rate of overall coal production will equal the forecasted rate of coal consumption in electricity generation (that is, an annual growth rate of 3.2% until 1980 and 1.3% thereafter), and (3) the proportion of coal produced by surface mining will continue at its current level (55%).[7] The forecasted series of annual tonnage ends in 1985, the year in which cumulative production will have approximately depleted the remaining strippable reserves.[8]

The second step in the estimation of the net benefits of development involves a computation of the net difference in the costs of production between deep and surface mining. For the purposes of this preliminary report, the cost calculation presented in Table 16.2 was based on the following assumptions: (1) all coal producers are in long-run competitive equilibrium such that price equals minimum long-run average cost, (2) coal deep mining is a constant cost industry such that industry output can be expanded by increasing the number of firms without affecting minimum long-run average cost, and (3) the potential influence of technological advance on deep and surface mining, respectively, is not considered. The observed prices are adjusted, in turn, to account for the increased costs of deep mining as a result of the 1969 Coal Mine Health and Safety Act and the increased costs of surface mining as a result of the 1972 Tennessee law regulating surface mining.[9]

Based on the information in Tables 16.1 and 16.2, the discounted present

Table 16.1 Estimated Annual Coal Surface Mine Production, Big South Fork Basin

Year	Tons (in thousands)
1973	1851
1974	1910
1975	1973
1976	2037
1977	2102
1978	2169
1979	2239
1980	2310
1981	2340
1982	2371
1983	2401
1984	2433
1985	2464
Total	28,600

Source: see text.

Table 16.2 Estimated 1973 Costs of Coal Deep and Surface Mining, Big South Fork Basin

(1) Deep Mining	
1970 F.O.B. mine value[a]	$5.07
Coal Mine Health and Safety Act[b]	1.01
Total	$6.08
(2) Surface Mining	
1970 F.O.B. mine value[a]	$4.71
Cost of reclamation[b]	.50
Total	$5.21
Net Difference: (1) Minus (2)	$0.87

Source: U.S. Bureau of Mines, *Coal—Bituminous and Lignite*, Preprint from 1970 Bureau of Mines Minerals Yearbook, Table 42, p. 59.
[b] See text.

value of the net benefits of development can be calculated using a revised form of Equation 16.6, that is,

$$b_d = \sum_{t=1}^{T} (C_{a0} - C_{d0})X_t/(1 + i)^t, \qquad (16.6')$$

where

b_d = discounted net benefit from development,
$C_{a0} - C_{d0}$ = \$0.87 (Table 16.2)
X_t = production in year t (Table 16.1),
i = rate of discount.

Using an interest rate (opportunity cost of capital) of 9%, with alternatives of 8 and 10% for purposes of comparison, the values of b_d are presented in Table 16.3. These values must now be compared with the discounted present value of the net benefits from preservation in order to obtain an indication of the preferred land use for the basin.

16.4.3. Net Benefits of Preservation

For the purposes of this paper, the calculation of the benefits of preservation of the Big South Fork basin will be limited to consideration only of the recreation benefits. In general, the value of the quantity of recreation consumed per unit time can be measured by the area under the demand curve (Krutilla and Cicchetti, 1972, pp. 9–13). Since a natural environment is a reusable, nondepreciating asset, then giving time the value of one year, the gross benefit can be approximated by summing the discounted benefits as represented by the area under the demand curve for each time period the basin is used. Moreover, since the supply of recreational services provided by the Big South Fork is not augmentable, the annual value of the services can be expected to grow, reflecting increases in demand as a result

Table 16.3 Discounted Present Value of the Net Benefit of Development by Coal Surface Mining

i	b_d
.08	\$14,732,000
.09	13,906,000
.10	13,172,000

Source: see text.

of changes in income and population over time. Following Krutilla and Cicchetti (1972), this growth can be visualized as a series of outward shifts in the demand curve, with the shift in the intercept of the price axis related to growth in real per capita income and the shift along the quantity axis related to the rate of growth in quantity demanded at zero price (Krutilla and Cicchetti, 1972, pp. 9–13). However, such growth cannot be allowed to occur without bound. Rather, in order to avoid the destruction of the environmental characteristics of the area, the existence of a specified recreational capacity must be recognized and rationing must eventually be imposed.[10]

Summarizing then, the rate of growth of annual recreational benefits s depends on the following variables:[11]

γ = annual rate of growth of quantity demanded at given price,
r_y = annual rate of growth of price per user day,
k = number of year after initial year in which carrying capacity constraint becomes effective,
m = number of years after initial year in which γ falls to the rate of growth of population.

Ideally, a complete estimation of the recreation benefits would involve the development of a demand schedule for each of the several recreational activities the Big South Fork could provide. Alternatively, a composite demand function might be constructed to represent the combination of these independent but related individual demand functions. However, since information sufficient to implement either of these approaches is not available, an indirect method is employed. A determination is made instead of what the initial year's benefit from preservation would need to be, growing at the rate s implied by the annual shifts in an assumed composite demand function, to be equal to the present value of development by surface mining. This "threshold value" of the required initial year's benefit can be calculated by dividing the present value of $1 growing at the rate s into the present value of development as presented in Table 16.3.

The determination of the annual rate of growth s requires the specification of a range of values for γ, r_y, k, and m. For the purposes of this paper, it is assumed that a sufficient similarity exists between the recreational opportunities of the Big South Fork and the Hells Canyon such that the values employed by Krutilla and Cicchetti are appliable to this study. In general,

Table 16.4 Initial Year's Preservation Benefits (Growing at the Rate s) Required in Order to Have Present Value Equal to Development

r_y	$i = 8\%$ $\gamma = 7.5\%$ $k = 25$ years	$m = 50$ years $\gamma = 10\%$ $k = 20$ years	$b_d = \$14,732,000$ $\gamma = 12.5\%$ $k = 15$ years
0.04	\$109,875	\$86,730	\$84,715
0.05	69,582	55,911	56,302
0.06	38,255	31,526	32,811
r_y	$i = 9\%$ $\gamma = 7.5\%$ $k = 25$ years	$m = 50$ years $\gamma = 10\%$ $k = 20$ years	$b_d = \$13,906,000$ $\gamma = 12.5\%$ $k = 15$ years
0.04	\$148,457	\$115,816	\$110,462
0.05	102,160	80,685	78,899
0.06	64,751	52,063	52,577
r_y	$i = 10\%$ $\gamma = 7.5\%$ $k = 25$ years	$m = 50$ years $\gamma = 10\%$ $k = 20$ years	$b_d = \$13,172,000$ $\gamma = 12.5\%$ $k = 15$ years
0.04	\$190,127	\$147,255	\$137,624
0.05	138,434	108,047	103,164
0.06	95,332	75,333	73,727

Source: see text.

these values are based on limited available evidence regarding ". . . the growth in demand for primitive recreation generally, and the income elasticity and related phenomena for this type service." (Krutilla and Cicchetti, 1972, p. 12.)

The required initial year's preservation benefits calculated according to the procedure outlined are presented in Table 16.4. Taking, for purposes of illustration, an i of 9%, γ of 10%, and k of 20 years, m of 50 years, and r_y of 0.05, that is, \$80,685, then in order to justify preserving the Big South Fork, the actual initial year's preservation benefits must equal or exceed this amount. This figure must be compared, then, with the results in Table 16.5, where quantifiable benefits from the existing recreational use of the Big South Fork are presented. The estimates of current visitor days are those contained in the Interagency Report (1970, p. 73). The dollar values attached to the respective visitor day estimates are essentially lower bounds on average willingness to pay for the types of recreational experiences provided in this environment (Fisher, Krutilla, and Cicchetti, 1972, p. 37).

The estimated actual initial year's preservation benefits in Table 16.5

Table 16.5 Current Recreational Use of the Big South Fork, Cumberland River

Type	Visitor Days	
General Use	40,000 at $5.00/day =	$200,000
Fishing	6,500 at 5.00/day =	32,500
Hunting	3,800 at 10.00/day =	38,000
Total Quantifiable Benefits		$270,500

Source: see text.

appear to be more than three times larger than necessary in order to have a present value equal to or exceeding that of the development alternative. This excess of benefits is, on economic grounds, apparently sufficient to justify the prohibition of coal surface mining in the Big South Fork basin in order to preserve the value of the gorge for outdoor recreation. It must be remembered, however, that the derivation of this result is based on a number of simplifying assumptions that may or may not be justified. Thus pending further analysis, these results are offered only as preliminary and illustrative of the analytical techniques involved.

16.5. Summary

In general, two land-use problems potentially are associated with coal surface mining: (1) external costs of production represented by adverse environmental impacts on neighboring sites, and (2) discrepancies between private and social evaluations of the opportunity costs of the development of land for coal surface mining. This paper addresses the latter problem, particularly as it may lead to the preservation of less than the socially optimal amount of natural environments.

Based on previous research on the evaluation of environmental resources, an analytical framework is presented by which the respective net benefits of alternative and incompatible land uses can be compared. Once developed, this framework is applied to the basin of the Big South Fork of the Cumberland River, an area possessing both strippable coal reserves and environmental features of value for scenic river recreation. Representing the net benefits of development by surface mining as the difference in cost between surface and deep mining, and the net benefit of preservation as the gross benefit of recreation, the analysis suggests preservation to be the preferred land use in the basin. This result is preliminary, however, pending the outcome of continuing research in progress.

Notes

1. For a fuller discussion of the potential benefits from the preservation of wilderness areas, see J. V. Krutilla (1967).

2. The arguments presented in this section are principally those as first developed by J. V. Krutilla and C. J. Cicchetti (1972).

3. See, for example, U.S. Department of the Interior (1967).

4. For example, the Tennessee Department of Conservation has imposed a moratorium canceling the issuance of coal mining permits in the watershed of the Obed River pending completion of a National Wild and Scenic River Study.

5. See Krutilla (1967) for examples of other preservation values that can be irreversibly lost by the development of natural environments.

6. For a more complete discussion of this point see Krutilla, ed. (1972, p. 6).

7. In support of assumption (1) it can be noted that 74 percent of the coal shipments in Tennessee went to electric utilities; see U.S. Bureau of Mines (1972, Table 41, p. 58). The growth rates employed in assumption (2) were computed from National Economic Research Associates (1972, Table 14, p. 29). The percent of coal produced by surface mining is that for the nine counties in the basin in 1970; see U.S. Bureau of Mines (1972).

8. The assumed depletion of strippable reserves is based on the assumption that no substantial changes occur in the economic stripping ratio used to compute the reserves. For a discussion of the general procedures used in computing strippable reserves as defined in this study, see U.S. Bureau of Mines (1971, pp. 1–13).

9. Little usable information is available on the actual effect on costs of either of these pieces of legislation. An increase in deep mining costs attributable to the Coal Mine Health and Safety Act of 20% is used in Table 16.2, a figure currently quoted in the trade press and privately communicated by government officials. The $0.50/ton increase in costs due to reclamation is similarly estimated, and is equivalent to an expenditure of approximately $300 per acre disturbed.

10. For a discussion of how the optimal capacity might be determined, see Fisher and Krutilla (1972, pp. 115–141).

11. For a detailed presentation of the functional relationship between these variables and the growth rate of annual preservation benefits, see Krutilla and Cicchetti (1972, Appendix B, pp. 25–29).

References

Brooks, D. (1966). "Strip Mine Reclamation and Economic Analysis," *Natural Resources Journal, 6,* pp. 13–44, January.

Crecine, John P.; Davis, Otto A.; and Jackson, John E. (1967). "Urban Property Markets: Some Empirical Results and the Implications for Municipal Zoning," *Journal of Law and Economics, 10,* pp. 79–99, October.

Fisher, A. C., and Krutilla, J. V. (1972). "Determination of Optimal Capacity of Resource-Based Recreation Facilities," in J. V. Krutilla, ed. (1972), *Natural Environments: Studies in Theoretical and Applied Analysis,* The Johns Hopkins Press, Baltimore, pp. 115–141.

Fisher, A. C.; Krutilla, J. V.; and Cicchetti, C. J. (1972). "Alternative Uses of Natural Environments: The Economics of Environmental Modification," in J. V. Krutilla, ed. (1972), *Natural Environments: Studies in Theoretical and Applied Analysis,* The Johns Hopkins Press, Baltimore.

Interagency Report (1970). *Big South Fork, Cumberland River, Kentucky and Tennessee,* Interagency Report to the Committee on Public Works, U.S. Senate by U.S. Army Corps of Engineers, U.S. Department of the Interior, and U.S. Department of Agriculture, U.S. Government Printing Office, Washington, D.C., pp. 68–69.

Krutilla, J. V. (1967). "Conservation Reconsidered," *American Economic Review, 57,* pp. 777–786, September.

———, ed. (1972). "Introduction," in *Natural Environments: Studies in Theoretical and Applied Analysis,* The Johns Hopkins Press, Baltimore, p. 2.

———, and Cicchetti, C. J. (1972). "Evaluating Benefits of Environmental Resources with Special Application to the Hells Canyon," *Natural Resources Journal, 12,* pp. 1–29, January.

National Economic Research Associates, Inc. (1972). *Fuels for the Electric Utility Industry, 1971–1985,* Edison Electric Institute, New York, p. 95.

Spore, R. L. (forthcoming 1973). "The Economic Problem of Coal Surface Mining," *Environmental Affairs,* Vol. II, No. 4.

———, and Nephew, E. A. (in progress). *Economic Alternatives in Landscape Modification by Coal Surface Mining,* ORNL-NSF Environmental Program Report.

U.S. Bureau of Mines (1971). *Strippable Reserves of Bituminous Coal and Lignite in the U.S.,* U.S. Government Printing Office, Washington, D.C.

——— (1972). *Coal—Bituminous and Lignite,* preprint from the *1970 Bureau of Mines Minerals Yearbook,* U.S. Government Printing Office, Washington, D.C.

U.S. Department of the Interior (1967). *Surface Mining and Our Environment,* U.S. Government Printing Office, Washington, D.C.

17 Coal's Role in the Age of Environmental Concern*

RICHARD L. GORDON†

In discussions of energy problems, it has been conventional for at least half a century to argue that coal is the only fuel amply available enough to cover future needs. Despite the continued failure of this faith to be justified, it has reemerged with renewed vigor. My contention here is that the argument remains as dubious as ever. A decline in coal output seems more likely than the explosive growth so widely forecast.

In viewing the conventional arguments about coal, it is well to recall Ambrose Bierce's definition of self-evident as "evident to one's self and to nobody else." The idea that it is self-evident that the world must turn to coal is, in fact, physical and economic nonsense. The physical supply of energy that we know how to use far exceeds the amounts humanity will consume throughout its existence. The choice among them is a simple question of comparative costs. In my technical innocence, I used to think that technical breakthroughs were needed before some of these sources could become physically available. Recent discussions of the technology make clear that the problem is entirely economic. Technologies exist but they cost too much.

We have been inundated lately with a plethora of suggestions about how to secure essentially unlimited energy supplies—be it directly from the sun, by thermonuclear power, the earth's internal heat, and ocean heat differentials, to name a few. To be sure, many of the concepts seem inherently absurd, and certainly the particular schemes proposed often are unappealing. I am not very excited about the idea of plantations to grow firewood or of giant satellite collectors to feed massive ground stations. Nevertheless, I suspect that ultimately, albeit later than some enthusiasts predict, one or more of these technologies will become economically viable as they are improved.

* Research support from the National Science Foundation and Resources for the Future, Inc., is gratefully acknowledged. The paper in part summarizes writings in various stages of completion. These include "Major Factors Affecting the Coal Market at Present," in *Symposium on Coal and Public Policies,* Center for Business and Economic Research, University of Tennessee, 1972, "Coal's Role in a National Materials Policy" (delivered at the University of Texas and to be published by it), and a book-length manuscript now under review. Other portions of the research discussed are still in progress. Useful comments on this manuscript were provided by John D. Ridge and John E. Tilton.

† Professor of Mineral Economics, the College of Earth and Mineral Sciences, The Pennsylvania State University.

From this analysis falls the conclusion that coal is not some fabulous race-horse against which no one will enter. We have a crowded field, and coal must beat all the rivals. Victory here means an ability to be the fuel cheapest to use.

Once we recognize these facts, the arguments necessary to make a valid case for coal become more apparent. We must be able to prove that coal—currently far more expensive to use than its rivals—will shortly become superior. Three types of supply developments may be distinguished: those of coal's present rivals, those for new rivals, and those for coal. The optimistic view of coal assumes a deterioration of the position of coal's present rivals long before any of the potential rivals can surpass coal.

The counterhypotheses that I would like to present are that a combination of long-continued superiority of the present rivals and ultimate development of new alternatives and continued deterioration in coal's position will produce the decline already mentioned. Following the standard economic principle of comparative advantage, I will dwell most heavily on the coal element of the argument. Professor Adelman (1972) has effectively characterized the petroleum side of the analysis and I shall rely heavily on his arguments for support.[1] The prospects for new technologies defy conclusive analysis; research is risky because we cannot be sure what it will produce. However, Professor Adelman's contentions about petroleum suggest it will remain the economically superior fuel long enough to permit an enormous amount of new technology to be developed. Given this time span and the attractiveness of the alternatives, it seems highly probable that several will become cheaper to use than coal.

My research of the last several years has provided insights into the coal situation. The work began with a study, for Resources for the Future, Inc., on coal's market position. This was followed by a still ongoing project on industrial fuel markets, supported by the National Science Foundation, which included an appraisal of coal supply trends and the development of computer models that among other things permit appraisals of the implications of my supply analysis to coal's position in the critical electric power market. Further insights are coming from another current project—an appraisal of the environmental impacts of energy production and use that my colleague John E. Tilton and I are conducting as part of Resources for the Future's study for the Energy Policy Project of U.S. energy supply prospects.

In his book on the world petroleum market (1972), Professor Adelman established that the ideal solution to the nation's energy problems would be a more competitive world oil market, that is, that the producing countries' efforts to raise prices must fail. He suggested, moreover, that such increases in competition were attainable if the consuming countries improved their oil policies. He has elsewhere noted that Arctic oil and gas developments might produce large amounts of low-cost fuels; others have been willing to speculate that these North American resources might make it economically unnecessary to rely on Eastern Hemisphere oil. From these arguments we deduce that oil will be a more attractive energy source than most observers contend.

The potential role of natural gas is more difficult to appraise. Just how responsive natural gas supply will be to price increases is unclear. However, it is quite apparent that the present shortage is severely aggravated by the field price controls administered by the Federal Power Commission. Many observers of gas prospects have confused the effects of these controls with basic changes in the market and at least exaggerated the inherent difficulties in securing gas.

It may be added that my current work suggests that more explicit consideration of environmental issues, if anything, strengthens the case for petroleum. Natural gas emerges from analysis as the fuel with by far the fewest environmental problems. Oil at worst is a good match environmentally for its closest rival—nuclear power. Environmental problems in oil production, distribution, and processing seem particularly susceptible to satisfactory control at lesser cost than comparable problems with other fuels. For example, the most significant oil spill problem—routine discharges from tankers—has been materially reduced by introduction of new techniques, and further improvements are feasible. The extensive environmental review of the Trans-Alaskan pipeline seems to have reduced the hazards to a tolerable level.[2]

In looking at problems of oil use, the most serious difficulty is with the automobile, and here rival fuels are unlikely to provide a superior alternative for some time. In uses in which other fuels can compete, oil has some significant advantages. It is very difficult to appraise the comparative environmental costs of oil and nuclear power, given our uncertainties about the optimal approaches to waste heat disposal, nuclear accidents, and the storage of nuclear wastes. The most reasonable conclusion, however, is that even if the nuclear problems can be resolved fairly cheaply, nuclear power would find it difficult to compete with cheap oil, even when both bear their full environmental costs.

With these points in mind, we may turn to coal's position. My work on coal has involved continued discovery of its enormous and growing drawbacks. It is hard to imagine how one fuel can have so many disadvantages to its competitors. Actually, all these problems seem to be reflections of one proposition: coal is a solid fuel containing large proportions of impurities. This makes coal harder to mine, transport, process, and use. Solidity also has proved to have increased coal's environmental problems.

Some rough perspective is in order here. First, environmentalism did not emerge recently full-grown as Athena from the head of Zeus. The concerns have been around for centuries. However, a quite mixed impact resulted from these concerns. Some problems were subjected to vigorous efforts at control while others were ignored. The determinants of choice were complex. By and large, stress was upon the most conspicuous and clearly damaging impacts. Presumably, the ease of abatement may also have had an impact in causing stress on more easily controlled problems. In any case, on balance, by the 1960s coal was way behind its competitors in facing controls. Although the defense in depth concept of nuclear power development has been overpublicized to a degree that often inspires glib cynicism, this should not obscure the substantial achievements. It is less well recognized that the petroleum industry also has been subjected to extensive environmental regulations.[3]

Second, in what follows, I act as an interpreter rather than an advocate of prevailing policies. I, in fact, do believe that additional environmental controls and other regulations of coal mining are desirable. However, it is by no means clear that the principles are being sensibly implemented.

Quite the contrary, those supporting environmental policies have quickly attained the incredible feat of matching the energy industries in inept analysis. This has led to excesses in many areas and threats of worse to come. For example, the attacks on strip mining go beyond cure of side effects to restoring the land to its original use even if this is unjustified. A critical overall error needs particular note. It is conventional for proponents of controls on coal to argue that the impacts will be minor. In fact, when properly costed, these regulations imply that coal use is so expensive to society that the industry may be regulated out of existence. This may be a correct appraisal but it is clearly one of such profound impact that it long since should have been explicitly stated instead of being continually denied.

Third, even granting some moderation in environmental pressures, coal will remain in trouble. Its nonenvironmental difficulties are likely to increase enough to more than offset any reasonable changes in public policy.

Coal's present problems arise from numerous sources, and it is extremely difficult to allocate responsibility among the various influences. The evidence suggests that the basic market conditions for coal mining were heading naturally toward sharp changes. Until the late 1960s, forces favorable to stable mining costs prevailed. Improved techniques in both underground and strip mining and the ability to shift into the more productive strip-mining method combined with steady declines in employment that may have moderated wage demands to keep costs level. The introduction of reduced rates for trainload shipments (the unit train) sharply lowered delivery costs.

Considerable evidence exists that market forces were arising to reverse these conditions. Concern was widespread that technology could no longer continue to improve as rapidly as it had up to the middle 1960s. The shift to more mechanized methods of coal cutting and loading had gone as far as practicable, and the need had arisen for another radically new technique such as a faster way to haul coal from underground mines. This particular development is crucial because mining machines can cut much faster than the coal can be removed. Strippable reserves near major markets were becoming more difficult to find. Death, retirements, relocation, and industry expansion had depleted the reserve of mine labor. Higher wages were needed to attract newcomers to the industry. The benefits of unit train rates were eroded by steady rail rate increases and unreliable service.

To all this was added the pressure of new regulations. Coal mining costs apparently have been markedly affected by the Mine Health and Safety Act of 1969. The various provisions of the Act require purchase of new equipment, radical changes in practice, and frequent inspections. Control of the other problems with mining, such as land scarring by strip mines, underground mine subsidence, and acid mine drainage, do not yet seem to have had significant influence. In part this is because thus far the controls have been less stringent, but it appears that within broad limits more severity might not have major impacts.

The critical measure of these impacts is the cost of steam coal F.O.B. mine sold on long-term contracts. The price of this coal largely reflects long-run trends and is less sensitive to short-run market fluctuations. East of the Mississippi (Eastern coal hereafter), a typical 1969 price level F.O.B. mine was around 18 cents a million Btu's; the 1972 level was probably close to 30 cents. This rise is the result of all the influences noted here and also perhaps of general labor unrest. For example, labor costs per ton of production have risen because wages have risen and output per man has declined. This decline involves the effects of worker inexperience, the loss of momentum caused by wildcat strikes, and the extra work needed to comply with the Health and Safety Act. No one has conducted controlled experiments to determine the relative contribution of each factor, and the allocations made by coal companies often are heavily influenced by the degree to which the company dislikes the Health and Safety Act. These pressures were aggravated by rising equipment and material prices and the equipment requirements of the Mine Health and Safety Act.

Moreover, the most reasonable prognosis is that Eastern coal prices will continue to rise substantially, at least in this decade. To be sure, some of the 1969 to 1972 cost rise is due to temporary factors. However, reversal of these pressures probably will be offset by other forces. Wages will continue to rise rapidly. Outputs per worker are likely to show a slow rate of increase in relation to 1969 levels. It will become increasingly difficult to find attractive strippable reserves. More stringent regulation would clearly worsen the situation.

To make matters worse, sulfur emission regulations at best will make use of Eastern coal extremely expensive at least in the 1970s. It has become widely recognized that perfecting a stack scrubbing process has proved extremely difficult. Many observers contend that the wrong route was chosen and we should shift to perfecting precombustion treatment such as by solvent refining or producing a low Btu gas at the power plant. It appears likely that stack scrubbing will increase the cost of Eastern coal use by 40 cents or more per million Btu's treated, and it is still not clear when a satisfactory process will be perfected. Pretreatment may be somewhat cheaper but is unlikely to be commercially available in this decade.

All this adds up to cost pressures that appear far greater than those faced in the use of rival fuels. It seems probable that nuclear plants will be built in preference to coal-fired plants. Indeed, the evidence suggests that movements in this direction are well under way. TVA and Commonwealth Edison long ago made heavy nuclear commitments albeit for reasons quite different from those that now suggest the choice was appropriate. A failure to anticipate massive increases in nuclear plant costs was offset by failure to anticipate coal's problems. More recently, we have seen companies with particularly favorable relative coal positions, such as Union Electric and Illinois Power, moving toward nuclear power.

This incidentally may prove to have been the incorrect choice. Oil may prove to have been the best alternative. Unfortunately, given present U.S. policies, the electric utilities cannot be confident that oil prices can be kept at levels that make it cheaper than nuclear power.

Given the long nuclear plant lead times, and the large amounts of coal-burning capacity existing or on order, nuclear power's contribution in this decade is severely limited. However, we need not conclude from this that high levels of coal use will continue. Sufficiently severe environmental standards will force conversion to oil. Indeed, we have already seen utilities in Illinois and Michigan announce oil-fired plants. Others may take similar steps and even convert old coal plants to oil. Thus, the prospects for Eastern coal in this decade appear dismal at least unless a massive backdown on environmental regulations occurs.

Western low-sulfur fuels could perhaps displace Eastern coals. My estimates show that it is quite conceivable that these coals could be burned more cheaply than oil in such states as Minnesota, Wisconsin, and Illinois. However, this requires that all the uncertainties are resolved in Western coal's favor. First, it is clear that expansion of Western coal output will raise prices above the extraordinarily low level—15 cents a million Btu's in some cases—prevailing for early developers of the most favorable deposits. The key question then is how high prices will rise. Second, considerable uncertainty prevails about the sustainable long-run cost of rail transportation to major markets. Third, while the best Western seams are so rich that reclamation that is highly costly per acre has a small impact on the cost per ton, a strip-mining ban would have catastrophic economic impacts. Fourth, should world oil prices decline and import controls be liberalized, oil could easily become the most attractive alternative.

I am presently directing work on computer programs designed to quantify the implications of my general conclusions for coal demand. Stress is upon the electric power market. We have chosen to substitute greater than usual disaggregation for complex analytic techniques. Aggregation often obscures what are essentially quite simple points. In the electric power industry, the plant decisions are made years in advance, and these choices constrain the short-run fuel choices that can be made. If we know these constraints and the fuel prices faced by a particular utility, specification of the fuel mix that will most efficiently meet given demand becomes a matter of mere arithmetic. Disaggregation also provides a simple way of handling capital investment. If we are willing to specify a given requirement for base-load plants and a probable operating pattern expectation, we can calculate the cheapest plant alternative for the utility (of course, problems do arise in ensuring consistency between actual and expected operating rates but this problem may be solvable by an iterative approach).

At present, we have been able to make only a few test runs with the model. Many further refinements are necessary. Nevertheless, the results of the test runs are so striking that they are included here (Tables 17.1 and 17.2). We contrast a "standard" forecast of electric utility fuel use—that is, one that assumes that coal will remain strong—to one based on the most unfavorable plausible assumptions about coal. Since these assumptions imply the oil will be cheaper to use in the Eastern and Midwestern markets examined in the runs, we naturally get a near disappearance in coal use.

Table 17.1 Projected Shares of Critical Fuels in 1980 Electric Power Generation (percentage of total)

Federal Power Commission Region	"Standard Forecast" Coal	Oil	Nuclear	Pessimistic Case for Coal Coal	Oil	Nuclear
I	7.2	36.4	51.9	0.1	43.0	52.4
II	79.5	1.8	17.4	5.5	71.3	21.8
III	56.1	5.4	33.8	3.9	50.0	41.1
IV	70.2	3.4	25.7	10.0	56.8	32.4

Table 17.2 Projected Coal Consumption in 1980 Electric Power Generation (million tons)

Federal Power Commission Region	"Standard Forecast"	Pessimistic Case for Coal
I	13.7	0.1
II	175.5	11.0
III	147.4	9.6
IV	106.2	14.1
Total	442.8	34.8

While neither figure is my last word in forecasts, the estimates give us another indicator of the uncertainties surrounding coal's future. Further runs on the model should provide a greatly improved evaluation of these problems, and I hope that such work will be completed rapidly.

It may be noted that these views are hardly controversial in the coal industry.[4] It generally expects Eastern coal to be squeezed out of the electric utility market and expects only modest shipment of Western coals to Midwestern markets. The industry, however, hopes for an eventual vast expansion because of widespread synthesis of oil and gas from coal and perhaps the development of pretreatment techniques to permit recapture of some sales to the electric power industry. It is clearly recognized, however, that the required new technologies will not be available commercially until the 1980s.

The critical question in all cases again relates to the comparative economics of the alternatives. At least six relevant cases might be considered. There are three basic options: gasification, liquifaction, and pretreatment for boiler fuel use, and analyses for Eastern and for Western coal should be made for each option.

We have already seen the differences between Western and Eastern coal supply conditions. Prevailing industry views suggest that the strength in the West is most dependent upon developments such as strip-mine regulations over which little control is possible, while improved underground mining technology might hold down the rise in Eastern costs. Even so, Western coal at the mine may be much cheaper. However, it will face two disadvantages. First, transportation costs to the East will be higher than from Eastern mines. Second, water costs may be much higher. (This is ineptly described in the literature as a problem of shortages. Economic theory suggests that a "shortage" is a reflection of excessively low prices, and the literature on water problems makes abundantly clear that low prices are precisely what are involved.)

In any case, the three new basic approaches will be viable only if their total costs are low enough. Gasification has been stressed because the cost of gas transportation gives coal gas a substantial transportation cost advantage over imports and Arctic gas. Synthesis faces potential strong competition from gas produced in the contiguous 48 states if field prices are deregulated. With sufficiently low oil prices, oil gasification would also be a strong competitor. This development could also produce additional competition for coal in power plants. Proposals have been made to build simple refineries whose products are essentially naphtha for gasification and fuel oil for electric utilities and other industrial fuel consumers. Even without such a development, low oil prices would constitute a threat to new processes for using coal in electric power plants. Nuclear power also remains a strong threat. Liquifaction's position obviously would be untenable if cheaper oil became available. Clearly then, the euphoria about new technology is at best excessive and may be misplaced.

It should be noted, however, that these radical differences with conventional view, in practice, may have only marginal impacts on my short-run policy proposals. I, in fact, concur with suggestions for significant increases in energy research including that related to fossil fuels. However, I believe the justification and the implementation should be quite different from that usually stated. We need the research to determine which alternative is best. A few million spent in comparing coal conversion to breeders could save billions in the choice of the inferior option, whatever it might prove to be. Information, even if it is bad news, is a valuable commodity. This stress on learning more about which the true winners and losers may be carries with it the implication that we should be prepared to stop research once the facts are sufficiently clear—a point I suspect we may have reached with the fast breeder reactor.

A more general observation seems appropriate here. We seem to have gotten our view of technology completely backward. Much of the current discussion assumes that we can secure short-run miracles and long-run fiascoes. The environmentalists think stack scrubbing can be produced overnight; the engineering community expects to solve the difficult coal problems just outlined. Those suggesting that someday one of what I called essentially inexhaustible energy sources will become available at a low price often are denounced as visionaries. It seems to me that the truth is that rapid breakthroughs are hard to achieve but long-term success is quite likely. It is this view of technology that causes me to reject the conventional views about coal.

Finally, I would like to reflect on where my work is heading. I hope this year to complete the first series of runs on the electric power model and to complete and manipulate a simple model of the regional distribution of other industrial fuel uses. The appropriate next steps would be first to integrate my submodels with each other and then tie them into one or more of the national energy models now under construction. At present the linkages to the rest of the energy economy are left implicit and a more specific connection would greatly improve the realism. From this point, I can proceed to examine more fully the consequences for the submarkets being studied of different energy developments.

Notes

1. His 1973 book (Adelman, 1972) provides the fullest presentation of his views; his (and other) appraisals of Alaskan oil, appear in Adelman, ed. (1971).

2. A fuller evaluation of Alaskan oil appears in a paper of mine (Gordon, 1972), delivered at the December 1972 meetings of the American Association for the Advancement of Science; the organizers of the session hope to have the papers published.

3. See National Petroleum Council (1972), especially Volume 2 for a review of the problems and potential solutions.

4. Much of the material here was produced after discussions with leading coal companies.

References

Adelman, M. A. (1972). *The World Petroleum Market,* The Johns Hopkins Press, Baltimore.

———, ed. (1971). *Alaskan Oil,* Praeger, New York.

Gordon, Richard L. (1972). "Alaskan Oil in the U.S. Energy Market," paper delivered at the December meetings of the American Association for the Advancement of Science.

National Petroleum Council (1972). *Environmental Conservation, the Oil and Gas Industries,* 2 Vols., Washington, D.C.

18 Recent Sulfur Tax Proposals: An Econometric Evaluation of Welfare Gains*

JAMES M. GRIFFIN†

There is considerable doubt concerning the appropriate policy tools for dealing with sulfur oxide emissions. Recent proposed legislation such as The Pure Air Tax Act of 1972 calls for a fixed tax rate.[1] Alternatively, several economists, including Baumol and Oates (1971), propose the use of both an ambient air standard and a tax, which the pollution control agency would adjust periodically to achieve target reductions in pollution. At least four major problem areas constrain meaningful comparisons of these alternative policy instruments. Applied welfare analysis is circumscribed both by the lack of reliable estimates of the damage function and by deficiencies in the estimation of control costs. Furthermore, Ayres and Kneese (1969) suggest that second-best considerations are likely to be important, implying an inherent weakness of the traditional partial equilibrium approach to welfare analysis. Finally, while the theoretical literature examines alternative policy instruments, it does so within a comparative static framework, thereby overlooking potentially significant dynamic problems in pollution control.

This paper seeks to demonstrate how deficiencies in control cost estimation and the restrictions of comparative static analysis can be remedied by an econometric model. Specifically, simulations with an econometric model of the electric utility industry (Griffin, 1973) are presented to assess the likely welfare effects of the Pure Air Tax Act of 1972 and the dynamic properties of the Baumol and Oates approach. Study of the electric utility industry is particularly appropriate because in 1970 power plants contributed 54% of the nation's sulfur dioxide emissions.[2]

Theoretically, one should be able to recommend a tax on pollutant emissions which would minimize the sum of the damage costs, the polluters' control costs, and the regulatory agency's administrative costs. Unfortunately, our understanding of these cost relationships is generally not sufficient to allow such a calculation. Considerable effort has been devoted toward determining the damage function of sulfur dioxide and particulate

* This project was partially financed by a Faculty Development Research Grant by the University of Houston. I wish to express thanks to F. G. Adams, George Daly, R. William Thomas, Tom Mayor, and Henry Steele for numerous helpful comments.
† Assistant Professor of Economics, University of Houston.

emissions through analysis of property-value data. These studies generally obtain the expected negative relation between property values and pollution levels,[3] but the estimated coefficients are highly sensitive to the specification of the model. There is also some question as to the shape of the marginal damage function. Ridker and Henning (1967) found the function to be flat while Anderson and Crocker (1970) indicate that it declines over low and intermediate ranges of pollution (Crocker, 1969, p. 70). While recognizing its importance, this paper does not attempt to add to the damage function literature, but rather draws upon existing studies when necessary. Fortunately, the policy conclusions of this study hold for a wide variety of assumptions about damage functions.

In contrast to the attention given the estimation of damage costs, the problems of estimating control costs have been generally overlooked. Perhaps because engineering data are a necessary input, the estimation of control costs has tended to be viewed as strictly an engineering question. For example, in the case of a sulfur tax, the estimation of control costs has been viewed merely as the cost of reducing a fuel of high-sulfur content to low-sulfur content after some specific period of transition.[4] Yet, economic theory would suggest the possibilities of factor input substitution and final-product substitution by the consumer. Since these substitution possibilities are likely to produce substantial and pervasive lagged effects, knowledge of these dynamic properties may offer insight into the problem of dynamic pollution control. Therefore, it is especially important that the econometric model include possibilities for both current and lagged period substitution effects.

Section 18.1 describes the model and modifications necessary to examine the sulfur issue. In Section 18.2, the probable welfare and distributional effects of the recently proposed sulfur tax are estimated for alternative technological assumptions and tax rates. A brief summary is given in Section 18.3.

18.1. The Model

While the electricity model is described in detail elsewhere (Griffin, 1973), nevertheless, some discussion of the model is instructive. In its present preliminary form, the model consists of six estimated behavioral equations, eighteen identities, two linking equations, and twenty-three exogenous variables. The model posits a recursive structure using ordinary least squares as the principal estimation approach, with Almon polynomial distributed lags utilized in three of the six behavioral functions. A 5% t test criterion was used for inclusion of variables that were first determined as relevant on theoretical grounds. In the estimation of the Almon distributed lags, the lag structure that maximized the coefficient of determination was generally selected. The data consisted of annual observations over the period 1950–1970. A complete equation list is given in Griffin (1973).

The *demand block* of the model separates residential from commercial and industrial demands by using two equations. In the first, residential electricity demand per capita is hypothesized to depend on the per capita stock of central air conditioners and on distributed lags of per capita real disposable income and the price of electricity relative to the GNP deflator, both of which lags were estimated using the Almon procedure. The lagged income and price effects were included to indicate that residential electricity demand is a derived demand emanating from the stock of appliances and that decisions to add to these stocks involve substantial lagged effects of income and price. The industrial and commercial demand for electricity in the second equation is cast in terms of kilowatt-hour requirements per dollar of real GNP. This ratio is assumed to depend on the stock of commercial air conditioning relative to GNP and a weighted average of past electricity prices relative to the GNP deflator. In both equations, the lagged price effects are substantial; when converted into mean long-run price elasticities, they are −.55 for residential and −.53 for industrial and commercial demand. These elasticities probably include the effect of interfuel competition between electricity and natural gas because due to multicollinearity it was not possible to separate price effects into interfuel substitution and substitution between electricity and all other goods except natural gas.

Given the two demand components, in equations (1) and (2), it is possible to calculate total kilowatt-hour generation by assuming an exogenous value for net imports, and by utilizing a simple linking equation, which relates transmission losses to the predetermined variables. Kilowatt-hour generation requirements to be met by fossil fuels in equation (8) are then calculated by subtracting from total electricity generation in (7) the exogenous values for net imports, internal generation by industries, and hydroelectric and nuclear generation.

In the conversion block of the model, two equations are estimated which allow the translation of kilowatt-hour requirements to be met by fossil fuels into Btu of fuel required by the electric utilities. The conversion rate is generally measured by the heat rate (HR), which gives the Btu required to generate a kilowatt-hour of electricity. Equation (3) is based on the hypothesis that the average heat rate is best viewed as a vintage capital stock problem in which technical progress is embodied. At any point in time, the heat rate can be changed through new, more efficient capacity and through the effects of relative capital and fuel costs. The effects of variations in capital costs can manifest themselves both in the determination of the design efficiency of new units and in the use rates of alternative vintages as peaking requirements affect the opportunity costs of the various vintages. These latter effects are included by using equation (4) and equation (10) to approximate capital costs, which appear as an explanatory variable in the heat rate equation (3). Fuel costs are calculated in equation (15) being based on exogenous price values for each particular fuel and the appropriate quantity weights, which must be determined simultaneously in equations (5), (6) and (12). Given capital and fuel costs together with exogenous values for capacity expansion, it is possible to determine the heat rate in equation (3) and consequently Btu fossil fuel requirements in equation (9).

The third block of the model determines specific fuel consumption patterns. Equation (5) states that natural gas consumption by electric utilities depends upon the exogenous supply of "low priority" gas, that is, the supply available after meeting basic residential and industrial demand for natural gas. Natural gas purchases are then subtracted from total Btu requirements in equation (11) to give the remaining requirements, which must be met by fuel oil or coal. This requirements variable together with a dummy sulfur control variable (1969 and 1970) and a distributed lag on the relative prices of coal and fuel oil determines coal consumption. The shape and length of the distributed lag indicates that while there is some short-run substitution for plants designed to burn both fuels, the longer-run substitution possibilities are much greater. With the determination of coal consumption, in equation (6), fuel oil purchases are determined, by difference in equation (12).

Over the sample period, it is implicitly assumed that regulatory agencies set rates to cover long-run costs. In forecasting and simulation exercises, this same assumption is made implicitly in equations (16) and (17), where the weights and changes in cost components are measured relative to 1970. This has the effect of endogenizing electricity prices and forcing consistency between factor prices and electricity prices. Also in simulation exercises, equation (14) is utilized to endogenize capacity expansions of fossil-fired plants by assuming they will proceed at the same rate as the output of these plants.

The modifications necessary to consider the sulfur tax question require the addition of only six exogenous variables and equations (19)–(26). In equations (19) and (20), the prices of fuel oil and coal are adjusted to include the cost of sulfur control and the tax paid on the remaining sulfur. These equations require exogenous values for the control costs for each fuel, the incremental damage of a ton of sulfur, the tax rate, and the percent sulfur content remaining in the fuel. Equation (21) calculates total control costs, while equation (22) calculates total taxes. Similarly, total sulfur emissions are estimated, in equation (23). The welfare gain from a policy is calculated in equation (24) as the incremental damage cost of sulfur emissions multiplied by the reduction in sulfur emissions from some control solution. The welfare loss from such a policy is then calculated in equation (25) by the incremental resources devoted to sulfur controls plus the loss in consumer surplus[5] by electricity users facing higher prices. The net welfare gain, which is the relevant variable from a policy perspective, is the difference between equations (24) and (25).

Implicit in the model are most of the substitution effects alluded to earlier. For example, in addition to the substitution of low-sulfur fuel for a high-sulfur fuel, a sulfur tax will affect the price differential between coal and fuel oil, causing an interfuel substitution effect. The model also includes the possibility of substitution between fuel inputs and capital inputs as the fuel conversion efficiency rate depends on the relative prices of capital and fuel. The lagged price variables in both the residential and industrial demand equations imply that consumers will respond to the higher electricity prices by substituting other goods for electricity-intensive goods. However, the treatment of substitution effects is incomplete since interfuel substitution between coal and fuel oil and the other power sources—natural gas, hydroelectric, and nuclear—does not occur in the model. In the case of natural gas and hydroelectric power, this omission is not serious, as supply conditions are likely to restrict their substitutability. Clearly nuclear power will compete with fossil fuel installations and should be endogenized despite the difficulty of doing so; subsequent research will attempt this. In the present application, the omission of this substitution possibility is probably not too serious because the welfare gains of the tax will tend to be underestimated as control costs are reduced by substitution to nuclear units.

18.2. The President's Sulfur Tax

In this section, we consider the welfare and distributional effects of the Pure Air Tax Act of 1972 as they relate to electric utility generation. It is particularly instructive to simulate the model over the period 1971 to 1985 with varying technological assumptions about sulfur control costs and with alternative sulfur tax rates and then to compare these results with a base case solution in which the air is treated as a free good. The relative differences between the alternative simulations are probably much more informative from a policy perspective than the absolute welfare losses and gains, which are conditional upon quite restrictive assumptions.

Estimates of welfare gains and losses, which are calculated in the manner suggested by Harberger (1971), are particularly dependent upon determination of the appropriate damage function. Welfare gains are defined as those resulting from the reduction in sulfur emissions, while welfare losses are the social costs in terms of control costs, losses in consumer surplus, and administrative costs. The net welfare gain is the difference between the two components. This analysis presumes that sulfur control is the least-cost alternative, for example, residents cannot relocate at lower costs. The administrative costs of the tax are excluded from the calculation both for lack of an adequate estimate and for their insignificant effects on the principal conclusions of this section.

Figure 18.1 depicts the welfare losses resulting from the tax, assuming a flat long-run marginal cost function. LRAC denotes the long-run average cost function for electricity generation before the imposition of the tax. After the imposition of the tax, the new long-run average cost function becomes LRAC + TAX + CC as the costs of the taxes (TAX) and control costs (CC) are internalized. While the industry's average costs increase by AC, the incremental welfare loss consists of the control costs (ABEF) and the loss in consumer surplus (DFG). These sources of welfare loss comprise the two shaded areas in Figure 18.1. Of course, these calculations presume that the full amount of the control costs and taxes are reflected in both residential and industrial electricity prices.

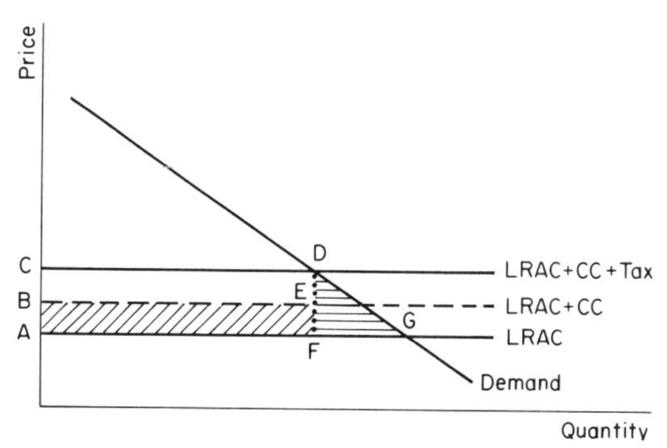

Figure 18.1. Welfare effects of sulfur tax.

Estimates of welfare gains resulting from reduced sulfur emissions are particularly suspect because of the uncertainty surrounding the damage function. Following Ridker and Henning's results, it is assumed that the marginal damage function is flat over the ranges considered in this analysis. Furthermore, a marginal damage cost of \$580 per ton of sulfur in 1971 dollars is assumed as it is the *average* social cost per ton of sulfur emissions implicit in the Environmental Protection Agency's estimate of \$16.1 billion damage from all forms of air pollution.[6] Because of the dubious methodological documentation for this estimate, the reader may want to substitute his own estimate of the damage cost per ton by scaling the calculated welfare gains up or down accordingly.

Before a comparison of sulfur emission profiles for various tax rates can be made, the cost curves for removing sulfur from fuel oil and coal must be specified. In this context, engineering cost data are a necessary input. The process technology for the desulfurization of fuel oil is well known, and the cost data are reliable. Detailed private estimates[7] of the cost function were used which postulate that the marginal control cost function for fuel oil desulfurization is a step function with two steps.[8]

The technology for economically reducing the sulfur content of coal is not well developed. By the process of wet cleaning, high-sulfur coals can be reduced to 1% sulfur, but this appears to have only limited applicability. Flue-gas desulfurization experiments are being conducted with a wet limestone scrubbing process as well as several others, but neither the exact cost nor effectiveness is known. For this reason, the model was simulated under a variety of assumptions, such as "not available" to costs ranging from 8.2 to 16.4¢/10^6 Btu ($2.00 to $4.00 per ton of coal).[9] The most likely case at this point appears to be the simulation at 12.3¢/10^6Btu($3.00 per ton). In all simulations, the process is assumed to remove 87% of the sulfur. While there is the possibility of opening new low-sulfur coal mines, the average sulfur content of coal before control is assumed to remain constant at its 1964 rate (2.3%).[10]

A base case solution to the model was calculated for the period 1971–1985 under the assumption that pollution controls or taxes would not be utilized. This simulation provides a benchmark by which to evaluate the effects of a sulfur tax. Of particular interest is the estimated annual emissions of sulfur in millions of tons based on the assumption of no sulfur taxes or controls.[11] Despite the rapid growth in nuclear capacity, sulfur emissions grow markedly as generation by fossil-fired plants continues to expand.

Given in Table 18.1 are the principal results of nine alternative simulations that vary in their assumptions about the tax rate and the cost of flue-gas desulfurization. With the exception of the tax adjustment and their effects on fuel and electricity prices, the same exogenous variables are utilized as in the base case simulation. In all cases, it is assumed the tax becomes operative in 1976. In addition to a 15¢/lb sulfur tax, alternative calculations are made for tax rates of 10 and 20¢/lb.[12] The cost and sulfur content (after control) for fuel oil desulfurization are similar except in cases 1 and 2, in which under the 10¢/lb tax only the use of the resid desulfurization process (which costs 8¢/lb of sulfur removed) is economic.

Table 18.1 reveals that in all cases the cumulative net welfare gains (*summed over the decade*) are not only positive but of substantial magnitude. The magnitude of these numbers is in large part due to the combination of (1) a large estimate of damages ($580 per ton) relative to the tax ($20 to $40 per ton), (2) the continued rapid growth of sulfur oxide emissions in the base case solution, and (3) the effectiveness of the sulfur-removal processes. The differences between the various cases depend primarily on the availability and cost of flue-gas desulfurization. Net welfare gains are the lowest in cases 1, 3, and 6, in which it is assumed that an economically feasible flue-gas desulfurization process is unavailable. In the cases in which this process is available, the average annual welfare gains over the 10-year period range from $6.5 to $7.7 billion. More important than the absolute size of these net welfare gains is the implication that even

Table 18.1 Simulation Results for Alternative Tax Rates and Control Costs

Assumptions:	Case 1	Case 2	Case 3	Case 4	Case 5	Case 6	Case 7	Case 8	Case 9
Tax rate (¢/lb)	10	10	15	15	15	20	20	20	20
Coal desulfurization (¢/10^6 Btu)	N.A.	8.2	N.A.	8.2	12.3	N.A.	8.2	12.3	16.4
Sulfur content (%)	2.3	.3	2.3	.3	.3	2.3	.3	.3	.3
Fuel oil desulfurization (¢/10^6 Btu)	6.6	6.6	11.9	11.9	11.9	11.9	11.9	11.9	11.9
Sulfur content (%)	.9	.9	.3	.3	.3	.3	.3	.3	.3
Results for 1976–1985 period:									
Cumulative effect (1976–1985)									
Welfare gain (10^9\$)	13.68	91.50	20.04	93.85	94.04	23.68	93.90	94.09	94.27
Welfare loss (10^9\$)	2.54	15.60	4.41	16.53	23.11	5.29	16.51	23.06	29.39
Net welfare gain (10^9\$)	11.14	75.90	15.63	77.32	70.93	18.39	77.39	71.03	64.88
Average annual sulfur emission (10^6 tons)	16.12	2.70	15.02	2.30	2.26	14.39	2.29	2.25	2.22
Percent sulfur reductions	12.7	85.4	18.7	87.5	87.8	22.1	87.6	87.8	88.0
Percent attributable to intrafuel substitution	6.5	81.9	9.8	83.5	82.3	10.8	83.2	82.0	81.0
Average percent increase in cost of electricity	10.8	6.5	15.6	7.2	9.3	19.4	7.9	10.0	12.0
Percent cumulative change in coal use	−7.1	−2.9	−11.2	−3.2	−5.4	−15.1	−3.7	−5.8	−8.1
Percent cumulative change in fuel oil use	3.7	−7.7	14.6	−8.9	−3.3	26.9	−8.5	−2.7	3.0

N.A. = Not available.

if the actual damage rate is one-third the assumed rate of $580 per ton, the net welfare gains in all cases would be positive.

In terms of the physical reduction of sulfur emissions, a comparison of the annual average sulfur emissions with the base case simulation levels indicates marked reductions. For example, if an economic flue-gas desulfurization process is available, the sulfur emission level from electric utilities in 1985 may be less than one-fifth of today's level.

To illustrate the point raised earlier with respect to the possibility of substitution between fuels, between capital and fuel, and between other goods and electricity-intensive goods, the percent of the sulfur reduction attributable solely to the substitution of low-sulfur for a high-sulfur fuel of the same type is reported. The differences between total sulfur reduction and that due to intrafuel sulfur substitution are attributable to the other forms of substitution typically overlooked in analyses of his type. The importance of these three substitution effects varies significantly depending on the technology. In cases 1, 3, and 6, which assume no economic flue-gas desulfurization process, these three substitution effects account for about half the change. In these cases, fuel oil has a decided competitive advantage over coal and induces a substantial interfuel substitution effect. This is apparent from the cumulative change in coal and fuel oil use given in the last two rows. In the remaining cases, the three substitution effects are less important as the reduction in the level of emissions is primarily attributable to intrafuel substitution. This seems plausible, because these substitution effects now have an impact on fuels with a much lower percentage sulfur composition so that their effects are mitigated.

While the net welfare gain may be the critical variable to the economist, the politician is likely to be more concerned with the distributional effects of a tax. Assuming these costs are fully passed forward to residential and industrial users, estimates of the increase in the average cost of a kilowatt-hour of electricity range from 6.5 to 19.4%. If an economic flue-gas desulfurization process is developed, this range is 6.5 to 12.0%. Despite significant distributional effects, the area representing consumer surplus (DFG) is generally small as at most it constitutes $1.18 billion of the $5.29 billion welfare loss in case 6.

As would be expected, sulfur taxes have a pronounced effect on both fuel oil and coal sales as revealed by comparing cumulative sales with the base case solution. The petroleum industry stands to gain a marked competitive advantage over coal only in the cases of high tax rates and the unavailability of flue-gas desulfurization (cases 3 and 6). In all cases, coal sales will deteriorate, but the coal industry is well aware, no doubt, of the long-run infeasibility of the present situation, which is implicit in the base case solution. Thus, from the coal industry's viewpoint, the critical variable is not so much the tax rate but the development of a cheap flue-gas desulfurization process.

18.3. Summary

The extension of econometric modeling to problems of environmental policy offers a potentially fruitful and relatively unexplored area of research. The dynamic framework, which includes various substitution possibilities, appears to offer some useful policy insights despite a number of restrictive assumptions. Over a wide range of marginal damage rates and technological assumptions, the Pure Air Tax Act of 1972 should result in a net welfare gain. However, it does not follow that the proposed tax rate is optimal.

Notes

1. Specifically, this proposal calls for a 15¢/lb tax on sulfur emissions beginning January 1, 1976, for those areas that do not meet the primary standards and a 10¢/lb tax effective in those areas that meet the primary standards but not the secondary standards as set forth in the Clean Air Act. (See U.S. Congress, Senate, February 8, 1972.)

2. For an informative review, see U.S. Congress, Senate (1970, pp. 172–188).

3. See Ridker (1967), Ridker and Henning (1967), Zerbe (1969), Anderson and Crocker (1970), Wieand (1970), and Crocker (1969).

4. For an example, see Hausgaard (1971).

5. Calculated by taking the shaded area DFG in Figure 18.1, which amounts to one-half the product of incremental price and output changes relative to the base case solution.

6. See Council on Environmental Quality (1971, pp. 103–107). This was based on a 1968 calculation that 16.6 million tons of sulfur did $8.3 billion damage, or $500 per ton, in 1968. The $8.3 billion was broken down as follows: $2.3 billion to residential property, $2.2 billion to materials, $3.3 billion to human health, and $13 million to vegetation. (See Chapman, Tyrrell, and Mount, 1971.)

7. Private estimates by a large petroleum refining firm are believed superior to public estimates such as in Hausgaard (1971).

8. On the first step, the resid desulfurization process can reduce the sulfur content of a barrel of fuel oil from 2.5 to .9% sulfur at a cost of 43¢/bbl or 8¢/lb of sulfur removed. To reduce the sulfur content of fuel oil from .9 to .3%, vacuum gas oil desulfurization must be utilized at a cost of 14.5¢/lb of sulfur removed. The process technology to reduce sulfur content below .3% apparently does not exist except by blending low-sulfur crude oils (<.3%), the supply of which is quite limited. Thus, the marginal cost curve is assumed to become vertical at the point where the sulfur content of fuel oil approaches .3%.

9. One of the more recent estimates of the flue-gas desulfurization costs is given in the N.E.R.A. discussion of electric power by the Council on Environmental Quality (1972). According to my rough calculations based on their data, the cost of sulfur removal they are implicitly assuming is $2.30 per ton of coal. This is based on taking 7% of average revenue per kWh (1.7¢/kWh), which is 1.2 mills/kWh. Of this 1.2 mills/kWh, fixed costs for cooling towers and flue-gas processes constitute .75 mill/kWh, the flue-gas component of fixed costs being .6 mill. Combined fixed and variable costs are .95 mill/kWh (.6 + .35 mill per kWh), a figure which implies a fuel cost of 9.5¢/10^6Btu, or $2.30 per ton of coal with standard Btu content.

10. See U.S. Bureau of Mines (1966).

11. The estimated figures are 9.7, 10.7, 11.7, 12.5, 13.3, 14.2, 15.3, 16.3, 17.1, 18.1, 19.0, 20.0, 20.7, 21.9, 22.2 millions of tons in 1971 through 1985.

12. The 10¢/lb calculation was made because, under the Nixon proposal (U.S. Congress, Senate, February 8, 1972), some areas that meet the primary standards would qualify for the lower rate so that on average the average effective tax rate will be between 10 and 15¢/lb. The 20¢/lb tax calculation is made to conform roughly to the Proxmire proposal that would begin with a 5¢/lb tax in 1972 and increment by 5¢/lb each year until it reaches a rate of 20¢/lb in 1975. The Proxmire proposal makes no provision for differential rates depending on the air quality of different geographic areas.

References

Anderson, R. J., Jr., and Crocker, T. D. (1970). *Air Pollution and Housing: Some Findings,* Paper No. 264, Institute for Research in the Behavioral Economic and Management Sciences, Purdue University, Lafayette, Indiana, January.

Ayres, R. U., and Kneese, A. V. (1969). "Production, Consumption, and Externalities," *American Economic Review, 59,* pp. 283–297, June.

Baumol, W. J., and Oates, W. E. (1971). "The Use of Standards and Prices for Protection of the Environment," *Swedish Journal of Economics, 73,* pp. 42–54, March.

Chapman, D.; Tyrrell, T. J.; and Mount, T. D. (1971). "Electricity and the Environment: Economic Aspects of Interdisciplinary Problem Solving," paper presented to National Institute of Social and Behavioral Sciences, Philadelphia, December.

Council on Environmental Quality (1971). *Environmental Quality 1971,* Environmental Protection Agency, U.S. Government Printing Office, Washington, D.C., August.

——— (1972). *The Economic Impact of Pollution Control,* Environmental Protection Agency, U.S. Government Printing Office, Washington, D.C., March.

Crocker, T. D. (1969). "Urban Air Pollution Damage Functions: Theory and Measurement," Report to the National Air Pollution Control Administration, CPA 22-69-52.

Griffin, J. M. (1973). "A Long-Term Forecasting Model of Electricity Demand and Fuel Requirements," Working Paper, University of Houston, March.

Harberger, Arnold C. (1971). "Three Basic Postulates for Applied Welfare Economics: An Interpretive Essay," *Journal of Economic Literature, 9,* pp. 785–797, September.

Hausgaard, O. (1971). "Proposed Tax on Sulphur Content of Fossil Fuels," *Public Utilities Fortnightly, 88,* pp. 27–32, September 16.

Ridker, R. G. (1967). *Economic Costs of Air Pollution,* Praeger, New York.

———, and Henning, J. A. (1967). "The Determinants of Residential Property Values with Specific Reference to Air Pollution," *The Review of Economics and Statistics, 49,* pp. 246–257, May.

U.S. Bureau of Mines (1966). "Sulphur Content of United States Coals," Information Circular 8312, Washington, D.C.

U.S. Congress, Senate (1970). *Air Pollution 1970, Part I,* Subcommittee of the Committee on Public Works, Hearings, 91st Congress, 2nd Session.

——— (1972, January 24). *Congressional Record,* Proxmire bill S 3057, p. S276.

——— (1972, February 8). Nixon Proposal, H 841.

Wieand, K. F. (1970). "Property Values and the Demand for Clean Air: Cross-Section Study for St. Louis," unpublished Ph.D. dissertation, Washington University, St. Louis, Missouri.

Zerbe, R. O., Jr. (1969). *The Economics of Air Pollution: A Cost-Benefit Approach,* Toronto.

19 The Northern Plains Coal Resource—Case Study in Public Nonpolicy

ERNST R. HABICHT, JR.*

19.1. Introduction

Energy scenarios for the future assume greatly increased exploitation of domestic fossil and nuclear fuel reserves. Coal is central to these predictions but has received an insufficient emphasis. Because of long lead times for nuclear development, uncertainty of imported fuel supply, or anticipated price increases in liquid and gaseous fuels, Western coal will be central in meeting new energy demands. A major shift is presently under way from high-sulfur Eastern deep-mined coal to low-sulfur Western strip-mined coal.[1] A sulfur dioxide effluent tax would, of course, greatly accelerate this trend.

Should there be a major slippage in anticipated energy supplies, particularly nuclear energy or oil imports, Western coal development will *have* to take up the slack. The major uncertainties derive from the timing and magnitude of changes in our oil import quota policy, the unpredictability of environmental control technology and regulations, and the arrival of new coal conversion and mining technology. Coal constitutes the bulk of U.S. known total fossil fuel reserves. Many recent energy scenarios, often formulated with the tunnel vision of industrial, project-oriented, regulatory, or environmental bias, have ignored coal development potential as a function of slippage in other fuel supplies. Such neglect is not in accord with the recent activities of the domestic energy companies.

The very name "energy company" has arisen because of the recent tendency of oil companies to buy coal companies and uranium futures. Of the ten largest U.S. domestic coal producers, four are subsidiaries of oil and gas companies; they accounted for 18.7% of our 1971 total coal production.[2] Most significantly, a number of oil companies and oil speculators, having no past association with coal development, figure prominently in recent acquisitions of coal leases, prospecting permits, and water options in the Northern Plains states, especially Montana and Wyoming. Much of this activity is said to be prompted by the intention to use these strippable coal reserves for conversion into other gaseous, liquid, and solid fuels, and appears to have been speculative in part (Department of the Interior,

* The Environmental Defense Fund, Inc., 162 Old Town Road, East Setauket, New York.

1972). Presently, the coal and water resources that have been leased or optioned are tied up in long-term (10- to 20-year) contracts. The energy companies have thus locked up the more attractive resources (especially water) until the 1980s.

I want to focus your attention on a relatively small geographic component of the coal development picture, but one which is large in terms of proved reserves. This is the portion of northeastern Wyoming and southeastern Montana in which repose the Powder River and Tongue River coal deposits. The amount of low-sulfur strippable coal reserves estimated to be at least 80% recoverable in this region is approximately 33 billion tons.[3] At least two-thirds of this is found in seams ranging from 20 to 250 feet thick; the overburden is from 10 to 200 feet thick. The most impressive concentration of this coal is found in the Wyodak Deposit, a 25- to 150-foot thick seam in Campbell County, Wyoming containing strippable reserves estimated to be 19 billion tons.

19.2. The North Central Power Study

With these reserves in mind, the Bureau of Reclamation in May of 1970 sought to hasten industrial development in the Northern Plains states by bringing together representatives of 35 electric utilities in order to probe the technical and economic realities of thermal generation in the Tongue and Powder River Basins. Their joint efforts culminated in the publication of the *North Central Power Study* (NCPS) (U.S. Bureau of Reclamation, 1971), which, after languishing some months, was allowed to sink into obscurity.

The study is of interest, not because it will soon be revived but because it forecast a number of developments that will be executed by individual utilities and energy companies. The study also constitutes a dismal paradigm for cooperation between the federal government and the private sector when truly massive investments are at stake; this will be discussed later in my paper.

Water is the key to industrialization of the Northern Plains; the supply and delivery of cooling water were crucial to the NCPS. The Bureau of Reclamation is the primary supplier of water in Montana and Wyoming. Having exhausted most of the hydropower, irrigation, and recreational potential of the Yellowstone and upper Missouri Rivers, the Bureau has turned to industrial water demands for its mission. The prevailing austere climate for public works projects has, if anything, hastened the Bureau's abandonment of agrarian and public power goals so clearly delineated in its charter, the Reclamation Act of 1902. Most of the developed and potential industrial water supply of the region is in the state of Montana at a distance of 30 to 200 miles from those coal fields capable of supporting large-scale mine mouth industrial development. Thus, the question of water delivery as well as water supply is of great interest to the Bureau of Reclamation.

Both Montana and Wyoming have water laws embodying the key concept that present water use establishes a legally recognized claim for continued use. This perpetuated water right has little or no relationship to stream location. It is also a strong motivation for rapid development and maximum use. Some industrial interests[4] have gone into intensively irrigated farm operations with marginal returns in order to preserve their water rights for the moment when technology and demand render coal gasification or mine mouth generation propitious.

As the legal concept of "first in time, first in right" applies to individuals, so does it apply to states. The slogan "use it or lose it" has been an important weapon in the arsenal of water project construction agencies in persuading Montanans to support a proliferating network of dams and diversions.[5] It is perhaps ironic that a substantial portion of this Montana agricultural water will be diverted into Wyoming for industrial development.

In brief, the North Central Power Study examined the resource potential for mine mouth coal-fired steam electric generation in the Northern Plains states. The preponderance of sites were in the Tongue and Powder River Basins and depended largely on subbituminous coal deposits. These were determined to be capable of sustaining 156,000 MW(e) for at least 35 years. The study detailed a 50,000 MW(e) generation complex operating at an 85 to 93% load factor in conjunction with 3000 MW(e) of pumped storage hydropeaking power. Cooling water consumptive use was estimated to be 28 cfs per 1000 MW(e), or 855,000 acre-feet per year at the lower load factor, assuming no losses in storage or delivery. The study also called for multipurpose transmission corridors requiring 4800 square miles of land.

The power was to be transmitted to load centers to the east, principally Omaha, Des Moines, Minneapolis–St. Paul, St. Louis, and Kansas City, on a largely self-contained grid. State-of-the-art technology was deployed throughout the study. Alternative uses of the coal, socioeconomic impacts, and environmental impacts were not discussed in any depth. The load growth projections of the participating utilities formed the basis of future demands. Water supply projections for the region were formulated by the Bureau of Reclamation.[6,7] Legal and institutional problems were never seriously considered. Liaison with other interests in the region were minimal or nonexistent. The press releases issued by the study group never discussed the project's consumptive water demands, air pollution problems, land requirements for transmission, and socioeconomic impacts. Even if the 50,000 MW(e) project met present federal and state standards, it would still emit annually 2,100,000 tons of sulfur dioxide, between 1,226,000 and 1,879,000 tons of nitrogen oxides, and well over 100,000 tons of particulate matter.

Reviewed by the individual participating utilities after its publication, the study fell from favor. Nonetheless, certain projects delineated in the study are now being implemented independently by some companies. Though some environmental spokesmen were stirred to frenetic dismay, it was not public opinion alone that defeated the study; it had been rejected by many of the participating utilities well before it was released. There are obvious reasons for this rejection of the study—the same reasons that will continue to militate against sound energy supply and environmental planning in the region:

1. The investor-owned utilities or IOUs are eager to acquire the small publicly owned systems in the region. It is senseless for the IOUs to enter into a cooperative venture that strengthens the hand of the very entities the IOUs wish to weaken. In fact, while the report was being prepared, the IOUs and several public systems were making other arrangements with the Bureau of Reclamation, the effect of which would be to deny power supply or wheeling facilities to numerous publicly owned systems in the Northern Plains.[8]

2. The electric utility industry seeks to minimize public scrutiny of its intentions. A cooperative venture with the Bureau of Reclamation presents the utilities with an array of carrots and sticks. The carrot of a firm water supply at a subsidized rate may, for the present, be outweighed by the stick of increased public scrutiny and possible litigation under laws applicable to federal agencies. This mitigates against cooperative ventures despite certain economies of scale. System reliability and supply are presently of insufficient concern to engender national or large-scale regional power grids.

3. While the NCPS was attractive to certain of the participants, some of the utilities in the region refused to participate or were excluded at the outset. Most notable was the unexplained absence of the Bonneville Power Administration. This led to abandoning any plans for an intertie to the Pacific Northwest despite obvious planning elsewhere for such an eventuality.[9] One participant in the study told me privately that a suggestion that the NCPS should be rewritten prior to release (so as to incorporate some environmental cosmetics) was not seriously considered. It almost seems as if some of the utilities sabotaged the study to make it appear to be *both* an economic and an environmental disaster.

4. Few state government representatives participated during the planning sessions or the preparation of the final study. Wyoming, the most development-oriented state in the impacted region, was heavily represented while other more apprehensive state and federal agencies were notable by their absence over the course of the study. Understandably, the study met with outrage from many state officials, irrigation interests, and stockmen in the area, all of whom had been strong supporters of past water reclamation projects. Suddenly, they faced an industrial consumptive use of some 2¾-million acre-feet of water per year, at 156,000 MW(e), in a region where the annual precipitation averages less than 15 inches. The state of Montana was particularly upset with the lack of communication afforded by the Department of the Interior.[10]

5. Though the concept of the NCPS originated in the Bureau of Reclamation, even a casual reading of the full report reveals that the agency rapidly lost control of the direction of the study. The agency's most innovative ideas were buried under the parochial interests of the influential utilities. The Bureau urged the logical incorporation of a northwestern intertie, but to no avail. Coordination with coal conversion technology fizzled as the large utilities began to dominate the study. The Bureau was, of course, primarily interested in the hydropower and water supply aspects. Such project orientation almost certainly caused it to become increasingly timid about influencing policy matters as the study progressed.[11]

I have painted a dismal picture of the joint planning effort of government and utilities; let me hasten to add that the energy companies, armed with coal leases and water options, offer no brighter outlook. Indeed, I look at the NCPS as a regrettable model for future planning and cooperation. An optimistic federal bureaucracy, seeking joint ventures with the energy industry in the West, is not likely to act in the public interest until a number of institutional reforms take place in both the private and the public sectors.

19.3. The Energy Companies in the Northern Plains

On September 25, 1972, the president of the National Coal Association, a former member of the Federal Power Commission, addressed an attentive audience in Billings, Montana (Bagge, 1972). He praised the big coal producers for their endeavors and said, "I am quite comfortable in representing the responsible segment of the coal industry which has not only endorsed but embraced conservationist ideals" (Bagge, 1972). Then, in what must be one of the most memorable understatements of the post-Coolidge era, he said of strip mining and reclamation, "On the sparsely-vegetated hills and plains of the West, we never promised a rose garden." At the same Billings meeting, one stockman remarked that, given the amount of fertilizer and water some coal companies were using on their reseeded reclamation demonstration plots, he could grow alfalfa in the bed of his pickup truck.

The prospect of coal mining certainly seemed to be no "rose garden" to those ranchers in the region whose mineral rights had been retained by the federal government when the Northern Plains were homesteaded. In a letter to one rancher, a Westmoreland Resources' legal representative had recently written (Ennis, 1972):

> $137 per acre for your entire surface with a leaseback on reasonable terms constitutes Westmoreland's final offer. This offer will terminate on March 3, 1972 and so you must take positive steps to accept it on or before that date.
>
> Absent such positive steps or acceptance, Westmoreland will have no alternative but to institute an action against you to condemn the surface necessary for its mining operations.

Some months later, Dell Adams, vice-president of the Western Division of the Consolidation Coal Company, wrote the Northern Cheyenne Indian tribe in the Tongue River Basin of southeastern Montana detailing Consol's offer to enrich the tribe with coal gasification plants (Adams, 1972):

> There are several reasons why we feel that it is proper and prudent for the Northern Cheyenne to negotiate a prospecting permit with option to lease to Consol instead of asking for competitive bids. The obvious problem to Consol of going to competitive bids would be the loss of several months that would be required to properly advertise and set up the sale. Due to pressure being exerted on Consol by gas users and transmitters, Consol

will be forced in the near future to locate the gas project, which has previously been outlined to the tribe, on a large body of coal. If Consol cannot conclude negotiations with the Northern Cheyenne Tribe at an early date, Consol will be forced to take this project elsewhere. If it becomes necessary to do this, this project will be lost to the Northern Cheyenne and it may be a long time before a project of this magnitude comes again, if ever.

The Indians have provided the traditional soft underbelly for industrialization of the West. The Navaho and Hopi served this function with the Southwest Energy Development, now slated to provide some 36,000 MW(e) of coal-fired capacity—and heavily dependent on cheap Indian land, coal, and water. Both the Northern Cheyenne and the Crow in southeastern Montana have been under great pressure to relinquish their coal and valuable water rights to the energy and primary metal companies. Perhaps in this context it is prophetic as well as historical to note the remarks of an anonymous Indian reflecting upon the white man (Brown, 1971):

They made us many promises, more than I can remember, but they never kept but one; they promised to take our land, and they took it.

The Northern Cheyenne Reservation is entirely underlain with coal, much of which is easily strippable. These Indians possess significant water rights (which have never been adjudicated) and, lacking any natural resources code or reclamation law, are a prime target for coal development. Of the reservation's 415,000 acres, some 227,000 acres are already under prospecting permits and 16,000 acres are under lease to Peabody Coal (a subsidiary of Kennecott Copper).

Suspicious of the intense interest in their coal and water rights, the Northern Cheyenne Tribal Council rejected Consol's bid and in late 1972 turned to the Native American Rights Fund and other counsel for legal assistance in preparing an adequate tribal natural resources code. This is an unprecedented decision for an Indian tribe and should afford them more protection than provided by the Bureau of Indian Affairs (BIA), another Interior agency that has been imbedded in some remarkable institutional conflicts of interest.[12]

19.4. The Federal Partners

The "pressure" of which Adams wrote, and which the Reddings and the Northern Cheyenne felt in 1972, had already struck receptive antennas. In

November of 1969, Harold Aldrich, the Regional Director of the Bureau of Reclamation in Billings, Montana, spoke of the demand for industrial water attendant to coal development (*Proceedings of the Montana Coal Symposium,* 1969, p. 89):

The Bureau is not in the synthetic fuel business, but it is well aware of the confidence existing among these companies. It is a fact that we have already executed option contracts or have received contracts and applications for 623,000 acrefeet of water available from the Bighorn River. The interest shown in obtaining Bighorn River water for industrial use, in connection with development of the coal resources of northeastern Wyoming and southeastern Montana is—to say the least—amazing.

* * * *

All of this development is not only possible, but quite probable. It can come about—and quickly. Undoubtedly, it is going to require unified action between private industry and State and Federal entities. Money and cooperation will be prerequisites—but it can and should be undertaken.

Fulfillment of Aldrich's vision might require substantial revision of the Reclamation Act of 1902. Alternatively, the Bureau could continue to ignore the law as it has done in the issuance of a flurry of industrial water option contracts. Some 400,000 acre-feet of Bighorn River water has been committed in a series of 10-year options to ten energy companies *since* the passage of the National Environmental Policy Act (NEPA). *Not one* of these contracts has been accompanied by an environmental impact statement. A total of 708,000 acre-feet of this water has been optioned by the Bureau (Armstrong, 1972), but the options have not yet been exercised, thus rendering the contracts susceptible to litigation under the NEPA as well as other federal statutes and state water laws. Indian water rights in the region have never been properly adjudicated. The low prices for this industrial water ($9.00 to $11.00 per acrefoot) are startling and indicate a need for substantial alterations in the Bureau's policy.

Noncompliance with NEPA seems to be characteristic of Interior agencies in the Northern Plains. The BIA has steadfastly refused to comply with the Act. It is a certainty that the coal- and water-rich Indians would be far better off had the BIA subjected Indian leases and permits to NEPA review.

At least eight coal leases have been granted by the Bureau of Land Management (BLM) since January 1, 1970. *None* of these has been accompanied by an environmental impact statement. The BLM has come under substantial criticism for its noncompetitive leasing practices and concomitant low royalties, rents, and bonus payments. Substantial reform of BLM and USGS practices has been urged by the General Accounting Office. Minimum coal bonus payments of $1.00 per acre have been the rule rather than the exception (although the most recent leases have seen a much healthier level of competition). The suggested reforms should be implemented, and NEPA compliance should commence at once. It is foolish for the Department of the Interior or its successors to continue to disregard the law when compliance will give rise to much needed data and sound program planning.

19.5. The Future

1. Though there is a clear need to centralize some authority in federal energy policy, overcentralization of decision making could be disastrous inasmuch as project-oriented thinking is likely to gain sway over problem-oriented thinking. Many intelligent and well-meaning scientists and engineers will put up with the most onerous or misguided leadership in order to work at the bench or the drawing board. The Soviet geneticists under Lysenko occasionally seem to have their contemporary parallel among the staff of some of our Department of the Interior agencies. As the President's Science Advisor recently stated:

> We should retain pluralism in both scientific decision-making and funding, so that entrenched viewpoints or incorrect judgments by a single agency or administrator can be offset by others in a system comparable to the checks and balances of the executive, legislative, and judicial systems. We should not create a czar for science and technology, nor should we put together easily understood organization charts just for the sake of neatness. (David, 1973.)

Inter- and intra-agency competition for funds and programs is essential for problem solving. After all, if we believe that competition is a healthy part of American creativity, should we allow it to lapse in both the private and the public sectors?

2. If we are reluctant to place the United States at the "mercy" of the OPEC Nations, why should we not be equally concerned about handing our future over to an even more powerful array of oligopolies? The competitive implications of horizontally integrated oil companies should be fertile ground for the economists. I would only suggest that there is every reason to believe that, given present trends, these corporations will to an increasing extent be equipped with the economic and political power to make self-fulfilling projections.

3. The only rational policy to preserve the Northern Plains in a state approximating their present beauty and agricultural productivity is to conserve energy demands. This will at least buy time until the arrival of less intrusive technology. None of the federal agencies I have discussed have given serious thought to demand conservation.

4. If any policy decisions are to be made in the federal government, should we not begin, at every opportunity, to segregate promotional from regulatory responsibilities? The Department of the Interior is often in the same position as the Atomic Energy Commission, which has been so sharply criticized for internal conflicts of interest. Recently, the AEC has provided a good example for NEPA compliance. The courts since the Calvert Cliffs decision have added to the substantive import of NEPA. In *Environmental Defense Fund, Inc.* v. *Corps of Engineers* (Eighth Circuit Court of Appeals, 1972), the Court of Appeals for the 8th Circuit held that

> The unequivocal intent of NEPA is to require agencies to consider and give effect to the environmental goals set forth in the Act, not just to file detailed impact studies which will fill government archives.
>
> * * * *
>
> Given an agency obligation to carry out the substantive requirements of the Act, we believe that courts have an obligation to review substantive agency decisions on the merits.* * * [t]he prospect of substantive review should improve the quality of agency decisions and should make it more likely that the broad purposes of NEPA will be realized.

I am not overly sanguine of outlook, but I do believe that compliance with existing law would alleviate many planning and implementation problems in the Northern Plains states. The Department of the Interior would best be advised to look forward from recent interpretations of NEPA. If the Department yearns for past practices, let it seek the biblical admonition taught me repeatedly by my mother, who was born and raised in a now defunct coal mining camp in West Virginia:

> As a dog returneth to his vomit, so a fool returneth to his folly. (Proverbs 26:11.)

Notes

1. See Montana Department of Natural Resources and Conservation (1973). Production in Montana has increased from 1 million tons in 1969 to 7 million tons in 1971; 16 million tons is projected for 1973 and 75 to 80 million tons for 1980. Total U.S. production in 1971 was 534.9 million tons.

2. These are by rank: No. 2 Consolidation Coal (division of Continental Oil); No. 3 Island Creek Coal (subsidiary of Occidental Petroleum); No. 8 Eastern Associated Coal (division of Eastern Gas and Fuel Associates); and No. 10 Old Ben Coal (subsidiary of Standard Oil of Ohio). By 1971, four others of the top ten coal producers were subsidiaries of primary metals companies and accounted for another 18.3% of total tonnage.

3. In terms of raw energy, this amounts to approximately 500 quadrillion Btu and is a most conservative estimate of regional recoverable strippable reserves. This coal is 80% subbituminous C, 17% lignite A, and 3% subbituminous B. These data are derived from the NCPS.

4. For example, Reynolds Metals has control of lands (with attendant water rights) around Buffalo, Wyoming. These lands are a major source of hay and alfalfa in the area.

5. For an instructive review of these activities, see Stern (1971). Stern makes a convincing argument for mine mouth coal-fired steam electric generation in place of further hydropower development in the Northern Plains states.

6. Water supply costs were determined assuming federal government construction and operation of the aqueducts; a 3.463% interest rate was assumed. See NCPS, that is U.S. Bureau of Reclamation (1971, Vol. II, p. V-1).

7. Details are available in U.S. Bureau of Reclamation (1972).

8. See, for example, the controversy surrounding the Mid-Continent Area Power Pool Agreement (MAPP) before the Federal Power Commission (1972).

9. See, for example, memoranda detailing meetings dating from June 24, 1971, between the Forest Service and the Bonneville Power Administration concerning electrical transmission through the Magruder Corridor in the Bitterroot chain of the Rocky Mountains.

10. Gary J. Wicks, Director, Montana Department of Natural Resources, in an anguished speech given in Washington, D.C. on July 27, 1972, identified this frustration and singled out the Department of the Interior and the Congress as most culpable: Requests by Montana Governor Anderson for "[A] coordinated study of the total impact of coal development" were met by ". . . silence (except for a visit by a political appointee of the Department of the Interior who expressed concern, promised action, delivered nothing). . . ." * * * "There seems to be a belief in Washington that problems will disappear if enough people express 'grave' concern . . ." Finally, after several false starts, Assistant Secretary of the Interior Larson promulgated a study outline on January 29, 1973, for the Northern Great Plains Resource Program. It is long overdue; however, the program recently has been diminished in both funding and duration and now promises to provide a benchmark for only a few of the many impacts of interest.

11. One of the participants suggested early in the Study that ". . . it would be necessary to 'optimize by brute force' many of the items to be considered." (NCPS Vol. II, p. I–6). This is precisely what happened.

12. For example, the Director of the BIA has had to report upward in the Department of the Interior through the Assistant Secretary for Public Lands, one of whose primary responsibilities is to raise revenues for the public treasury.

References

Adams, Dell (1972). Letter to the Northern Cheyennes, July 7.

Armstrong, Ellis (1972). Letter from the Commissioner of Reclamation to the Honorable Lee Metcalf, United States Senate, June 20.

Bagge, Carl E. (1972). An address to the Western States Water and Power Consumers Conference, Billings, Mont., September 25.

Brown, Dee (1971). *Bury My Heart at Wounded Knee,* Holt, Rinehart & Winston, Inc., New York.

David, Edward E., Jr. (1973). "Toward a National Science Policy," *American Scientist, 61*(1), p. 20.

Department of the Interior (1972). *Improvements Needed in Administration of Federal Coal-Leasing Program,* Department of the Interior B-169124 Comptroller General's Report to the Honorable Lee Metcalf, United States Senate, March 29.

Eighth Circuit Court of Appeals (1972). *Environmental Defense Fund, Inc.* versus *Corps of Engineers,* No. 72-1326, 4 ERC 1721 (BNA), November 28.

Ennis, Bruce L. (1972). Letter to Mr. and Mrs. J. T. Redding of Hysham, Mont.

Federal Power Commission (1972). *Mid-Continent Area Power Pool Agreement (MAPP)*, Docket No. E-7734, Washington, D.C., filed May 23.

Montana Department of Natural Resources and Conservation (1973). *Coal Development in Eastern Montana,* Helena, Mont. 59601, January.

Proceedings of the Montana Coal Symposium (1969). Eastern Montana College, Billings, Mont., p. 89, November 6 and 7.

Stern, Carlos D. (1971). "A Critique of Federal Water Resources Policies: Hydroelectric Power versus Wilderness Waterway on the Upper Missouri River," Ph.D. dissertation, Cornell University, Environmental Sciences; University Microfilms, Ann Arbor, Mich., ref. #71-18,911.

U.S. Bureau of Reclamation (1971). *North Central Power Study,* Vols. I and II, Montana Office of the Bureau of Reclamation, P.O. Box 2553, Billings, Mont. 59103, October.

——— (1972). *Appraisal Report on Montana-Wyoming Aqueducts,* Washington, D.C.

20 Institutional Design for Energy Systems/ Environmental Decision Making

GLENN BUCHAN*

Power plant siting questions have become increasingly controversial in recent years as demands for more electricity and simultaneously for improved environmental quality have put electric power companies and local environmentalists on a collision course. The work of Gage and Jopling (1971) in analyzing selected nuclear power plant siting controversies suggests that these conflicts frequently lead to irrational outcomes in which nobody's interests are adequately served. In such instances, rational planning is typically replaced by crisis management, which in turn generally yields less than satisfactory results. What is required, then, is an institutional decision-making system designed to identify and protect the interests of all parties concerned with power plant siting questions.

One way to approach this problem is to view society as a cybernetic system whose state is defined by a set of quality parameters including, among other things, electric power availability, environmental quality, and the cost of electricity. The installation of a power plant "disturbs" the system by moving it to a new state. The problem is to design a cybernetic regulator that keeps the new state within acceptable bounds (after defining what "acceptable" means in this context). By analyzing a variety of economic, political, and legal mechanisms, this study will propose a design for such a societal regulator to resolve specific power plant siting questions and to promote rational energy/environmental planning. Among the specific issues to be examined are the appropriate functions of regulatory agencies, the value of local referenda, and effective mechanisms for generating and resolving conflicts.

20.1. The Inherent Limitations of the Regulatory Process

The difficulty in adequately considering environmental quality in power plant siting questions derives from the economic nature of pollution. This subject has been rather thoroughly analyzed (Buchan, 1972; Dolan, 1971; Kneese, 1964; Walker, 1969). Briefly, power companies, along with other potential polluters, can reduce their production costs by discharging various kinds of waste into the environment, thereby reducing the cost of electricity to the consumer. The difference between the true cost of produc-

* University of Texas at Austin.

ing the electricity and the price paid by consumer must be borne by the society at large. Thus, pollution is a negative externality. From a societal point of view, an excessive quantity of pollution is produced since the price of electricity is artificially low, leading to excessive power production.

Clearly then, the first requirement for any environmental decision-making system is to provide a mechanism to compel electric power producers to internalize their external diseconomies. This task, however, is not quite as straightforward as some analyses suggest. For example, the most common response to this type of problem is the creation of some type of regulatory agency empowered to etablish and enforce some type of environmental quality standards, levy fines on waste dischargers, or provide subsidies to polluters who reduce their waste discharge. In general, the economic literature favors the effluent charge approach primarily due to the potential for marginal adjustments to waste discharge levels and the lower information requirements inherent in the system.

Unfortunately, this approach fails to account for the inherent limitations of the regulatory process. The limitations include both operational problems and excessive erosion of the theoretical basis of the regulatory process. The contrast between the ideal and the actual functioning of the regulatory process is illustrated in Figure 20.1. Figure 20.1a schematically depicts the ideal regulatory process in which a negative externality constitutes a "signal" to the public. In the ideal case, the signal is transmitted undiminished around the closed control loop shown in Figure 20.1a and the negative feedback effectively eliminates the externality.

To observe that the actual process does not function as effectively as this is almost to belabor the obvious. An examination of Figure 20.1b suggests some of the difficulties with the idealized model. To begin with, the initial signal is subjected to noise, filtering, and distortion as it travels around the loop. Distortions can occur in the initial public perception of the signal (that is, the ecological effects of subtle pollutants—waste heat, for example—can be difficult to predict or perceive). Substantial damping and filtering of the signal are virtually inevitable in the process of translating public preferences into legislative or other political action (Bobrow, 1970; Buchan, 1972). Policy is typically made according to various elite preferences and merely reacted to by the public. Signals emanating from the public are filtered accordingly.

Further distortion of the signal derives from the nature of regulatory

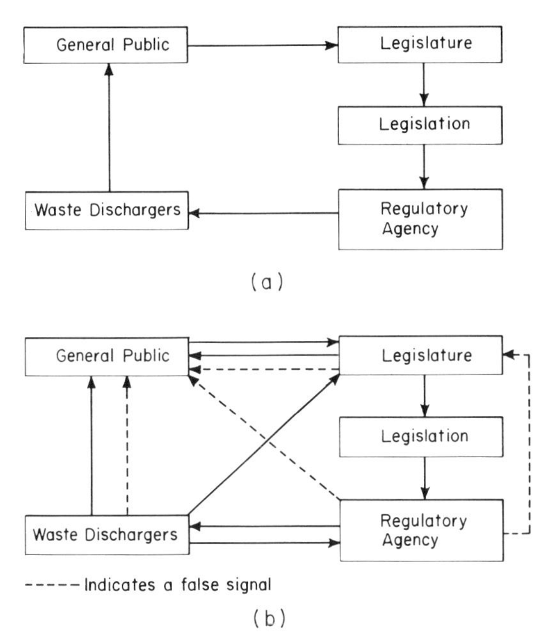

Figure 20.1. Contrast between (a) the idealized regulatory process and (b) the actual regulatory process.

agencies. Most idealized regulatory schemes implicitly assume that regulatory agencies are both benign and omniscient. It seems much more reasonable to assume that regulatory agencies are self-interested groups which conform to the "laws" governing bureaucracies (Downs, 1967). Among other things, this suggests that regulatory agencies become very rigid, hierarchical, and unadaptive (Downs, 1967). They perform routinized tasks adequately but cannot adapt to changing conditions. They collect data effectively but analyze it poorly (Bobrow, 1970). They concentrate on the present at the expense of the future (Bobrow, 1970). They tend to be more of a nuisance in terms of required reports, and so on, than a controlling force to those whom they are supposed to regulate (Downs, 1967). Finally, the regulatory agency's decisions are influenced primarily by its own interests, which are basically continued existence and ever-expanding authority (Downs, 1967). Thus, the already distorted signal is distorted even more as it passes through the agency hierarchy in order to make it conform more closely to the preference ordering of agency officials.

At this point, the possibility of effective regulation is rather small, but even more serious problems exist. Figure 20.lb suggests the possibility of false signals and of closed internal loops that effectively exclude the public entirely from the decision-making process. The first type of false signal involves the symbolic use of the regulatory process to reassure the public that its interests are being protected rather than to act effectively. The effect of these false signals is to add "noise" to relevant signals, thus complicating public perception of real changes in the initial signal. A second kind of false signal can result from dual roles of some agencies such as the AEC as both regulators and promoters of particular kinds of projects or technologies. The inevitable conflict of interest inherent in this dual role can lead to biased analysis and to the transmission of false signals to both the public and political decision makers (Fabricant and Hallman, 1971; Hoehn, 1969).

Two kinds of closed internal loops are suggested by Figure 20.lb. First, waste dischargers may have more influence with government decision makers than does the less organized, more amorphous public. Second is the familiar "captive" argument in which the regulator becomes more responsive to the interests of the regulator than to the interests of the public. This situation is most likely to occur when the regulatory agency has substantial discretionary power in establishing quality standards, effluent charge schedules, and so forth. In both cases, the role of the public in decision making is substantially diminished.

Beyond these operational problems (which are in themselves quite substantial), traditional approaches to environmental regulation suffer from a significant erosion of their theoretical foundations. This is particularly true of the effluent charge system favored in much of the economic literature. The effluent charge system requires the quantification of all damage costs associated with various levels of waste discharges. These costs are then levied as a fine against the waste discharger to force him to internalize the costs. There are several problems with this approach, however. The first is the entire philosophy of quantifying and comparing all costs and benefits on a single cardinal utility scale (dollars, in this case). This is a prima facie violation of welfare economic constraints that allow only ordinal utility scales and forbid interpersonal utility comparisons but is generally rationalized as being necessary for applying welfare economic ideas to real-world problems. Similarly, the difficulty in measuring environmental costs has been widely recognized but has generally been viewed as a methodological problem rather than as a fundamental deficiency in the basic approach (Kneese, 1964; Walker, 1969).

In fact, it is neither necessary nor desirable to quantify and compare all costs and benefits in terms of dollars. Implicit in that approach is at least a tacit assumption about the existence of a unique social welfare function or an inherently correct set of values with a benefit-cost model playing the role of surrogate decision maker. A more sophisticated view of benefit-cost analysis (Hoehn, 1969; Watt, 1970) requires only that costs and benefits be identified and measured if possible but not necessarily compared on the same utility scale. Thus, the results of such an analysis might appear as a set of parameters: monetary costs, monetary benefits, aesthetic alterations, physical and biological changes in the environment, and so forth.

Such an analysis would be difficult for a regulatory agency to apply for obvious reasons, but it is entirely appropriate for a different style of societal decision making. In particular, humans may largely "satisfice" rather than optimize in their own decision making (Simon, 1955). That is, individuals may fill a variety of different needs up to approximately fixed levels prescribed by their own valuations rather than elaborately optimizing some single-valued welfare function. Even if individuals attempt to optimize, their actual decision-making process may closely approximate the "satisficing" mode due to the limits imposed by decision costs. Therefore, the role of analysis should be to present individuals with a matrix of costs and benefits for various alternate futures. A societal decision-making system must then be designed to allow individuals to state their preferences for goods, that is, to choose among the alternate futures presented to them.

There are other important objections to narrowly based economic approaches to environmental regulation. Economic analysis cannot, by definition, treat questions of equity and these can be important. Also, as suggested earlier, economic regulatory schemes fail to provide adequate means for individuals to reveal their preferences and valuations on environmental and energy-related questions.

In summary, regulatory agencies cannot be depended upon to perform adequately in making trade-offs between environmental quality and energy demands. In general, regulatory agencies can perform adequately when their functions are limited to the following: (1) data collection; (2) environmental monitoring; and (3) enforcement of standards devised by somebody else. On the other hand, pollution abatement schemes that require regulatory agencies to be adaptive, discretionary, analytical, and sensitive to broad and diverse interests of various public groups are almost certainly doomed to failure or, at the very least, suboptimal operation. Specifically, this casts doubt upon the feasibility of the widely recommended effluent charge system. This conclusion obviously begs the question of what kind of control system can be designed to replace the regulatory agency approach. The remainder of this study will address that question.

20.2. The Political Process—Local Referenda

Separation of political and economic systems is somewhat artificial in this context since political economy, the relevant discipline, obviously encompasses both areas. This section deals with voting as a means for determining the ideal mix of public goods (which are analytically equivalent to externalities). In particular, it will examine the role of local referenda in power plant siting disputes.

The voting process has been subjected to positive analysis by Downs (1957), Buchanan and Tullock (1969), Arrow (1963), and others. The object of voting is to minimize the sum of external costs plus decision costs. The latter include both the actual costs of voting and the costs of obtaining information. The magnitude of these decision costs becomes crucial in analyzing the rationality of voting and the effectiveness of referenda in revealing public preferences. Two other points are of particular interest. First, no single voting rule (majority rule, for example) can always be relied upon to minimize costs. Second, any voting rule other than unanimity (which is subject to other objections) subjects a minority to political externality costs.

For a variety of reasons, the voting process cannot function as efficiently in revealing demand for public goods as the market can in dealing with private goods (Buchanan, 1968; Buchanan and Tullock, 1969; Dolan, 1971; Downs, 1957). Thus, as a mechanism for conflict resolution, voting is generally suboptimal. Nevertheless, a requirement for local referenda to approve power plant siting can prove most useful in performing two other functions. The first is the generation of low-cost information. This is important because excessive information costs can discourage rational individuals from voting or, if they do vote, reduce the probability of their thoroughly evaluating the alternatives. The second is the generation of conflict and the early identification of the various interests involved.

Conflict inspiration essentially involves the identification of all significant effects of a particular action and the subsequent identification of all affected interests. Suppose, for example, that a local referendum were required to approve the construction of any electric power plant. In terms of conflict resolution, such a referendum would not generally be sufficient to protect all individuals against the imposition of external costs or to assure an economically optimal or equitable decision. On the other hand, the referendum would have a clearly positive effect due to its potential for conflict inspiration. First, by providing for local veto power, a referendum requirement reduces the possibility of an entire community's being subjected to excessive political externality costs. This provides power companies and their political supporters with strong incentives to provide information to citizens to convince them that the power plant is in their interest. Similarly, the process tends to prompt individuals and groups to seek more information about the effect of the proposed plant on their particular interest.

When negative effects are discovered, opposition tends to catalyze in the manner described by Gage and Jopling (1971). The fact that the issue will be decided by referendum gives both proponents and opponents of the power plant strong incentives to provide low-cost information to voters and provides political entrepreneurs with opportunities to generate alternatives and supply still more information to the public. Therefore, the existence of a referendum causes more information to be generated and more interests to be defined and represented than would otherwise be the case. Use of the political system in this fashion, coupled with analysis of citizen feedback, opinion polls, and so forth, provides a reasonably reliable means of determining societal preferences for public goods, a task that the economic system failed to accomplish.

This process can also improve the efficiency of normal political channels by improving feedback between citizens and elected officials. First, elected officials (and their potential opponents) receive information about voter preferences at low cost. That makes it easier and more rational for them to consider their constituents' views in making policy decisions. If they do not, political entrepreneurship becomes easier and more rewarding, because political opponents can easily identify areas of divergence between representatives and their constituents. Under these conditions, political entrepreneurs will find it to their advantage to supply more information to the public on issues where incumbents have performed "poorly" (that is, not in accord with constituents' preferences), and the incumbents will be similarly compelled to provide information to defend their actions. The net effect is that voters receive more information at lower cost and can afford to invest more resources in analyzing policy alternatives. This in turn has a positive feedback effect on individual preference selection and leads to a general increase in the rationality of the decision-making process.

Thus, the political system provides the potential for generating information, for revealing and measuring individual preferences for public goods, and, to a lesser extent, for making broad policy decisions. On local issues such as power plant siting, a required referendum can both protect a local community from political externalities imposed by larger political units (state governments, for example) by guaranteeing local veto power and provide power companies, elected officials, political entrepreneurs, and special-interest groups with incentives to provide the public with information (and, in the case of the opposition, substantive alternatives) to support their positions. The role of the referendum in generating low-cost (at least from the public viewpoint) information and forcing power companies to consider local preferences in their decision-making process is more important than the outcome of the referendum itself, since no political decision rule can be constructed which can guarantee optimal decisions. The possibility of political externalities and inequity in distributing costs and benefits still exist, however. Marginal adjustments of this type can be made by use of the legal system.

20.3. The Legal System—Public Advocates

Legal remedies to environmental problems are becoming more fashionable (Dolan, 1971; Katz, 1971; Landau and Rheingold, 1971; Mishan, 1969). The general approach is legally to ban negative externalities through various types of tort actions. It is probably worth pointing out that the simultaneous generation of positive externalities does not justify ignoring or discounting negative externalities. In general, the victims of negative externalities are not exactly the same group as the beneficiaries of the positive externality. Also, the magnitudes of the opposing externalities are generally different. Thus, equity demands that victims of negative externalities be compensated and that positive externalities are accounted for by subsidies financed through appropriate taxation measures (granting the difficulties inherent in any subsidy approach). Such subsidies would then have to be rationalized and approved through the political system. The point is that positive and negative externalities are separate issues and should be treated as such.

Various types of reforms have been proposed to make the legal system function better in these cases, the most important being the establishment of standing for citizens to bring class action suits against nuisances. Unfortunately, even the existence of adequate legal doctrine in the environmental area is not sufficient to assure adequate performance of the system due to the practical barriers encountered by individual plaintiffs. Individuals rarely have adequate money, time, or expertise to undertake litigation effectively against a large polluter (a power company, for example), especially in the usual case where the potential damages to be collected are relatively small (or nonexistent where only an injunction is sought). Even under the best conditions, where lawyers and expert witnesses agree either to donate their services free or accept only nominal fees, the incidental costs are likely to exceed the benefits received by an individual plaintiff (Landau and Rheingold, 1971). Where class actions are employed, the free rider problem reduces individual incentives to act still further. By contrast, defendants such as power companies have a considerable private stake in the outcome of environmental suits, so they can afford to invest more resources in winning these cases. Also, of course, polluters generally have more access to resources and expertise than do pollution victims. Thus, the legal system is clearly biased in favor of the polluters.

An effective public advocate system can change that by redistributing power and expertise. These public advocates' sole reason for existence would be to represent directly individual citizens in disputes with polluters. They would have no role in either enforcing environmental standards or participating directly in political decision making. Rather, they would be lawyer-scientist types (probably lawyers with a technical staff to do analytical work) who would represent citizens both in legal actions against polluters and in an ombudsman capacity, if citizens desire, in early planning stages of power plants, and so forth.

The power inherent in the public advocate's office makes its other role as citizen ombudsman and surrogate representative much more viable. The fact that it has the resources and expertise available to take effective legal action against polluters provides the polluters with strong incentives to communicate and negotiate informally with the public advocates. The power of the public advocates is further enhanced by the incentives that they have to take an activist role. Since they can justify their existence (and continued tax support) only by satisfying individual citizens and demonstrating that they provide a worthwhile service not available elsewhere, the public advocates must maintain reasonably high visibility and effectiveness. This practical requirement for activism, in turn, enhances the credibility of the public advocates as negotiators and surrogate representatives.

This negotiation process is the crux of the entire decision-making process. In particular, the process will improve the planning process in general and serve the interest of power companies as well as citizens since later crises of the type described by Gage and Jopling (1971) can be averted. Furthermore, since power plant siting is a relatively slow and deliberate process, the added decision costs should not be excessive. The net effect should be a rational planning process that adequately accounts for both energy and environmental quality demands of local communities.

20.4. Summary and Conclusions

The most important aspects of the institutional decision-making system for balancing energy and environmental quality demands are (1) limitation of regulatory agency functions to the collection and publication of data and the enforcement of appropriate quality standards established by someone else; (2) local referenda to approve power plant siting; and (3) the creation of a public advocate system to represent citizens in both legal actions against and negotiations with power companies, thus facilitating rational planning.

References

Arrow, K. J. (1963). *Social Choice and Individual Values,* Wiley, New York.

Baumol, W. (1965). *Welfare Economics and the Theory of the State,* Harvard University Press, Cambridge, Mass.

Bobrow, D. B. (1970). "Computers and a Normative Model of the Policy Process," *Policy Sciences, 1,* No. 1, pp. 123–134, Spring.

Buchan, G. C. (1972). *A Cybernetic Aproach to Thermal Pollution Decision-Making,* Ph.D. dissertation, University of Texas at Austin, August.

Buchanan, J. M. (1968). *The Demand and Supply of Public Goods,* Rand McNally, Chicago.

———, and Tullock, G. (1969). *The Calculus of Consent,* University of Michigan Press, Ann Arbor.

Dolan, E. G. (1971). *TANSTAAFL: The Economic Strategy for the Environmental Crisis,* Holt, Rinehart, and Winston, New York.

Downs, A. (1957). *An Economic Theory of Democracy,* Harper & Row, New York.

——— (1967). *Inside Bureaucracy,* Little, Brown and Company, Boston.

Fabricant, N., and Hallman, R. M. (1971). *Toward a Rational Power Policy: Energy, Politics, and Pollution,* Braziller, New York.

Gage, S. J., and Jopling, D. G. (1971). "The Pattern of Public Political Resistance," *Nuclear News, 14*(3), pp. 32–35, March.

Hoehn, William E. (1969). "Economic Analysis in Governmental Decisionmaking," RAND P-4222, Santa Monica, Calif., October.

Katz, M. (1971). "Decision-Making in the Production of Power," *Scientific American, 224*(3), pp. 191–200, September.

Kneese, A. V., ed. (1964). *The Economics of Regional Water Quality Management,* Resources for the Future, Washington, D.C.

Landau, N. J., and Rheingold, P. D. (1971). *The Environmental Law Handbook,* Ballantine, New York.

Milsum, J. H. (1968). "Technosphere, Biosphere, and Sociosphere: An Approach to their Systems Modeling and Optimization," *General Systems Yearbook,* pp. 37–48.

Mishan, E. J. (1969). *Technology & Growth,* Praeger, New York.

Olson, M. (1968). *The Logic of Collective Action,* Schocken, New York.

Simon, Herbert A. (1955). "A Behavioral Model of Rational Choice," *Quarterly Journal of Economics, 69*(1), pp. 99–118, February.

Stevens, C. H. (1971). "Citizen Feedback: The Need and the Response," *Technology Review, 73*(3), pp. 39–45, January.

U.S. House of Representatives, Committee on Science and Astronautics (1969). *Technology: Processes of Assessment and Choice,* Report of the National Academy of Sciences, Washington, D.C., July.

Walker, W. R., ed. (1969). *Economics of Air and Water Pollution,* V.P.I. Press, Blacksburg, Va., October.

Watt, K. E. F. (1970). "A Model of Society," *Simulation, 14*(4), pp. 153–164, April.

IV Supply and Demand

21 The Rational Allocation of Natural Gas under Chronic Supply Constraints

THOMAS R. STAUFFER* AND JAMES T. JENSEN†

21.1. Introduction

The gas gap that had hitherto represented a purely *potential* excess of demand over supply is now an actuality and no longer merely an academic prospect. This paper will focus upon the economic implications of eliminating excess demand for natural gas via administrative measures, namely rationing, as distinct from market mechanisms. Since excess demand is expected to grow over the next years, stimulated partly by environmental protection regulations, increasing numbers of users of natural gas will perforce be obliged either to shift to some alternative fuel or cease operations. Acceptable alternate fuels are almost without exception more expensive than natural gas—up to triple its price or more, and fuel switching also frequently entails sizable supplementary investments and incremental expenses. Consequently, the total economic cost of coping with the gas gap is also sizable, and it is relevant indeed to determine whether the total incremental economic burden can be minimized.

Section 21.2 of this paper explores alternative mechanisms for equilibrating natural gas supply with demand, concluding that rationing is inevitable in the short run unless the prices of *all* gas, vintage supplies as well as newly discovered volumes, are decontrolled and rise to a level approximating fuel oil parity prices. Section 21.3 analyzes fuel conversion economics, contrasting the new situation of long-term gas supply deficiency with the prior, more conventional case of short-term, cold-weather interruptions to which the gas industry and its customers had long since effectively accommodated themselves. Finally, Section 21.4 compares the economic implications of the several gas allocation or rationing schemes that have either been advanced or are presently being implemented on an ad hoc basis.

21.2. Dimensions of the Gas Gap

While the present gas shortage has aptly been characterized as regulation-induced (MacAvoy, 1971), we shall propose here that still further regulation, in the form of an administrated rationing system, must be imposed in order to balance supply and demand. Since the Federal Power Commission

* Center for Middle Eastern Studies, Harvard University.
† Arthur D. Little, Inc., Cambridge, Mass.

has failed in its original mandate, one may well ask why it is either desirable or necessary to invoke still another form of regulatory apparatus as an antidote for the symptoms of that same regulatory body's past failures.

Rationing now is an unfortunately inevitable expedient because the supply and demand for natural gas are so far out of balance that no incremental adjustments can redress that imbalance. Injudicious price regulation has distorted both sides of the market equation. Low prices have severely inhibited new supply while simultaneously overstimulating demand. The price disparity between natural gas and alternate fuels is so large that rehabilitation of the market, which otherwise would have reallocated gas supplies anonymously and efficiently, would itself be seriously disruptive if rapid. Transitional therapy is therefore needed.

The available measures that might induce new gas supply or curb existing demand differ considerably in their impact, particularly in terms of the likely speed with which the system might respond to even the most enlightened of remedial steps.

21.2.1. Gas Supply

Supply responds both to increased prices and to increased offerings of federal lands for leasing, but in both cases the lead time for exploration and development is at least two years in known areas and up to five to seven years under less favorable conditions. Several years at the minimum must lapse before any significant impact upon reserve additions can result, yet the rate of additions to gas reserves of the past 3 to 4 years must be doubled merely to maintain the status quo. Greatly increased exploratory activity is needed merely to hold our own, and increased production, even after that 3- to 5-year lag, would involve a still more massive effort.

Although increased field prices for *new* domestic natural gas appear to be clearly desirable, whatever new supply might be ultimately possible cannot be elicited rapidly enough to solve the shortages over the next 3 to 5 years. Similarly, supplementary supplies—liquid natural gas (LNG) or syngas plants—if not already designed and certificated, cannot come on stream in much less than that amount of time.

21.2.2. Gas Demand

On the demand side, most of the deregulation options also involve considerable time lags, so that any reduction in demand due to higher, deregulated prices would be equally slow. The principal options are the following:

1. *Deregulation of Newly Produced Gas*[1]
Since any supplies of new gas at higher prices would be "rolled" into or averaged with the historically low prices of vintage gas, the prices perceived by consumers would rise only asymptotically to a market-clearing level. Impact upon demand would thus be imperceptible in most markets for 3 to 8 years, depending upon actual circumstances.
2. *Deregulation of All Gas*
If prices of vintage gas also were allowed to seek their own level, in abrogation of existing long-term supply contracts, demand could be reduced rapidly as fuel switching became financially attractive to certain gas users. This would imply politically conspicuous windfalls to producers, even if such extensive violations of contracts were otherwise certified as constitutional.
3. *Restructuring of Gas Rates*
New gas costs could be charged exclusively to the cost of service for industrial users, rather than being spread over all users, or other changes in rate formulas can be devised that shift more of the fixed charges on to industrial or large-volume users. This case is intermediate between the first two but concentrates increased costs into the most price-sensitive sector of gas consumption.[2]

Simulation studies indicate that only the extreme measure of complete price deregulation, where vintage gas prices also increase by 300% or more, can effect a significant reduction in demand in less than 3 to 4 years. Under all other circumstances, the "salutary" impact of higher prices is thwarted by rate structure provisions, and price signals are seriously muffled.

The extreme case of complete deregulation appears to be politically improbable; thus, excluding that contingency, we conclude that no market-based measure provides adequate relief in the short or medium run. Even if new supplies are priced at 80¢ or more per MCF (thousand cubic feet), there remains a gap between available supply and prospective demand which must somehow be closed via nonmarket measures.[3]

The time dimensions of this gap are illustrated in Figure 21.1, where the solid lines represent potential demand and potential supply in the absence of any fundamental price reform. The dashed lines show directionally the reduction in demand and the increase in supply that might ultimately result

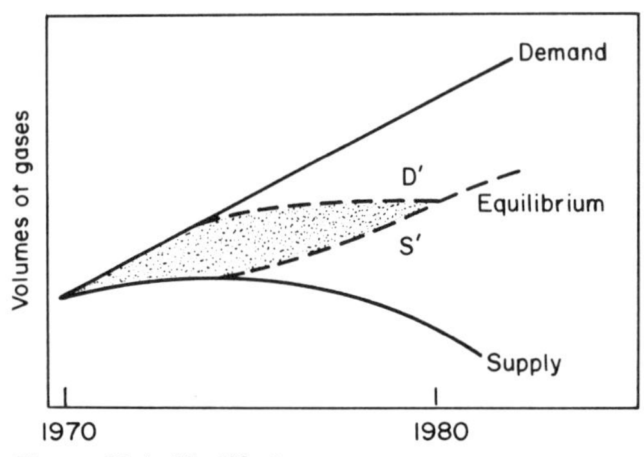

Figure 21.1. Modified gas gap.

from the immediate deregulation of *new* gas as described by the FPC, Bureau of Natural Gas (1972). This report contains one version of the "unmodified" gas gap. We note again that both supply and demand respond slowly, so that the emancipation of market forces could suffice to restore market equilibrium no earlier than the late 1970s. In the interim, the cross-hatched area reflects the unsatisfied, excess demand that must be selectively allocated to alternative fuels by some rationing system.[4]

Finally, we also observe from Figure 21.1 that natural gas rationing and some version of price deregulation are entirely complementary remedial measures. Deregulation permits the ultimate restoration of market balance over the longer period, while a judiciously designed rationing system prevents disruptive imbalances in the short run. It is crucial that the short-run allocation system, however, ensure a smooth transition into the ultimate equilibrium. Thus, the set of users who are compelled to switch from gas to alternative fuels under the rationing program should also be the same users of natural gas who would otherwise have switched under market equilibrium conditions. An important criterion is that the hierarchical ranking of priorities should duplicate as closely as is administratively feasible the actual workings of the market.

21.3. Fuel-Switching Economics

Industrial and commercial uses of natural gas differ greatly and, thus, there is a wide spectrum of costs involved in converting diverse applications to alternate fuels. The rationing system that successfully selects the lowest-cost uses for earliest curtailment can therefore effect significant economic savings.

Four principal determinants of switching cost may be isolated, and we shall examine these in turn:

1. Duration of gas supply curtailment.
2. Type of gas-using facility.
3. Size of unit (scale).
4. Operating load factor.

21.3.1. Duration of Curtailments

It is crucial to distinguish between two quite distinct types of gas supply shortage—episodic interruptions versus chronic curtailments—because industry copes with each type in a totally different fashion. Episodic interruptions in natural gas service, as a consequence of short periods of cold weather or *force majeure* conditions, are the conventional "shortage" problem. The operating procedures of all gas pipeline and distribution companies embody end-use priorities whereby precedence is given to residence and commercial users during such short-term shortfalls. The most frequent standby fuel for such circumstances is propane (LPG) which can be gasified and used directly in the plant's gas mains as a direct replacement for flowing natural gas. Although propane itself is significantly more expensive than natural gas, this has proved to be the minimum cost tactic for coping with episodic interruptions in natural gas service.[5]

Chronic curtailment, however, is an entirely new prospect for most industrial users of gas located east of the Rocky Mountains, and sustained curtailment of gas supply involves a radical rearrangement of fuel use patterns.[6] While propane is logically the alternate fuel of first recourse for an episodic supply interruption, it is not an alternate under the present circumstances of a chronic gas supply deficiency. Propane represents the transformation of part of the natural gas supply into a form that is more amenable to storage.

Some 75% of the domestic supply of propane is extracted from natural gas in the field, reducing the Btu value of the gaseous stream committed to the pipelines. The remaining 25% is a minor by-product of petroleum refinery operations, for which the supply is virtually inelastic and which is used in the petrochemical industry as a feedstock.

Since gas shortages are anticipated to be chronic, rather than transient, the conventional approach to supply interruptions—propane standby—

is irrelevant.[7] Whereas prior interruptions were almost always brief and occasional, many users hereafter will be forced to switch completely or, at best, for much longer periods. Converting from natural gas in the broadest sense—that is, switching away from methane *plus* propane—involves two types of additional costs: (1) provision of the alternate fuel and (2) modification or replacement of gas-using equipment to accommodate some new fuel.

21.3.2. Type of Facility

The costs of "conserving" a unit of gas by converting to another fuel depend intimately upon the type of facility. The first cost component relates to the fuel itself. In most instances tankage, piping, and pumps for liquid fuels will be required, but some electric utilities might be obliged to rehabilitate or extend coal yards (Office of Emergency Preparedness, 1973; U.S. Senate, 1972). New burners and related instrumentation will need to be installed. The additional costs, measured in terms of the volume of natural gas that is displaced or "conserved," comprise the annualized capital charges for the incremental investments, plus related incremental operating costs.

A major item is the price differential between natural gas and the alternative fuel itself. In some cases, for example, in the conversion of a gas turbine, only a more costly premium fuel such as No. 2 distillate can be used. In utility boilers, however, the cheaper residual oil is acceptable. Additional operating and maintenance costs will ordinarily also be incurred; a coal-burning electrical plant, for example, typically requires almost triple the labor per kilowatt-hour as compared to a gas-fired plant.

The second category of fuel-switching costs is related to the process itself, as distinct from the fuel-handling and combustion equipment. Some processes can be adapted to alternative fuels more economically than others. In the case of boilers, for example, sootblowers must be installed because alternative fuels produce more ash and particulate matter than gas.

In other cases, much more extensive modifications and supplementary investments prove necessary. If an ammonia plant is converted to use naphtha as a feedstock, say, in place of natural gas, the overall material balances are upset and systemwide debottlenecking must be undertaken. The additional cost per unit of gas displaced is four to seven times higher than for a large boiler. In glass furnaces the greater luminosity of an oil flame radically alters the temperature profile so that the furnace can safely be operated only at reduced capacity. In extreme cases conversion is not technically feasible at all, and the gas-burning piece of equipment must be completely replaced by a different process. Examples that involve complete replacement of otherwise serviceable equipment are the fiber-formation stage in the manufacture of fiberglass, direct-flame makeup air heaters, or certain types of radiant-tube furnaces for metallurgical applications.

Furthermore, even where *conversion,* as distinct from *replacement* is feasible, additional process-related operating costs arise. The quality of the product is impaired by contamination from combustion products or from temperature fluctuations due to the inherently greater difficulty in controlling liquid burners as compared to gas burners. Further, capacity reductions (deratings) imply higher unit costs and/or increased investment in order to maintain plant output levels. This effect might be particularly important for the electric utilities where the allegedly higher forced-outage ratios for oil-fired plants, plus the increased downtime for scheduled maintenance, narrow an already tight capacity margin.

The variations in conversion costs are thus quite large when all components are considered: fuel quality premium, handling facilities, process conversion or replacement, and additional operating costs.

21.3.3. Economies of Scale

For any given type of facility—whether a steam boiler, chemical plant furnace, refinery heater, gas turbine driven compressor, or a soaking pit in a steel plant—an additional factor is the size of the individual unit. Technical-economic studies (U.S. Senate, 1972) indicate that significant economies of scale in fuel conversion are possible. Indeed, since industrial gas-burning equipment spans more than three orders of magnitude in size —from heat input ratings of 6 billion Btu per hour down to burners of 1 million Btu per hour or less—one finds considerable scope for selectively exploiting economies of scale. These are illustrated in Figure 21.2.

Illustrative of the range of conversion costs for boilers is the difference between the cost of converting a large electric utility boiler to oil versus a household gas furnace. The conversion cost in the former case amounts to about 5 to 7 cents per thousand cubic feet of gas (MCF) conserved, whereas switching the residential user to oil costs \$1.10 to \$1.50 per MCF.

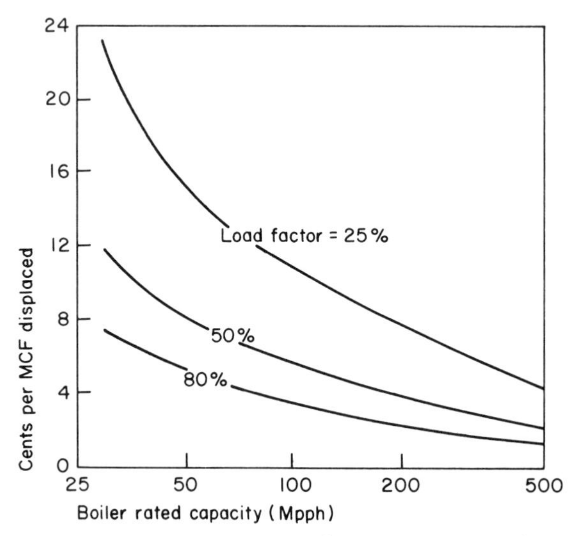

Figure 21.2. Conversion of gas-fired boilers to fuel oil (capital charge per MCF displaced).

21.3.4. Load Factor

Load factor also influences the economic cost of fuel switching because the investment costs of converting a unit of any given size will be spread over a larger volume of gas in such measure as the load factor is higher. Load factors run as high as 85 to 90% for some utility boilers or industrial furnaces, while 20 to 25% is more typical of the smaller boilers used for space heating alone. Broadly speaking, higher load factors are correlated with larger units, so that the effect of economies of scale is reenforced.

The estimated distribution of fuel-switching costs is displayed in Figure 21.3. It is to be noted that almost 2 trillion cubic feet (TCF), equivalent to about 11% of the total 1969 consumption, or some 18% of industrial consumption, can be switched at "zero" incremental cost. This represents known large electric utility boilers, already capable of burning either coal or fuel oil, for which the only additional costs would be the fuel price differential that is otherwise common to any user curtailed.[8]

The conversion costs rise rapidly for other applications, but full coverage of industrial gas consumption at the upper end was not possible. Ever smaller volumes of gas are involved in the proliferation of higher-priority

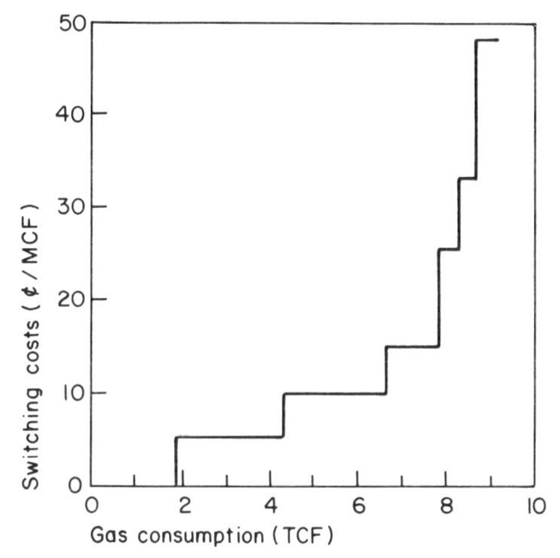

Figure 21.3. Distribution of curtailment costs.

applications that could not be surveyed, and there is an additional total volume of 2 to 4 trillion cubic feet consumed in applications where conversion costs are well in excess of 60 cents per MCF of gas displaced. Direct-flame makeup air heaters, for example, involve conversion costs of $.70 to $1.35 per MCF of gas, while some metallurgical processes imply conversion penalties above $2 per MCF. The cost distribution curve in Figure 21.3, therefore, rises still more steeply if one were to attempt to convert the remaining several trillion cubic feet of industrial gas consumption.[9]

It is clear that quite significant savings can be realized if any given gas shortfall is met by first obliging users in the lowest "cost" categories to switch before forcing conversion of gas uses that entail a differential conversion cost of, say, 60 cents per MCF. The unnecessary cost burden of switching the latter, rather than the former, to meet only a 5% level of curtailments would amount to some $600 million annually, which is a quite substantial and quite unnecessary extra burden. Since much larger curtailments are envisaged—10 to 15% of present demand, or perhaps more—an injudiciously designed rationing system could impose important extra costs upon our economy and require the unnecessary redeployment of important economic resources. Given the magnitudes at stake, the rationing system warrants careful analysis.

21.4. Comparison of Gas Allocation Systems

Three principal systems have been proposed for allocating scarce natural gas:

1. Contractual form—"firm" versus "interruptible."
2. Pro rata curtailments (industrial load).
3. End-use priorities.

Since curtailment of gas deliveries has important repercussions upon all affected parties—both an increased financial burden upon the individual customers who must switch to a more expensive and less convenient fuel and an economic burden upon the national economy as a whole—it is important to examine the economic implications of each of these allocation proposals.

Common to all systems is an explicit set of partial "end-use" priorities in that all gas use, except for minimum volumes required to protect industrial plant and equipment, is ultimately subordinated to requirements for private residences, hospitals, and certain other designated preferred applications. Otherwise, the tariff sheets of each gas utility stipulate where each class of customer fits on the rationing totem pole in event of supply interruption. The quite different systems of curtailment priority in effect today have evolved consonant with the historical patterns of gas supply and the customer characteristics that are peculiar to each gas pipeline or distribution system.

21.4.1. Contractual Basis

Priority rankings here are based upon whether the customer has contracted for "firm" or "interruptible" service. An "interruptible" customer is thus curtailed first, "firm" industrial customers are next, and only then, in the most dire emergency, might customers in the highest "end-use" category—residential users—be curtailed.

The conventional procedure for handling gas supply interruptions is really a hybrid system and combines "end-use" priorities, based upon social and economic considerations, with a subset of priorities related strictly to contractual form. The contractual distinction between "firm" and "interruptible" service was vitally important in the historical development of the natural gas industry, since "interruptible" sales were originally off-peak deliveries that permitted part of the system's peak-day capacity costs to be borne by other users who did not contribute to that peak demand and who bought only "valley" or "off-peak" gas. A lower gas price was the "interruptible" customer's compensation for accommodating himself to possible loss of his supply.

This traditional approach to curtailment evolved when natural gas was in surplus, and supply interruptions occurred only on peak cold days when delivery capability was limited. The situation now has been altered basically, perhaps irreversibly. Now, supply itself is the binding constraint, and not surprisingly, the well-established solution to yesterday's problem of *balancing* loads is no longer applicable to the problem for today and tomorrow of *shedding* load. Currently, where supply is chronically deficient, reliance upon contractually determined curtailment priorities poses certain disadvantages:

1. The form of a gas service contract is entirely uncorrelated with the uses of that gas. In particular, high conversion cost uses might be on "interruptible" service while electric utilities can burn gas for boilers under "firm" contracts. The economic impact of curtailing on that basis is therefore capricious.
2. The alternate fuel systems for coping with a chronic curtailment are quite different than the standby fuel suitable for the hitherto customary, short-duration, cold-weather interruptions.
3. "Interruptible" gas service has become much too imprecise to be functional for regulatory purposes. In some areas "interruptible" users have historically switched quite regularly to alternate fuels for months at a time, while in other areas "interruptible" service has been *de facto* firm.

21.4.2. Pro Rata Curtailments

A second basic proposition is to spread the shortfall evenly among all industrial users of gas in proportion to their consumption in some base period. Pro rata assignment of the curtailment obviously treats "like" users in a like fashion, but it also has the equally obvious disadvantage that it also treats *un*like users in just the same "like" fashion, requiring every industrial user to provide for alternate fuels. Formally, pro rata rationing is nondiscriminatory among users, but it is economically highly discriminatory among uses.

21.4.3. End-Use Priorities

Uses of gas are ranked in terms of the relative cost of switching each to the least expensive alternate fuel, making all due allowances for the several associated costs discussed in Section 21.2. In the event of a gas deficiency those uses are curtailed first for which the total conversion cost is least, working up the distribution curve until the curtailed volume equals the gas supply shortfall.

Theoretically, implementation of a system of end-use priorities would entail surveying all uses of natural gas and carrying through technical-economic analyses of the fuel conversion costs. Administratively, such a survey would become an impossible quagmire, analogous to but even greater than the problem created by the effort undertaken in the mid-1950s to apply a cost-of-service analysis to gas producers in the field. However, it is possible to devise an administratively workable approximation to this classification, without introducing significant economic inefficencies, by (1) identifying nonpremium uses of gas and (2) explicitly recognizing economies of scale. One set of priorities based upon these principles is tabulated now, ranked in descending order of precedence:

1. Residential and small commercial users.
2. Commercial and industrial users below a *de minimis* level of annual consumption.
3. Feedstock, "process," and "form-value" uses.
4. Boilers, gas turbines, indirect-flame furnaces, and equivalent uses:
   a. Smallest units
   b. Medium-sized units
   c. Largest units.

Whereas a hierarchical classification of all uses of gas is infeasible, it is possible to identify about a score of types of use for which the conversion costs are the lowest and comparable for given rated heat inputs. These identifiable uses for which conversion costs are minimum—the "nonpremium uses"—are explicitly assigned to the lowest priority category, which is further subdivided in order to recognize economies of scale. The largest units are assigned the very lowest priority, the next largest into the next higher priority class, and so on. The uses in the bottom category would be curtailed in inverse order in order to meet a gas supply deficiency of any specified level. Small plants, irrespective of how gas is used, would be assigned to priority category 2, reflecting the diseconomies of scale associated with conversions at a small location, as well as to reduce further the administrative burden. All remaining uses of industrial or commercial gas would fall into category 3. This approach, which avoids the difficult question of identifying premium use by addressing itself to the easier but complementary question of recognizing nonpremium uses, requires no complicated engineering studies, tailored to each individual user.

Several other such rationing programs, involving end-use priorities, have been advanced. The Federal Power Commission (1973a) itself quite recently announced one version of such a scheme where it argued that natural gas for boiler fuel should be cut off first, reserving the highest priority among industrial uses to feedstocks and "process gas." [10] The FPC proposal focuses upon one use alone as the lowest-priority application; another system, advanced by a related group of gas pipeline and distribution companies, divides uses into premium and nonpremium uses, expanding the restrictive FPC designation of "boiler fuel" to include soaking pits, reheat furnaces, cement kilns, and so forth.

The economic impacts of the several rationing procedures differ markedly. End-use priorities minimize by construction the total economic costs of realigning natural gas usage. Pro rata curtailments, however, entail much higher costs, since a proportional volume is cut from all uses, thereby requiring much more extensive conversions, as well as ignoring the large volumes consumed where curtailment is virtually costless.

The impact of curtailments based upon contractual form—that is, curtailing "interruptible" users first—is not calculable, as there is no relationship whatsoever between the form of the contract and the relative costs of switching fuels. Indeed, about one-half of the gas burned by large electric generating plants is classed as "firm" service, while a significant fraction of the metallurgical process gas for which conversion costs can exceed $2.00 per MCF is classed as "interruptible." The economic impact of contractually based curtailments is therefore as capricious as pro rata rationing. The extra economic burden of allocating gas using either basis, as distinct from an end-use system, would be of the order of $1 billion annually, and involve an additional investment outlay for conversion equipment of several billion dollars over and above the base case.

21.5. Summary

It is concluded that there exist only two basic remedial strategies for the growing gas gap:

Option One. Deregulate all gas, both vintage and new supplies.

Option Two. Partially deregulate gas prices *and* implement a formal rationing system to reallocate excess demand during the transitional period.

Table 21.1 Differential Impacts of Natural Gas Rationing Systems

Characteristics	Pro Rata and Contractual	End-Use Priorities
Selectivity	Capricious	Identifiable
Economic burden	High	Minimum
Administrative requirements	Routine	Classifications needed
Financial impacts	Pro rata: diffuse Contractual: focused	Focused

Since the first option involves an abrupt tripling of the prices of gas for interstate pipelines, as well as proportionate windfalls for gas producers, we have concentrated on the second option, deeming it to be the more probable of implementation.

Three model types of gas rationing schemes have been proposed as devices to bridge the 4- to 8-year period until market forces can reestablish equilibrium. The trade-offs inherent in the different rationing systems are summarized in Table 21.1. The price of gas is below that of alternate fuels, and whoever must switch fuels is penalized. It is therefore useful to distinguish between financial effects and economic impact. The user who must replace 40-cent gas with 80-cent fuel oil, *plus* incurring conversion costs, will perceive a significant financial burden, which he may or may not be able to pass forward to his customers. Curtailments reduce the financial impacts upon individual users by distributing the incremental burden quite widely, at the expense of greatly increasing the aggregate costs, thereby implying an unquantifiable trade-off between considerations of equity and economic efficiency.

However, end-use priorities concentrate the burden of natural gas curtailments on that subclass of users which is potentially best able to bear the extra costs. The largest users of low-priority gas are the electric utilities —for boiler and peaking turbine fuel—and the natural gas pipelines themselves—for compressor station fuel. If their rates include automatic adjustments for purchased fuel costs, that financial burden can be redistributed over the broader class of consumers who in prior years were implicitly subsidized by artificially low gas prices.

More generally, however, insofar as the burden of switching must be borne by customers who are not natural or designated monopolies, the financial effects are large but their location is less predictable. Since some financial impact is inevitable, it is more effective to minimize the aggregate of all financial impacts by minimizing the total economic burden, since this has the collateral advantage, cited earlier, of selectively shifting the financial burden upon those parties best able to pass it on. End-use priorities thus achieving the economic objective while greatly reducing the otherwise inevitable inequities.

We conclude, therefore, that rationing is inevitable unless the nation is willing to accept all the implications of retrospective increases in the prices of all natural gas. Barring that expedient, a two-part package for restoring equilibrium is advisable:

1. Deregulation of prices for "new" gas, permitting gradual restoration of market balancing forces.
2. Implementation of end-use priorities for curbing excess demand during the transitional period.

The implementation of end-use allocation priorities minimizes the necessary total economic costs of readjusting national patterns of fuel utilization and also facilitates the transition toward a longer-run equilibrium. Indeed, since end-use priorities are close surrogates to the market mechanism itself, those priorities offer an additional important advantage because the "control" system discretely "self-destructs" in such measure as it restores the targeted market equilibrium. It thus fulfills the criteria both for economic and regulatory efficiency.

Notes

1. See the forthcoming paper by Jensen and Stauffer (1973).

2. The first is a variation upon the concept of "incremental pricing" as advanced by FPC in Opinion 622, 28 June 1972, while the second proposal was advanced officially in the FPC's statement of Policy (1973b). The inversion of block rates is still another variation upon this theme. Under conditions where existing customers can no longer be served with gas, this variation of incremental pricing in its conventional interpretation, as applied to new demand, is of reduced relevance.

3. A field price of 80¢ per MCF approximates fuel oil parity, allowing for gas's intrinsic form value. If the free field price were determined by the wellhead netback equivalent of substitute or synthetic gas, it would be still higher—up to $1.20/MCF, or almost six times the present average for interstate sales.

4. "Full deregulation" could significantly discourage demand and reduce low-priority consumption. However, freeing the prices of old gas necessarily involves widespread

abrogation of existing long-term contracts and probably requires congressional action in order to revise quite radically the Natural Gas Act. Lacking such draconian measures, however, rationing is by default the sole alternative for coping with the problem over the next few years.

5. Engineering analyses conclude that the trade-off between the high fuel cost but nominal supplementary investments for propane standby clearly favor reliance upon propane for short-duration gas shortages.

6. The demarcation between the two polar cases of curtailment versus interruption is less polar in practice. Broadly speaking, however, large industrial gas customers on the West Coast generally are accustomed to sustained curtailment and have functioning alternate fuel capability. The opposite is true elsewhere, and chronic seasonal cutoffs of gas are very much less frequent, even for users whose service is legally "interruptible."

7. The prospective gas shortages exceed manyfold even the most sanguine estimates of potential propane supply, even allowing for refinery modifications and disruption of supplies of propane feedstocks to the petrochemical industry.

8. As discussed earlier, since the cost of an alternate fuel is common to all curtailed customers, we concentrate here strictly upon those additional cost components that differ among the potential candidates for curtailment.

9. The striking disparities in conversion costs among different users of natural gas are increased still more if we introduce the costs of pollution abatement systems that would be needed to minimize the environmental impacts of widespread fuel switching. This equipment, too, is characterized by strong economies of scale; see *Natural Gas Policy Issues* (U.S. Senate, 1972).

10. The FPC later reversed itself in a series of inconsistent opinions.

References

Federal Power Commission (1972). *National Gas Supply and Demand 1971–1990, Staff Report No. 2,* Bureau of Natural Gas, Washington, D.C., February.

——— (1973a). Opinion 643, Docket No. RP-71-122, Washington, D.C., January 8.

——— (1973b). Order No. 467: "Statement of Policy," Docket R-469, Washington, D.C., January 8.

Jensen, James T., and Stauffer, Thomas R. (1973, forthcoming). "An Economic Rationale for Rationing Gas Supplies in the United States," Council on Economics, American Institute of Mechanical Engineers, New York.

MacAvoy, Paul (1971). "The Regulation-Induced Shortage of Natural Gas," Reprint No. 124, The Brookings Institution, Washington, D.C.

Office of Emergency Preparedness (1973). "The Potential for Energy Conservation: Substitution for Scarce Fuels," January.

U.S. Senate (1972). *Natural Gas Policy Issues,* Part I, Serial No. 92-22, pp. 426–509.

22 Market Structure and Regulation: The Natural Gas Industry

ROBERT S. PINDYCK*

22.1. Introduction

If there really is or soon will be an energy crisis in the United States, the first outward sign of it will probably be a shortage of natural gas. It has often been said that there is already a growing shortage of natural gas in this country, and that this shortage is at least partly the result of the Federal Power Commission's regulatory policies over the past decade. It is thus not surprising that there has recently been a growing interest in modeling and understanding the natural gas industry and the impact of FPC regulatory policy on that industry.

Several econometric studies have appeared in recent years which attempt to measure the impact of alternative ceiling prices on the production and sale of gas. Erickson and Spann (1971), for example, developed a multi-equation model of the supply of gas that related the ceiling price of gas to the number of wells drilled and in turn relates the number of wells drilled to eventual new discoveries and production of gas. And a recent study by Khazzoom (1971), which looked at the supply of gas by estimating equations for new discoveries and extensions and revisions using cross-section and time-series data, indicated that the supply of new reserves responds very elastically to the wellhead price, that is, a small increase in the ceiling price of gas would result in large increases in new discoveries in the future.

The problem with studies such as these is that they investigate either the demand for or the supply of gas but neglect the simultaneous interaction of these variables. Also, even if they did deal with demand and supply simultaneously, they ignore two important aspects of the natural gas industry that are critical to understanding its performance. The first of these is that there are actually two markets for natural gas: a field market in which producers sell gas from the ground to pipeline companies and a wholesale market where pipeline companies sell gas to residential and commercial buyers. These two markets are distinct, and they have their own demand-supply equilibria, even though they are certainly interrelated. The second characteristic of the natural gas industry which is very important is that it is a regional industry whose spatial characteristics are critical to its performance. The natural gas field market, for example, actually consists of several

* Massachusetts Institute of Technology.

regional field markets at different points in the country. These regional field markets correspond to areas of the country where gas producers sell gas to pipeline buyers. Similarly, the wholesale market actually corresponds to several regional wholesale markets, where pipeline companies all compete to sell gas to consumers. Thus, when one speaks of a shortage of natural gas, one must keep in mind that there may be a shortage of gas in one market region but an excess in another market region.

This paper will discuss briefly some of the important aspects of the market structure of the natural gas industry. We will show how the mechanisms that determine the production and sale of natural gas depend on the spatial characteristics of these two distinct markets, and how demand and supply disequilibrium is the result of constraints across both markets. This discussion will also provide a conceptual introduction to the econometric policy model of the natural gas industry that is being constructed at M.I.T. (MacAvoy and Pindyck, 1972), and we will conclude by summarizing some of the initial results of that study and their implications for natural gas regulatory policy.

22.2. Regional Field Markets and Regional Wholesale Markets

There exist two distinct but interrelated sets of regional markets for gas. The first of these consists of regional field markets, where reserve commitments for gas are sold by producers to pipeline companies. There are probably about six or eight such market regions in the country. The south Central Plains region, for example, would contain northern Arkansas, Kansas, Oklahoma, and Texas Railroad Commission District No. 10. This represents a geographical region of market exchange on the field level.

There also exist about five or six regional wholesale markets around the country, where pipeline companies sell gas to both final consumers and to public utilities for resale. Note that any single regional wholesale market might receive gas from pipelines originating in two or more different regional field markets. The spatial interrelationships of regional markets can thus be somewhat complicated, and this in turn is what complicates any attempt to measure the "excess demand" for gas. We will discuss these interrelationships in more detail later, but first we will briefly outline the functioning of each market separately.

22.2.1. Field Markets

In the field market the supply variable is the change in reserves, ΔR (these new reserves can then be "committed" to pipelines). The change in reserves comes about from additions to reserves of both nonassociated and associated gas that result from new discoveries and extensions and revisions, less subtractions from reserves due to actual production (a function of final consumption and past commitments of reserves) and also any net changes in underground storage. The accounting structure for reserve and production levels is actually somewhat complicated because of the existence of repressuring (which adds to reserves), field use, transportation losses, and so on.

The demand variable in each regional field market is the wellhead price that pipelines are willing to pay for new reserves. After 1961, however, this price variable became an exogenous ceiling price when and where it was lower than the price that would have resulted from a market equilibrium. (In fact, regulation of field markets became effective in 1961, the first year of "area ceiling prices," because these ceilings were then lower than existing market prices.)

The major component of additions to reserves is new discoveries of both nonassociated and associated gas (associated gas includes gas dissolved in oil that is recovered from wells that are drilled primarily for oil production as well as free natural gas in immediate contact with crude oil). The process of discovering gas begins with, and is driven by the drilling of wells.[1] Drilling in a particular area depends on costs per foot drilled in the previous period as well as economic incentives such as the price of oil [2] and the price of gas that producers expect to receive (which in turn are weighted distributions of past prices they did receive), as well as the success ratio and the average discovery size.

In the case of nonassociated gas, new discoveries of gas are primarily dependent upon the number of wells drilled, but they are also a function of the wellhead price of gas. The wellhead price affects new discoveries through the drilling of wells, which is already accounted for by the wells equations, but also because that price has a direct influence on the discovery levels that are *reported*. A lower wellhead price may result in fewer discoveries being reported.

Depending on the relative prices of gas and oil, associated gas can be viewed as an input factor for producing oil or as a separate output. As Khazzoom (1971) indicates, if the price of oil is considerably greater than the price of gas, then gas is to some extent an input for producing oil from reserves (since gas pressure forces the oil to the surface, and the higher the gas pressure, the higher is the oil extraction rate). In many regions, however, the supply of gas is regulated by state conservation laws that constrain the gas-to-oil production ratio to be within some range so as to conserve the "gas-input" resources. This effectively puts upper bounds on production. In this case, the supply of associated gas will be linked to the production and the price of oil. On the whole, one would expect new discoveries of associated gas to be relatively inelastic with respect to the price of gas but more a function of the price of oil.

Extensions and revisions for both nonassociated and associated gas are determined by previous new discoveries, the wellhead price, and the price of oil, but extensions and revisions for associated gas should also depend on the rate of production of oil, particularly in those regions that constrain the gas-to-oil production ratio.

The supply of new reserves, then, is just equal to total new discoveries plus total extensions and revisions. Subtracting from this marketed production, losses, and the change in underground storage yields the net change in reserves.

The variable that specifies the demand for new reserves by pipelines is the wellhead price P. Some time after 1961, however, the regulated ceiling price had to be below the price that would have resulted from a supply-demand equilibrium, since otherwise the ceiling would not prevail. When ceilings became effective, excess demand for reserves was a necessary result, and it is no longer possible to observe the demand curve after that point. This is shown graphically in Figures 22.1a (before ceilings) and 22.1b (after ceilings). Before ceilings are in effect, the endogenous wellhead price is a function of the change in reserves (it is this change in reserves that is

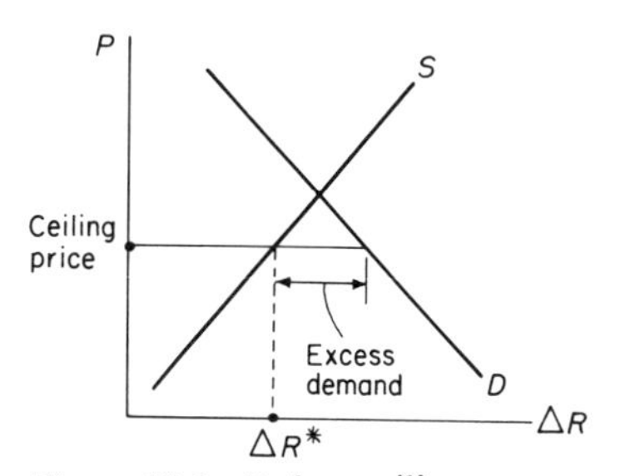

Figure 22.1a. Before ceilings.

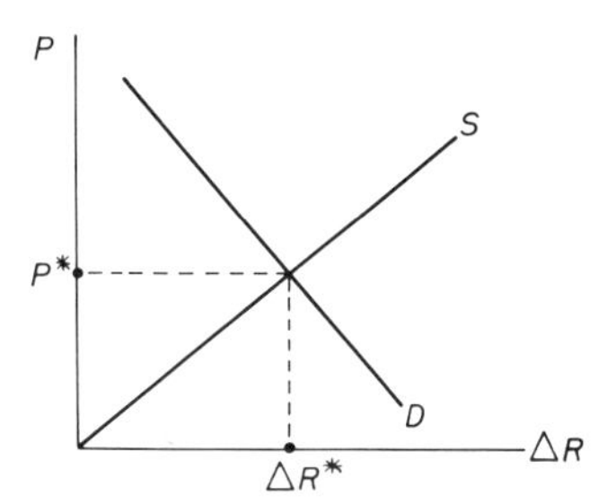

Figure 22.1b. After ceilings.

"demanded" as security reserves by the pipelines), the average mileage between the center of the production region and the distribution area, and the price of oil.

22.2.2. Wholesale Markets

In the wholesale market the demand for gas by public utilities and industrial consumers is a function of prices, the prices of alternative fuels, and exogenous variables such as population and income. Because the wholesale market is oligopolistic, there is no explicit supply curve, but wholesale prices are endogenous and are determined by a markup on the field price that depends on the marginal costs of transmission. Until recently it was safe to assume that wholesale markets cleared and there was no excess demand for gas by consumers. Thus, the equilibrium quantity of gas sold by pipelines (and therefore, excluding losses and inventory change, produced from reserves) is determined by the intersection of the consumer demand curve with a curve that specifies the pipelines' markup on the field price.

When the wholesale price is regulated to be a markup over the wellhead price and the marginal cost of transmission (as it is in the case of residential and commercial gas), excess demand can exist in the wholesale market. This is shown in Figure 22.2. The marginal cost curve for new production at the field level becomes almost vertical when reserve-production ratios become too low and rises to infinity at $\Delta Q_{\max}$. Thus, when demand increases from *DD* to *D′D′* and the wholesale price is a fixed markup over the regulated wellhead price, there will be excess demand for new production, as shown in the figure.

Note that excess demand for reserves *need not imply* excess demand for production. During the middle 1960s production demand was largely satisfied even though there was excess demand for reserves; what occurred was a

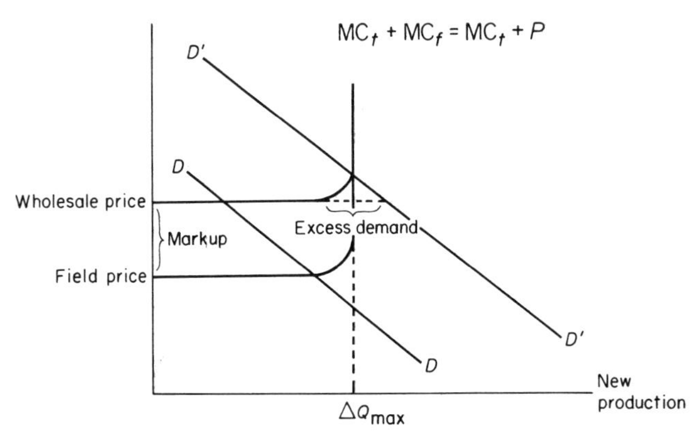

Figure 22.2. Excess demand in wholesale markets.

decrease in the reserves-to-production ratio. By 1970, however, reserve-to-production ratios were so low that resulting high marginal production costs resulted in excess demand for production as well.

22.3. Spatial Characteristics of the Gas Industry

As may have become clear in Section 22.2, equilibrium in the gas market requires joint equilibrium in the two distinct but interrelated sets of regional markets, namely field markets and wholesale markets. The way these markets are related depends largely on their relative geographical locations, as well as the degree of potential monopoly and monopsony power potentially afforded to pipelines in different regions. A hypothetical example of how these regional markets might be connected is shown in Figure 22.3. Approximately twenty production districts (indexed "*j*") determine the supply of gas reserves in six or so regional field markets (indexed "*i*"), while the supply and demand of gas on the wholesale level is determined in about five *different* regional markets (indexed "*k*"). As can be seen in the figure, a single field market may supply gas to two or more wholesale markets, and a single wholesale market may receive gas from two or more field markets.

As one would expect, the identification of regional field markets is not a straightforward matter. The AGA Reserve Committee has identified ten regional field markets, but these markets do not seem to be distinct, that is, there are reserve commitments across markets. One may also identify field markets by "eyeballing" the FPC map, which divides the United States into

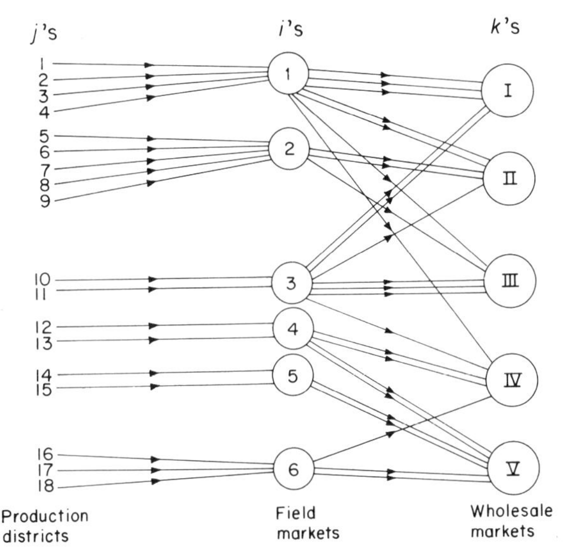

Figure 22.3. Hypothetical example of connection of regional markets: *j* production districts, *i* regional field markets, and *k* regional wholesale markets.

production districts and also indicates the placement of major pipelines. However, new pipelines continue to come into existence, and also the geographical location of production districts is not the sole factor determining markets.

Regional wholesale markets are defined and identified in much the same way. We define a "wholesale market" as a distinct geographical region where pipeline companies (in competition with each other) sell gas to public utilities and industrial consumers (who also are in competition with each other as buyers). The identification of these markets, also not straightforward, can be based on an examination of pipeline locations and regional data on sales of gas.

The difficulty in determining excess demand on the wholesale level (that is, where it affects consumers) may perhaps now be made clear. One would like to know at what point $\Delta Q_{\max}$ does the marginal cost curve in Figure 22.2 become vertical. This occurs at the point at which new production becomes constrained by the existing level of new reserves needed to support (on a security basis) that new production. Generally, people discussing the "gas crisis" have used an aggregate "guideline" constraint of the form

$$\Delta R \geq \alpha \, \Delta Q_{\max}, \tag{22.1}$$

with α approximately equal to 15 or 20. This, unfortunately, does not take into account the spatial interconnections between regional field markets and regional wholesale markets. One must instead consider a separate constraint for each wholesale market region k:

$$\Delta Q_k \leq \Sigma \frac{1}{\alpha_{i,k}} \beta_{i,k} \, \Delta R_i, \tag{22.2}$$

where this summation is taken over those i's that feed into the particular k, and $\beta_{i,k}$ is the fraction of ΔR_i going to region k. Thus, the change in consumption ΔQ for wholesale market II would be constrained by additions to reserves in production districts $j = 1, 2, \ldots, 12$ that feed into field markets 1, 2, and 3.

22.4. Some Initial Results from the M.I.T. Econometric Policy Model

The econometric policy model of the natural gas industry that is being built at M.I.T. is an attempt to provide a quantitative tool for examining the possible impacts of alternative regulatory policies. In developing the model, cross-section and time-series data were pooled and used to estimate equations that basically characterize the relationships described in Section 22.2.[3]

The model, at least in its preliminary form, does not however attempt to describe the spatial characteristics of the gas industry that we discussed earlier. One reason for this is that it is difficult to identify regional markets exactly, and in fact several alternative regional breakdowns could be specified which all seem to meet the criterion of representing actual economic markets. But this is a relatively small problem, since econometric tests could be used to compare the "market quality" of the alternative breakdowns. A larger problem is specifying the interconnections between field markets and wholesale markets. The difficulty here is that these interconnections change over time as new pipelines are built or as the capacities of existing pipelines change.

Nonetheless, the model seems to be useful in providing some rough indications of what we might expect from alternative ceiling price policies. Six simulations were run using the preliminary version of the model. In the first, the ceiling price of gas was fixed at its 1970 values (on a region-by-region basis). In the second, the ceiling price was increased by 1¢ per year, so that in 1980 it is 10¢ higher than its 1970 level in each production district. In the remaining simulations, it increases by 2¢ per year, 3¢ per year, 4¢ per year, and 5¢ per year. The results are summarized graphically for several key aggregate variables in Figures 22.4 through 22.6.

With no change in the ceiling price of gas during the decade 1970–1980, the number of wells drilled declines slightly, total new discoveries remain about constant, and extensions and revisions decline rapidly. Total demand for gas (that is, production), however, increases by about 11% per year, for a total increase of 150% over the decade. If this demand were to be somehow satisfied (by depleting reserves), we would run out of reserves by 1980. This demand, of course, would not be satisfied. By the time the reserves-to-production ratio dropped to about 7 or 8 (that is, by 1974), considerable rationing would occur, and many consumers would simply not be able to purchase gas.

Even with a ceiling price increase of 2¢ every year, demand for reserves would increase faster than the supply, and the reserves-to-production ratio would drop to about 2.3 by 1980. Many more wells would be drilled in this case, and both new discoveries and extensions and revisions would increase accordingly, but this increase in reserves would be outstripped by the more rapidly growing increase in production. Even with a 2¢ per year increase in the price of gas, increases in the prices of other fuels, in GNP, and in capital expenditures would result in a doubling of wholesale gas demand by 1980.

A ceiling price increase of 4¢ per year or 5¢ per year would result in increases in reserve additions that would be larger than the increases in wholesale demand, so that the reserves-to-production ratio would increase over the decade, reaching, by 1980, 23 for a 4¢ annual price increase and 41 for a 5¢ increase. This is the result of a dramatic increase in well drilling together with an increase in wholesale demand of only 4 to 6% per year.

If FPC objectives are to keep the reserves-to-production ratio approximately constant at around 12, then these results indicate that the ceiling price should be increased by about 3 to 3.5¢ per year. In this case, additions to reserves would be commensurate with increases in production demand, and the reserves-to-production ratio would remain about the same over the long term. There may, however, still be some excess demand during the first 3 or 4 years after the policy is instituted.

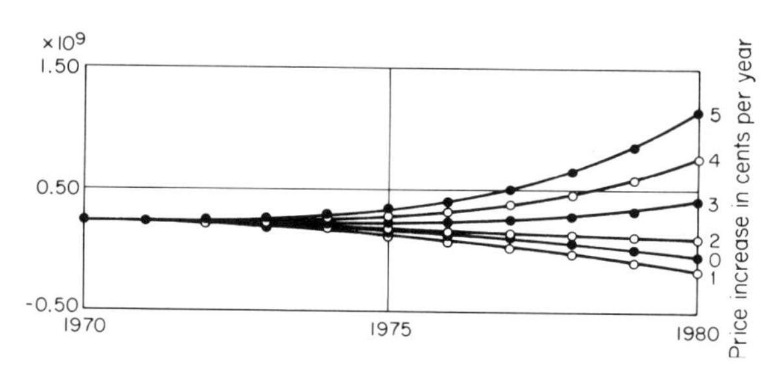

Figure 22.4. Total year-end reserves.

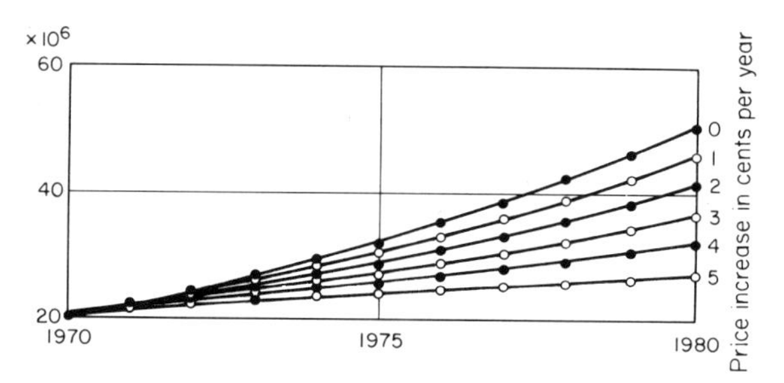

Figure 22.5. Total production.

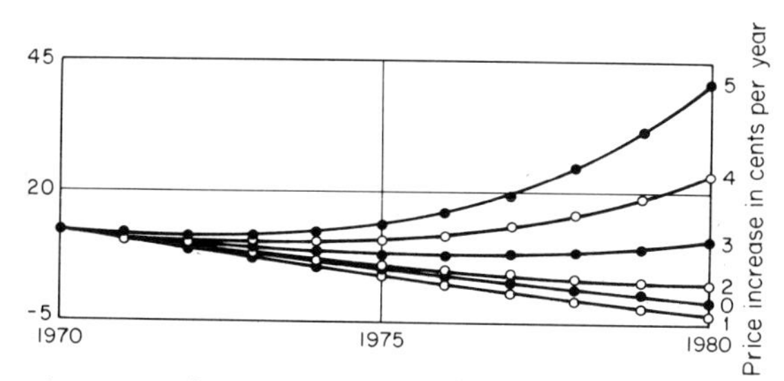

Figure 22.6. Reserves-to-production ratio.

It is interesting that the initial version of the model indicates that complete deregulation is not necessary to stabilize the growing shortage in gas. Rather, a *controlled* increase in the ceiling price could result in a constant reserves-to-production ratio. An unanswered question, of course, is whether the resulting stabilization of the *aggregate* reserves-to-production ratio is consistent with the spatial constraints described in Section 22.3. Perhaps this question will be answered by continued research into the structure of the gas industry.

Notes

1. It is useful to distinguish between four types of wells, all of which can be classified according to intent when they are drilled or by the result after drilling is completed (or abandoned). The classification according to intent by the American Association of Petroleum Geologists (AAPG) consists of drilling for new fields (field wildcats), drilling for new pools (new pool wildcats, shallower-pool tests, deeper-pool tests) or to extend presently known pools (outposts). After drilling, wells are classified as "dry holes" if they are unsuccessful, or as new field or new pool-discovery wells if a new field or pool is discovered, or extension wells if they extend the size of presently known pools. These wells can be further subdivided into oil or gas wells.

2. As Khazzoom (1971) points out, the price of oil has two effects: directionality, that is, gas discovered in the search for oil, as well as discovery of associated gas.

3. For a full description of the model, see MacAvoy and Pindyck (1972).

References

Alchian, A. (1958). "Costs and Outputs," in *Allocation of Economic Resources*, M. Abramowitz (ed.), Stanford University Press, Stanford, Calif.

Balestra, P. (1967). *The Demand for Natural Gas in the United States*, North Holland Publishing Company, Amsterdam.

———, and Nerlove, M. (1966). "Pooling Cross-Section and Time Series Data in the Estimation of a Dynamic Model: The Demand for Natural Gas," *Econometrica*, *34*(3), pp. 585–612, July.

Breyer, S., and MacAvoy, P. W. (1973). *Energy Regulation by the Federal Power Commission*, The Brookings Institution (Washington, D.C.), forthcoming, Spring.

Erickson, Edward W., and Spann, Robert M. (1971). "Supply Price in a Regulated Industry, The Case of Natural Gas," *The Bell Journal of Economics and Management Science*, Vol. 2, No. 1, pp. 94–121.

Garfield, P. J., and Lovejoy, W. F. (1964). *Public Utility Economics,* Prentice-Hall, Englewood Cliffs, N.J.

Houthakker, H. S., and Taylor, L. D. (1966). *Consumer Demand in the United States, 1929–1970,* Harvard University Press, Cambridge, Mass.

Khazzoom, J. Daniel (1971). "The FPC Staff's Econometric Model of Natural Gas Supply in The United States," *The Bell Journal of Economics and Management Science,* Vol. 2, No. 1, pp. 51–93.

MacAvoy, P. W. (1962). *Price Formation in Natural Gas Fields,* Yale University Press, New Haven, Conn.

——— (1971). "The Regulation-Induced Shortage of Natural Gas," *Journal of Law and Economics, 14*(1), pp. 167–199, April.

———, and Pindyck, R. S. (1972). "An Econometric Policy Model of Natural Gas," Sloan School of Management Working Paper #635-72, Massachusetts Institute of Technology, Cambridge, Mass., December.

Wellisz, Stanislaw H. (1963). "Regulation of Natural Gas Pipeline Companies: An Economic Analysis," *Journal of Political Economy, 71*(1), pp. 30–43, February.

23 Cost Trends and the Supply of Crude Oil in the United States: Analysis and 1973–1985 Supply Schedule Projections

HENRY STEELE*

It is useful to begin by conceding that it is impossible to make truly defensible long-run supply schedule projections for oil and gas in the United States. However, some aspects of the projection apparatus are more impossible to implement than others. The major problems are well known. First, there is no way to allocate truly joint costs of oil and gas exploration which is consistent with the analytical conscience of the microeconomic theorist. Second, there are the familiar limitations in the available statistical data. Although these data have steadily improved in both quantity and quality over the years (and particularly since 1966), too little detail is available with regard to geographical breakdowns for other outlays than drilling, and the highest-quality, consistently defined time series usually go back no earlier than 1966. Third, the statistical data that are most interesting from the point of view of long-run supply schedule analysis are usually not available at all. We have data on total industry expenditures for a number of cost categories, but nowhere is there to be found any consistently organized statistics that would permit computation of long-run, or even short-run, marginal costs—again, with the notable exception of drilling costs. Fourth, even if marginal production costs were to be known, there are the further difficulties of translating cost-per-barrel data into the supply-price versus production-rate calculus of the supply schedule. The complexities of tax accounting and tax treatment in the oil and gas industry (consisting as it does of many firms that differ greatly in terms of size, degree of vertical integration, diversification, reliance upon foreign operations, and the like) make it quite difficult and uncertain to determine what portion of a given expenditure is capitalized, to say nothing of the complexities of estimating the time profile of exploration, development, and production in a given reservoir, and the assigning of an equilibrium rate of return on investment.

It is certainly interesting to study past trends in the real costs and alleged money costs of such "separate" activities as predrilling exploration, exploratory and development drilling, and field production, but it is quite another matter to attempt to combine the available data into a rigorously defensible long-run supply schedule along theoretical lines. Can it be done? In his re-

* University of Houston.

cent book on the world petroleum market, Professor Adelman (1972) states: "In my opinion, without information on the distribution of fields and wells by capacity and cost, no long-run incremental cost function can be reckoned." I entirely agree. And yet, even in the absence of such information, essential as it is, there is increasing activity in the field of systematic investigation of future petroleum supply.

The existence of long-run supply-and-demand projections for petroleum may or may not demonstrate the property of foresight on the part of the forecasters, but it does prove the existence of a definite supply of, and demand for, oil and gas market forecasts. Supply forecasts, in particular, appear to be demand determined. Formerly there was curiosity about future supply; now it is closer to apprehension if not anxiety. Under these circumstances, what economists call Say's law is reversed: demand creates its own supply. Where there is effective demand for projections, the supply will be forthcoming. For example, the National Petroleum Council (1972) has recently presented a long-range study on the future energy market which promises to provide much interesting and potentially valuable new material for supply projections, even though the NPC did not adopt the long-run marginal cost supply schedule frame of reference. And my own studies of petroleum cost trends, originally motivated by analytical curiosity, have been increasingly directed into areas less rewarding in terms of neat theoretical formulations and more demanding in terms of the provision of long-run forecasts. In 1969 I was requested by the Senate Subcommittee on Antitrust and Monopoly to present them with data on petroleum supply and on the effects of conservation regulation on the supply schedule. Due to two fortunate accidents—the conducting of a study by the U.S. Bureau of Mines (1967) with regard to oil production by depth of well and rate of well production, and the publication of a table by Eggleston of well operation costs by production rate[1]—I was able to develop a rough approximation of the short-run marginal cost curve for oil lifting costs for 1965, but I could see no basis for estimating any long-run supply relationships. Even though I was strongly urged by the Senate Subcommittee to take some stand on long-run supply, I concluded in my statement to them that the difficulties of so doing were prohibitive (Steele, 1969).

In the next few years, however, I had some second thoughts. First, better data were being made available by the American Petroleum Institute (API) in regard to costs and reserves. Second, long-run supply schedules were in fact being drawn up for crude oil. Most notably, the Cabinet Task Force on Oil Import Control (1970), drawing upon substantial amounts of data provided to them by individual oil companies, had prepared a plausible long-run supply projection to 1980. When I was asked by Resources for the Future, Inc. in 1972 to make long-run supply schedules through the year 2000 for oil and gas and other energy resources, I rather rashly agreed. I later came to realize that, having been identified as an habitual offender in the petroleum cost study area, I was in real danger of receiving a life sentence. The remainder of this paper is a report on the research that I have done for Resources for the Future, Inc. in recent months. Although tentative results have been reached, the work continues to remain in progress.

Like other economists, I have been somewhat unhappy with the way in which the joint cost problem is often apparently rather casually resolved by some arbitrary allocation of joint costs between oil and gas on the basis of relative prices or relative Btu content. (American Petroleum Institute economist Tony Copp aptly refers to this allocation process as "the theory of administered cost.") I was determined to do better and began by formulating my analysis in an explicit joint production and joint supply framework. In the long run, oil and gas are certainly joint products with variable proportions, and in theory separate marginal cost schedules can be drawn up for each product. Thus, interdependent supply functions could be established. Under these circumstances, the supply of oil would depend not only upon the price of oil but also upon the price and production level of gas. Appraising the impact on supply of a price change for either product would require the sorting out of the substitution and output effects as they affected both products.

The outcome of these efforts may be readily summarized: complete failure. No statistically significant relationships could be discovered. There are probably two major reasons for this disappointing result. First, the hypotheses suggested by theory had to be considerably blunted in order to accommodate the limited quality of available data. Second, and probably more important, during the entirety of the period for which the best data are available, gas prices were under FPC control. It appears that there was not enough price change for gas to allow observation of the presumed underlying joint supply mechanism. Also, the reaction of producers to price control is difficult to appraise through time. The time lag in supply response is complicated here because it appears that for some period of time the producers frequently felt that price control would be repealed or, even if not repealed, would prove incapable of implementation. In this respect, some producers were more optimistic than others, but it is impossible to ascertain the rate at which gas-finding efforts dropped off in direct relation to price control.

Even after I joined the ranks of the cost administrators, no better results were obtainable. The employment of various "directionality" assumptions for cost allocation, in conjunction with the study of time trends (or trends in cumulative discoveries, cumulative production, or cumulative drilling) for the major per barrel cost components of finding, developing, and producing oil and gas did not permit the development of any statistically significant relationships that could be used for long-run projections. Typically, projections based on cumulative drilling or discoveries tended to increase explosively as the time horizon progressed through the 1980s. Projections based merely on time trends also increased implausibly rapidly before 1985. One reason for this phenomenon is of course known to everyone in this field: the tendency of proved reserves assigned to new discoveries in the early years of field development to be extremely conservative. Unfortunately, no one agrees as to what sort of adjustment can be made for this largely unavoidable underestimation. I have, however, observed one expedient employed by some analysts in the oil industry. When making projections of this sort, instead of completely disregarding the most recent years, they tend to fit linear curves to the data even where a curvilinearly increasing form would provide a better fit, and there may be a sort of wisdom in this practice.

It is perhaps not too surprising that statistically significant projections of past oil and gas cost trends remain elusive. At a minimum, the money cost data must be expressed in constant dollars, and available deflators are at best imperfect. In general, the wholesale price index of industrial commodities (that is, exclusive of farm products and foodstuffs) seemed to give better results than the implicit GNP deflator or other indices. For joint cost "administration" purposes, different bases seemed preferable for different applications. In the allocation of total drilling costs, for example, dry hole costs were allocated between oil and gas on the basis of the ratio of oil or gas footage. Thus, it became possible to distinguish between those dry holes which did not find oil and those dry holes which did not find gas. For estimating field production costs of oil and gas, however, it seemed preferable to rely upon technical information gleaned from trade journals, such as the generalization that it costs about three times as much to produce (from the wellhead) the average Btu from oil as from gas.

The main difficulty in charting cost trends is "matching" oil and gas discovered and/or developed with the relevant costs of discovery and development. Ideally, one would like to match all oil ultimately producible from a new reservoir against all the costs, direct and indirect, of discovering that reservoir. Matching new field and pool discoveries in a given year against total "finding costs" incurred in that year is an extremely poor approximation, for well-known reasons: conservatism of original reserve estimates, the fact that exploration expenditures are typically incurred over a period of several years prior to the discovery of a given reservoir, and all the inherent faults in the arbitrary allocation of the total of many types of exploratory costs (both successful and unsuccessful) against the total of proved reserves from a wide variety of reservoirs. Matters are improved only slightly by revising the denominator to include an estimate of future ultimately recoverable reserves from the newly found reservoirs. The latter estimate is subject to wide margins of error until quite a few years of development have occurred. Furthermore, subsequent costs of an exploratory nature may be required in order to confirm the magnitude of the reserves "discovered" by the initial-year finding efforts.

Industry analysts often try to avoid this difficulty by matching finding costs against the total of new discoveries, extensions, and revisions during the current year. Here, too, everyone is aware of the problems involved in resorting to this expedient: extensions and revisions relate to past rather than current finding activity, and are primarily the result of current development efforts instead of current finding efforts—to mention only the most important objection to this procedure.

A standard objection to naïve trend analysis projections relates to the implicit assumption that the underlying structure of the investigated relationships remains constant over time. Many factors influencing the structure of the petroleum industry have varied significantly during this century, including not only technology but also such regulatory arrangements as taxation, conservation regulation, and import controls. (To these we may soon add environmental measures, already allowed for in the National Petroleum Council projections.) In terms of long-run trend analysis, the petroleum industry appears to have a longer "memory" than most industries. There appears to be a genuine continuity in trends over a period of decades, and this is particularly true in the analysis of finding costs. In large part this stems from the circumstance that exploration in the United States during the period of historical data availability has consisted largely of the ever more intensive exploitation of a fixed universe of discovery possibilities. Another way of saying this is to note that there has been no fundamental breakthrough in oil-finding methods since the introduction of intensive geophysical methods in the mid-1930s. Improvements and refinements are continually made, and we are repeatedly promised that in a few years the perfection of existing techniques will reveal the fugitive secrets of the earth's crust, but so far no new breakthrough of the type that would open new horizons and significantly extend the universe of discovery possibilities has been made. Hence, to the extent that the data can be relied upon, it is useful to make finding-cost trend projections based on time series extending back to the 1930s.

Trends in development costs are almost equally difficult to appraise with precision. The costs in money terms of development activities undertaken during a given year are probably more representative of all the development activities which contribute to newly developed reserve during that year, than in the case of a parallel computation of finding costs. The time profile of development activities in connection with a given field or reservoir is more regular and concentrated in time than in the case of the exploration process. On the joint cost allocation side, the magnitude of this issue in development is less than in exploration, although one must still come to terms with the allocation of the costs of development dry holes. Matching expenditures and reserves, however, is again a major perplexity. Even if we assume that the annual total of development costs in dollars is a good measure of the activity resulting in the development of those reserves made additionally producible during the year, we are hard put to find a precise measure of those barrels of reserves developed in that year. Successful development wells typically extend the proved areal extent of a reservoir, and the "extensions" category of reserve additions should measure the success of this aspect of developmental activity. However, development costs include not only the drilling of development wells but also the costs of equipping wells and leases not only for primary but also for additional (secondary and tertiary) recovery. The "revisions" category of reserve additions is designed to measure this phenomenon, but only imperfectly. The American Petroleum Institute (1972) (API) Reserves Volume for 1971 defines Revisions as follows:

REVISIONS. Both development drilling and production history add to the basic geological and engineering knowledge of a petroleum reservoir and provide the basis for more accurate estimates of proved reserves in years following discovery. Changes in earlier estimates, either upward or downward, resulting from new information (except for an increase in proved acreage) are classified as "revisions." Revisions for a given year also include (1) increases in proved reserves associated with the installation of improved recovery techniques; and (2) an amount which corrects the effect on proved reserves of the difference between estimated production for the previous year and actual production for that year.

It would appear that only those reserves included in category (1) should be credited directly against the costs of installation of secondary and tertiary recovery facilities. I am not aware of any information that would allow one to estimate how large a fraction of total revisions stem directly from additional recovery installations, but the wording of the API definition suggests that it may be less than half of the total. Assuming discrepancies between actual and estimated production during a given year to be minimal, the major fraction of revisions would appear to be due to reevaluations of the probable primary recovery potential from a field or reservoir, apart from the extension activities of the typical development well. This increase in reserves could not logically be attributed to the development activities of the current year only but would rather be a function of the cumulative production history of the reservoir. Recently, however, the NPC has adopted the practice of considering all of the revisions category as resulting from installation of improved recovery techniques, arguing that only a very small fraction of revisions reflects improved primary recovery. (This decision was not adopted without some dissent on the part of the members of the oil supply task group of NPC.) This treatment is rather surprising, but if justified it implies that a reasonably good estimate of annual per barrel development costs could be obtained by dividing total development costs by the sum of extensions and revisions.

The time trend of development costs per barrel has been influenced by structural changes in the petroleum industry. Technological changes have been pronounced in the postwar period, with increasing emphasis on additional recovery installations, as higher finding costs caused companies to substitute development for exploration at the margin. More distressing from the statistical point of view, however, was the influence of increasingly strict conservation regulation in the 1950s and 1960s. As is well known, the emphasis on per well allowables in market-demand prorationing states stimulated the drilling of many unnecessary wells in order to obtain more production, and this considerably increased the cost of field development. The amount of the cost increase is impossible to quantify, but estimates in the range of 50¢ per barrel have been cited.[2]

Per barrel production costs are more readily computed on a time trend basis since both the costs and the production relate to the same time period. Hence, the problems of matching costs and barrels are greatly reduced. Nevertheless, the data on production costs, as currently reported, are far from satisfactory. The Joint Association Survey publishes national totals, but there is no breakdown between oil and gas, even where it is feasible, particularly in the case of severance of production taxes. Furthermore, although the data would be capable of collection, there is no information relating to production costs by well depth and production rate. This is the sort of information that is crucial in the estimation of short-run marginal costs and hence supply relationships.

Production costs, like development costs, have been considerably influenced in the 1950s and 1960s by the impact of conservation regulation in increasing the number of producing wells and reducing the production rate of flowing wells greatly below the optimum. In appraising the historical record, it is difficult to allow for the impact of this factor on producing costs, although during the period of stringent regulation it was substantial. As mentioned earlier, only for 1965 are data available that permit an estimate of the cost impact of prorationing. On the basis of these data it would appear that the marginal cost of lifting some 2.5 billion barrels of crude oil under prorationing was about 74¢, while it could have been reduced to about 36¢ in the absence of prorationing. (The gap between average costs with and without prorationing would of course be less per barrel.) Fortunately, this sort of distortion in the efficiency of resource allocation is largely a thing of the past now that excess producing capacity has been virtually eliminated.

In the following paragraphs, an attempt is made to estimate long-run supply for oil by combining the foregoing trend analysis (which is of course expressed in terms of average costs) with data relating as closely as possible to the presumed lifetime exploration, development, and production pattern of a "typical well." Industry analysts sometimes use the concept of the typical well in order to determine the profitability of development of a known reservoir or to estimate the rate of return on an exploratory venture over its entire life, assuming a typical pattern for costs of, and production from, exploratory and developmental wells. My model for the typical well has been developed from a variety of sources, including discussions over a period of time with economists and production engineers from several major integrated oil companies and one large independent producer, and from a study of the voluminous literature on production patterns from various types of fields. Data from the typical well model are used to estimate the extent to which the indicated average per barrel costs of the various stages are capitalized, so that the total outlay can be reduced to its capitalized component. The capitalized cost is then multiplied by a factor that places the outlay on a discounted cash flow basis over the entire time horizon of the typical well, from initial predrilling exploration (in the case of finding costs) to final depletion of all production. This capitalized cost is expressed in 1972 dollars per barrel.

The development of this hybrid approach involves the computation of per barrel finding and development costs, using time-series data for available years from 1936 through 1971. The long-term trend in production costs is then computed on a current basis without capitalization. Although the observations are on an annual basis (with the exception of some years for which no cost data were available), the basis for the trend projection is cumulative crude oil discoveries since 1859, as reported by API and the U.S. Bureau of Mines, in the case of finding and development costs, and cumulative production since 1859, in the case of production costs.

The total minimum supply price of crude oil finding, development, and production can then be obtained by adding up per barrel finding, development, and production costs for a given level of cumulative discoveries and production. The price can then be related to the production rate by relating production to reserves through the decline rate built into the typical well calculation. Thus, the pattern of prices over time will depend upon the rate of increase in annual production and consumption, with import levels capable of entering the analysis as a policy variable.

As thus conceived, the long-run supply schedule is based on annual trends in average costs but as modified in the direction of marginal considerations by use of typical well data to compute necessary supply prices for the required new investment in each time period. The supply projection is thus hybrid in nature, mixing elements of average and marginal cost. Crucial to the computation is the representativeness of the "typical well" model, and the assumption that the characteristics of the typical well have been substantially constant over the entire time period. It would be more desirable to have a model of the typical reservoir, rather than the typical well, since the reservoir is the logical unit of production. Ideally, data should be obtained which would make possible the construction of models of the typical well, typical reservoir, and typical field, with the necessary interrelationships elucidated. Until such data become available, however, it may be instructive to do what may be done within the more restricted confines of the typical well analysis.

Limitations on available space dictate a rather abridged description of the analysis. The data are basically production and reserves statistics from API and cost data taken from the Joint Association Surveys and prior cost studies done by the oil industry done during World War II for the Petroleum Administration for War. The data cover the United States, excluding only the Alaskan North Slope, for which it is preferable to make separate estimates.

The series on finding costs is analyzed in the following manner. First, take total finding costs as reported in each year, including such rent elements as lease bonuses, and attempt to adjust the earlier year costs in such a way as to make the division between exploration and development costs consistent with the data since 1965. Prior to 1966, all dry hole costs were deemed exploratory costs, and the costs of all productive wells were allotted to development. To remedy this rather undesirable cost division, a reallocation of costs was performed such that the costs of both dry and productive exploratory wells were included in finding costs, and similarly for development wells and costs. The allocations were made on the basis of numbers and footages drilled of dry and successful exploratory and development wells. Next, divide costs between oil and gas wells on the basis of relative total footage drilled in successful oil and gas wells. Then multiply oil finding costs by the fraction of finding costs expensed for tax purposes; the remainder represents the capitalized finding cost. This finding cost is then reduced to a per barrel basis by dividing total capitalized finding cost by the most recent (1971) estimate of ultimately recoverable reserves currently credited to fields discovered in each year. Data for the last 6 years were adjusted upward by a factor derived by analysis of the trend over time in the expansion of initial estimates of reserves as revealed in later reestimates. Finding costs per barrel of reserves discovered were then converted into finding costs per barrel of oil ultimately produced by reference to the characteristics of the typical well. For such a well, predrilling exploration is conducted at varying levels of intensity over a period of 7 years prior to drilling. Production commences shortly after drilling, and continues at an assumed constant decline rate for 25 years. If we assume that exploration is risky enough to require a 12% rate of return on investment, then the capitalized finding cost per barrel is multiplied by a factor of 3.09 in order to convert dollars per barrel found to dollars per barrel eventually produced.[2] The current dollar finding cost is then deflated by the wholesale price index for industrial commodities, with the base converted to 1972 = 100, in order to obtain a long-term trend in per barrel finding costs versus cumulative discoveries, expressed in constant 1972 dollars.

Rather curiously, a linear regression line provides the best fit. The finding cost schedule is

$$Y = -43.22 + 1.456X \qquad (Y \text{ in cents; } X \text{ in billion barrels}),$$

where Y is finding costs in 1972 cents per barrel, and X is cumulative discoveries (production plus proved reserves) since 1859. The coefficient of determination is .646, and the regression coefficient is significant at the 5% level.

Thus, it would appear that finding costs increase 1.46¢ per barrel for every additional billion barrels of cumulative discoveries. As of the end of 1971, cumulative discoveries totaled 125.1 billion barrels, so that the regression-derived relationship would indicate a finding cost of $1.39 per barrel, while the computed actual cost of finding oil in 1971 was $1.45. What implications does this have for the future? The NPC study, in its most optimistic domestic oil supply projection, projects total oil consumption in the United States at about 65 billion barrels from 1972 through 1985. If 15 billion barrels were to come from the North Slope (which of course is not included in the above regression), then 50 billion more barrels would have to be found in the contiguous United States if total reserves were not to decline. The need to discover 50 billion more barrels by 1985 would increase finding costs to $2.12 per barrel by the end of the period.

The same process is repeated in order to estimate the trend in per barrel development costs. Here, the factor applied to capitalized development investments is smaller than for finding investments because of the absence of significant predrilling development expenditures and the greater proximity in time between development drilling and the depletion of the additional reserves developed. The factor involved is 1.72, in contrast to the 3.06 factor used for finding costs. Since development is a less risky investment than exploration, a rate of return of 10%, rather than 12%, is used for this computation.

The long-term trend of development costs for the entire period 1936–1971 did not produce a statistically significant regression. Matters were improved slightly by limiting the period to the years 1959–1971, where the statistical series is more consistently defined. Even here, the relationship derived was at best marginal. Again, a linear fit was obtained. (It might be observed that a linear trend over cumulative discoveries translates into a curvilinearly increasing trend over calendar time.) The equation obtained is

$$D = 27.2 + .518X,$$

where D is discovery costs in 1972 cents per barrel and X is cumulative oil discoveries since 1985 (excluding the Alaskan North Slope). As of 1971, the regression value for development costs was 92.0¢ per barrel, while the actual value for 1971 was 90.5¢.

The time trend for production costs is computed by taking total producing costs (including taxes), dividing them up between oil and gas, and then matching the current oil production costs against annual oil production. The resulting per barrel costs are then converted into constant 1972 dollar costs by use of the wholesale price index of industrial commodities. Again, no significant trend can be plotted for the entire period, and it is necessary to use the more restricted period 1959–1971 where the statistics are more comparable in quality. For this period, there is a slow decline in the production cost of oil, and a significant increase in the production cost of gas. It is quite likely that the decline in oil production costs resulted largely from the phasing out of excess capacity and the resultant easing of prorationing of flowing well output. There would be no correspondingly notable effect for gas; hence, its increasing cost trend might reflect its growing physical scarcity.

The trend in oil production costs is

$$P = 108.3 - .2733Q$$

where P is production costs in 1972 cents per barrel and Q is cumulative production since 1859 (excluding North Slope) in billions of barrels. The coefficient of determination is .898, and significance is at the 1% level. The computed production cost for 1971 is 82.9¢ per barrel, and the actual value for 1971 is 82.4¢.

It thus appears that producing costs have been declining at the rate of about ¼¢ per barrel since 1959. It is possible that the decline may cease when excess capacity is completely eliminated and all prorationing is at an end. In any event, it does not make too much difference whether we assume constant production costs or a decline of .273¢ per barrel per billion barrels of cumulative production.

Combining these three trends presents something of a problem. Until more is known about the development of the typical reservoir, as distinct from the economics of the typical well, synchronization in time of finding and development costs will be imperfect. If we simply add together the 1971 values for the three cost components, we obtain a supply price of $3.14 per barrel to yield 12% on finding investment and 10% on development investment. If we assume that another 55 billion barrels of crude oil will be discovered between 1971 and the end of 1985, the supply price increases to $4.09 per barrel in 1972 dollars. This would probably translate into a 1985 production rate of about 12.5 million barrels per day from the contiguous United States, which contrasts with 1971 production of 8.9 million barrels per day at an average wellhead price of $3.31. A mechanical application of the formula for computing supply elasticity yields a coefficient of 1.13, which should probably be accepted only with reservations. The computed 1985 price is probably somewhat too low, for two principal reasons. First, some additional time lag should be incorporated in the capitalization of development costs in order to reflect the average time period between discovery and development, weighted by consideration of the ratio between those reserves producible from discovery wells and those producible only through additional development wells. A second reason is the omission of royalties from the analysis. There is a fairly simple way of making an adjustment for royalties paid, at least to a first approximation. The total royalty share in production is estimated today at 14.5 to 15% of all oil and gas produced. Some of these royalties, however, represent interindustry payments. The NPC study estimates that royalty payments net of interindustry transfers is about 12.2% of production. Hence, it could be argued that the reserves discovered, developed, and produced should be multiplied by .878 in order to obtain net reserves in terms of oil operator working interest. Hence, all computed costs should be multiplied by 1/.878, which would increase costs by 13.9%. This adjustment would increase the 1985 price to $4.66 and reduce the computed elasticity coefficient to 0.99, or almost precisely unity. There is, however, an approximate offset, in that natural gas liquids have been excluded from consideration, although in the past decade their production and discoveries have been on the average more than one-seventh that of crude oil. Hence, the inclusion of natural gas liquids would cut unit costs by about the amount of the net royalty payments on total liquid hydrocarbons.

In conclusion, it should be repeated that these supply schedule computations do not distinguish between production costs and rents but include all rent elements except royalties on production. Hence, the projection assumes the general continuation of existing conditions and does not attempt to predict what might happen if, for example, lease bonuses or production taxes were to be reduced. From the point of view of a discounted cash flow analysis, reduction in lease bonuses would not have a major impact on finding costs because the great majority are soon expensed. Elimination of production and severance taxes, however, would have a more appreciable impact—the NPC study projects constant future *ad valorem* and production tax rates at 5.6% of oil prices and 9.3% of gas prices.

Notes

1. W. S. Eggleston, unpublished communication to M. A. Adelman. See Adelman (1972, p. 286).

2. See Stauffer (1972) for a more general computation of a finding cost capitalization factor covering an idealized production pattern from an individual well. Using the same 12% rate of return, Stauffer arrives at a factor of 3.82, using a longer life (25 versus 30 years) and a lower decline rate (5 versus 7.4%) than is employed in the "typical well" calculations.

References

Adelman, M. A. (1972). *The World Petroleum Market,* for Resources for the Future, Inc., The Johns Hopkins Press, Baltimore, p. 68.

American Petroleum Institute (1972). *Reserves of Crude Oil, Natural Gas Liquids, and Natural Gas in the United States and Canada as of December 31, 1971,* Washington, D.C., p. 16.

Cabinet Task Force on Oil Import Control (1970). *The Oil Import Question,* Washington, D.C., p. 240.

National Petroleum Council (1972). *U.S. Energy Outlook—A Summary of the National Petroleum Council,* Washington, D.C.

Stauffer, T. R. (1972). "Estimated Economic Cost of U.S. Crude Oil Production," Paper SPE 4129 presented before the October 1972 Meeting of the Society of Petroleum Engineers, San Antonio, Texas.

Steele, H. B. (1969). "Government Intervention in the Market Mechanism," Testimony in U.S. Senate, Committee on the Judiciary, Subcommittee on Antitrust and Monopoly, Washington, D.C., pp. 439–444.

U.S. Bureau of Mines (1967). *Information Circular 8362: Depth and Producing Rate Classification of Oil Reservoirs in the Fourteen Principal Oil Producing States,* Washington, D.C.

24 Electricity Demand in the United States: An Econometric Analysis*

T. D. MOUNT, L. D. CHAPMAN, AND T. J. TYRRELL†

24.1. Introduction

The quantity of electricity demanded in the United States has grown consistently since the end of World War II. If past trends are extrapolated to the year 2000, the quantity demanded will increase to at least six times the 1970 level. It is questionable, however, whether these trends will remain unchanged in the future. In economic theory, demand is related to various causal factors, and if the direction of change of one or more of these factors is reversed, demand will be affected. For example, the relative price of electricity has decreased in the past but will almost certainly increase during the next decade. This increase will retard the growth of demand for electricity unless price is an unimportant determinant of demand. The main objective of our analysis is to determine the magnitude of the relationships between the quantity of electricity demanded and causal factors such as price. The five factors identified are population, income, the price of electricity, the price of substitute fuels such as gas, and the price of complementary products such as household appliances. The demand elasticity[1] of each factor is estimated, and its magnitude indicates how important the factor is as a determinant of demand.

Annual values of the explanatory factors (particularly population, income, and the relative price of electricity)[2] exhibit very strong trend components between 1946 and 1970 and slightly less since 1970. Consequently, the correlation between factors is high and estimation of the elasticities is difficult. Even though most of the variation of demand can be explained by a variety of simple models, the fact that a model provides a close fit to the data need not imply that the estimated elasticities or projections of future

* Research sponsored by the National Science Foundation RANN Environmental Program at the Oak Ridge National Laboratory, and by the New York State College of Agriculture and Life Sciences at Cornell University. A longer version of this paper, which contains more information on the statistical aspects of the analysis, has been published by the ORNL-NSF Environmental Program (ORNL-NSF-EP-49), Oak Ridge National Laboratory, Oak Ridge, Tenn. 37830, June 1973. The authors wish to thank M. E. Czerwinski, M. M. Slanger, H. Wang, N. L. Brown, J. M. Ostro, and J. K. Baldwin for their assistance in conducting the analysis and preparing this manuscript.

† T. D. Mount and L. D. Chapman are assistant professors at Cornell University, Ithaca, N.Y. 14850, and T. J. Tyrrell is a statistical programmer at the Oak Ridge National Laboratory, Oak Ridge, Tenn. 37830.

demand derived from this model are reliable. Results from some preliminary analyses suggest that accurate estimators of the elasticities cannot be obtained from a single time series of observations. In fact, with this type of data, it is not always possible to show that any factor other than autonomous growth is statistically significant. On the other hand, it is doubtful that estimates derived from a single cross section of observations are satisfactory for making projections over time. Consequently, we rely on pooling both cross-section (states) and time-series (years) observations to provide a suitable data base for estimation.

In many empirical studies of the demand for electricity, a constant elasticity model (CEM) is specified. However, the use of cross-section data from different states introduces an additional problem into our analysis. Estimates of the CEM are statistically sound only if the elasticity for each factor is the same in each state. For this reason, a variable elasticity model (VEM) is used to permit some degree of heterogeneity between states. In the VEM, the value of the elasticity for a particular factor depends on the level of that factor. In addition, models for residential, commercial, and industrial consumer classes (R, C, and I) are estimated separately.

The correct specification of a model should also allow for the gradual adjustment of demand through time in response to changes in the causal factors. This lagged response reflects the relationship between the use of electricity and existing stocks of electrical equipment and appliances. The size of these stocks depends on past as well as current decisions and, consequently, on past and current levels of the explanatory factors. A geometric lag structure is specified in our model, and both short-run elasticities (percentage change of demand in the current time period) and long-run elasticities (percentage change of demand after the response is completed) are estimated. It is the long-run elasticities, however, that have the most direct bearing on the future growth of demand for electricity.

The exact specifications of the CEM and VEM are summarized in Appendix 24.A. Both models can be linearized, and standard linear regression procedures may then be used to estimate the elasticities. With a geometric lag specified, one of the variables used to predict the current quantity demanded is the quantity demanded last year.[3] The presence of this variable implies that the familiar ordinary least-squares (OLS) estimator may be statistically unreliable. For this reason a more reliable alternative, an instrumental variable (IV) estimator, is used for the VEM as well as the OLS estimator.

It is convenient at this point to summarize other empirical analyses of electricity demand. In Table 24.1, seven recent studies are cited, and information on the source of data, the factors considered, the type of model and the estimation procedures used is provided. However, no attempt to appraise the specific empirical results in these studies is made until the final section.

24.2. The Empirical Results

Annual observations were obtained for 47 contiguous states[4] in the United States from 1947 to 1970. The VEM was estimated using two alternative procedures (OLS and IV) for each of the three consumer classes (R, C, and I). In addition, the OLS estimates of the CEM were computed for comparative purposes, as this model is the most widely used in other studies (see Table 24.1). The estimated coefficients that are summarized in Appendix 24.A refer to the final form of each model used in the analysis.

The major conclusion is that the price of electricity is a more important determinant of demand than either population or income in terms of the long-run elasticities (LRE). Demand for all three consumer classes is generally elastic with respect to price and inelastic with respect to income. The LRE of population is close to one, as would be expected, and the LRE for the price of gas is consistently very small. The LRE for the appliance price index is inelastic for class R, and the LRE for the industrial electrical machinery and equipment price index was constrained to zero for C and I due to the incorrect sign and lack of statistical significance of the initial estimates. Estimated LRE's for two contrasting states (New York and Tennessee) are summarized in Table 24.2.

The estimated LRE's for each of the three models are surprisingly consistent with one another. The major difference between the IV and OLS estimates of the VEM is that the implied rate of adjustment is faster using IV, particularly for the C and I classes. Hence, even though the LRE's are similar, the short-run elasticities are generally larger with IV. Attempts to determine which model is most appropriate by comparing actual 1971 demand with predicted demand for each state (recall 1971 data are not used for estimation) suggest that the OLS models are better, particularly for classes C and I.

The LRE's for population, income, and the price of electricity depend on the level of the corresponding factor for both the OLS and IV estimates

Table 24.1 Summary of Recent Studies of Electricity Demand

Authors Date of Publication Class of Consumer	Type of Data	Variables Included[a]						Type of Elasticity Models	Type of Distributed Lag	Estimation Procedure
		P_E	Y	N	P_F	T	Others			
1. Fisher and Kaysen (1962) R and I	Annual time series for each of 47 U.S. states. (1946–1957)	X	X	[b]		X		CEM for each state and industry	None	OLS on first differences of variables
2. Baxter and Rees (1968) I	Quarterly time series for industries in U.K. (1954–1964)	X			X	X	Industrial output, Temperature, Wage rate	CEM and a linear function for each industry	Geometric	OLS
3. MacAvoy (1969) Combined	Pooled quadrennial time series for 9 regions in U.S. (1958–1972)	X	X	X				CEM	None	OLS
4. Wilson (1971) R	Cross section of 77 cities in U.S. (1963, 1967)	X	X	[b]	X		Housing unit size, Temperature	CEM and a linear function	None	OLS
5. Halvorsen (1972) R	Pooled annual time series for 47 states in U.S. (1961–1969)	X	X	[b]	X		Temperature, Urbanization	CEM, hyperbolic and linear functions	None	Simultaneous model using two stage least squares
6. Anderson (1972) R	Cross section of 47 states in U.S. (1969)	X	X	[b]	X		Housing unit size, Temperature, Urbanization	CEM	None	OLS
7. Griffin (1972) R and (C + I)	Annual national totals for the U.S. (1950–1970)	X	X	[b]	X		Stock of air conditioners	CEM	Almon	OLS

[a] P_E = Price of Electricity, Y = Income, N = Population, P_F = Price of Alternative Fuel, T = Trend, Price of Appliances not included in any study.
[b] Quantity variables specified on a per capita or per household basis.

Table 24.2 Estimated Long-Run Elasticities Evaluated at 1971 Factor Levels, 1970 Dollars

		New York				Tennessee				Mean Level of All States			
Factor	Consumer Class	Level of Factor	CEM OLS	VEM OLS	VEM IV	Level of Factor	CEM OLS	VEM OLS	VEM IV	Level of Factor	CEM OLS	VEM OLS	VEM IV
Population (thousands)	R	18391	.94	1.00	.96	3990	.94	.99	.95	4365	.94	.99	.95
	C	18391	.98	1.04	.99	3990	.98	1.02	.98	4365	.98	1.03	.98
	I	18391	1.09	.99	1.02	3990	1.09	1.01	1.05	4365	1.09	1.01	1.05
Income (thousands of dollars per capita)	R	4.81	.30	.19	.17	3.19	.30	.21	.25	3.72	.30	.20	.21
	C	4.81	.80	.93	.87	3.19	.80	.81	.89	3.72	.80	.86	.88
	I	4.81	.72	.40	.50	3.19	.72	.60	.76	3.72	.72	.51	.65
Price of Electricity (average mills/kWh)	R	29.39	−1.21	−1.24	−1.34	12.13	−1.21	−1.10	−0.96	21.39	−1.21	−1.20	−1.24
	C	29.29	−1.60	−1.65	−1.50	16.01	−1.60	−1.12	−1.40	20.26	−1.60	−1.36	−1.45
	I	12.20	−1.79	−1.89	−1.81	7.70	−1.79	−1.53	−1.46	10.89	−1.79	−1.82	−1.74
Price of Gas (average dollars per thousand therms)	R	137.70	.21	.19	.13	89.12	.21	.19	.13	117.89	.21	.19	.13
	C	127.07	.05	.06	.04	72.35	.05	.06	.04	90.84	.05	.06	.04
	I	74.56	.00	.00	.06	36.50	.00	.00	.06	47.22	.00	.00	.06
Price of Appliances[a] (index of appliance prices)	R	1.00	−0.36	−0.42	−0.74	1.00	−0.36	−0.42	−0.74	1.00	−0.36	−0.42	−0.74
	C	1.00	.00	.00	.00	1.00	.00	.00	.00	1.00	.00	.00	.00
	I	1.00	.00	.00	.00	1.00	.00	.00	.00	1.00	.00	.00	.00

[a] Index of industrial electrical equipment and machinery wholesale prices for classes C and I.

of the VEM. The LRE's for the price of gas and the price of appliances, on the other hand, are constant in all models. The exact relationships between the variable LRE estimates and the factor levels are illustrated in Figure 24.1 for all three consumer classes. The LRE of population is similar in all classes and is close to one in all but the smallest states. In contrast, the income and price LRE's vary considerably over the observed range of factor levels. The LRE of income increases (except for class C using OLS) as income decreases and, in fact, is elastic for class I at low income levels. However, income is expected to rise in the future, and consequently, its LRE will become increasingly inelastic and approach zero in classes R and I and will remain fairly stable at slightly less than unit elasticity in class C. On the other hand, the LRE of price is inelastic at low price levels and becomes increasingly elastic as price increases. The relative price of electricity is expected to rise in the future for all consumer classes, and as a result, price will become a more important determinant of electricity demand. This effect is particularly important in class I, as the LRE of price is more elastic at high prices in this class than in R or C. In spite of this, the price is generally lower for class I than for R and C, and consequently, the industrial LRE of price may currently be more inelastic in some regions relative to the other consumer classes. If all prices increase, however, this situation will change, and eventually class I may exhibit the highest LRE for price in all regions.

24.3. Conclusion

The estimated long-run elasticities summarized in Table 24.2 demonstrate that electricity demand is generally price elastic for all three consumer classes and becomes increasingly elastic as prices rise. In contrast, demand is generally inelastic with respect to income, and for residential and industrial classes, approaches zero as income increases. The income elasticity for commercial demand is, however, only slightly inelastic over a wide range of income levels. Population exhibits approximately unit elasticity for all classes, and the elasticities for both the price of gas and the price of appliances are consistently found to be inelastic.

Generally, demand is found to be income inelastic in other studies, but the nature of the price elasticity is more controversial.[5] In particular, Fisher and Kaysen (1962) and Griffin (1972) conclude that the long-run price

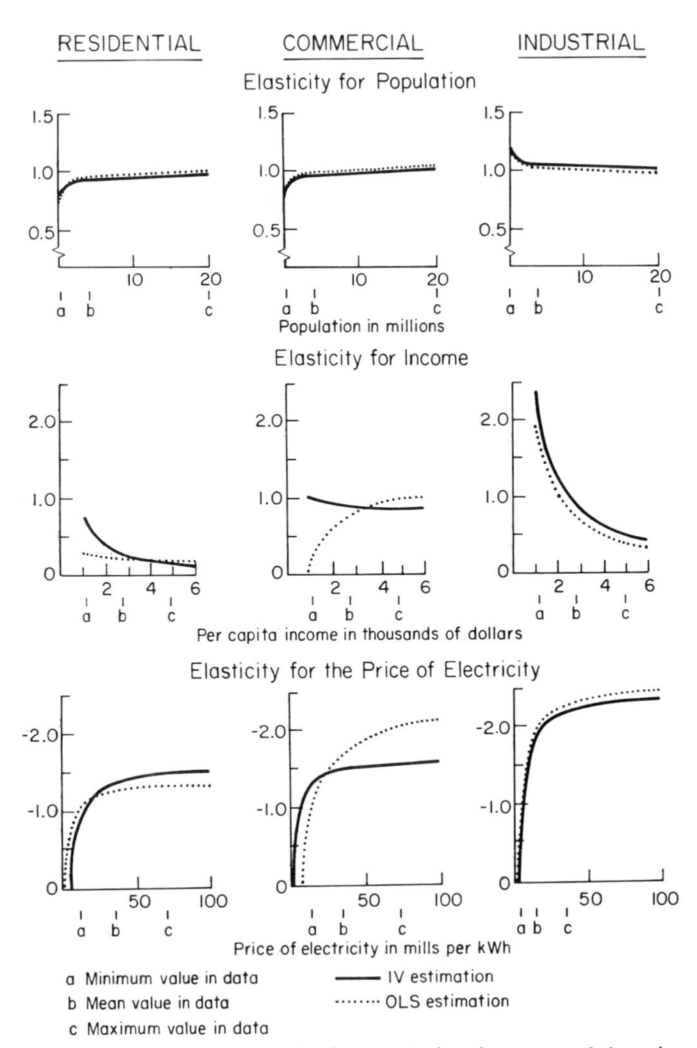

Figure 24.1. Demand elasticity for population, income, and the price of electricity.

relationship is clearly inelastic. On the other hand, Wilson (1969), MacAvoy (1969), and Halvorsen (1972) find the relationship elastic. The results of Baxter and Rees (1968) and Anderson (1972) are less clear cut. While it is difficult to make any direct comparison between these analyses due to the wide variety of models and data sources, one general comment is appropriate. Does the stock of electrical appliances and machinery respond to changes in the price of electricity? Fisher and Kaysen conclude that it does not, and Griffin implicitly assumes that price does not influence the stock, as the number of air conditioners is identified as a separate exogenous variable in his demand model. Hence, price influences the intensity of use of existing stocks but not the size of the stock itself. In contrast, our results and those of Wilson, MacAvoy, and Halvorsen suggest that price also plays a role in determining the life-style of residential consumers (for example, whether or not to install air conditioning) and the types of facilities and production methods employed by commercial and industrial consumers. However, it should be remembered that demand responds relatively slowly to changes of the causal factors, and consequently, any adaptation of present life-styles or of existing facilities and production methods will also be very gradual.

The relative importance of price as a determinant of demand has implications for the future need of generating capacity. If prices increase over the next few years in response to increased fuel costs etc., the growth of electricity demand will gradually decrease from the present rate. No accelerated growth of population or income is expected to offset this price effect. We have discussed the broad implications of this conclusion in an earlier article (Chapman, Tyrrell, and Mount, 1972). In summary, we consider that planning new generating capacity by extrapolating past trends significantly overestimates the future need for additional capacity.

Appendix 24.A. The Demand Models

Constant Elasticity Model

$Q_{it} = Q_{it-1}{}^{\lambda} V_{1it}{}^{\beta_1} \ldots V_{Nit}{}^{\beta_N} \exp(\alpha_j),$

Short-run elasticity for V_n is β_n,
Long-run elasticity for V_n is $\beta_n/(1-\lambda)$,

Table 24.A Parameter Estimates for the Constant and Variable Elasticity Models (Estimated t statistics are given in parentheses.)

Explanatory Factor	Units of Measurement	Parameter Estimated	Residential: Constant Elasticity OLS	Residential: Variable Elasticity (Model A) OLS	Residential: Variable Elasticity (Model A) IV
1. Lagged demand	Million kilowatt hours	λ	.8859 (136.5)	.8837 (119.2)	.7177 (33.0)
2. Population	Thousands	β	.1075 (17.0)	.1172 (15.3)	.2703 (13.3)
		γ	.0000 —	9.7821 (3.3)	8.3737 (2.3)
3. Income	Thousand dollars per capita (deflated)	β	.0343 (4.2)	.0195 (0.6)	.0000 —
		γ	.0000 —	−.0139 (0.2)	−.2246 (6.8)
4. Price of electricity	Mills per kilowatt hour (deflated)	β	−.1385 (12.5)	−.1552 (7.7)	−.4524 (10.4)
		γ	.0000 —	−.3304 (0.9)	−2.1965 (4.2)
5. Price of gas	Dollars per thousand therms (deflated)	β	.0238 (6.7)	.0225 (6.2)	.0370 (7.7)
6. Price of appliances	Price index (deflated)	β	−.0408 (2.4)	−.0486 (2.5)	−.2094 (7.5)
7. Region	Northeast	α	.4620	.4986	1.7699
	Mid-Atlantic	α	.4612	.4917	1.7564
	East North Central	α	.4802	.5117	1.7818
	West North Central	α	.5056	.5404	1.8319
	South Atlantic	α	.4858	.5181	1.7669
	East South Central	α	.4844	.5150	1.7443
	West South Central	α	.5175	.5492	1.8190
	Mountain	α	.4837	.5151	1.7641
	Pacific	α	.4679	.5020	1.7445
Estimated residual variance			.001335	.001324	.002042
Sum of squared residuals			1.4229	1.4071	2.1725

Table 24.A (continued)

	Commercial			Industrial		
	Constant Elasticity	Variable Elasticity (Model A)		Constant Elasticity	Variable Elasticity (Model A)	
Explanatory Factor	OLS	OLS	IV	OLS	OLS	IV
1. Lagged demand	.8735	.8724	.1843	.8869	.8765	.2167
	(75.7)	(75.7)	(2.4)	(79.9)	(76.5)	(2.7)
2. Population	.1244	.1333	.8105	.1237	.1220	.7970
	(10.6)	(10.2)	(10.4)	(8.5)	(7.5)	(9.1)
	.0000	10.5007	32.9935	.0000	−11.7364	−97.7104
	—	(1.6)	(2.4)	—	(1.0)	(3.5)
3. Income	.1011	.1486	.6825	.0817	.0000	.0000
	(5.3)	(2.5)	(4.9)	(3.3)	—	—
	.0000	.1432	−.1387	.0000	−.2358	−1.8921
	—	(1.0)	(0.5)	—	(3.6)	(8.0)
4. Price of Electricity	−.2030	−.2925	−1.3242	−.2021	−.3097	−1.8867
	(9.6)	(6.2)	(8.9)	(8.1)	(8.9)	(9.1)
	.0000	−2.4014	−2.8931	.0000	−.9290[a]	−5.7016
	—	(2.0)	(1.2)	—	(8.9)	(5.4)
5. Price of gas	.0068	.0082	.0305	.0000	.0000	.0475
	(1.0)	(1.1)	(2.0)	—	—	(2.4)
6. Price of appliances	.0000	.0000	.0000	.0000	.0000	.0000
	—	—	—	—	—	—
7. Region	.5960	.8115	3.4981	.4897	1.1280	6.4089
	.5897	.7958	3.5180	.4363	1.0766	6.0087
	.5930	.8021	3.5006	.4814	1.1190	6.2585
	.6180	.8340	3.6629	.4374	1.0640	5.8895
	.6088	.8178	3.5222	.4764	1.1052	6.1519
	.5619	.7692	3.3090	.4995	1.1335	6.3657
	.6381	.8484	3.6860	.4639	1.0876	6.0289
	.6392	.8493	3.6691	.4635	1.0875	6.0307
	.5688	.7862	3.3586	.3586	1.0050	5.5643
	.006834	.006795	.029547	.022271	.022034	.090169
	7.2917	7.2295	31.4384	23.7852	23.5100	96.0305

[a] The constraint $\gamma = 3\beta$ is imposed for the industrial price of electricity in this model to ensure that the estimated price elasticity is negative for all observed price levels.

Variable Elasticity Model

$$Q_{it} = Q_{it-1}^{\lambda} V_{1it}^{\beta_1} \ldots V_{Nit}^{\beta_N} \exp(\alpha_j + \gamma_1/V_{1it} + \ldots + \gamma_N/V_{Nit}),$$

Short-run elasticity for V_n is $(\beta_n - \gamma_n/V_n)$,
Long-run elasticity for V_n is $(\beta_n - \gamma_n/V_n)/(1 - \lambda)$,

where

i denotes the ith state,
j denotes the jth region,
t denotes the tth year,
Q is the quantity of electricity demanded,
V_n is the level of the nth factor such as income,
λ, β_n, γ_n, α_j are unknown parameters.

Notes

1. The demand elasticity is defined as the percentage change in the quantity demanded associated with a 1% increase in a particular factor.

2. In these three cases, the direction of trend is logically consistent with an increase of the demand for electricity.

3. The coefficient for this variable represents the proportion of the adjustment of demand that is unfinished at the end of the current time period, assuming causal factors remain constant at the changed levels.

4. North and South Carolina are combined in some of the data sources, as are Maryland and the District of Columbia.

5. The use of the average price of electricity has been criticized by two authors. However, Wilson, who uses marginal prices, and Halvorsen, who specifies that price and quantity are determined simultaneously, both find that demand is price elastic.

References

Anderson, K. P. (1972). *The Residential Demand for Electricity: Econometric Estimates for California and the United States,* RAND R-905-NSF, Rand Corporation, Santa Monica, Calif., January.

Baxter, R. E., and Rees, R. (1968). "Analysis of Industrial Demand for Electricity," *Economic Journal, 78,* pp. 277–98.

Chapman, L. D.; Tyrrell, T. J.; and Mount, T. D. (1972). "Electricity Demand Growth and the Energy Crisis," *Science, 178,* pp. 703–708.

Fisher, F. M., and Kaysen, C. (1962). *The Demand for Electricity in the United States,* North Holland, Amsterdam.

Griffin, J. M. (1972). *A Long Term Forecasting Model of U.S. Electricity Supply and Demand,* unpublished manuscript, University of Houston, Tex.

Halvorsen, R. (1972). "Residential Electricity: Demand and Supply," presented at the Sierra Club Conference on Power and Public Policy, Vermont, January.

MacAvoy, P. W. (1969). *Economic Strategy for Developing Nuclear Breeder Reactors,* MIT Press, Cambridge, Mass.

Wilson, J. W. (1969). "Residential and Industrial Demand for Electricity," unpublished Ph.D. thesis, Cornell University, Ithaca, N.Y.

——— (1971). "Residential Demand for Electricity," *Quarterly Review of Economics and Business, 11,* pp. 7–22, 1971.

25 Electricity Growth: Economic Incentives and Environmental Quality*

CHARLES J. CICCHETTI† AND WILLIAM J. GILLEN‡

According to the Federal Power Commission's National Power Survey (1970), the average annual growth in consumption of electric power has been on the order of 7.0% per year since the early 1880s and is expected to continue at this rate through 1980. A slightly lower rate of growth, 6.4%, is forecast between 1980 and 1990. The capacity of the system then would expand from 1642 million megawatt-hours in 1970 to 3202 million megawatt-hours in 1980, and 5978 million megawatt-hours in 1990 (Federal Power Commission, 1970). If the forecast is borne out, the electric power system will expand 364% between 1970 and 1990. This rate of expansion in the industry has been cause for concern for both consumer groups and environmental organizations. While these groups have considerable mutuality of interest, residential rate payers may be presumed to be most concerned with rising real prices for electricity, and environmentalists with the proliferation of generating stations, transmission lines, and power plant effluents. In several instances these groups have intervened in regulatory proceedings, both rate and facilities cases, and urged the alteration of electric utility pricing and promotion policies. These efforts are aimed at reducing or eliminating price discrimination and dampening the rate of growth in production and consumption of electric power.

In this paper, we first will consider the public utility pricing question from the economist's point of view. Second, we will discuss recent trends and near-term estimates of cost curves facing the industry. Third, we will discuss recent econometric evidence of the demand for electricity. The objectives of economic efficiency, social equity, and environmental quality underlie the following analysis.

25.1. Pricing and Costs

Between 65 and 75% of the electric generating capacity in the United States is privately owned but publicly regulated. This segment of private industry has been brought within the regulatory sphere so that the public may capture the efficiency benefits of so-called "natural monopolies" while

* This work was supported by the Environmental Defense Fund.

† Visiting Associate Professor of Economics and Environmental Studies, University of Wisconsin, Madison.

‡ Staff Economist, Environmental Defense Fund, Washington, D.C.

avoiding the less desirable tendencies of monopolists to limit output and charge higher prices than the competitive norm. As Kahn (1970, p. 17) put it, "The single *most widely accepted* rule for the governance of the regulated industries is [to] regulate them in such a way as to produce the same results as would be produced by effective competition, *if it were feasible* [emphasis added]." In general, the economist advises that the way to achieve the objectives of equity and efficiency in regulated markets is to require that the single producing unit produced equals the highest price that consumers are willing to pay for this level of production. This is marginal cost pricing and is precisely the perfectly competitive equilibrium solution.

Historically, one of the characteristics of the electric utility as "natural monopoly" was that the long-run marginal cost (LRMC) curve of the firm has decreased throughout the relevant range. In this instance the economist's prescription of setting price equal to LRMC may not cover the total costs of production, and some form of subsidy was required. Regulators, however, are either unable or unwilling to use this means to achieve the efficient allocation of resources. Consequently, a price structure was developed that took advantage of differing elasticities of demand among various categories of consumer. Price discrimination permits the utility to recover its full costs plus its allowed rate of return. The regulating authority may require that these price differentials not be "unduly discriminatory," that is, that they be based on differential costs of serving the particular category of customer.

In addition to price discrimination, electric utility tariffs are characterized by the so-called declining block rate. That is, for each class of customer (for example, residential, commercial, or industrial) a tariff schedule is established that is a decreasing step function of quantity consumed. Thus, average price decreases as consumption increases.[1] It is price discrimination and declining block rates that have recently come under attack. The former are said to be unfair and the latter to promote more rapid growth of the system. The response of the industry is mixed but may be generally characterized as rejecting these contentions on grounds that (1) the demand for electric power is generally inelastic, except perhaps for the large industrial and the residential space-heating demands, and therefore rates do not significantly influence consumption; and (2) price differentials are based on cost of service and are therefore not unduly discriminatory.

Consideration of the first defense of declining block rates, inelasticity of demand, we defer for the moment. The second defense, that rates are based on costs of service, is doubtful. According to Russell E. Caywood (1972),

> Although most rate forms follow cost patterns in one way or another, costs have been looked upon as only one of many tools used in the pricing process and their use has been rather limited.

The actual *de facto* relation of costs to prices aside, costs remain the most significant consideration in the economic analysis. Moreover, the electric power industry is widely regarded as being a "decreasing cost industry." We should consider then the various ways in which costs may be said to be decreasing. First relates to short-run decreasing costs, which is the familiar case of spreading the overhead. These may be sufficiently important to justify different prices for what may be the same service. If new users pay the incremental or marginal cost of service and make some contribution to capacity costs, differential pricing may be "efficient" in the sense of a move in the direction of Pareto-optimality.

Second concerns long-run decreasing costs. A utility is said to confront long-run decreasing costs (LRDC) when at any given point in time and for a given level of technology the unit costs of a larger plant are less than those for a smaller plant. Such a cost curve would exhibit "economies of scale" or increasing returns to scale. These also may be sufficiently important to justify the provision of otherwise similar service to new users at reduced rates as long as they cover the incremental cost of supplying them the service.

Finally, technological change may occur, in which case it is again possible to find costs decreasing over time. This is distinguished from the previous situation because improved technology may not be dependent upon the level of output. Utility growth may therefore be keyed to changes in technology and result in reduced customer costs.

Let us consider the historical relevance of costs in the electric power industry. Kahn points out that between 1950 and 1963 the electric generating units constructed in Great Britain increased from 30 MW to 550 MW, and at the same time the capital costs per kilowatt declined from £67 to £37. In the United States in 1956 the largest generating unit in operation was 260 MW, and in 1965 the largest was 1000 MW. Over this same time the annual operating costs of privately owned electric utilities decreased from 8.95 mills per kWh in 1948 to 6.56 mills per kWh in 1966 despite an overall price inflation of 17% (Kahn, 1970, p. 128). During this period of declining costs electric utilities priced electricity in a manner to encourage system growth. Some smaller customers paid more per kWh of energy and per kW of capacity than larger customers; however, rate increases were uncommon and price decreases in real terms were impressive.

For reasons not having to do with the economic principles involved, regulatory commissions have generally rejected the use of a subsidy to make up for losses that would result from the implementation of long-run marginal cost pricing. Instead, regulators and the power industry have made use of a "value of service" pricing concept. Under this principle differential rates based upon "willingness to pay" were established by public utilities and generally approved as recommended by regulatory commissions. As a result, a practice has evolved akin to the pricing scheme a discriminating monopolist might employ for different classes of customer.

One current problem facing the public utility commissions is that although the electric utility industry may no longer be operating with decreasing costs, the companies, like any rational monopolist, are unwilling to give up their discriminatory pricing arrangements. Regulatory commissions around the nation are, therefore, being inundated with rate increase requests from companies continuing to use declining block rate pricing and trying to keep up with increasing costs.

Granting the historical decline in capacity costs just discussed, environmentalists are wont to point out that not all costs have necessarily been decreasing; to wit, a broad spectrum of external costs, that is, costs not included in the industry's calculations. In addition, short-run costs may also be increasing. Since electric utilities operate under the principle of minimizing the unit costs of operation, the newest and most efficient plants are used by the utility to meet the demands put on it first. However, when peak demands have to be met, all plants in the system might have to be utilized, thus requiring the use of older and more costly plants. Users who contribute to peak demand should be required to pay the costs imposed on the system from utilizing these older, more expensive plants. In other words, promotional rates to off-peak users may not only be fair but also efficient; on the other hand, promotional rates to peak users are likely to be both discriminatory and inefficient.

The logic of separate pricing for on-peak and off-peak use generally breaks down in practice for two related reasons. First, it is not typical in this country to distinguish between on-peak and off-peak use. Instead users are placed in categories because they are more likely to use electricity on or off the peak period. Second, there is usually no restriction placed on the off-peak-user category when they demand service on the peak period. When this occurs and they are paying a lower price for service similar in every respect to the higher-priced user, then the objectives of economic efficiency and social equity are violated. Accordingly, while in practice a fuller utilization of capacity should be encouraged off the peak (we are excluding external and energy conservation concerns), if there is no way to restrict or identify these same users, who either come into the system or expand their use, when they demand service at the peak, the off-peak incentive system breaks down. The key word is "identify." Without metering, the use of differential rates to flatten the peak is likely to result in uneconomic use of resources and an inequitable or discriminatory pricing system. Exceptions for such items as hot water heaters or winter space heating have sometimes been utilized.

Finally, although in the past, peak periods usually occurred during the winter, in recent years due to increased air conditioning the nation's peak electric periods have been in the summer. Electric utilities seek to fill in the difference between these peaks by encouraging the use of winter electricity rather than to encourage the conservation of energy in the summer. It should be noted, as discussed by Hirst (1972), that there presently exists a wide range of variance in the efficiency of air conditioners of comparable cooling capacity. Accordingly, energy conservation in the summer does not imply turning off the air conditioner. For units equal to about 8000 Btu (a standard room-size unit), the efficiency range in terms of Btu/watt-hour ranges from below 6 to almost 12. If the more efficient units were to replace the more inefficient units, energy needs would be cut in half for the same amount of cooling.

Utilities, however, think in terms of their capital investment. For utilities to increase allowable income, they must increase the rate base, assuming regulatory bodies ensure proper accounting of operating expenses and the so-called fair rate of return exceeds the cost of capital. A philosophy that is based upon retarding the growth in the rate base even if it reduced operating expenses is irrational for utility executives under such circumstances.

The curtailment of the summer peak rather than the expansion of the winter peak as a way of flattening the peaks is, therefore, unlikely, given present institutional incentives.

A further argument for promoting the use of electricity is based on the second concept of decreasing costs, that is, long-run decreasing costs or economies of scale. But the notion of economies of scale in the long run here is not exactly what economists mean by that expression. To economists the economies of scale means that at any point in time and for a given state of technology, the unit costs of a larger plant are lower than those of a smaller plant. In the electric power industry confusion sometimes is present because the term has been given another dimension, that is, time.[2] Not only are the unit costs of larger plants lower than smaller plants at a point in time, but at successive points in time the costs of a plant of any given size are lower than the costs of the same size plant at an earlier point in time. This latter intertemporal effect is not an economy of scale in the economist's sense.

In order to facilitate the analysis and accommodate both concepts, we will distinguish the "intermediate run" as a point in time where plant size is variable. This may be regarded as a planning curve that confronts the firm contemplating new capacity and seeking to determine the optimum size of the new plant. On the other hand, the "long run" becomes a dynamic concept where both plant size and time are variable. If costs are decreasing over time and economies of scale are present, then utility customers are better off in the long run if the utility can effectively promote its service.[3]

Let us now consider some current evidence on economies of scale and long-run cost trends in the electric power industry. The once generally accepted proposition that both operating and capital costs of larger plants are lower than smaller plants has been challenged by utility executives and industry observers. According to former Federal Power Commissioner Charles Ross (1971):

> [T]he record in this proceeding reveals a distinct flattening of the economies of scale curve in generation at the present levels. . . . Furthermore, I know of no great projected breakthrough that would provide additional generation economies of scale *per se*. . . . The higher outage rates associated with increasing unit size also serves to severely reduce, if not entirely eliminate, initial capital cost savings. . . .

In the same proceeding economist James R. Nelson (1971) testified that

> . . . the 600 MW, 800 MW, and 1300 MW units built and to be built between 1966 and 1973 will all have the same heat rates because of technological factors. In fact heat rates (a measure of efficiency) of the larger units may even be increased in an effort to improve reliability. So, economies of scale due to improved heat rates from larger fossil-fueled thermal plants (or, for that matter, other sources of improved heat rates) seem to be mainly in the past—resulting from changes from much smaller units and more primitive technologies.

The trend to lower rates of unit availability was also recorded in *Electrical World's* 17th Steam Station Cost Survey (1971). The average unavailability of the 20 newest stations in the survey was 18.6%. In the previous survey the outage rate had been 11.3% for the same group (*Electrical World,* 1971, November 1). Evidence that higher outage rates are associated with larger units comes from an Edison Electric Institute report (1972) on "Equipment Availability for the Twelve Year Period 1960–1971."

The problem of forced outage is particularly acute with nuclear stations that make up an increasing share of planned new capacity. In an address to the 1972 Atomic Industrial Forum, Louis H. Roddis, Jr., President of Consolidated Edison Company of New York, noted that the average plant factor for domestic reactors is 60.9%. This contrasts sharply with the 80% plant factor usually assumed for new plants. The general point of his remarks was that nuclear generation of electrical power has failed to live up to expectations, and that a major contributing cause is higher than anticipated forced outage. Mr. Philip Sporn, former president of the American Electric Power Company, also noted this fact as long ago as 1969 in a report prepared for the Joint Committee on Atomic Energy, U.S. Congress (Sporn, 1969). He stated that

> . . . [w]ith the growth of atomic power which will take place between now and 1980 no atomic plant can, except for the shortest time, be expected to operate at a capacity factor as high as 80% and that, therefore, a more rational capacity factor is . . . 75%.

If, as it appears, we have for the present reached a limit to long-run decreasing costs, it is more to the point to consider the time trend of costs of new generating capacity. Promotional practices seek to make future construction necessary.

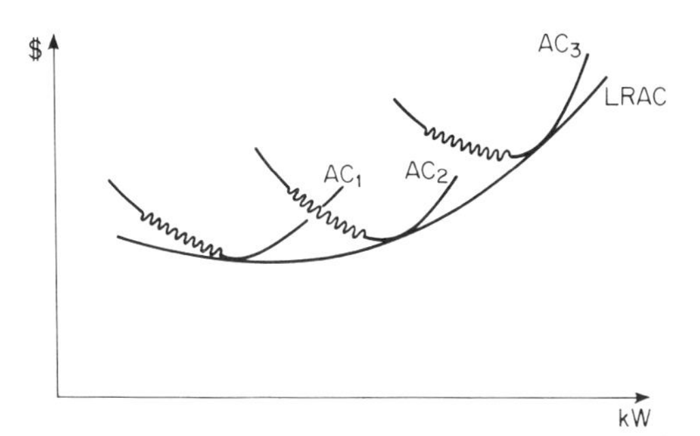

Figure 25.1. Long-run average cost and average intermediate costs.

Even promotional policies that are intended only to improve load factors, that is, to relieve temporary excess capacity, may reasonably be expected to have future impact because these new demands also continue to grow. Only a portion of new demand is a "once only" increase in consumption.

We will designate the period during which generating capacity is added as the "intermediate long run." The LRAC curve in Figure 25.1 is the envelope of intermediate-run average cost. The situation depicted by the hashed segments of AC_i curves identify situations of "economy of scale" at a point in time. Note that even in the instance of AC_i decreasing over the relevant range, the LRAC function may still be increasing. In other words, if future capacity costs for units of the same size are expected to be higher in the future, then even with "economies of scale" promotional policies will be perverse if present use is encouraged to utilize fully the expanded capacity of and thus require a more rapid installation of new capacity at higher unit costs AC_{i+1}.

The actual recent record on capital cost trends is a tale of woe told by utility executives in rate proceeding after rate proceeding. "Plant capital costs [are] spiraling upward." *Electrical World* (1971, July 1, p. 36) "Nuclear plant costs . . . have not merely evolved in recent years; they have exploded" (McTague et al., 1972, p. 31). An 800-MW fossil fuel plant that in 1967 may have come on line at $85 to $125 per kilowatt would presently cost almost twice that. By 1980 the cost may have quadrupled (Roe and Young, 1972, p. 40). While mere extrapolation of these trends is not particularly useful, we know of no observers of the industry who anticipate a full reversal in the direction of future costs.

25.2. Demand and Growth

We turn now to consideration of whether declining block-rate pricing is, in fact, promotional. Generally, those who contend that high rates of growth in the power industry are socially deleterious, and those concerned that growth is being spurred to ever more costly levels contend that this pricing is promotional. The electric industry generally responds that the pricing is not promotional. The argument is whether the quantity of electricity is significantly affected by the price charged; that is, whether the elasticity of demand is significantly different from zero.

Studies that were done about 10 years ago on the demand elasticity of electricity showed the quantity consumed to be basically insensitive to price changes. However, an inverse relationship different from zero was found even in these early studies. This is quite logical as electricity consumption at that time was primarily for lights and appliances—items that consumed relatively little electricity and for which good substitutes were unavailable. In addition, these studies were usually based upon time-series data, and thus relatively short-run estimates of price elasticity were determined. We usually expect the short run to be the more inelastic.

More recent studies, however, show the demand for electricity to be more responsive to changes in price. This is also logical as electricity has come to be used for heating, air conditioning, water heating, and clothes drying. Not only do these uses of electricity require larger amounts of power, there are also competitive sources of energy to meet these demands. Moreover, as the percentage of weather-sensitive electricity demand increases, quantity purchased is likely to become more sensitive to price. This is true for residential as well as commercial and industrial customers. As substitutability increases, price responsiveness or elasticity increases.

To the extent that consumers consider the price of electricity in purchasing appliances and comfort-conditioning equipment, the demand for electric energy will be responsive to the price charged. Because the use of electricity for heating and cooling tends to fall in the tail-blocks of the currently used declining block-rate structures, it is the rates charged in these blocks that the potential heating and cooling equipment customers consider. The lower the price of electricity in the tail-block, the less careful will customers of heating and cooling equipment be in shopping for the most energy-efficient units, or in making sure that their homes are properly insulated, or in taking other steps to minimize power usage. This lack of attention to efficient power use due to a high price elasticity for such uses and low price will also stimulate wasteful consumption.

One of the recent studies was presented by John W. Wilson, an economist with the Federal Power Commission, who found the residential demand elasticity for electricity to be −1.33 (Wilson, 1971, pp. 7–22). That is, a 1% *increase* in the price of electricity causes a 1.33% *decrease* in the demand for electricity, other things remaining equal. Wilson draws the following conclusions:

> My study demonstrates that price is the major determinant of intermarket variations in the average volume of residential electricity consumption. Concomitantly, residential demand for electricity is price elastic and is influenced substantially by the price and availability of natural gas. These conclusions remain basically the same whether we are concerned with aggregate demand or specific appliances.

Work done by Duane Chapman, Timothy Tyrrell, and Timothy Mount of Cornell University confirms the Wilson research (Chapman, Tyrrell, and Mount, 1972).[4] Their estimates of average price elasticity were −1.3, −1.5, and −1.7 for residential, commercial, and industrial consumers, respectively. Again similar results were found by Kent P. Anderson in a Rand Corporation Paper (1972). Anderson's price elasticity estimate was −.91, indicating that quantity taken is proportionately less (but not by much) than price changes.

25.3. Policy Implications

In the preceding discussion several interrelated factors were analyzed separately. First, it was pointed out that electric utilities price in a declining block-rate fashion. The smaller user pays more per unit than the larger user, and the more each user consumes, the less additional "blocks" of electricity cost. Such practices were said to be discriminatory and inefficient when the costs of providing the electricity are increasing. However, in situations in which costs are decreasing as a utility expands its service and where a subsidy policy (preferred by economists) is rejected, then charging different users different prices for the same service may not be discriminatory or inefficient in the sense that all users would find their electric bills declining when the utility expands its service.

Several empirical analyses of price elasticity have been performed recently, and the results were outlined earlier. All of these studies point to the conclusion that electricity consumption is sensitive to price changes. If prices are changed because of rising costs to either a "flat" rate (everyone pays the same price per kWh) or an "inverted" rate (the bigger users pay more since they are "causing" the growth and the resultant higher costs), the rate of growth in consumption will be dampened.

Policy makers must decide whether costs are in fact increasing. Most companies have recently filed petitions for rate increases, in many instances after two decades of no rate increases. In these petitions they have outlined recent changes that have resulted in increasing costs. Since whatever merit discriminatory and promotional pricing may have is clearly contingent on a downward slope in long-run costs, public policy makers should scrutinize utilities' requests for rate increases for consistency in rate structure.

Notes

1. Tariffs may be further divided into "demand" and "energy" components, based on kW and kWh consumption, respectively. In addition, there may be special rates for designated purposes, for example, residential space heating; or for types of service, for example, interruptible or standby.

2. More precisely, there are several variables subsumed under "time," for example, technological progress, different relative returns in other sectors, inflation or deflation, changing factor prices, and so forth.

3. Some obvious forms of promotion are advertising and incentives to home builders to install electricity-consuming devices. A pricing policy may be said to be promotional whenever prices charged on-peak consumers are less than short-run marginal cost (SRMC). Declining block rates that do not reflect so-called coincident peaks are almost inevitably below SRMC.

4. See Chapman, Mount, and Tyrrell (1972); also see Chapman, Tyrrell, and Mount (1972).

References

Anderson, Kent P. (1972). "The Residential Demand for Electricity: Econometric Estimates for California and the United States," RAND R-905-NSF, Rand Corporation, Santa Monica, Calif., January.

Caywood, Russell E. (1972). "Electric Utility Ratemaking—1972," *Public Utilities Fortnightly,* p. 16, August 3.

Chapman, D.; Mount, T.; and Tyrrell, T. (1972). "Electricity Demand Growth: Implications for Research and Development," prepared for the Subcommittee on

Science, Research and Development of the Committee on Science and Astronautics of the United States House of Representatives, June 16.

Chapman, D.; Tyrrell, T.; and Mount, T. (1972). "Electricity Demand Growth and the Energy Crisis," *Science, 178,* pp. 703–708, November 17.

Edison Electric Institute (1972). "Equipment Availability for the Twelve Year Period 1960–1971," Publication No. 72-44, November.

Electrical World (1971, July 1). Vol. 175, p. 36.

——— (1971, November 1). Vol. 176.

Federal Power Commission (1970). *1970 National Power Survey,* Washington, D.C., pp. I–3–3 and I–3–4.

Hirst, Eric (1972). "Electric Utility Advertising and the Environment," Oak Ridge National Laboratory, Oak Ridge, Tenn., April.

Kahn, Alfred (1970). *The Economics of Regulation,* Vol. 1, John Wiley & Sons, New York, p. 17.

McTague, Peter J.; Davidson, G. J.; Bredin, R. M.; and Herman, A. A. (1972). "The Evolution of Nuclear Plant Costs," *Nuclear News, 15*(2), pp. 31–35, February.

Nelson, James R. (1971). Testimony, "In the Matter of American Electric Power Company," File No. 3-1476, Securities and Exchange Commission, Washington, D.C.

Roe, Kenneth A., and Young, William H. (1972). "Trends in Capital Costs of Generating Plants," *Power Engineering, 76*(6), p. 40, June.

Ross, Charles (1971). Testimony, "In the Matter of American Electric Power Company," File No. 3-1476, Securities and Exchange Commission, Washington, D.C.

Sporn, Philip (1969). "Developments in Nuclear Power Economics, January 1968–December 1969," Hearings before the Joint Committee of Atomic Energy, Congress of the United States, U.S. Government Printing Office, Washington, D.C., November 18–20.

Wilson, John (1971). "Residential Demand for Electricity," *Quarterly Review of Economics and Business, 11*(1), pp. 7–22, Spring.

26 Projections of Electricity Demand *

T. J. TYRRELL†

With the arrival of the "Energy Crisis" the need for accurate predictions of electricity demand becomes crucial. For decades reasonably accurate results have been obtained by simple extrapolation of historical growth rates. Recently, however, changes have occurred in some of the underlying economic factors, increasing the desirability of a more sophisticated predictive approach. Recently developed economic models describe electricity demand as a function of population, per capita income, and prices of electricity, gas, and electrical appliances. In this paper we examine some features of one such model and illustrate the projections that can be made using it. These projections are found to be generally lower than other well-known forecasts and, depending on assumptions, show a wide range of possible future electricity demand levels.

26.1. Introduction

For decades electricity demand growth has consistently followed an exponential rate of growth. Thus, accurate predictions have been made using simple extrapolation techniques. A recent study of electricity demand and the factors contributing to its growth has shown that such techniques may not continue to provide accurate predictions of the future.

In an article written for *Science* (Chapman, Tyrrell, and Mount, 1972), preliminary findings of the study were discussed. Electricity demand growth was described as a function of four causal factors: population, per capita personal income, the price of electricity, and the price of natural gas. Using an econometric model of this demand relationship and likely futures of the causal factors, electricity demand projections were made and compared to other government and industry estimates. Similar comparisons are shown in Table 26.3.

Since that article, models have been refined to reflect more accurately the behavior of each class of electric customers. Refinements include the

* ORNL-NSF Environmental Program, work supported by the National Science Foundation, Interagency Agreement No. AEC 40-237-70 and NSF AG 398.
The author wishes to express his appreciation to Ms. Patricia Tyrrell and Ms. Margie Adair for their assistance with the figures and tables. The original conference article was coauthored by R. S. Carlsmith, Oak Ridge National Laboratory.
† Oak Ridge National Laboratory, Oak Ridge, Tenn. 37830, operated by Union Carbide Corporation for the U.S. Atomic Energy Commission, Contract No. W–7405–eng–26.

Table 26.1 Long-Run Electricity Demand Elasticities (evaluated at 1971, average U.S. factor levels)

	Residential	Commercial	Industrial
Population	0.99	1.03	1.01
Income	0.20	0.86	0.51
Price of Electricity	−1.20	−1.36	−1.82
Price of Gas	0.19	0.06	0.00
Price of Appliances	−0.42	—	—

addition of the price of electric appliances as an explanatory variable, the formulation of a variable-elasticity model, and the use of additional statistical techniques. These models and their development were described in a recent ORNL report (Mount, Chapman, and Tyrrell, 1973). Several alternative models were presented, discussed, and compared to electricity demand models of other studies. Finally, elasticity estimates were given for each model, for the three major consumer classes (residential, commercial, and industrial). One set of these is given in Table 26.1.

The elasticity of an explanatory variable can be thought of as the percentage change in electricity demand which would result from a 1% increase in the variable.

The foregoing estimates are from "Variable Elasticity Model A" (VEMA) of "Electricity Demand in the United States" (Mount, Chapman, and Tyrrell, 1973) as estimated by ordinary least squares (OLS). Two other sets of elasticities were also presented: (1) a constant elasticity model (CEM) estimated by OLS and (2) the VEMA again, this time estimated by the Instrumental Variables (IV) technique. For projections we have chosen the VEMA, since it allows for heterogeneity between states, and the OLS estimates, since projections of 1971 values were generally closer to actual values using these.

In this paper we have projected electricity demands out to the year 2000. While the model we have used is different and some of the assumptions about the future of causal factors have changed, the range of resultant demand projections is consistent with results presented in *Science* (Chapman, Tyrrell, and Mount, 1972).

26.2. The Electricity Demand Model

The estimated Variable Elasticity Demand Model (VEMA) is of the form:

$$\mathrm{DE}_{i,j,t} = \mathrm{RC}_j + \prod_{k=t}^{t-n} \lambda^{t-k} \Big[\mathrm{PE}_{i,j,k}{}^{\beta_1} \mathrm{N}_{j,k}{}^{\beta_2} \mathrm{Y}_{j,k}{}^{\beta_3} \mathrm{PG}_{i,j,k-1}{}^{\beta_4} \mathrm{PA}_{1,k-1}{}^{\beta_5} \exp\Big(\frac{\alpha_1}{\mathrm{PE}_{i,j,k}} + \frac{\alpha_2}{\mathrm{N}_{j,k}} + \frac{\alpha_3}{\mathrm{Y}_{j,k}}\Big)\Big],$$

where

$\mathrm{DE}_{i,j,t}$ = electricity sales (demand) by consumer class i, in state j, in year t,

RC_j = a regional constant, the same for each state j within a census region,

$\mathrm{PE}_{i,j,k}$ = the average price of electricity for class i, in state j, in year t, deflated by the national consumer (or wholesale) price index for year k,

$\mathrm{N}_{j,k}$ = population of state j in year k,

$\mathrm{Y}_{j,k}$ = per capita personal income for state j in year k, deflated,

$\mathrm{PA}_{1,k-1}$ = the national BLS (Bureau of Labor Statistics) index of the wholesale price of electric household appliances for residential consumer class only, for year $k - 1$, deflated. The national index of electrical machinery and equipment was eliminated from commercial and industrial models on statistical grounds.

i = the consumer class: (1) residential, (2) commercial, or (3) industrial,

j = the state: one of 47 states areas (Maryland and District of Columbia are combined; so are North and South Carolina),

t = the year,

k = an index of the n previous years that influence electricity demands of year t,

λ = a lag coefficient weighting the influence of previous years,

$\beta_1, \ldots, \beta_5, \alpha_1, \alpha_2, \alpha_3$ = OLS estimates of the coefficients of each of the variables, for each of the three consumer classes. $\beta_5 = 0$ for commercial and industrial consumers.

In this form the expression for a single $\mathrm{DE}_{i,j,t}$ is a nonlinear product of many factors. For estimating coefficients and making projections, a simpler, reduced form is used:

$$\mathrm{LDE}_{i,j,t} = \mathrm{LRC}_j + \lambda \mathrm{LDE}_{i,j,t-1} + \beta_1 \mathrm{LPE}_{i,j,t} + \beta_2 \mathrm{LN}_{j,t} + \beta_3 \mathrm{LY}_{j,t} + \beta_4 \mathrm{LPG}_{i,j,t-1} + \beta_5 \mathrm{LPA}_{1,j,t-1} + \frac{\alpha_1}{\mathrm{PE}_{i,j,t}} + \frac{\alpha_2}{\mathrm{N}_{j,t}} + \frac{\alpha_3}{\mathrm{Y}_{j,t}},$$

where we have taken the natural logarithm of the equation (note the prefix L), eliminated the need for n past values of each variable by using the lagged dependent variable ($LDE_{i,j,t-1}$) and all other expressions have the same meaning as before.

Two important features of this model are illustrated in Figures 26.1 and 26.2.

26.2.1. The Variable Elasticity

The elasticities of electricity price, population, and income of this model are functions of the level of these variables. The elasticity of a variable (say V) equals $\beta_V - \alpha_V/V$, that is, customer response to percentage changes in V varies according to the initial value of V. For example, the short-run residential electricity price elasticity is $-0.16 - (-0.33)/PE$ (see Mount, Chapman, and Tyrrell, 1973, Table B1). This implies that residential consumers are more sensitive to price changes when prices are relatively high to begin with. At the 1971 average residential price of 21.39 mills/kWh, the short-run residential price elasticity is $-0.16 - (-0.33)/21.39 = -0.145$. Such variations are also accounted for in population and income elasticities. The case of the variable residential price elasticity is illustrated in Figure 26.1.

26.2.2. The Lag Structure

The geometric lag structure of the model is represented by λ, where

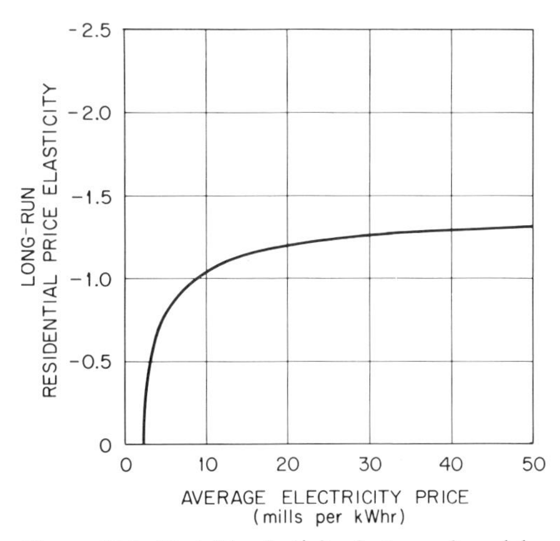

Figure 26.1. Variable elasticity feature of model.

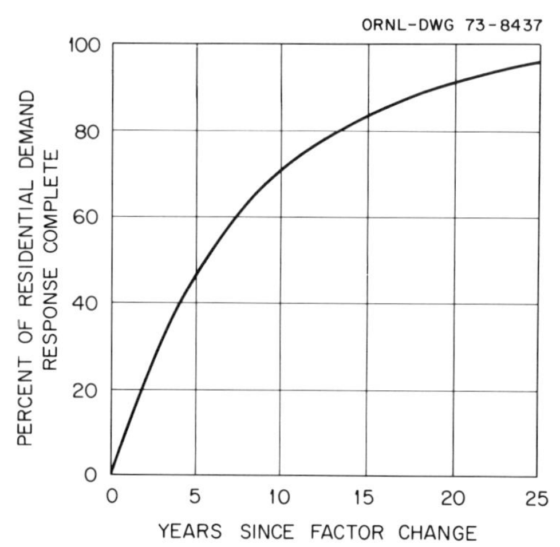

Figure 26.2. Lag structure feature of model.

$(0 < \lambda < 1)$. Thus λ^m $(m = 0,n)$ weights the past n values of the explanatory variables according to a geometrically declining factor. In this way the nature of electricity demand is a function of past as well as current values of causal factors. An obvious example of this delayed response to factor changes is the purchase of electrical appliances. Stocks of appliances currently contributing to electricity consumption are the result of many past purchases and thus past values of prices, incomes, and so on. Similarly, projections of future demand will be a decreasing function of current changes in causal factors. Long-run elasticities are related to short-run elasticities via this lag factor. Thus, the long-run U.S. average residential price elasticity equals $-0.145/(1 - 0.88)$ equals -1.20 (see Table 26.1). The estimated lag in residential customer response is illustrated in Figure 26.2.

26.3. Projections

As a point of reference and a typical example, we have made a "base case" projection from the model and one set of reasonable assumptions about the future of our five causal factors. These are taken primarily from independent forecasts and are shown in Table 26.2.

Table 26.3 compares our base case results with other predictions of residential demand and total generation requirements. Our results for 1990

Table 26.2 Assumed Future of Five Causal Factors

Causal Factor	Assumption[a]	Source
Electricity Price	Regional cost increases averaging 19% for the United States by 1990 (from 1968)	FPC (1971), *1970 National Power Survey*
Population	State estimates totaling 270 million for the United States	Graham, Degraff, and Trott (1972)
Per Capita Income	State estimates averaging 2.9% annual growth for the United States to 1990	Graham, Degraff, and Trott (1972)
Gas Price	U.S. estimate of 6.7% annual growth	Mean of NPC (1971) range
Appliance Price	U.S. estimate of 2.83% annual decline	Average annual decline of the past 15 years

[a] The next section discusses these assumptions more fully.

Table 26.3 Electricity Demand Predictions for 1990 (trillion kWh)

Source	Residential Demand (1970 = 0.45)	Total Generation (1970 = 1.52)
Electrical World (Olmsted, 1972)	1.79	5.93
Federal Power Commission (1971)	1.41	5.83
Cornell-NSF Workshop (U.S. Congress, Senate, 1972)	—	5.38
Our Base Case	1.20	3.88[a]

[a] Our electricity demand models do not account for sales to other than residential, commercial, or industrial customers, or for transmission and distribution losses. To compare projections we have made the following assumptions. In the "other" consumer category we have assumed a constant percentage equal to the 1971 value for each state. For "losses" we have assumed a single estimate for the nation of 9.9% of total generation.

are lower than the other three and significantly lower than some. Further details are given in Appendix 26.A.

The first comparative prediction is from the "23rd Annual Electrical Industry Forecast" of *Electrical World* based on an industry study of factors contributing to electricity demand growth (Olmsted, 1972). The second is from the Federal Power Commission's *1970 National Power Survey*: a summation of the work by the FPC's six Regional Advisory Committees (Federal Power Commission, 1971). The third is the base case estimate from the Cornell-NSF Workshop on Energy and the Environment (U.S. Congress, Senate, 1972): an extrapolation of a National

Petroleum Council estimate. [It should be noted that the workshop also presented an alternative "Total Generation" estimate of 4.48 trillion kilowatt hours which was derived from a model that included electricity price as a causal factor (1972).]

26.4. Model Sensitivity

In this section we apply the models to a range of future assumptions about the explanatory variables to produce a range of projections around our base case. Appendix 26.B summarizes these alternative total generation results.

In general, we have used independent estimates of a likely range. When these were not available (for income and appliance prices), we have used the highest and lowest annual rates of change of the past 15 years.

26.4.1. The Price of Electricity

The price of electricity has shown a relatively consistent decline over the past 25 years, until 1971, when we experienced the first increase in "real" prices. Several factors account for this. Increases in thermal efficiency, which were an important cost factor in the early days of the utility industry, came to an end in 1964. Since then the average system efficiency has actually dropped slightly. Reductions in unit construction costs through economics of scale came to an end in the mid-1960s with the development of both fossil-fueled and nuclear stations of approximately 1000 MW(e). Since that time the size of individual generating units has not increased rapidly. In the meantime there have been sharp increases in virtually every component of electricity costs: construction costs have risen due to higher labor costs, lower labor productivity, higher interest rates, licensing problems, and more stringent safety and environmental regulations. Fossil fuel costs have also shown dramatic increases.

In June 1972, Aubrey J. Wagner, chairman of the Tennessee Valley Authority, published estimates of maximum environmental protection costs for strip mine regulation, mine safety, fly ash and sulfur removal, a sulfur tax, waste heat control, and licensing delays. To meet all of these requirements, he stated: "TVA would be faced with an approximate doubling of its revenue requirements" (Wagner, 1972).

The Federal Power Commission predicted average U.S. electricity prices will increase 19% by 1990 from 1968 prices (Federal Power Commission, 1971).

On the other hand, *Electrical World,* in its "23rd Annual Electrical Industry Forecast" (Olmsted, 1972), also predicted price rises, but only until 1974, when prices will "reestablish the long-term down trend."

In 1972, real prices were about 7% below what they were in 1968 and about 3% above what they were in 1970.

In view of the foregoing we have used the FPC prediction for our base case and taken as two extreme assumptions: (1) constant prices at the 1970 level and (2) a doubling of prices by 2000.

Figure 26.3 shows the effect of electricity prices on electricity demand projections. The calculations include the estimated lag in response time and the variation in elasticities with level of population, income, and electricity price levels. The base case is the middle projection; high and low projections are made by changing the electricity price assumption and

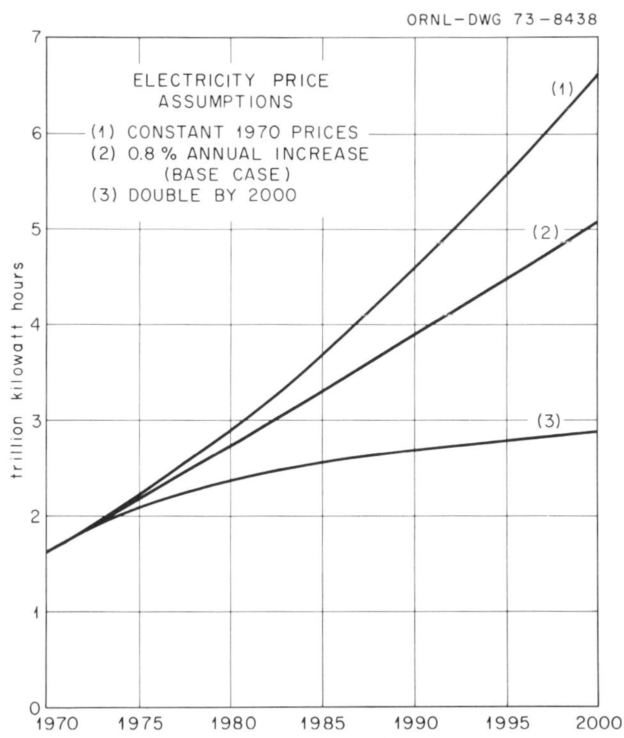

Figure 26.3. Electricity demand projections from alternative electricity price assumptions.

keeping the other variables at the base case assumptions. Holding prices at their 1970 level we get a projection 30% above the base case by the year 2000. If we assume prices double, we get a projection 44% below the base case by the year 2000.

26.4.2. The Price of Appliances

The "real" price of appliances has decreased at an average rate of 2.8% per year for the past 15 years. For our base case we have assumed a continuation of this trend.

Figure 26.4 shows the base case with the effect of varying appliance price projections. The long-run residential appliance price elasticity is −0.42. When we assume constant 1970 prices (in 1959 the growth rate was virtually 0) we get a projection for the year 2000 that is 9% below our base case. When we assume appliance prices decrease by 4.2% per

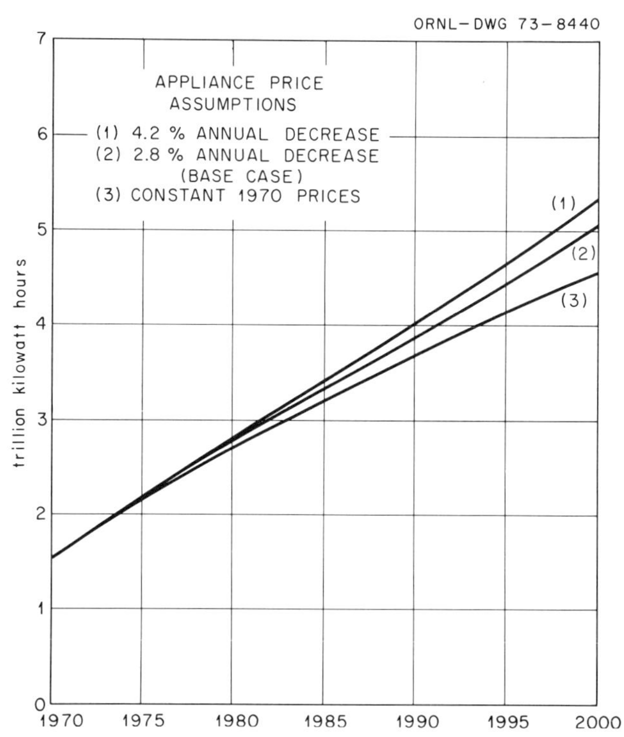

Figure 26.4. Electricity demand projections from alternative appliance price assumptions.

year (the 1958 rate) we get projections of demand 6% above our base case.

26.4.3. The Price of Gas

Recent gas shortages have put great pressure on its regulation (*Public Utilities Fortnightly,* 1972). As prices move above the established ceilings we should see some considerable increases. A chemical processing industry analyst predicts natural gas prices on the Gulf Coast to rise from 15 to 20¢/million Btu up to 50¢/million Btu by 1980 or earlier (Spitz, 1972). The National Petroleum Council (1971) estimates that well-head prices will need to increase 80 to 250% by 1985.

Our base case assumption is the mean of these NPC "required price" increase estimates: 165% in 15 years.

Figure 26.5 shows demand variations due to alternative gas price as-

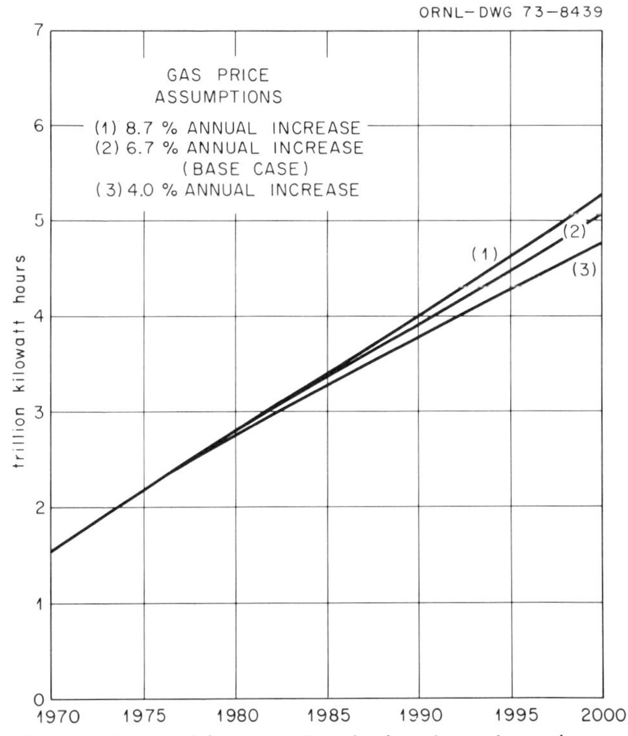

Figure 26.5. Electricity demand projections from alternative gas price assumptions.

sumptions. The long-run gas price elasticity is low for each consumer class. Although the future price of gas is very uncertain, the effect on demand projections of a wide range of projections is small due to this low cross-elasticity.

26.4.4. Population and Income

In April 1972, the *Survey of Current Business* printed "State Projections of Income, Employment, and Population," findings of a joint effort by the Bureau of Economic Analysis of the Department of Commerce and the Economic Research Service of the Department of Agriculture (Graham, Degraff, and Trott, 1972). These projections take into account geographically shifting population and income distributions as well as characteristics of specific industries. We used these projections for our base case estimates of both population and income.

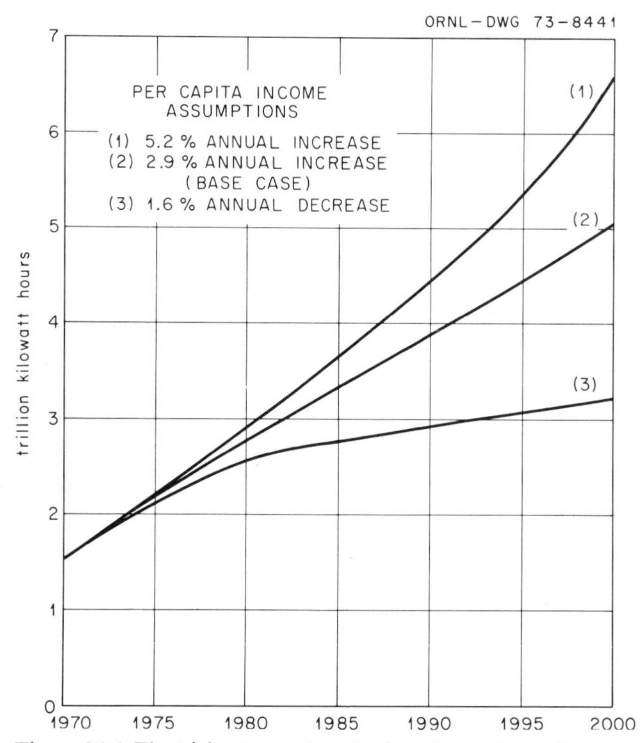

Figure 26.6. Electricity demand projections from alternative per capita income assumptions.

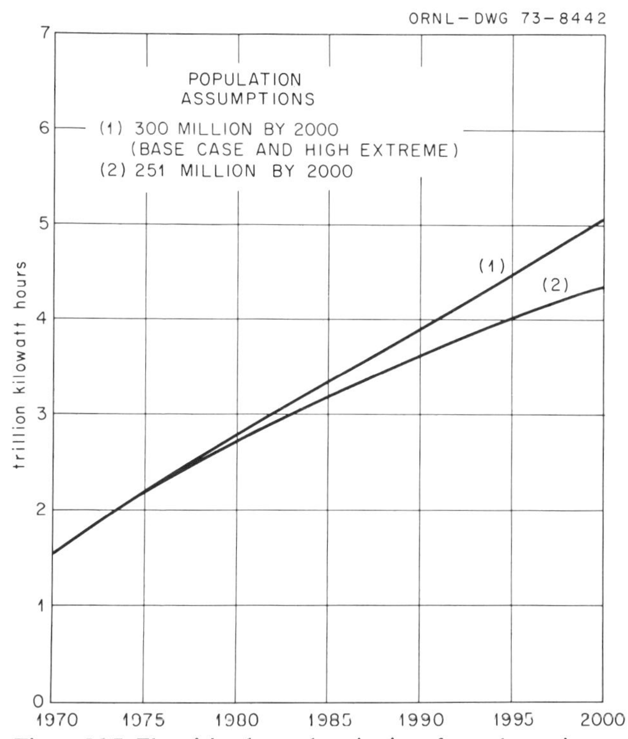

Figure 26.7. Electricity demand projections from alternative population assumptions.

Income effects are shown in Figure 26.6. Long-run income elasticities at current income levels are 0.20, 0.86, and 0.51 for our three consumer classes. Assuming the extreme range of −1.6 (1958) to 5.2 (1965) annual per capita income growth, we get projections for the year 2000 that are 37% below and 30% above our base case.

The range of alternative population forecasts for our analysis is the same as the range adopted by the Bureau of the Census (1972). In 1972 the actual fertility rate dropped below the replacement level (2.1) for the first time (National Center for Health Statistics, 1972), and the Census Bureau issued a new low set of population projections based on a fertility rate of 1.8 children per family. A high set was eliminated (Bureau of the Census, 1972), and our base case population assumption is coincident with the highest census projection as shown in Figure 26.7.

26.5. Summary

Long-run projections of electricity demand using the described econometric model show significant growth in the future but not as fast as some predictions. Our base case indicates trends that may occur. However, there is considerable uncertainty due to the sensitivity of the model to alternative assumptions about explanatory variables:

1. The future of the price of electricity seems very critical to demand projections. This results from its high elasticity.
2. The prediction of future appliance prices is important due to the relatively high elasticity and our uncertainty about its future.
3. Even with the wide range of gas price projections, demand projections are not strongly affected by this variable, because of the low cross-elasticity.
4. Alternative predictions of income and population seem of moderate importance. Only by taking an extreme range of income growth rates do we see a significant change in demand. The relatively narrow range of projected population growth rates results in only minor variations in demand projections.

Appendix 26.A

State and Regional Projections (Base Case Assumptions)

	Residential Demand (millions of kilowatt hours)						
	1970 (actual)	1975	1980	1985	1990	1995	2000
Maine	1722	2421	3127	3870	4678	5570	6566
New Hampshire	1462	1885	2332	2828	3388	4022	4741
Vermont	1157	1540	1933	2351	2809	3317	3885
Massachusetts	8910	12324	15868	19679	23894	28605	33911
Rhode Island	1366	2002	2647	3314	4027	4804	5664
Connecticut	6283	8558	10981	13653	16663	20072	23945
New England	20900	28729	36888	45693	55459	66390	78711
New York	25212	34561	44104	54207	65246	77484	91188
New Jersey	12121	16412	20766	25447	30675	36554	43199
Pennsylvania	22376	28892	35695	42976	50924	59717	69534
Middle Atlantic	59709	79865	100565	122629	146845	173754	203920
Ohio	21170	28643	36362	44602	53635	63651	74851
Indiana	11899	15690	19593	23825	28554	33861	39841
Illinois	20152	27263	34906	43438	53190	64386	77277
Michigan	16878	23615	30477	37703	45557	54214	63859
Wisconsin	9588	13095	16873	21095	25923	31466	37850
East North Central	79687	108306	138211	170662	206858	247597	293678
Minnesota	8001	11867	16021	20589	25736	31589	38291
Iowa	6262	8558	10879	13370	16154	19298	22870
Missouri	9729	13584	17743	22233	27137	32575	38671
North Dakota	1319	1865	2394	2920	3462	4035	4653
South Dakota	1476	2065	2628	3188	3769	4387	5058
Nebraska	3598	5360	7156	9015	10996	13156	15552
Kansas	4954	6865	8711	10627	12730	15069	17695
West North Central	35339	50163	65533	81941	99984	120108	142791
Delaware	1168	1700	2258	2852	3500	4218	5020
Maryland & D.C.	8269	12821	17561	22561	27992	33981	40667
Virginia	11280	16104	21245	26860	33131	40191	48187
West Virginia	3327	4819	6248	7607	8934	10283	11701
N. & S. Carolina	21304	29344	37613	46454	56196	67046	79226
Georgia	12607	17833	23315	29321	36099	43782	52522
Florida	23538	29620	36643	45065	55190	67133	81064
South Atlantic	81493	112240	144883	180719	221043	266633	318387
Kentucky	7148	9905	12541	15240	18160	21365	24920
Tennessee	19247	25477	32248	39896	48697	58802	70392
Alabama	11141	15101	19081	23300	27947	33118	38915
Mississippi	6252	8387	10490	12673	15030	17613	20475
East South Central	81493	112240	144883	180719	221043	266633	318387
Arkansas	4183	6139	8090	10089	12213	14517	17058
Louisiana	9097	12856	16702	20759	25175	30056	35512
Oklahoma	5834	8176	10543	13053	15823	18915	22397
Texas	28883	42359	56525	71891	89059	108405	130352
West South Central	47997	69530	91861	115792	142269	171893	205319
Montana	1521	2069	2558	3008	3447	3892	4360
Idaho	2406	3297	4090	4839	5592	6375	7213
Wyoming	607	874	1122	1353	1580	1811	2055
Colorado	3488	4949	6425	7968	9635	11464	13494
New Mexico	1360	2041	2696	3327	3954	4600	5286
Arizona	4050	5345	6703	8204	9907	11837	14024
Utah	1630	2541	3464	4395	5358	6379	7486
Nevada	1915	2794	3765	4896	6240	7811	9629
Mountain	16997	23910	30823	37989	45711	54169	63546
Washington	16226	22443	28756	35423	42699	50744	59725
Oregon	9389	12461	15628	18996	22670	26728	31251
California	34556	50665	67441	85462	105394	127610	152534
Pacific	60171	85569	111826	139881	170763	205082	243510
Continental U.S.	446061	617179	794948	986415	1198765	1436505	1704563

	Commercial Demand (millions of kilowatt hours)						
	1970 (actual)	1975	1980	1985	1990	1995	2000
Maine	970	1412	1835	2264	2716	3186	3675
New Hampshire	593	935	1272	1613	1969	2339	2721
Vermont	479	781	1078	1372	1671	1975	2286
Massachusetts	7211	10663	14077	17541	21143	21143	24878
Rhode Island	1125	1783	2425	3063	3720	4395	5089
Connecticut	4265	6919	9600	12336	15194	18170	21266
New England	14643	22493	30287	38190	46413	54943	63778
New York	24874	34160	43192	52384	61986	71953	82255
New Jersey	10185	14432	18600	22911	27506	32350	37415
Pennsylvania	13427	21452	29394	37202	45013	52872	60823
Middle Atlantic	48486	70043	91186	112497	134505	157174	180493
Ohio	14399	21750	29045	36365	43845	51476	59259
Indiana	6268	9682	13048	16436	19934	23530	27217
Illinois	17791	24572	31781	39763	48714	58592	69365
Michigan	10505	15785	20981	26170	31460	36842	42316
Wisconsin	4948	7198	9563	12149	15029	18195	21641
East North Central	53911	78986	104418	130883	158982	188635	219796
Minnesota	3228	5785	8551	11491	14665	18076	21737
Iowa	3344	5029	6655	8330	10137	12075	14143
Missouri	6037	9472	13121	16910	20836	24908	29140
North Dakota	862	1193	1503	1817	2149	2493	2848
South Dakota	754	1038	1297	1552	1819	2095	2378
Nebraska	2966	4637	6282	7974	9785	11716	13770
Kansas	4215	6011	7722	9478	11366	13375	15493
West North Central	21406	33165	45129	57552	70757	84738	99509
Delaware	822	1339	1871	2419	2993	3591	4213
Maryland & D.C.	8388	12119	15933	19967	24312	28935	33807
West Virginia	2095	3326	4494	5565	6554	7481	8363
N. & S. Carolina	11888	19312	26910	34827	43288	52266	61746
Georgia	8037	11141	14329	17856	21852	26259	31032
Florida	11896	15910	20140	24986	30525	36947	43867
South Atlantic	50356	74691	99809	126657	155960	187488	221075
Kentucky	3285	4713	6045	7388	8808	10290	11822
Tennessee	3266	6054	9133	12411	15914	19635	23580
Alabama	4434	5842	7274	8841	10592	12497	14532
Mississippi	3164	3651	4216	4895	5688	6563	7501
East South Central	14149	20260	26668	33535	41002	48986	57436
Arkansas	2818	4147	5419	6710	8075	9508	11005
Louisiana	5443	8383	11332	14361	17553	20906	24419
Oklahoma	4484	6740	8976	11288	13755	16374	19140
Texas	23137	36187	49663	64069	79877	97033	115492
West South Central	35882	55457	75391	96428	119260	143820	170057
Montana	1188	1804	2336	2801	3226	3618	3985
Idaho	2311	3076	3734	4371	5029	5699	6372
Wyoming	1112	1406	1678	1944	2213	2481	2748
Colorado	4243	6355	8440	10539	12699	14912	17172
New Mexico	2018	2712	3356	3994	4646	5303	5962
Arizona	3989	5579	7141	8779	10547	12420	14379
Utah	1502	2554	3613	4649	5676	6697	7719
Nevada	2004	2757	3587	4561	5703	6983	8376
Mountain	18367	26243	33885	41638	49738	58112	66713
Washington	7753	11638	15692	20011	24670	29633	34869
Oregon	5406	7177	9045	11096	13363	15806	18397
California	41277	56561	72592	90017	109096	129522	151055
Pacific	54436	75376	97329	121125	147129	174961	204320
Continental U.S.	311636	456714	604102	758505	923745	1098858	1283176

	Industrial Demand (millions of kilowatt hours)						
	1970 (actual)	1975	1980	1985	1990	1995	2000
Maine	1968	3194	4224	5014	5599	6016	6305
New Hampshire	1455	2049	2521	2885	3162	3366	3512
Vermont	688	965	1187	1356	1483	1575	1640
Massachusetts	7519	11963	15486	17979	19663	20753	21430
Rhode Island	1242	1900	2416	2786	3042	3211	3318
Connecticut	5289	8532	11169	13103	14482	15414	16050
New England	18161	28602	37003	43123	47421	50334	52255
New York	27306	46112	61347	72011	78994	83339	85911
New Jersey	15396	19777	22800	24885	26357	27362	28017
Pennsylvania	38140	42421	45689	48104	49785	50839	51392
Middle Atlantic	80842	108310	129836	145000	155136	161539	165321
Ohio	46045	62752	75223	83987	89953	93808	96129
Indiana	18022	22834	26252	28648	30315	31400	32040
Illinois	25678	41123	54275	64657	72776	79148	84239
Michigan	24277	33184	39590	43884	46647	48291	49146
Wisconsin	9373	12847	15746	18148	20159	21834	23234
East North Central	123395	172739	211085	239324	259850	274481	284788
Minnesota	8481	9349	10253	11180	12095	12947	13719
Iowa	5209	6590	7638	8475	9174	9748	10218
Missouri	9854	11493	12942	14138	15080	15802	16342
North Dakota	354	466	551	613	656	684	700
South Dakota	345	611	837	1001	1113	1185	1228
Nebraska	1797	3096	4223	5097	5747	6222	6569
Kansas	4674	6095	7081	7779	8299	8676	8944
West North Central	30703	37700	43525	48283	52164	55264	57720
Delaware	2635	2967	3279	3568	3828	4050	4232
Maryland & D.C.	10631	15178	18632	21129	22926	24188	25055
Virginia	7594	14759	21623	27213	31465	34587	36856
West Virginia	9195	10738	11859	12575	12951	13072	13014
N. & S. Carolina	26371	39256	49947	58329	64758	69534	73012
Georgia	11121	18288	24522	29552	33534	36608	38958
Florida	10358	18407	25722	31755	36604	40404	43356
South Atlantic	77905	119592	155583	184122	206067	222444	234483
Kentucky	20540	28670	34818	39363	42736	45116	46711
Tennessee	31289	41557	50546	58346	65077	70738	75430
Alabama	19208	27545	34429	39898	44192	47453	49876
Mississippi	5574	9146	12213	14570	16283	17457	18221
East South Central	76611	106919	132006	152176	168289	180764	190238
Arkansas	5857	7852	9458	10739	11754	12528	13102
Louisiana	12896	18224	22652	26189	28977	31124	32758
Oklahoma	4647	7653	10202	12173	13661	14760	15564
Texas	40098	58369	73638	86111	96355	104657	111364
West South Central	63498	92097	115951	135212	150747	163068	172788
Montana	6120	7767	8970	9815	10383	10727	10902
Idaho	4950	5911	6525	6931	7199	7348	7403
Wyoming	1427	1355	1328	1318	1312	1302	1287
Colorado	2355	4662	6752	8296	9322	9949	10298
New Mexico	1401	2308	3026	3507	3789	3926	3967
Arizona	4707	5821	6643	7272	7759	8117	8364
Utah	1790	2517	3072	3461	3719	3876	3959
Nevada	1537	2689	3815	4872	5848	6708	7442
Mountain	24287	33028	40130	45471	49331	51952	53620
Washington	25255	30056	34776	39389	43799	47849	51473
Oregon	10995	14289	17137	19569	21617	23297	24652
California	39053	51074	59921	66382	71138	74506	76782
Pacific	75303	95419	111834	125340	136553	145652	152907
Continental U. S.	570705	794408	976952	1118050	1225556	1305497	1364117

Appendix 26.B

Alternative National Total Generation Requirements Projections
(trillion kWh) 1970 = 1.52

Base Case Assumption Change	1975	1980	1985	1990	1995	2000
None (base case)	2.17	2.76	3.32	3.88	4.46	5.05
Electricity Prices:						
Double by 2000	2.06	2.37	2.55	2.67	2.76	2.84
Constant at 1970 levels	2.21	2.92	3.70	4.55	5.51	6.57
Gas Prices:						
Increase 8.7% per year	2.17	2.78	3.37	3.97	4.59	5.26
Increase 4.0% per year	2.16	2.73	3.26	3.77	4.28	4.78
Appliance Prices:						
Constant at 1970 levels	2.15	2.71	3.21	3.68	4.14	4.58
Decrease 4% per year	2.17	2.78	3.38	4.00	4.65	5.34
Income:						
Increases at 5.2% per year	2.19	2.87	3.60	4.42	5.38	6.56
Decreases at 1.6% per year	2.08	2.56	2.72	2.41	3.06	3.21
Population:						
300 million by 2000	2.17	2.78	3.34	3.90	4.46	5.06
251 million by 2000	2.15	2.70	3.18	3.61	4.01	4.34

References

Bureau of the Census (1972). "Projections of the Population of the United States, 1972 to 2020," Department of Commerce, GPO P-25, No. 495, U.S. Government Printing Office, Washington, D.C.

Chapman, D.; Tyrrell, T.; and Mount, T. (1972). "Electricity Demand Growth and the Energy Crisis," *Science, 178,* pp. 703–708, November 17.

Federal Power Commission (1971). *The 1970 National Power Survey,* part I, U.S. Government Printing Office, Washington, D.C., pp. I-3-12, I-19-10.

Graham, R. E., Jr.; Degraff, H. L.; and Trott, E. A., Jr. (1972). "State Projections of Income, Employment, and Population," *Survey of Current Business, 52,* Bureau of Economic Analysis of the Department of Commerce and the Economic Research Service of the Department of Agriculture, p. 22, April.

Mount, T.; Chapman, D.; and Tyrrell, T. (1973). "Electricity Demand in the United States, An Econometric Analysis," ORNL-NSF-EP-49, Oak Ridge National Laboratory, Oak Ridge, Tenn.

National Center for Health Statistics (1972). "Monthly Vital Statistics Reports," Department of Health, Education, and Welfare, Washington, D.C.

National Petroleum Council (1971). *U.S. Energy Outlook,* Vol. 2, Committee on U.S. Energy Outlook, Washington, D.C., pp. 7, 10, 11.

Olmsted, Leonard M. (1972). "23rd Annual Electrical Industry Forecast," *Electrical World,* p. 61, September 15.

Public Utilities Fortnightly (1972). "AGA Speakers Call for Bold Policies on Gas Shortage," Vol. 90, No. 11, p. 48, November 23.

Spitz, Peter H. (1972). "Raw Material and Energy Challenges," *Chemical Engineering, 79*(1), pp. 77–81, January 10.

U.S. Congress, Senate (1972). *Summary Report of the Cornell Workshop on Energy and the Environment,* sponsored by the National Science Foundation RANN Program, Committee on Interior and Insular Affairs, 92nd Congress, 2nd Session, Chapter III, pp. 123–137.

Wagner, Aubrey S. (1972). "Power, Environment, and Your Pocketbook," *Public Utilities Fortnightly, 89*(13), pp. 27–31.

27 An Econometric Model of the Demand for Energy in Canada*

J. DANIEL KHAZZOOM†

27.1. The Economic Background

In this study, I estimate for each province the demand for energy in the industrial, residential, and residential-commercial markets. I focus on the four major sources of energy: gas, oil, coal, and electricity.

This paper lays out the background and illustrates briefly some ramifications of the basic ideas. I will also report in this paper some estimation results for the industrial demand for gas and discuss the use of the model for simulation. A more extensive paper on the subject (Khazzoom, unpublished) is available from the author upon request.

The demand for the four major sources of energy may be thought of as made up of two parts: a "captive" and a "free" part. As the names suggest, captive demand is that part of demand which is immobilized by past commitments. It exists because of past investment in appliances that use the commodity in question, and it is generally immune to the influence of economic stimuli. Free demand is that part of the demand which is free from past commitments and which may be reasonably expected to be responsive to changes in the economic conditions.

For those commodities whose use requires a substantial investment in a stock of appliances (as is generally the case with gas, oil, coal, or electricity), the captive demand makes up the major component of the total. To give the reader a concrete idea about the magnitudes involved, I have prepared Figure 27.1. Six provinces are covered: British Columbia, Alberta, Saskatchewan, Manitoba, Ontario, and Quebec. (In estimating the industrial demand for these provinces, I constrained the depreciation rate to zero.) As the diagram shows, the captive component constitutes typically between 85 to 90% of the total industrial demand for gas.

* This study originated several years ago when I served as Chief Econometrician with the U.S. Federal Power Commission (FPC). I am indebted to the Honorable Lee C. White, former Chairman of the FPC and Haskell P. Wald, Chief, Office of Economics of the FPC, for sponsoring the initial stages of my work. Subsequent work was also supported by the Sub-Committee on Research, Social Science Division of McGill University, and the Department of Energy, Mines and Resources. My research on natural gas was supported by a grant from the Canada Council. Among those who assisted me, Allan Gruchy, Jr., Robert Stonebreaker, Susan Case-Schwartz, and Julia Klincsek deserve special mention for their substantial contribution.

† Associate Professor of Economics, McGill University.

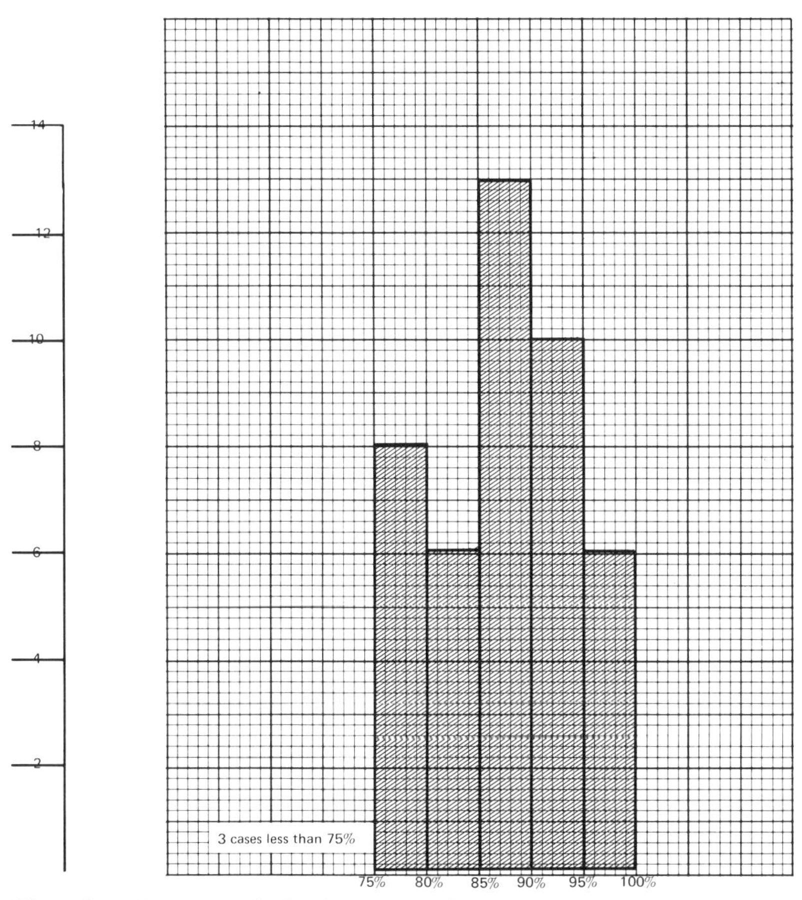

Figure 27.1. Frequency distribution of the ratio of captive industrial demand to total industrial demand in Canada, 1961–1968.

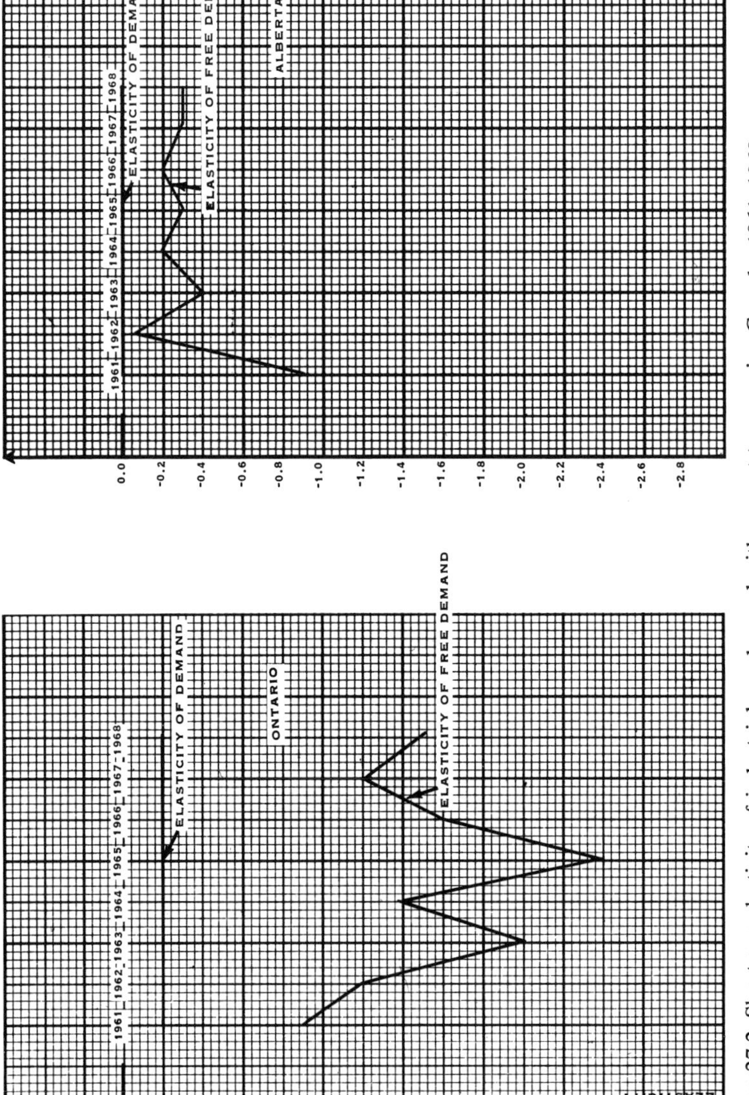

Figure 27.2. Short-run elasticity of industrial gas demand with respect to gas price, Canada, 1961–1968.

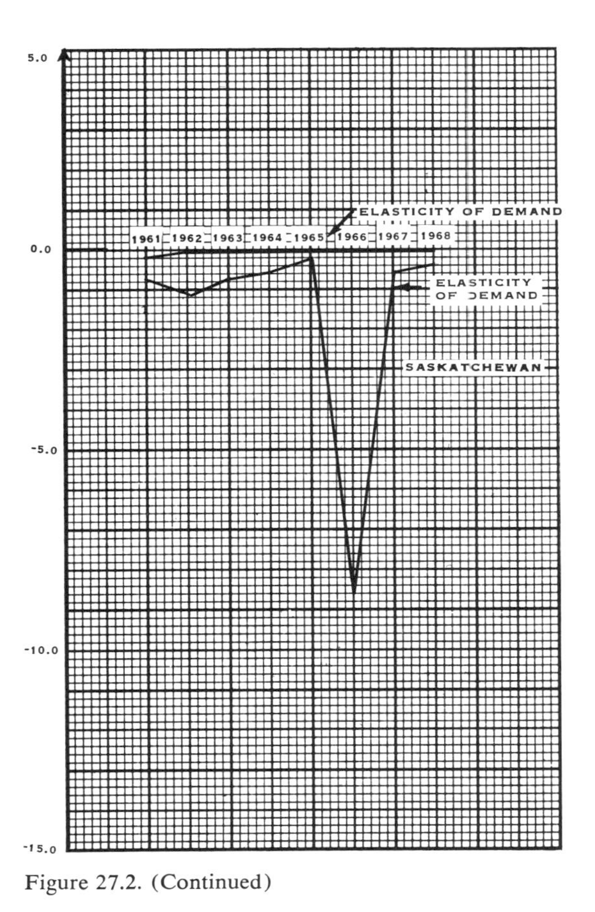

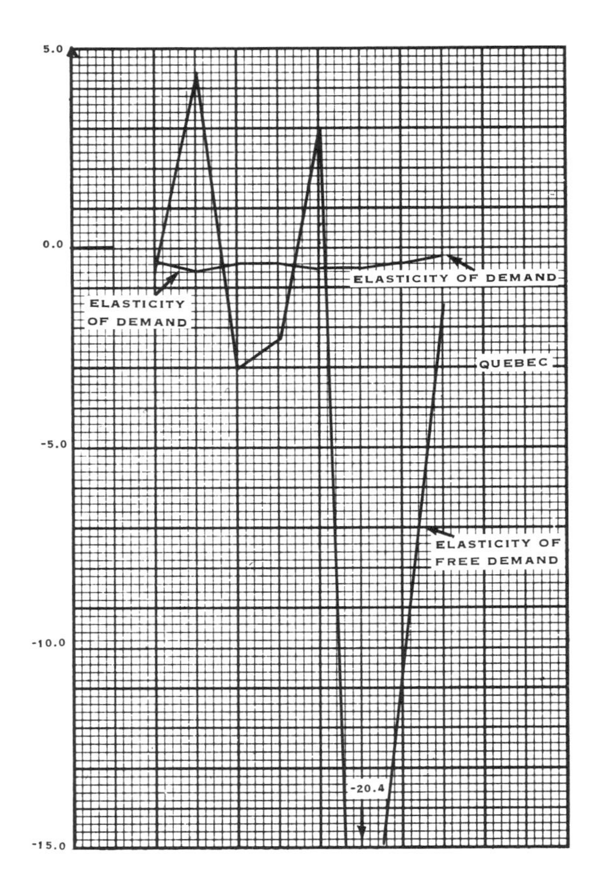

Figure 27.2. (Continued)

The predominance of the captive component in the total results in a reduction in the consumer's agility (or the producer's agility, if the commodity in question is a factor of production), and the overall response of total demand to, say, price variations, will be much more restrained than economic theory would lead one to believe. To measure the impact of an economic stimulus on demand, one necessarily has to focus on the free demand. Otherwise, the estimates will be swamped by the rigidities in the captive part. In fact, one may not be able to discern any response in the total demand, even though the free component may be extremely sensitive to the stimulus. This point is illustrated in Figure 27.2. Note the *stability* of the elasticity of total demand as opposed to the *variability* (almost volatility) of the elasticity of the free demand.[1] Note also how much larger the elasticity of the free demand tends to be compared with the elasticity of total demand.

The first statement about the need to isolate the free demand from the rest (at least in the field of energy) appeared about a decade ago in a series of testimonies by Sherman Clark before the Federal Power Commission (1963). Clark, who testified on the relationship between gas demand and gas price, coined the term "incremental demand" (which corresponds to the "free" demand in this study).

Various versions of this idea were implicit in some recent work in demand analysis (Fisher and Kaysen, 1962; Houthakker and Taylor, 1966) and investment decision theory (Drymes and Kurz, 1964). It remained, however, for Balestra and Nerlove (1966) to propose a mathematical formulation for Clark's incremental demand of gas. Nerlove's and Balestra's formulation did not express correctly the constituents of the incremental demand. This led to logical difficulties that required the authors to postulate subsequently a constant utilization rate for both gas- and all other energy-using appliances. This is not a realistic assumption when applied to a whole market. Other difficulties in the specification and estimation results of Nerlove's and Balestra's model are discussed in Khazzoom (1970; unpublished).

I used a diagram to identify the free and captive constituents of demand. The step requires a simple transformation from a one- to a two-dimensional space. This is a commonly used procedure. Often the problem is simplified by transforming it to a higher-dimensional space, solving it in there, and then bringing it back to the lower-dimensional space.

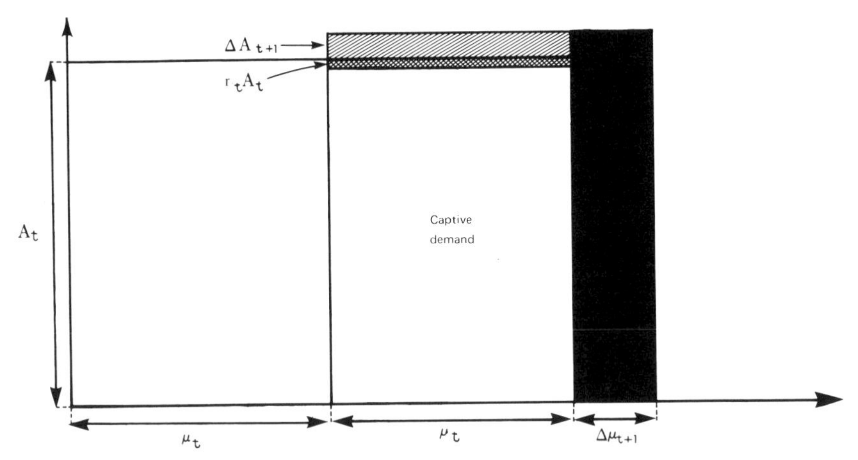

Figure 27.3. A diagrammatic derivation of the free and captive demands.

For our purpose, we measure the stock of an appliance in terms of the maximum amount of Btu that may be consumed if this appliance were fully utilized. We can then write the demand for the commodity as the product of the stock of appliances that use that commodity times the stock's utilization rate:

$$D_t = u_t A_t, \tag{27.1}$$

where D, u, and A stand for demand, utilization rate, and stock of appliances, respectively. (The subscript t, denoting time, is attached to u, to indicate that the utilization rate is not constrained to a constant.) The transformation, Equation 27.1, simply means that instead of thinking of demand as a scalar, we will henceforth think of it as a two-dimensional vector identified by a stock of appliances and a utilization rate. In Figure 27.3 the demand at time t, D_t, is seen to represent the area defined by the rectangle in the left-hand side of the diagram.

It can be shown (for space limitations, we will not go into the details here) that the free demand corresponds to the sum of three areas in the diagram, defined by the shaded, the crosshatched, and the solid black rectangles. Hence, denoting the free demand by D, we have from the diagram

$$\mathsf{D}_{t+1} = u_t r_t A_t + u_t \, \Delta A_{t+1} + A_{t+1} \, \Delta u_{t+1}. \tag{27.2}$$

Shifting Equation 27.2 back by one period and making use of Equation 27.1, we have

$$\mathsf{D}_t = D_t - (1 - r_{t-1})D_{t-1}. \tag{27.3}$$

Note that to derive Equation 27.3 we did not have to constrain u_t to a constant.

As a fraction of observed demand, the free demand is

$$(\mathsf{D}_t/D_t) = 1 - (1 - r_{t-1})(D_{t-1}/D_t). \tag{27.4}$$

This is the expression we used in calculating the frequencies shown in Figure 27.1, except that for the particular case of the industrial demand for gas in Canada, we constrained r_t to zero. (See Section 27.2.)

From the transformation, Equation 27.1, we were able to wring out several other useful results. To give one example: let S and De denote the sales of an appliance (for example, a coal-using appliance) and the total depreciation of the appliance, respectively, measured as before in terms of Btu equivalent. Then we have

$$\Delta A_t = S_t - \mathrm{De}_t. \tag{27.5}$$

Assuming that the depreciation rate for the appliance in question can be reasonably approximated by a constant, we have

$$\mathrm{De}_t = rA_{t-1}. \tag{27.6}$$

Substituting from Equation 27.6 into Equation 27.5, rearranging terms, and using E^{-1} to denote the (one-period) backward shift operator, we have

$$A_t = S_t/[1 - (1 - r)E^{-1}]. \tag{27.7}$$

Substituting from Equation 27.7 into Equation 27.1, clearing the denominator, and rearranging terms, we have

$$D_t - (1 - r)(u_t/u_{t-1})D_{t-1} = u_tS_t. \tag{27.8}$$

In the special case where we are dealing with an appliance whose utilization rate is known to be constant, Equation 27.8 simplifies to

$$D_t - (1 - r)D_{t-1} = uS_t, \tag{27.9}$$

where the left-hand side of Equation 27.9 is the same as the right-hand side of Equation 27.3 with $r_{t-1} = r$. In other words, when we are dealing with an appliance whose depreciation and utilization rates are reasonably constant, the free demand simply reduces to a fraction of the sales of the appliance. This is also as we would expect since the variation in the utilization rate (that is, the short-run effect) is then equal to zero, and we are left only with the long-run effect.[2] Hence, if data on the demand for the commodity that uses that appliance are not available, while data on the sale of the appliance itself are available, then the right-hand side of Equation 27.9 provides a useful alternative to the right-hand side of Equation 27.3 for expressing the (unobservable) free demand in terms of an observable variable. A good example is the free demand for electricity used in home electric refrigerators.

27.2. The Industrial Model and Its Estimation Results

The next step is to specify a relationship between the free demand and its determinants. This will vary with the markets. Here we confine ourselves to the industrial demand for gas.

The industrial demand for gas is a demand for a factor of production. As such, we would expect it to depend on a production and a price vector. In specifying the relationship, we have followed the tenets of the neoclassical theory of factor demand. Our only departure was in writing the free demand instead of the total demand as the dependent variable. Letting D denote in this section the free industrial demand for gas, we have

$$\mathsf{D}_t = f[\mathrm{PG}_t/\mathrm{PS}_t, \mathrm{MA}_t], \tag{27.10}$$

where PG denotes the industrial price of gas, PS denotes the price of gas substitutes, and MA denotes (real) manufacturing production.

In estimating Equation 27.10 we have substituted for D_t from Equation 27.3 but constrained the depreciation rate to zero, since the industrial demand for natural gas in Canada is a relatively new phenomenon.[3] Hence, our dependent variable reduces in this case to ΔD_t.

Several problems have to be dealt with prior to estimation: the operational definitions of PG and PS; the problem of joint determination of demand and price; and so on. We will not deal with these problems here.

Since the form of $f(\cdot)$ in Equation 27.10 is not known, it is necessary to aproximate the function by a Taylor's series for estimation. Equation 27.11 shows the results for the model we estimated for the six provinces: British Columbia, Alberta, Saskatchewan, Manitoba, Ontario, and Quebec for '60 through '68. These are the only provinces in which there was industrial consumption of gas during the sixties.

$$\begin{array}{ll} \Delta D_t = & 16.3 \text{ (British Columbia)} + 11.6 \text{ (other provinces)} - 50.7(\text{PG/PS})_t \\ t \text{ ratio} & (4.4) \qquad\qquad\qquad\qquad (5.5) \qquad\qquad\qquad\qquad (-4.2) \\ & + 57.6(\text{PG/PS})_t^2 + .7\text{MA}_t + .004\text{MA}_t^2 - 1.3(\text{PG/PS})_t \cdot \text{MA}_t. \\ t \text{ ratio} & \quad (3.3) \qquad\qquad (2.3) \qquad (2.3) \qquad\qquad (-2.1) \\ & R^2 = .51;\ \text{d.f. (degrees of freedom)} = 47;\ 1960 \text{ to } 1968. \end{array} \tag{27.11}$$

We have subjected this model to several predictive and nonpredictive tests, and it performed well.

Gas demand appears to have a tendency to increase at a slightly increasing rate (rather than at a constant rate) with the increase in manufacturing production. See the quadratic term in Equation 27.11. This may reflect waste in the use of gas; it may also reflect a tendency for the increase in MA to be weighed in favor of those industries that use gas rather than other sources of energy.

Figure 27.4 shows the impact multipliers computed from the foregoing model. It is interesting to note that Ontario has the highest impact multiplier with respect to gas price. Our estimates also show a clear tendency for the industrial gas demand in this province to become more responsive to gas price.[4] The diagram also shows that the impact effect of manufacturing production in this province has been rising during the same period. The same tendency appears to be present in Manitoba. On the other hand, in Quebec and British Columbia, the tendency seems to be in the opposite direction.

27.3. Simulation

The model we estimated is shown in Figure 27.4. For clarity's sake, we write PG_t in terms of what it stands for in the model we estimated, namely, $\text{PG}_t = (\text{pg}_t + \text{pg}_{t-1})/2$, where pg_t denotes industrial gas price in the year t.

We can now initiate the model at $t = 0$. Shifting Equation 27.11 successively forward by one period at a time and adding the resulting equations, we arrive at the following general result (where G denotes the demand for gas):

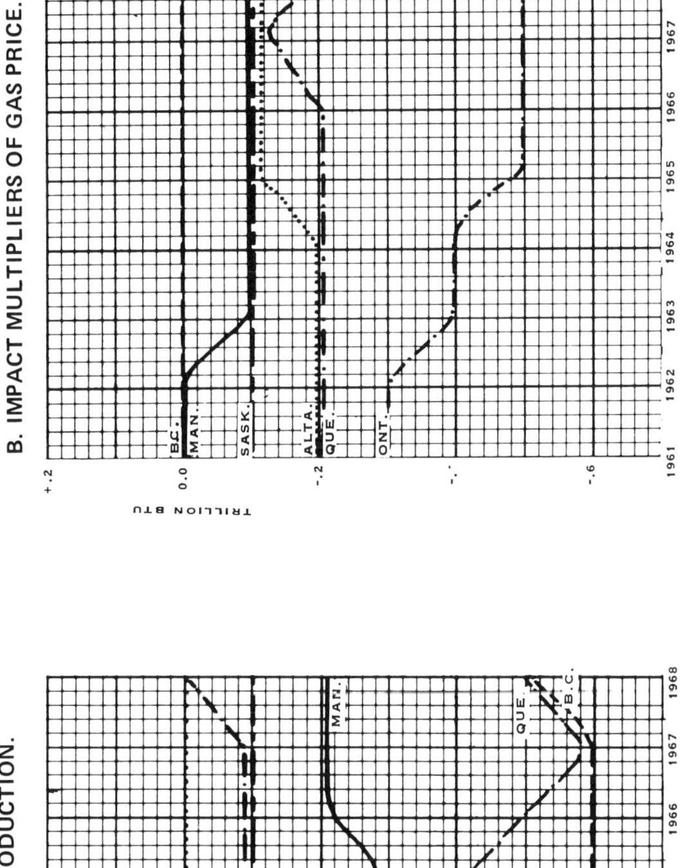

Figure 27.4. Impact multipliers for the industrial demand of gas in Canada, 1961–1968.

$$G_T - G_0 = T\alpha_0' + \sum_{t=1}^{T} \{-\beta_1'[(pg_t + pg_{t-1})/2PS_t] + \beta_2'[(pg_t + pg_{t-1})/2PS_t]^2 + \gamma_1'MA_t + \gamma_2'(MA)_t^2 - \delta'[(pg_t + pg_{t-1})/2PS_t] \cdot MA_t\}. \quad (27.12)$$

Recall we estimated two intercepts: one for British Columbia and another for the rest of Canada, so that there are really two different terms in $T\alpha_0'$, depending on the province. Otherwise, we estimated $\beta_1' = 50.7$; $\beta_2' = 57.6$; $\gamma_1' = .7$; $\gamma_2' = .004$; and $\delta' = 1.3$ (Equation 27.11).

The general expression of the multipliers with respect to prices over consecutive periods can be derived by differentiating Equation 27.12 with respect to pg_t, $t = 1, 2, \ldots, T$. We have:[5]

$$\frac{\partial G_T}{\partial pg_t} = \begin{cases} \frac{1}{2PS_t}\{-\beta_1' - \delta' MA_t + \beta_2'[(pg_t + pg_{t-1})/PS_t]\} & t = T \\ \sum_{i=0}^{1} \frac{1}{2PS_{t+i}}\{-\beta_1' - \delta' MA_{t+i} + \beta_2'[(pg_{t+i} + pg_{t-1+i})/PS_{t+i}]\} & 1 < t < T \\ \frac{1}{2PS_t}\{-\beta_1' - \delta' MA_t + \beta_2'[(pg_t + pg_{t-1})/PS_t]\}. & t = 1 \end{cases} \quad (27.13)$$

As gas price is allowed to increase (holding MA_t and PS_t fixed), the response (Equation 27.13) decreases numerically from one period to the next. This is also implied in the quadratic relationship estimated for ΔG_t. With pg_t rising, the cumulative response *a fortiori* reaches a maximum (numerically) after a finite number of steps beyond the initial period. Subsequently, the cumulative response begins actually to decrease (numerically). The use of the model for simulation purposes should therefore be done with a great deal of care. Equation 27.13 provides, in effect, a control mechanism that could be used to determine the horizon over which a particular simulation experiment can be carried out meaningfully. The danger signal is flashed when the model begins to yield positive values for any one of the terms in Equation 27.13. Of course, if PS, the price of substitute, is assumed to increase at the same time, the time it takes any interim response to reach zero will be accordingly longer. The same is true when MA is allowed to increase. In fact, with sufficiently high rate of growth of MA, Equation 27.13 will remain negative as long as that rate (of growth of MA) is maintained.

Hence, to simulate meaningfully with the model, one would want first to examine the results derived from Equation 27.13 for any combination of postulated changes in the predetermined variables. Certainly, the simulation should not be carried out beyond the time period T when $\partial G_T/\partial \mathrm{pg}_T$ ceases to be negative.

Exactly the same remarks apply to the various multipliers of G_T with respect to MA_t, $t = 1, \ldots, T$. From Equation 27.12 we find

$$\partial G_T/\partial \mathrm{MA}_t = \gamma_1' + 2\gamma_2'\mathrm{MA}_t - \delta'[(\mathrm{pg}_t + \mathrm{pg}_{t-1})/2\mathrm{PS}_t]. \quad 1 \leq t \leq T \quad (27.14)$$

For meaningful simulation results, we want Equation 27.14 to be positive for each province over the whole time period for which the simulation is carried out. Here again, with pg_t rising (holding MA and PS fixed), the cumulative response reaches a maximum after a finite number of steps beyond the initial period. Subsequently the cumulative response begins actually to decline. The time it takes this to happen may be prolonged, and in fact the reversal of sign in Equation 27.14 may never take place if MA or PS (or both) increase at the same time as gas price increases.

Taken together, Equations 27.13 and 27.14 determine the maximal time horizon over which we could simulate meaningfully the behavior of industrial gas demand under alternative assumptions about the behavior of the predetermined variables. The maximal horizon is the smaller of the two time horizons during which Equation 27.13 remains negative and Equation 27.14 remains positive. For a concrete example on how these rules were applied in this study, the reader is referred to a paper by Khazzoom (unpublished, pp. 93–97).

With these guidelines in mind, we have simulated the behavior of the industrial demand of gas under alternative assumptions about the behavior of the predetermined variables. The horizon extended beyond the year 2000 for most alternatives we attempted, but there were instances in which the simulation had to be constrained to a shorter horizon determined by the guidelines discussed earlier.

Note that our model was estimated for a period in which there was hardly any replacement market (at least, in any significant sense) for gas demand. Hence, our simulation results must have a tendency to overestimate the future industrial demand for gas in the face of rising gas price. On the other hand, the period for which we estimated was one in which environmental considerations played practically no role. The shift toward cleaner sources of energy will inevitably lead to a greater demand for gas in the future. Hence, to the extent that our simulation results tend to overestimate the future demand for gas, this tendency may be mitigated or even more than offset by the greater role that environmental factors will come to play in the future in enhancing the demand for gas.

Briefly, our results indicate a tremendous growth in the industrial demand for Ontario in the future. As it is, Ontario is the largest industrial user of gas, followed closely by Alberta. The gap in industrial demand between this province and Alberta is forecast to widen substantially over time. Indeed, even under conservative assumptions about the growth of manufacturing production, the consumption in Ontario is forecast by 1978 to exceed the combined industrial consumption in all other provinces taken together.

Gas price influences substantially the pattern of simulated demand. The relative position of the provinces vis-à-vis one another changes radically as gas price is allowed to increase, holding other predetermined variables fixed. Indeed, the impact of gas price on future demand in some provinces is so drastic that gas pricing policy, whatever its purpose may be, has very much to take into account this tremendous regional difference in the response of gas demand to gas price.

In closing, I would like to observe that most studies of the future outlook for energy in Canada that I have encountered focus primarily on the question of how to get more supplies to meet the forecast demand. Once the demand is determined (or is assumed to be determined), the question is simply reduced to one of how to get the necessary supplies to match the expected demand. What seem to be given only perfunctory treatment are the following:

1. Is the price necessary to elicit these supplies one that the public can afford or is prepared to accept? In my testimony on the supply of gas in the United States in 1970 (Federal Power Commission, 1970, p. 225) I pointed out that even an increase of 13¢ per thousand cubic feet in south Louisiana may not do the job of eliciting adequate supplies from south Lousiana. But it was unthinkable at that time to advocate such a price hike. An average increase of about 6¢ was sanctioned. The shortage became evident not long after. It is not clear even today that, in spite of the constant talk about a crisis in energy, the same problem does not lurk in the background. These remarks are equally applicable to the Canadian scene. Which brings me to the second point.

2. Is it really necessary to keep chasing that demand curve? Might it not be more profitable to direct funds into extensive research efforts to find ways and means of curbing the rate of growth of gas consumption (and energy consumption in general) without adversely affecting economic growth? There are potentially substantial savings that can be accomplished both at the technological as well as institutional levels. I am aware that efforts are being made to conserve energy, but these efforts are nowhere commensurate with the problem on hand.

Notes

1. In the diagram, the price elasticity of total demand in Ontario and Alberta looks completely flat, because I plotted the diagram with one decimal point. With two decimals, some ripples could have been observed, but the size of the diagram would have been unmanageable. Note also that the price elasticity of the free demand can be positive (since the free demand may be negative). I have included Quebec in the diagram to illustrate this point.

2. If we can reasonably assume also that the dollar value of an appliance is a good reflection of its Btu capacity, we may replace the Btu equivalent of the sales by their dollar value. This was implicit in Fisher's estimation of the long-run demand for electricity. Fisher used the sales of appliances as the dependent variable. See Fisher and Kaysen (1962).

3. For completeness, I have estimated each version I considered (not reported here) with a variable depreciation rate as in Equation 27.3, as well as with a constant depreciation rate. As expected, the results, though by no means uniform, did not indicate that there was much of a replacement market for gas to start with during the sampled period. Interprovincial gas did not start flowing until 1958 or 1959. Our sampled period ends in 1968, and it would probably be too much to expect the replacement market to be an important component of the free demand by the end of the sixties. Note also that the amount of gas consumed in the early sixties was very small compared to the later part of the sixties. Hence, even if we assume that part of the appliances were to be replaced after six or seven years of use (due to, say, technological obsolescence), this would hardly have any effect on the industrial demand for gas since the whole stock in the early sixties must have been very small.

4. Do not confuse response with elasticity. There is no evidence that the industrial demand for gas in Ontario is becoming more price elastic. See Figure 27.2.

5. When MA_t, PS_t, and pg_t are held fixed at their level at $t = 1$, Equation 27.13 simplifies greatly, and then we have $\partial G_T/\partial \mathrm{pg}_T = \partial G_T/\partial \mathrm{pg}_1 = \frac{1}{2}\partial G_T/\partial \mathrm{pg}_t$ for $1 < t < T$. This means that the impact multiplier with respect to gas price at time $t = 1$ (and $t = T$) is half the size of each of the successive multipliers. The unchanging pattern of the multipliers for $1 < t < T$ should be evident also from the fact that when the explanatory variables are fixed (Equation 27.11) simplifies to $\Delta G_t = k$, where k is a constant. The general solution of this equation is $G_t = G_0 + kt$, and $\partial G_t/\partial t = k$. The divergence of our results from this expression for $t = T$ and $t = 1$

stems from the fact that gas price enters (Equation 27.11) as a two-year moving average.

References

Balestra, P., and Nerlove, M. (1966). "Pooling Cross Section and Time Series Data in the Estimation of a Dynamic Model: The Demand for Natural Gas," *Econometrica, 34,* pp. 585–612, July.

Dhrymes, P. F., and Kurz, M. (1964). "Technologies and Scale in Electricity Generation," *Econometrica, 32,* pp. 287–315, July.

Federal Power Commission (1963). *A Study of the Demand of Gas to 1970,* Testimony by Sherman Clark in the Matter of Area Rate Proceedings, et al., FPC Docket AR61-2. See also various other testimonies by Clark before the FPC beginning with Permian Basin Case in 1961.

——— (1970). *An Econometric Model of U.S. Natural Gas Supply,* Testimony of Dr. J. Daniel Khazzoom, Docket No. AR69-1, Southern Louisiana Area, August 31.

Fisher, F. M., and Kaysen, C. (1962). *The Demand For Electricity in the U.S.,* North Holland Publishing Co., Amsterdam.

Houthakker, H. S., and Taylor, L. D. (1966). *The Consumer Demand in the U.S., 1959–1970, Analysis and Projection,* Harvard University Press, Cambridge, Mass.

Khazzoom, J. Daniel (1970). A review article of P. Balestra's, "The Demand for Natural Gas in the U.S.," *Econometrica 38,* pp. 946–947, November.

——— (unpublished). *An Econometric Model of the Demand For Energy in Canada, 1960–1968,* working paper in Econometric Studies.

——— (1971). "The FPC Staff Econometric Model of U.S. Natural Gas Supply," *The Bell Journal of Economics and Management Science,* pp. 51–93, Spring issue.

28 Residential Demand for Electricity*

JOHN TANSIL† AND JOHN C. MOYERS‡

The growth of residential electricity use for the period 1950 to 1970 is examined from the standpoint of increases in the number of households, appliance saturations, and the average annual electricity consumption per appliance. Growth patterns are defined which illustrate the factors accounting for the increase from 1800 kWh per household in 1950 to 7000 kWh per household in 1970. Space heating, water heating, and air conditioning have small saturations, large average annual consumptions, and the greatest growth potentials for contributing to the residential load. Energy conservation is stressed through (1) the importance of housing insulation, (2) more efficient room air conditioners, and (3) the substitution of heat pumps for electric resistance heating. The number of households, appliance saturations, and average annual electricity use per appliance are projected to 1990 to obtain the total electricity consumption per appliance. The sum of the disaggregated projections is compared to other independent projections based on extrapolation and econometric methods.

28.1. Summary of Past Use and Growth

The post-World War II exponential growth rate of electricity consumption in the United States is a well-documented phenomenon. The growth has been universal, encompassing all sectors: residential, commercial, and industrial. This article will summarize the historical patterns of residential electricity use, discuss the potential for electricity conservation in the home, and project residential electricity demand to 1990.[1]

Table 28.1 gives total electricity sales, residential sales, population, and the number of wired households from 1950 to 1970 (Edison Electric Institute, 1971; U.S. Bureau of the Census, 1971). Since 1960 total electricity sales have grown at the approximate annual rate required for a doubling time of 10 years. However, residential sales have grown somewhat faster, constituting about one-third of total sales in 1970.

* Research sponsored by the National Science Foundation RANN Program under Union Carbide Corporation's contract with the U.S. Atomic Energy Commission. Views expressed herein are those of the authors and not necessarily of the Oak Ridge National Laboratory.

† Shawnee College, Ullin, Ill. 62992.
‡ ORNL-NSF Environmental Program, Oak Ridge National Laboratory, Oak Ridge, Tennessee 37830.

Table 28.1 Electricity Sales, Population, and Wired Households

	1950	1960	1970
Total Electricity Sales (10^9 kWh)	281	683	1391
Residential Electricity Sales (10^9 kWh)	70.1	196	448
Residential Sales/Total Sales	25%	29%	32%
Resident Population (10^6)	152	180	204
Wired Households (10^6)	38.9	51.4	64.0
Residential Sales per Capita (kWh)	461	1089	2196
Residential Sales per Household (kWh)	1800	3820	7000
Average Size of Household	3.37	3.33	3.17

The growth in the residential use of electricity can be related to an increase in the number of households and a concomitant greater use of electricity per household. In 1970 the average American home used 7000 kWh, four times more electricity than in 1950. This came during a time when the average number of people per household showed a slight decline.

Table 28.2 shows the saturation and average annual electricity consumption for household uses of electricity from 1950 to 1970 and projected to 1990 (Tansil, 1973; *Merchandising Week,* 1972; U.S. Bureau of the Census, 1950, 1960, 1970; Edison Electric Institute, 1950–1969). The saturation is defined to be the fraction of wired households having one or more of the appliance. With the exception of clothes washers, which declined in saturation, no appliance had a decrease in either saturation or average annual use from 1950 to 1970.

The projections of appliance saturations are based on either the 1960 to 1970 saturation growth rates or current trends of appliance installations in all new homes. The projections of average annual electricity use per appliance are based on the 1960 to 1970 shifts to more energy intensive appliances, that is, quick-recovery water heaters, frost-free refrigerators, color television, and so on.

Figure 28.1 illustrates the allocation of the residential consumption of electricity to the various household uses. From 1950 to 1970, the large contributors to the growth of electricity use per household were space heating (20%), refrigerators (19%), air conditioning (17%), and water heating (14%). Of the big contributors to growth in this 20-year period, refrigerators and water heaters were also large consumers at the beginning and end of the period; however, space heating and air conditioning (A/C)

Table 28.2 Summary: Saturation and Average Annual Electricity Consumption for Household Uses of Electricity

	Saturations (%)					Average Annual Use for Households Having the Appliance (kWh/household)				
	1950	1960	1970	1980	1990	1950	1960	1970	1980	1990
Refrigerators	83	98	100	100	100	350	800	1,300	1,600	1,800
Air Conditioning: Room	1	11	27	36	41	1,400	1,700	1,900	1,950	1,950
Central	0	2	11	18	26	3,600	3,600	3,600	3,600	3,600
Lighting	100	100	100	100	100	500	600	750	900	900
Space Heating	1	2	8	16	27	10,000	13,000	15,000	15,000	15,000
Water Heating	11	21	25	32	41	3,700	4,000	4,500	4,800	4,800
Clothes Drying	1	12	29	41	49	500	950	1,000	1,000	1,000
Cooking	16	32	40	54	74	1,200	1,200	1,200	1,200	1,200
Television	13	90	95	100	100	300	350	400	450	450
Food Freezers	6	19	28	34	39	600	900	1,400	1,500	1,600
Clothes Washers	74	76	70	70	70	50	50	100	100	100
Dishwashers	2	7	19	28	34	100	350	350	350	350
Irons	80	86	100	100	100	100	150	150	150	150

ORNL-DWG 73-1256

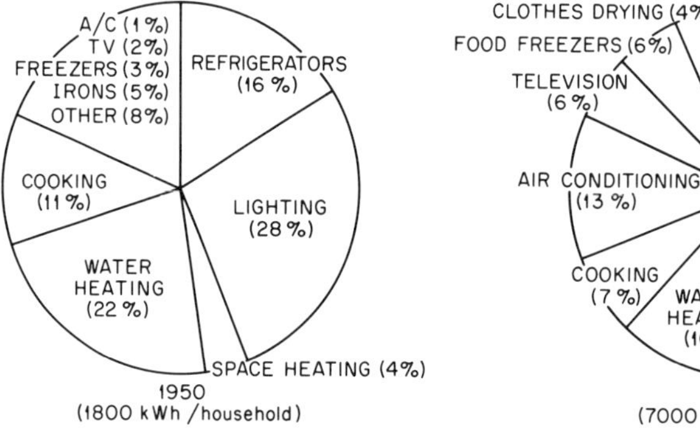

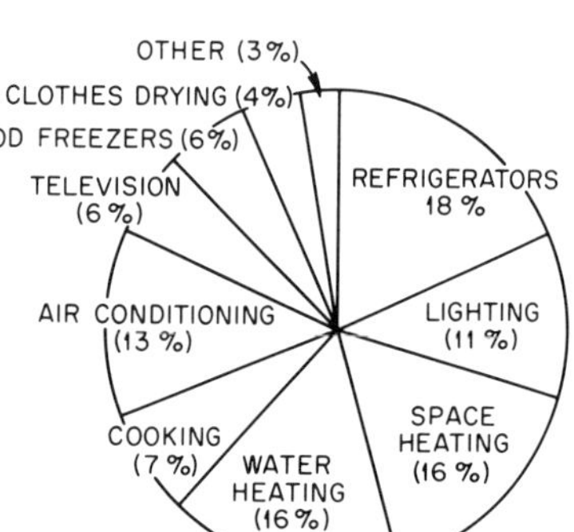

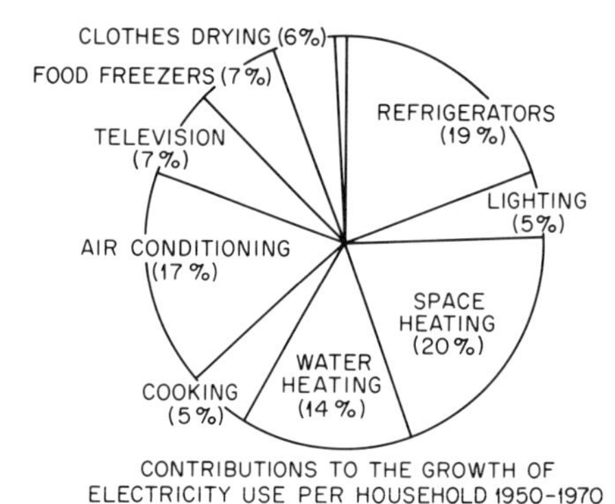

Figure 28.1. Household uses of electricity.

constituted new uses of electricity, being relatively unknown in 1950. Surprisingly, of the big consumers of electricity in 1970, only refrigerators, lighting, and television had saturations greater than 50%.

28.2. Potential for Energy Conservation

There are several possible ways to reduce electricity use in the home without a change of life-style. Many of these ways could be economically advantageous in the long run. These conservation measures include better-insulated housing, a shift to electrically driven heat pumps from resistance heating, more efficient air conditioners, a shift to fluorescent lighting from incandescent, and improved insulation for refrigerators, freezers, and water heaters. Three of these, heat pumps versus resistance heating, more efficient air conditioners, and better housing insulation, will be discussed in more detail.

There are, in general, two types of electric heating: (1) resistance heating and (2) electrically driven heat pumps. Resistance heating, which depends on resistive wire heating, includes electric furnaces, baseboard heating, ceiling cable, wall units, and duct heaters. The fraction of electrically heated households with heat pumps decreased from 15% in 1964 to 11% in 1970 (*Electric Heat and Air Conditioning,* 1956–1971; Air-Conditioning and Refrigeration Institute, unpublished data). The decline in the fraction of electrically heated homes with heat pumps is unfortunate from an energy resource conservation viewpoint: heat pumps are more efficient than resistance heating. The efficiency of a heat pump depends on the indoor-outdoor temperature differential, and, as a nationwide average, heat pumps are twice as efficient as resistance heating (Chiles, unpublished). In 1970 if all electrically heated households had heat pumps, then the electricity consumed for residential space heating would have been reduced by about 50% and total residential sales by 8%. The possible future reduction will be much larger because of the increasing share of the residential load which goes to electric heating.

Another energy savings would result from the use of more efficient air conditioners. There is a wide variation of efficiencies among different models of room air conditioners (Hirst and Moyers, 1973). Figure 28.2 shows the efficiencies of all room air conditioners having ratings up to 24,000 Btu/h (Association of Appliance Manufacturers, 1971). The efficiencies of these air conditioners range from 4.7 to 12.2 Btu of cooling per watt-hour of electricity consumed. Thus, for the same amount of cooling, the least efficient model uses 2.6 times as much electricity as the most efficient one. In 1970 the average room air conditioner had an efficiency of about 6 Btu per watt-hour. If the average efficiency had been 10 Btu per watt-hour, then the electricity consumed by room air conditioners would have been reduced by 40%, and total residential sales by 3%. In addition to the potential savings in energy consumption and operating costs, more efficient air conditioners would significantly reduce the summertime peak power demands that have resulted in "brownouts" in many major Eastern cities.

In 1970 residential space heating and air conditioning accounted for

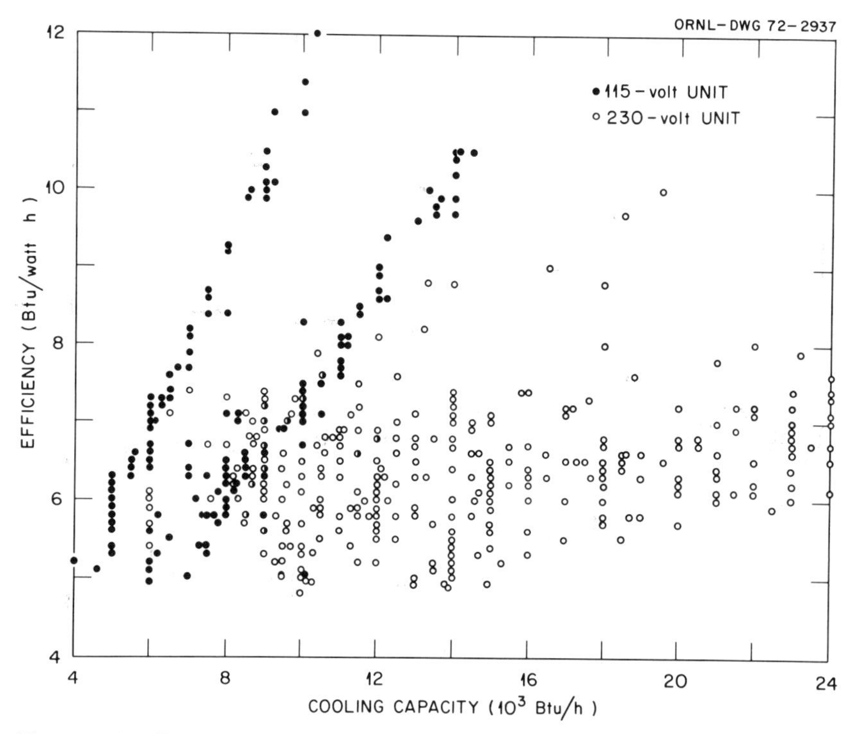

Figure 28.2. Efficiency of room air conditioners as a function of unit size.

almost 30% of household electricity use. A significant fraction of this could have been saved through the use of better housing insulation. The closest thing to a national standard for thermal insulation is incorporated in the Minimum Property Standards (MPS) of the Federal Housing Administration (FHA) (Federal Housing Administration, 1966 and 1971). The thermal insulation requirements, as outlined in these standards, apply to all federally financed housing, regardless of the type of heating fuel used. In response to the president's energy messages to Congress in 1971 and 1972, the thermal insulation requirements were upgraded for both homes and apartment housing.

Although the upgrading of thermal insulation requirements will reduce heating/cooling bills and conserve energy, the question arises as to whether additional housing insulation might be justified. A study of the use of additional thermal insulation in single-family homes concluded that significant savings in cost and energy consumption would result from the *economically optimum* amount of insulation rather than that required by the upgraded FHA standards (Moyers, 1971).

The fraction of all new homes and apartments that are federally financed is increasing, and, recently, federal financing has been made available for the purchase of mobile homes. No federal standards for mobile home construction exists; however, the FHA requires compliance with a production code entitled *Standard for Mobile Homes,* giving thermal insulation requirements for mobile home construction (Mobile Home Manufacturers Association, 1971). This standard specifies that for the common size 12 foot by 60 foot gas- or oil-heated mobile home, the maximum allowable heat loss is *55% greater* than the permissible heat loss specified in the revised FHA Minimum Property Standards for conventional single-family dwellings. For the same size electrically heated mobile home, the maximum allowable heat loss is *25%* greater than the permissible loss specified in the federal standards. Unless more stringent thermal insulation standards are instituted, inadequately insulated mobile homes will constitute a growing source of energy waste with unnecessarily high heating and cooling bills for mobile home dwellers.

28.3. Residential Electricity Demand of the Future

Table 28.3 gives the present and future residential demand for electricity (Tansil, 1973; Federal Power Commission, 1971; *Electrical World,* 1971; Chapman, Tyrrell, and Mount, 1972). The first part of the table gives the consumption by individual appliances. These figures are based on Bureau of the Census household projections combined with projections of appliance saturations and annual electricity use per appliance from Table 28.2. The projections are based on past trends and do not take into account any future potential energy conservation measures, that is, a shift to heat pumps from electric resistance heating, more efficient room air conditioners, improved housing insulation, and so on.

Also given in Table 28.3 is an estimate of the contribution from new users of electricity. The electric automobile has often been mentioned as a distinct possibility by 1990. One reference estimates that if the electric car is accepted by 10% of all families in 1990, it would contribute about 22 billion kWh to the residential load (*Electrical World,* 1971). Other new users might add between 5 to 15 billion kWh, approximately the amount contributed by all small users of electricity in 1970.

The second part of Table 28.3 compares the summed total of consumption per appliance with other estimates based on extrapolation of past growth and econometric modeling. These independent estimates are illustrated in Figure 28.3. The summed result predicts that residential demand in 1990 will be 2.7 times the 1970 usage, whereas the highest projection is 3.8 times the 1970 usage. The difference between these two projections is 500 billion kWh, more than the 1970 residential demand.

The extrapolation of past trends of electricity growth to the future gives a projection of about 1700 billion kWh in 1990. However, the residential demand for electricity is dependent on household size and occupancy: apartments and mobile homes generally use less electricity than homes. In the past, the trend has been an increasing number of homes at the relative expense of multifamily dwellings. However if current trends persist, by 1990 mobile homes will constitute 10% of the housing market, up from 3% in 1970 (Tansil, 1973). This increase will occur at the relative expense of conventional single-family homes, with the fraction of apartments remaining approximately constant. The following quotation from W. B. Shenk

Table 28.3 Present and Future Residential Electricity Demand

	1970	1975	1980	1985	1990
	(billions of kilowatt-hours)				
Refrigerators	83	—	122	—	158
Air Conditioning: Room	33	—	53	—	70
Central	26	—	50	—	83
Lighting	48	—	68	—	79
Space Heating	71	—	188	—	360
Water Heating	72	—	115	—	173
Clothes Drying	19	—	31	—	43
Cooking	30	—	49	—	78
Television	25	—	33	—	41
Food Freezers	28	—	39	—	54
Other (Clothes Washers, Irons, Etc.)	13	—	20	—	27
New Users (Electric Car, Etc.)	—	—	10	—	30
Total Residential Use (Summed Result)	448	—	778	—	1196
Federal Power Commission Estimate	448	—	775	—	1409
Electrical World Estimate	448	648	930	1205	1700
Econometric Model High Estimate	448	680	978	1331	1727
Econometric Model Low Estimate	448	613	711	749	751

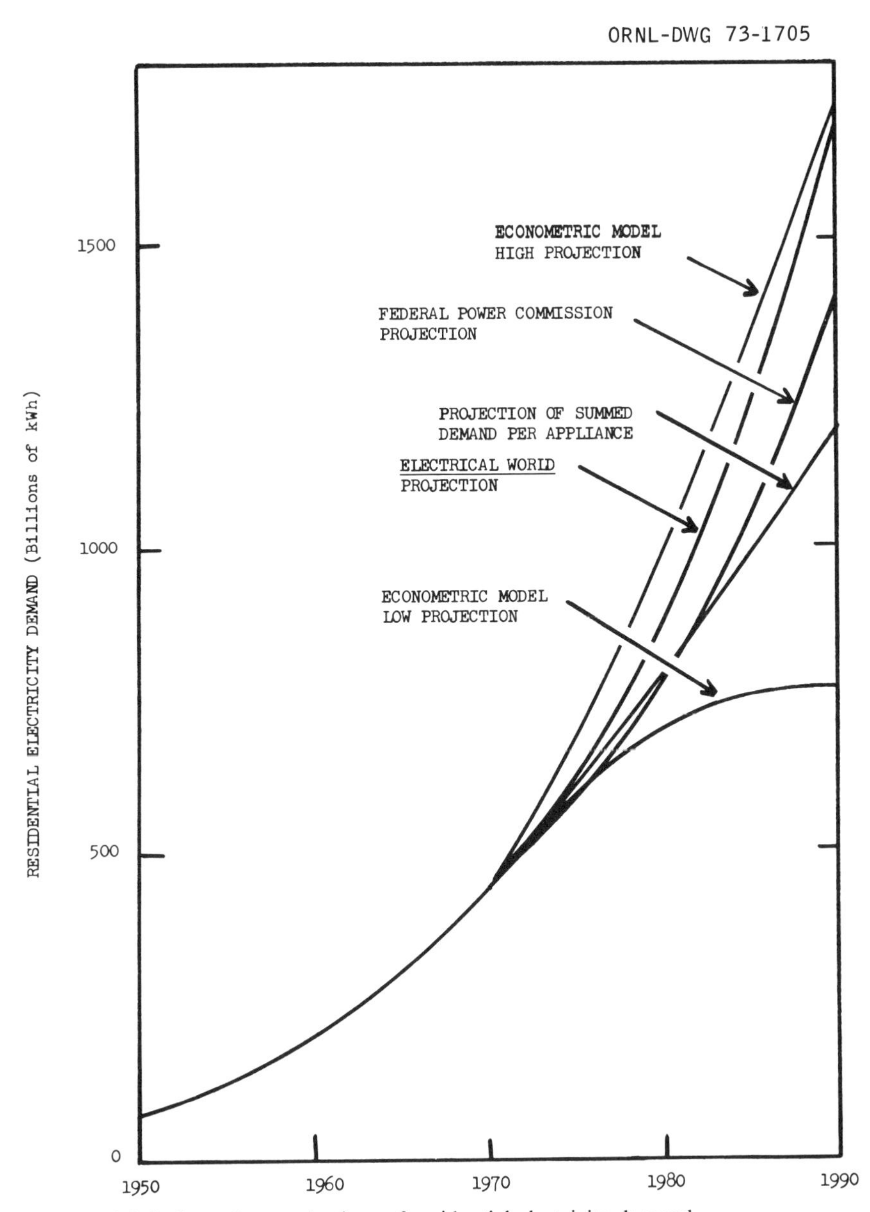

Figure 28.3. Independent projections of residential electricity demand.

of the Florida Power Corporation illustrates the net effect of these housing shifts on residential electricity use (*Electrical World,* 1969):

> The average use in the existing mobile home is 4,000 kwhr per year, or 34% less than the residential kwhr average. The total-electric mobile home falls short of today's conventional all-electric home by 29%.

In addition to the reduction in demand resulting from these housing shifts, the increasing national concern with fuel shortages, oil imports, coal strip mining, storage of radioactive wastes from nuclear power plants, and air and water quality standards have made the public aware of the need for conservation measures. The state of California recently passed a law requiring that *all* new residential units must meet or exceed the thermal insulation requirements for federally financed housing (*Electrical World,* 1973).

It can be concluded that the projection of residential electricity use, based on the consumption by individual appliances, yields significantly less demand in 1990 than extrapolation of past electricity growth. Housing shifts from standard homes to apartments and mobile homes will intrinsically reduce the growth rate of residential electricity use. The implementation of conservation measures such as better-insulated housing, a shift to heat pumps from resistance heating, more efficient air conditioners, and improved insulation for refrigerators, freezers, and water heaters would further reduce the growth rate *and* save the homeowner money.

Note

1. Portions of the material in this paper are excerpted from Tansil (1973).

References

Air-conditioning and Refrigeration Institute (unpublished data). Unitary Statistics Committee, Arlington, Va.

Association of Home Appliance Manufacturers (1971). *1971 Directory of Certified Room Air Conditioners,* Chicago.

Chapman, L. D.; Tyrrell, T. J.; and Mount, T. (1972). "Electricity Demand Growth and the Energy Crisis," *Science, 178,* pp. 703–708, November 17.

Chiles, J. Hunter (unpublished). "A Study of the Effect of the Heat Pump on the U.S. Total Energy Situation," Power Systems Planning, Westinghouse Corporation, Pittsburgh, Pa.

Edison Electric Institute (biennial issues 1950–1969). "Approximate Wattage Rating and Estimated Average Annual kwhr Consumption of Electric Household Appliances Assuming Normal Use," New York.

——— (1971). *Historical Statistics of the Electric Utility Industry,* Statistical Committee, New York.

Electrical World (1969). "Housing Shifts Will Alter Sales Methods Used," p. 79, November 10.

——— (1971). "22nd Annual Electrical Industry Forecast," September 15.

——— (1973). "California Ups All Residential Insulation Levels," p. 63, January 15.

Electric Heat and Air Conditioning (annual issues 1956–1971). "Annual Market Analysis of Electric Heating."

Federal Housing Administration (1966). *Minimum Property Standards for One and Two Living Units,* FHA No. 300, Washington, D.C.

——— (1971). *Minimum Property Standards for Multifamily Housing,* FHA No. 2600, Washington, D.C.

Federal Power Commission (1971). *1970 National Power Survey,* part I, p. I-3-13, Washington, D.C.

Hirst, Eric, and Moyers, John C. (1973). "Efficiency of Energy Use in the U.S.," *Science, 179,* pp. 1299–1304, March 30.

Merchandising Week (1972). "Fifty Years of Statistics and History," Vol. 104, No. 9.

Mobile Homes Manufacturers Association (1971). *Standard for Mobile Homes,* Chicago.

Moyers, John C. (1971). *The Value of Thermal Insulation in Residential Construction: Economics and the Conservation of Energy,* Oak Ridge National Laboratory Report ORNL-NSF-EP-9, Oak Ridge, Tenn.

Tansil, John (1973). *Residential Consumption of Electricity: 1950–1970,* Oak Ridge National Laboratory Report ORNL-NSF-EP-51, Oak Ridge, Tenn.

U.S. Bureau of the Census (decennial issues for 1950, 1960, and 1970). *Census of Housing,* Vol. I, U.S. Summary, Washington, D.C.

——— (1971). *Statistical Abstract of the United States: 1971,* 92nd edition, Washington, D.C.

29 Electricity Demand—One Utility's Econometric Model

R. B. COMERFORD* AND W. G. MICHAELSON*

Economic growth and the use of energy are inexorably tied together. To plan for future energy needs, one must closely examine the underlying economic forces that create energy demand. During the past decade econometric methods have been progressively inserted into the study of economic relationships.

29.1. Electric Energy and Demand Model

At PSE&G of New Jersey, a small annual model has been constructed to assist in the forecasting of future electric peak and energy demands. The model, which contains only 25 equations, provides a vehicle with which we can examine the effects of the economic sector upon our future energy consumption.

Generally this model postulates each sector i's electric energy consumption E_i as a function of some individual measure of real economic demand D such as personal income, investment, or manufacturing production. Thus,

$$E_i = F(D). \tag{29.1}$$

The relative simplicity of the model allows the easy testing of many sets of alternative economic factors. The trite equations, often found in larger models, which have little bearing on results, are omitted. Limiting the model to only essential equations leads to an easier understanding and a readier acceptance by those whose real-life decisions may be influenced by the output.

The econometric model that was formulated is designed to explain the behavior of the endogenous variables and to determine how well each of these variables can be forecast. The approach to the model was to specify a set of hypotheses that can be tested in a regression framework. Each hypothesis, formulated as a single equation, is examined separately. Ordinary least squares (OLS) is used to determine the parameter estimates. In matrix notation, the model can be written

$$\mathbf{By} + \mathbf{Tz} = \mathbf{u}, \tag{29.2}$$

$\mathbf{y}$ is a vector of 25 endogenous variables; $\mathbf{B}$, the coefficient matrix and $\mathbf{u}$,

* Public Service Electric and Gas (PSE&G) of New Jersey.

the vector of random disturbances. A 95% level of significance was used to establish the statistical significance of the coefficients.

We shall first examine what type of economic sector simulation was selected to relate with the electric energy sector. From there the direct energy-economy relationships will be looked at. And finally we can look at the uses of, testing of, and results from the model.

29.2. Economy Sector

PSE&G's electric service territory, like many utilities, is restricted to only one state—New Jersey. In actuality, PSE&G serves only 25% of the area of the state, but 80% of the state's economic activity takes place in PSE&G territory. As utilities serve specific, limited geographic areas, relationships may very often be ineffective if energy demands are related to the many national econometric models. Most areas' energy needs are peculiar to their specific economic environment. For many years the industrial sector growth in New Jersey closely paralleled that of the nation. However, during the past several years the industrial growth has not grown with the nation while the service sector has been booming. And energy needs for industrial and commercial sectors are quite different. Thus, it becomes very important to identify the areas' economic growth in each sector.

Therefore, the basic model of the economic sector uses state data. However, as will be shown, some national economic influences also enter into the model. The economic sector of the model revolves around the Gross State Product, or GSP, which is analogous to the Gross National Product. The historical GSP was constructed using a method similar to that developed by Kendrick and Jaycox (1965). Basically, the method employs the application of ratios of state to national economic series applied to national income accounts in each private nonfarm sector of the economy.

This method is described by the following:

$$\mathrm{GSP}_{ij} = \frac{\sum_{i=1}^{50} (\text{State Income}_j)}{\text{National Income}} \cdot (\text{National Product}_j), \tag{29.3}$$

where i represents the state and j the GSP sector. For example, if 5% of all manufacturing employment was located within a state, it would be assumed that 5% of the income originating in the manufacturing sector would also be from that state. This procedure is based upon the assumption that the structure within each private nonfarm industry group of the state is similar to that of the nation, or that any divergence is offsetting. Corrections for capital consumption allowances and indirect taxes are also made for each sector. For the more economically developed and diversified the state—as New Jersey is—the better the assumption is. The GSP is defined for the following sectors:

Manufacturing and Mining
Finance, Real Estate, Insurance
Construction
Trade
Services and Communications
Transportation and Utilities
Government
Agriculture

Figure 29.1 shows the growth of various sectors of New Jersey's economy from 1950 to 1972. It is readily apparent that the complexion, and therefore, the corresponding energy needs of the economy have changed considerably.

The relationships of the economic sector of the model are best described by Figure 29.2. The theoretical framework of the state's economy is developed along Keynesian lines. Basically, this economy sector describes consumption, investment behavior, levels of taxation, and the effect of the national economy. The model is similar in nature to one developed for Ohio (L'Esperance, Nestel, and Fromm, 1969).

The model's basic equilibrium condition can be described by $Y = C + I$, where Y = total income, C = consumption expenditures, and I = investment.

All variables representing sectors of the GSP are endogenous except the agricultural sector, which must be separately estimated. This sector is relatively unimportant, constituting less than 0.5% of the total GSP. The main exogenous variables include population, tax rates, finance rates, and the GNP. The reason the GNP forecast is input into the model is in order to insert the influences of national economics into the model. This is basically due to the fact that New Jersey is a net exporter of manufactured goods to other states.

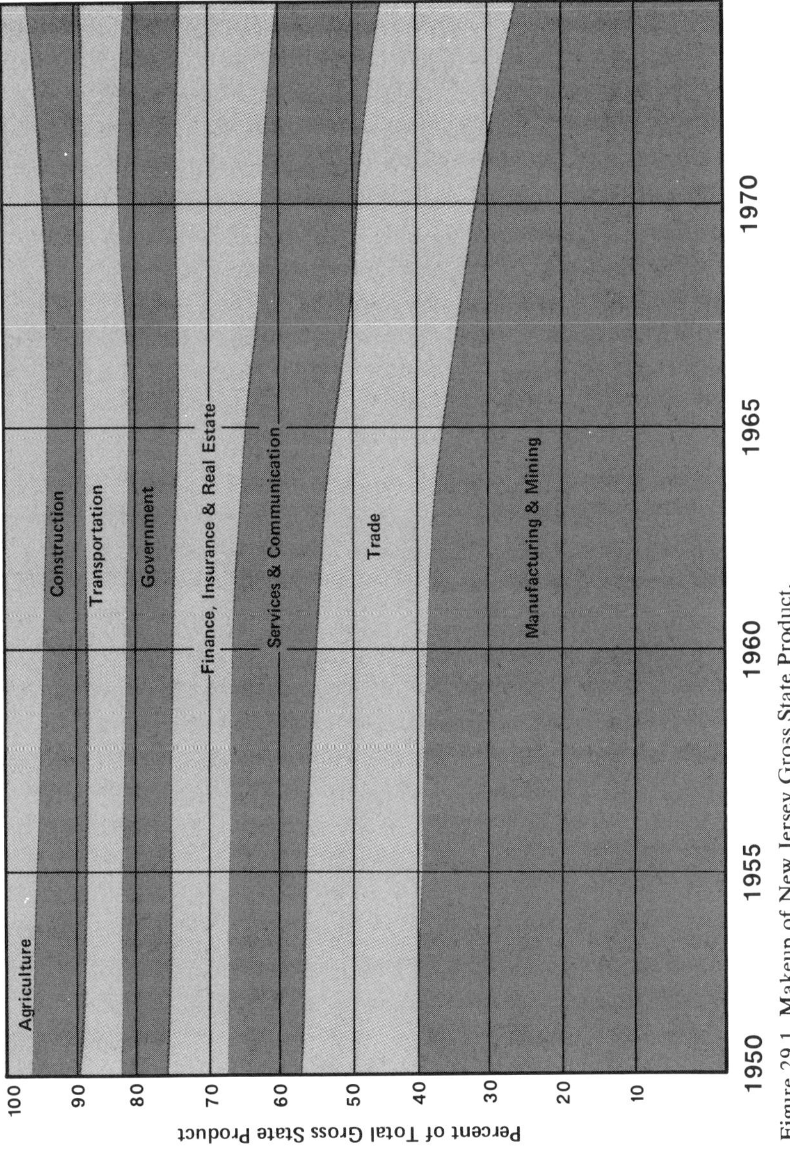

Figure 29.1. Makeup of New Jersey Gross State Product.

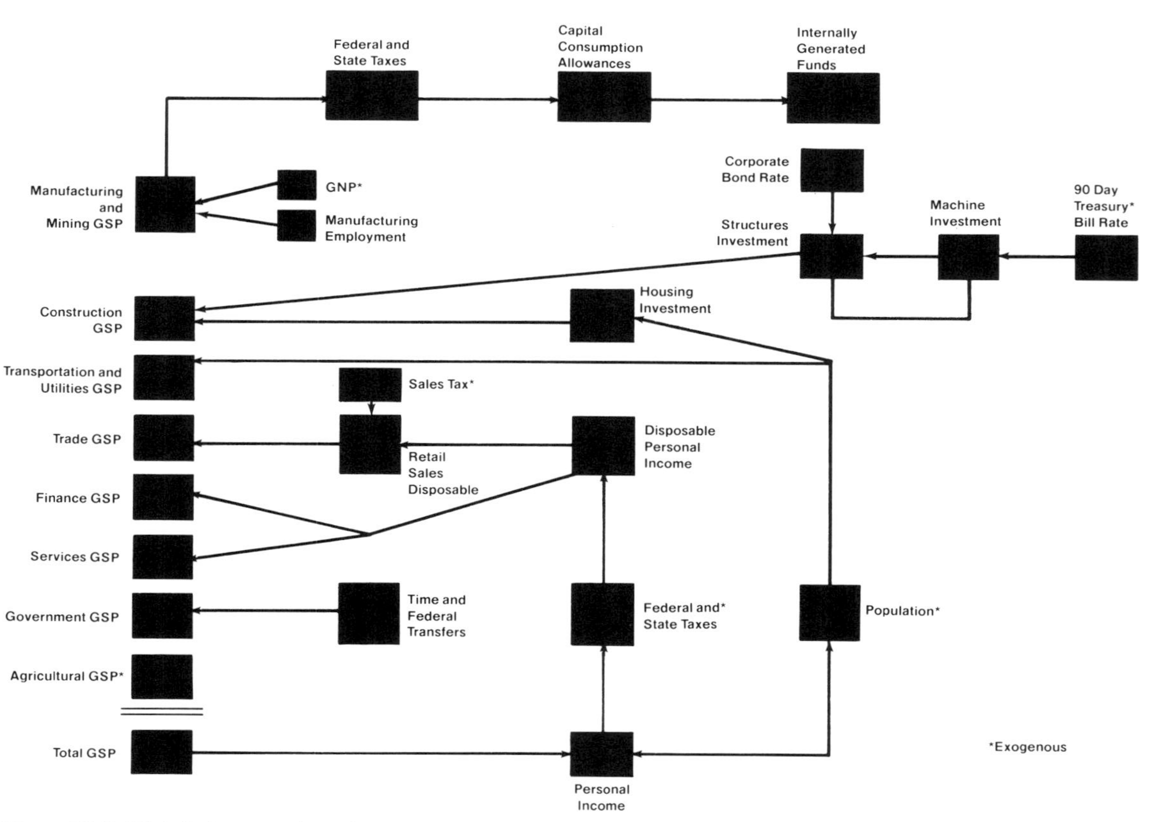

Figure 29.2. Model of economic sector.

29.3. Energy and Economy Relationships

After determining the various sectors of the GSP, the next step was to estimate the parameters that relate them to electric energy and demand needs. Electric energy sales are divided into three major groups: industrial, commercial, and residential.

29.4. Industrial Sector

Industrial sales in the model are related to the GSP sectors of manufacturing, mining, and construction. However, at this point not only must the demand be considered but also the effect of the price of electricity and fuels competitive to electricity and their respective prices. Thus, price elasticity relationships for the sector were also developed using econometric methods similar to work done at Cornell (Chapman, Mount, and Tyrrell, 1972) and the FPC (Kline, 1969).

The price elasticity is represented as

$$\gamma = \frac{\Delta\,\mathrm{RE}}{\mathrm{RE}} \Big/ \frac{\Delta P}{P}, \tag{29.4}$$

where γ is the elasticity, RE is the demand for electricity in the residential sector, and P is price. A more exact statement would be

$$\gamma_i = \frac{d(\mathrm{RE})}{d(X_i{}^*)} \cdot \frac{X_i{}^*}{\mathrm{RE}^*}, \tag{29.5}$$

where γ is the elasticity with respect to factor i when evaluated at some values $X_i{}^*$ and RE* and where RE is a function of n explanatory factors $X_1, \ldots, X_n$.

The equation for industrial sales

$$\mathrm{IS} = f(\mathrm{MGSP}, P_e, P_o) \tag{29.6}$$

must be converted to the logarithmic form in order to test the hypothesis via regression. The form then becomes

$$\mathrm{IS}_t = (a \cdot \mathrm{MGSP})(A_t), \tag{29.7}$$

where

$$A_t = \lambda(P_{et}/P_{e0})^{\gamma}(P_{ot}/P_{o0})^{B} + (1 - \lambda)A_{t-1}, \tag{29.8}$$

and where

IS_t = industrial kWh sales,
a = energy to manufacturing sector multiplier,
MGSP = manufacturing gross state product,
P_e = price of electricity,
γ = industrial elasticity for electricity,
P_o = price of oil,
B = industrial elasticity for oil,
λ = lag factor to take into account the time required to change from an oil use piece of equipment to an electric use type, or vice versa.

Thus the industrial energy submodel looks something like Figure 29.3.

29.5. Commercial Sector

Electric energy sales in the commercial sector are directly related to the GSP from six sectors that readily describe where commercial sector kWh are consumed: transportation, trade, finance, services, government, and construction. Again, we must consider price elasticities in the commercial sector. Elasticities in this sector are somewhat lower than those estimated for the industrial sector; however, they still show a reaction to price changes. Figure 29.4 shows the basic econometric relationship for the commercial sector.

29.6. Residential Sector

The submodel for the residential sector is more complex and is divided into

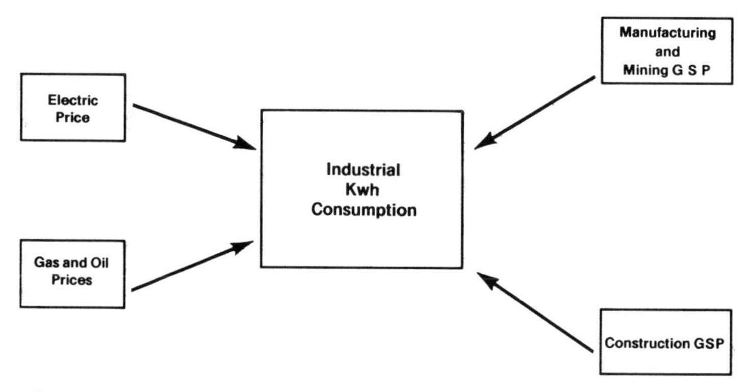

Figure 29.3. Industrial model.

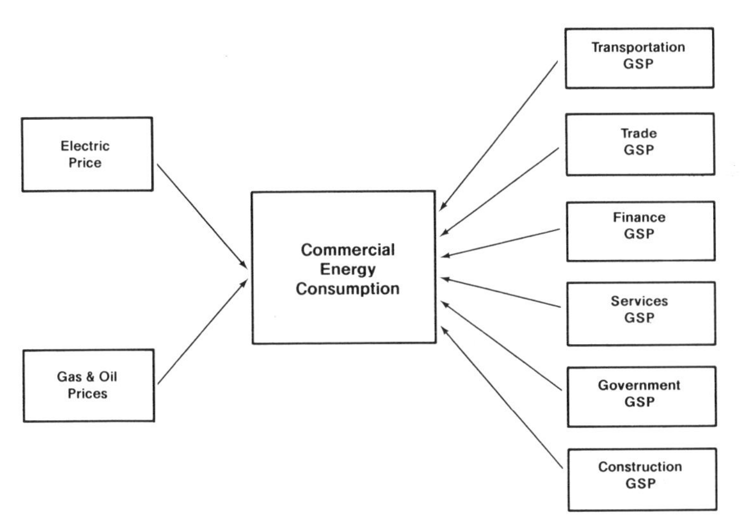

Figure 29.4. Commercial model.

several parts. Residential kWh use is divided into: (1) use per residential customer and (2) the number of residential customers. At this point another exogenous variable must be added to the model: air-conditioning saturation. Air-conditioning use now constitutes about 20% of the PSE&G average annual residential use of electric energy. The basic or non-air-conditioning kWh use per customer was found to be a direct function of disposable personal income. The number of residential customers is directly correlated to the population input. Thus, the schematic of the overall residential submodel might look something like Figure 29.5.

Very important in this is the air-conditioner saturation where the actual purchase of air conditioners is related to several sectors in the model.

$$\mathrm{ACS}_t = (a \cdot \mathrm{ACS}_{t-1}) + (b \cdot \mathrm{PAC}) - (c \cdot \mathrm{OAC}), \tag{29.9}$$

where

ACS = air conditioner saturation,
PAC = deflated average cost of an air conditioner,
OAC = price of electricity required to operate an air conditioner,
a, b, c = coefficients.

29.7. Peak Demand Sector

The final subsector of the model relates the economic relationships to peak

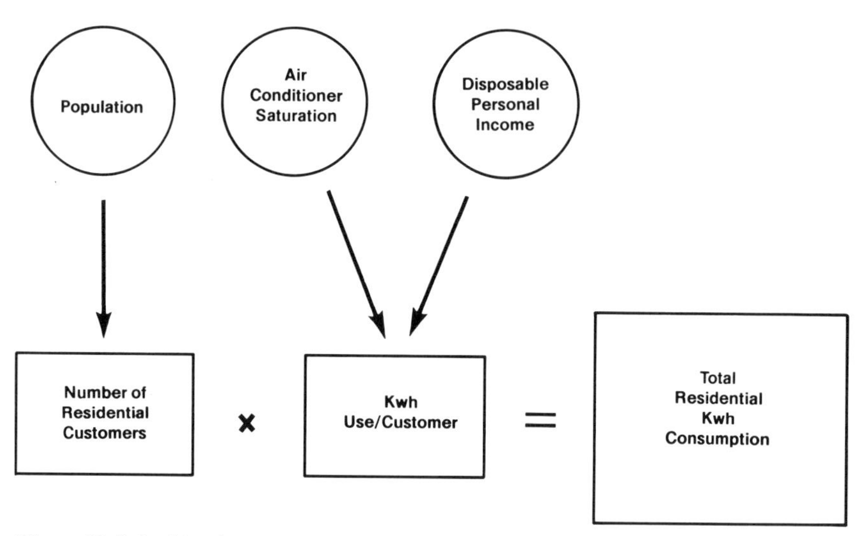

Figure 29.5. Residential model.

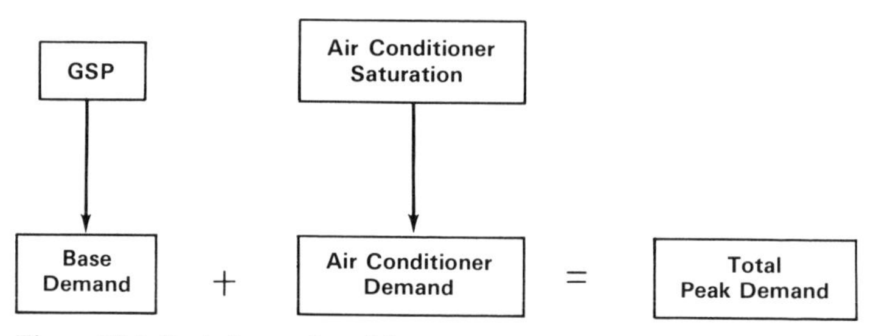

Figure 29.6. Peak demand model.

demand. Peak electrical demand is divided into two parts: (1) base, or non-weather sensitive and (2) weather sensitive. At this point due to the relatively greater magnitude of air conditioners versus electric heat in the PSE&G system, weather-sensitive demand is defined as cooling load.

The base demand that is constituted from factories, store and office lighting, home appliance, and so on, is correlated to total GSP. The cooling demand is correlated to air conditioner saturation. See Figure 29.6.

29.8. Uses—Alternative Case Simulation

After the observable relationships between the economy and energy sector have been established via statistical paths, the model was activated for use. The size of the PSE&G model readily lends itself to time-share computer simulation.

Once the assumptions with respect to the input variables have been defined, the model can be run to obtain an energy and peak forecast. However, while obtaining trial forecasts is an important application, the testing of the impact of alternative policy scenarios is the primary use of the PSE&G model.

The impact of studying hypothetical policies and changes on the endogenous variables in the model in year t can be represented by the vector

$$P_{it} = Y_{it} - \overline{Y}_{it}, \tag{29.10}$$

where $\overline{Y}_{it}$ is the value of the ith endogenous variable in year t generated by the model, given the assumption that the change in policy or economic situation was carried out and Y_{it} is the value before the introduction of the change. Complete development of the mathematics of this representation can be found in Kendrick and Jaycox (1965).

For example, after a base case has been run, simulations can be carried out for numerous other possibilities:

1. What effect would a 50% increase in electric rates and a 100% increase in gas and oil rates by 1980 have on total energy sales? (Figure 29.7)
2. What effect would zero population growth have on residential energy sales? (Figure 29.8)
3. What effect would the introduction of a state income tax in New Jersey have on the commercial electric energy sector? (Figure 29.9)
4. What effect does investment tax credit and ADR have on the industrial electric energy sector? (Figure 29.10) Here we introduced an earlier model developed for us by a Princeton economics student. Much of his work was based on studies by Hall of MIT and Jorgenson of Harvard (Hall and Jorgenson, 1967).
5. The model has also been used in very specialized cases. For example, the effects of the then impending 1970 steel strike on electric sales were analyzed (Figure 29.11). The direct and indirect impact on each of the eight major sectors of the GSP was first estimated by analysis of the effects of previous steel strikes. And then these reductions in GSP were factored into kWh consumption.

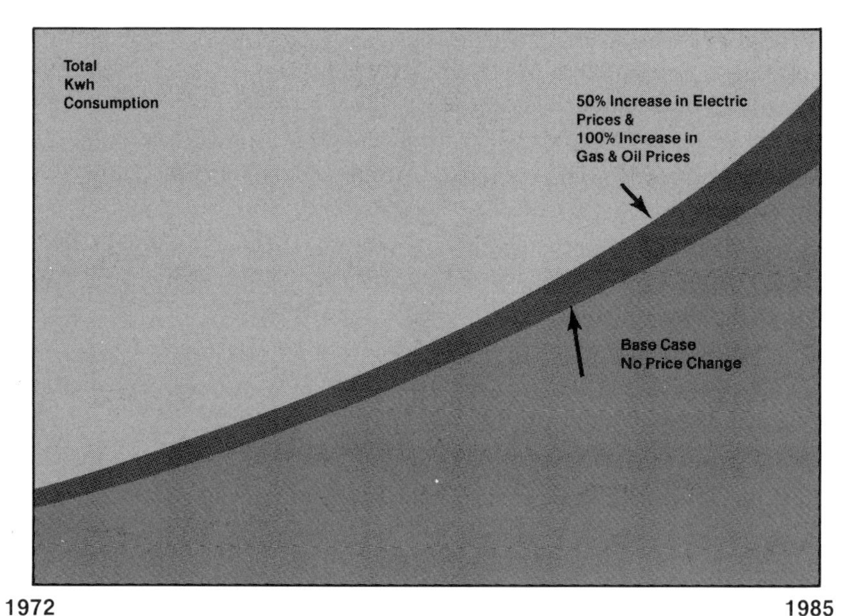

Figure 29.7. Effects of price increases.

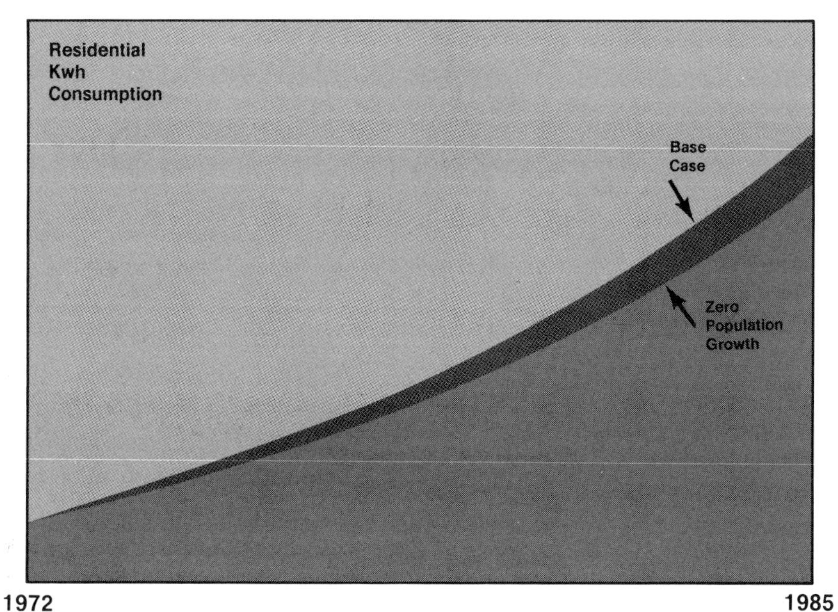

Figure 29.8. Effects of zero population growth.

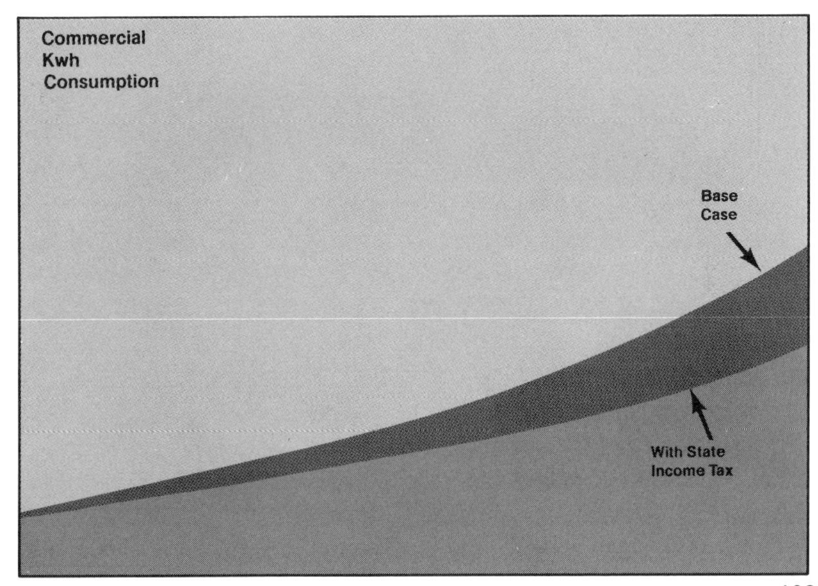

Figure 29.9. Effects of introduction of state income tax.

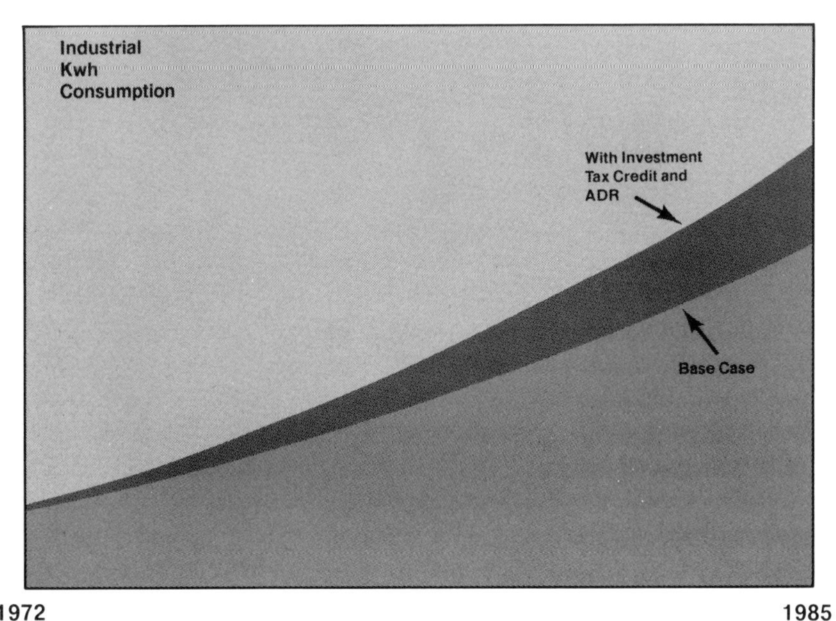

Figure 29.10. Effects of investment tax credit.

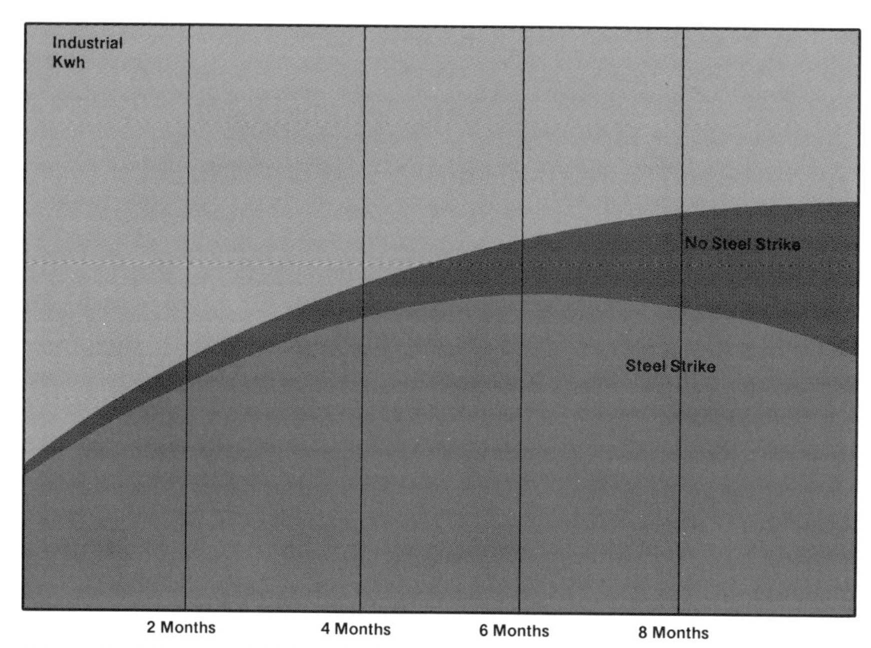

Figure 29.11. Impact of 1970 steel strike.

Therefore, at any one time any one or set of assumptions may be studied to find the type and magnitude of response. It is also possible with this type of model to test the internal properties of the model itself, that is, equation parameters. For example, the coefficient that relates use per residential customer and income might be varied to test its effects.

29.9. Model Use in Reverse

Not only can the growth of the economy and its effects on energy demand be analyzed, but the questions of the effects of limited or curtailed available energy and demand upon the economy can also be tested. For example, the question of what happens to New Jersey's economy if no additional electric generating capacity is added can be answered. If this became a reality, what would happen to the GSP, how many jobs would be lost, what would happen to state tax revenues? These are all questions that can and have been tested using the PSE&G Electric Energy and Demand Model.

29.10. How the Model Has Scored

The PSE&G Energy and Demand Model can be evaluated in several ways. One way is via a comparison of actual and simulated historical energy consumption figures. Figure 29.12 shows the actual (solid line) and the simulated (dashed line) values for total kWh consumption for 1960 to 1971. During this period the absolute error was 2.3%.

For those interested in the jargon of econometrics, the average R^2 for the model's 25 equations was .97. The lowest R^2 was .89. All the equations also passed the rigors of tests of T-statistics and the Durbin-Watson "d" statistic.

29.11. Summary

The application of econometric methods to planning for future energy needs has proved to be a valuable tool in studying these needs. And it should be emphasized that models are just that—tools, not the full and complete answer. The major strengths of econometric techniques in energy forecasting are that they allow the user to impute an economic/energy event to its various causes and allow the prediction of components on a consistent

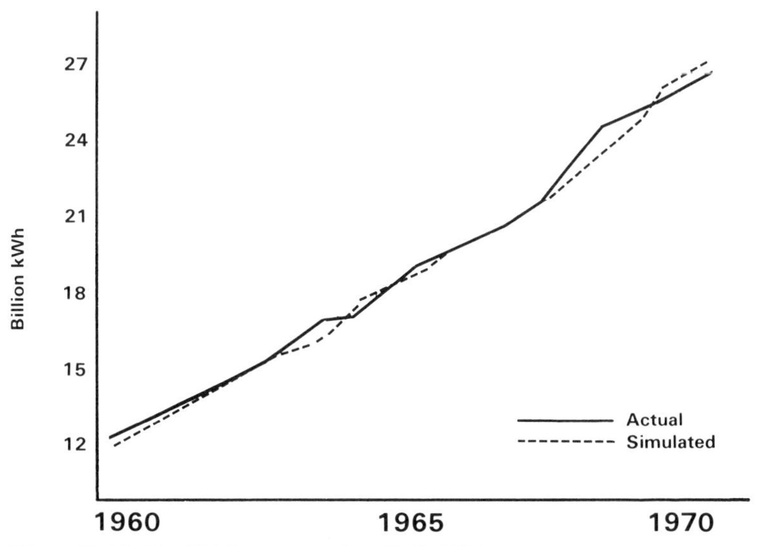

Figure 29.12. Total kWh consumption 1960–1971.

basis. The PSE&G Electric Energy and Demand Model is used to examine and study situations, but in the end human judgment must be applied to draw the final conclusions.

References

Chapman, D.; Mount, T.; and Tyrrell, T. (1972). "Predicting the Past and Future in Electricity Demand," Cornell Agricultural Economics Staff Paper, Cornell University, Ithaca, N.Y., February.

Hall, Robert E., and Jorgenson, Dale W. (1967). "Tax Policy and Investment Behavior," *American Economic Review,* Vol. 57, pp. 391–414, June.

Kendrick, John W., and Jaycox, C. Milton (1965). "The Concept and Estimation of Gross State Product," *The Southern Economic Journal,* Vol. 32, No. 2, pp. 153–168, October.

Kline, Phyllis (1969). "An Econometric Model for Residential Electricity Demand," *The Methodology of Load Forecasting,* Appendix B, prepared by the Technical Advisory Committee on Load Forecasting Methodology for the National Power Survey.

L'Esperance, W. L.; Nestel, G.; and Fromm, D. (1969). "Gross State Product and an Econometric Model of a State," *American Statistical Association Journal, 64*(327), pp. 787–807, September.

V Transportation

30 Some Problems and Prospects for Marine Transportation of Oil in the 1970s*

ZENON S. ZANNETOS†

30.1. Introduction

One of the most neglected and also misunderstood elements of the whole energy supply system is that of ocean transportation. At a time when outcries of impending and existing energy crises abound, not sufficiently serious thought is given to how the energy sources that are bound with geography can be brought effectively to the potential marketplace. In the final analysis the oil companies may find, if they continue their present policies, that the financial problems of oil production during the seventies may be less thorny than those of transportation. The latter, which at present is at best viewed as an ancillary evil deserving contempt and neglect, does now and will continue in the 1970s to provide one of the best opportunities for profit enhancement in the oil industry. I use the term profit enhancement in its broad sense. Later on I will point out that effective management in the transportation area also provides profit protection.

Because of the structure of the petroleum industry, the organization of the international oil companies, and the accounting systems of the latter, the impact of transportation on profitability is for the most part indirect. As a result, unless one looks carefully for these profit-making and profit-protection opportunities they tend to go unnoticed.

Another and more important reason for the relative neglect of transportation can be found in the lack of clear understanding by the oil industry as to what makes tanker rates fluctuate so wildly. And what a person cannot understand he naturally tends to ignore, because he does not know how to control through planning.

The amount of capital required for transportation and the conditions determining the availability of such funds cannot provide logical support to the attitude of the oil producers toward transportation. Unlike some commonly held beliefs, it is much easier for the integrated producers to find external capital for transportation than for exploration and production. As I have pointed out elsewhere (Zannetos, 1972), the oil companies provide either

* The underlying research was partly supported by NASA Contract No. NGL-22-009-309, Integrated Planning and Control Systems.

† Professor of Management, Alfred P. Sloan School of Management, Massachusetts Institute of Technology.

directly or indirectly almost all the credit support behind the capital that flows into the tanker markets. Furthermore, the amount of financial resources required for exploration and production activities is orders of magnitude greater than that required for transportation.

In trying to understand why the major oil producers have tended to ignore marine transportation, one cannot help but wonder how much those who are in charge of marine operations for the oil industry have contributed to the present state of affairs. A close look will show mainly two major classes of people at the helm. One group is transient, placed in transportation temporarily for training before reassignment to other "downstream" operations. By the time these managers have learned something about transportation, they are moved to "greener pastures." The other group is more permanent, it is hard working and in the final analysis more influential in guiding the activities of the transportation departments. These are the people who normally collect statistics, are reactive, and tend to be exclusively intuitive operators. Consequently, in an industry where scientific talent has indelibly left its mark on exploration, production, and refining, most transportation departments succeeded in the main to coexist with their "brethren" unaffected by management science techniques.

The end result of such a situation is that transportation has been dominated by other operations. The planning and anticipatory actions taken by the oil industry have been almost exclusively in the areas of exploration, production, refining, and distribution. Ocean transportation has been for years relegated to the status of a second-class citizen, and allocated resources either because of budgetary surpluses or because of crises.

The point that should not be lost is that this relative neglect of transportation by the integrated producers and their reactive behavior are not benign, if I am to use a now famous expression. In fact, herein lies the greatest cause of the wild fluctuations in both tanker rates and shipbuilding costs.

The purpose of my presentation is to look at the problems associated with, and the financial resources required for ocean transportation of petroleum in the 1970s. In the process I will also attempt to explain why I believe that in the future the oil industry and the producing countries cannot afford to ignore ocean transportation as much as they did in the past.

30.2. Some Background

30.2.1. Spot-Rate Fluctuations

If we look at the time series of spot rates for tankers (that is to say over time the current round-trip cost for delivering one ton of oil for a given route),[1] we will find that the rates fluctuate over a wide range. Forgetting the 1967 disturbance, in the post-1967 period the Persian Gulf/U.K. Continent rates reached a peak of Worldscale 297 in October 1970 and a low of Worldscale 25 in April–May 1972. This represents a fluctuation of twelve times from low to high. As Table 30.1 shows, if we take the Kharg Island/Philadelphia run, Worldscale 297 results in a transportation cost of $3.80 per barrel of crude oil delivered and the low of Worldscale 25 in a transportation cost of only 32 cents per barrel of oil. If we look at it in another way, at the high rate *the spot-rate cost of transportation alone was greater than the total value of the oil delivered during periods of low rates.*

Table 30.1 Cost of Transportation

	Worldscale[a]			
	297[b]		25[b]	
	Per Ton	Per Barrel[c]	Per Ton	Per Barrel[c]
1. Kharg Island/Philadelphia	$28.51	$3.80	$2.40	$0.32
2. Kharg Island/Rotterdam	27.65	3.70	2.33	0.31
3. Kharg Island/Yokohama	15.21	2.03	1.43	0.19
4 Aruba/Philadelphia	5.44	0.73	0.46	0.06
5. Aruba/Rotterdam	11.46	1.54	0.97	0.13

References: Conrad Boe Ltd. A/S Shipbrokers, *Estimated Tanker Market Rates Single Voyages 1947–1972*, Oslo, Norway, 1972.
John I. Jacobs & Co., Ltd., *World Tanker Fleet Review*, 30th June 1972, London, England.
Worldwide Tanker Nominal Freight Scale Applying to Tankers Carrying Oil in Bulk (for Flat Rates).

Notes:

[a] Flat Rate: Kharg Island/Philadelphia $9.01 + .59 = $9.60
Kharg Island/Rotterdam 8.72 + .59 = 9.31
Kharg Island/Yokohama 5.13 + .59 = 5.72
Aruba/Philadelphia 1.83
Aruba/Rotterdam 3.86

[b] The spot rate of Worldscale 297 was reached most recently in October 1970 and the low of Worldscale 25 in May 1972. Both rates were recorded for the Persian Gulf/U.K. Continent.

[c] We assume that there are 7.5 barrels to a long ton of crude oil of 34° API (American Petroleum Institute degrees).

I hasten to warn at this point that the spot market does not handle much carrying capacity. Over 80%[2] of all the oil shipped is transported in vessels that are either owned by the oil companies or chartered by the latter on a long-term basis. Although the long-term rates do fluctuate sympathetically with the short-term (spot) rates, the fluctuations of the former are more tempered. From the economic point of view, however, the spot rate is very important because it represents the short-run opportunity cost of transportation. It also affects the expectations of those in the industry (Zannetos, 1966), and brings about an overall impact that far transcends the percentage of tonnage involved in spot market activities.

In addition to affecting the long-term or time-charter rates, spot rates also influence the investment patterns in ocean transportation capacity. In the latter case not only the shipbuilding costs are affected by the level of spot rates in transportation but so is the amount of orders placed for new tonnage.

The relationships between spot rates for transportation and orders for new vessels create a complex network of dynamic interdependencies which can cause cyclical patterns and the "feast or famine" situations that we have been observing over the years in the price of ocean transportation. One significant consequence of these observations is that we do not necessarily have to have cyclical demand for transportation in order to observe cyclical price patterns. The forces operating on the supply side are sufficient to generate them without any aid from the factors affecting the demand for transportation. In practice of course both types of impacts are manifested.

Another important conclusion that we can draw is that the vital interrelationships among the various time periods (in terms of the spot rates, construction costs, orders placed for new tankers, deliveries of tankers, and eventual retirement of such) although complex, provide those in the industry with enough information on which to make rational plans regarding chartering and building of tankers and reduce the price fluctuations of transportation capacity. Such a reduction with its concomitant consequences will result in significant cost savings for the industry.[3] Ironically, these observations are not new. They were first expounded in their general form by Jan Tinbergen back in the 1930s (Tinbergen, 1958), later on by Tjalling Koopmans, whose work in the area of freight rates published in 1939 is now classic (Koopmans, 1939), and more recently by myself in the late 1950s and early 1960s (Zannetos, 1966, 1967). And here I am once again feeling the necessity to talk about these conclusions because they are still valid and as yet have not extensively influenced the oil industry.

30.2.2. Who Absorbs the Rate Fluctuations?

I see compelling reasons for change within the tankship markets. What the industry failed to do voluntarily through planning and rational anticipatory action in the past, it will be forced to do in the future as a matter of necessity. As the margins on production operations are reduced by the ever-increasing demands of the producing countries and the elimination of some of the special taxation benefits that are presently enjoyed by most international oil companies, it will be relatively more difficult for the latter to guarantee delivered prices and absorb large fluctuations in transportation costs. Transportation will now come to merit consideration as a profit center, not as a cost center. It will, therefore, be rationalized to the point where it is as efficient an operation as can be. The profits and losses from transportation will no longer be buried in other upstream and downstream operations. In short, I foresee that ocean transportation will come to maturity. In the future it will neither be able to hide its inefficiencies under the average profitability of other operations, nor will it subsidize others.[4] It will have to "stand on its own two feet."

Admittedly, the integrated oil companies may choose other alternatives such as

1. Pass along the increasing "costs" (including the transportation inefficiencies) to the consumer as they have done in the past. This, however, will bring increasing resentment from the consuming countries with deleterious long-run effects. The larger the price increases, the more vocal the complaints will tend to be. If we look very carefully, we will find out that this policy has never *fully* worked consistently. The existence of discounts from posted prices and the absorption of transportation fluctuations adds credence to the arguments of those who try to convince us that in reality there is no shortage of oil *in the extended short run,* at least not in the sense and to the degree that we are led to believe.[5]

2. Sell oil on an F.O.B. basis. This will imply a complete reversal of past practices of selling oil on a delivered basis,[6] and abandonment of a very powerful instrument for control of long-term contracts. Under a strict F.O.B. pricing scheme and an independent (more or less perfectly competitive) market for transportation, crude oil will be reaching the refineries at different prices because of the fluctuations in transportation costs. Similarly, the

Table 30.2 Impact of Distance on C.I.F. Prices

	December 1972 F.O.B. Price/barrel	Transportation Cost	C.I.F.	Adjustments[a]	Equiv. Cost[b]
A. Iranian, Light 34° API ex. Kharg Island	$2.467				
Transportation to Philadelphia					
W-297		$3.80	$6.267		$6.267
W-75		0.96	3.427		3.427
W-25		0.32	2.787		2.787
B. Venezuelan 34° API	$3.3138				
Transportation to Philadelphia					
W-297		0.73	4.044	(.07)	3.974
W-75		0.18	3.494	(.07)	3.424
W-25		0.06	3.374	(.07)	3.304

[a] For quality of crude. If we exclude it, equalization occurs at Worldscale 81.
[b] Excluding 10¢/bbl duty.

landed cost of crudes that have different distances to travel to the same market will be fluctuating with transportation and upsetting the delicate balance between the F.O.B. prices of the various producing centers. There will be no way of equalizing crude prices. For example, my calculations show that the posted prices at Kharg Island, Iran, and Aruba for 34° API crude will result in the same C.I.F. (Cost Including Freight) cost (adjusted for refinery values) if transportation is around Worldscale 75. (See Table 30.2.) At a lower rate the advantages shift in favor of the Persian Gulf crude and vice versa. Such shifts in the comparative advantages upset both the producing countries and the international oil companies. If the spot rates are on the high side, then little pressure is exerted on the producers for two reasons. First of all the new agreements provide for automatic escalation in the posted prices of producing countries nearer to the consumption centers, and second the producers whose oil is more transportation intensive have an option as to whether they wish to revise upward their C.I.F. price to reflect the higher spot rates. In most cases they choose not to, so that they do not upset the goodwill of their customers and the long-term contracts. Failure to take advantage of an opportunity to raise C.I.F. prices allows the producers, furthermore, to appear magnanimous and socially responsible. When the spot rate is very low, however, the pressure for discounts is really on, and cannot be ignored.

It has not escaped some producing countries that transportation differentials will be putting pressures on the F.O.B. prices. As we have already mentioned, the 1971 agreements provide for adjusting the posted prices, upwards with spot rates, in countries such as Venezuela, Libya, and Algeria which are closer to the major consuming centers.

The conclusion that we reach, therefore, is that it will be to the advantage of both the producing countries and the international oil companies if they were to control delivered prices so as to equalize in the marketplace the cost of the oil flowing from different geographic areas. In order to achieve this equalization, they must either fully control transportation or absorb freight differentials. Small fluctuations can be condoned. Wild fluctuations cannot be condoned, however, because these put pressure on the C.I.F. system which in turn strains the F.O.B. price structure. Of these pressures, of course, those that tend to raise prices are not very damaging because they afford the producers a choice. In fact they allow the producers to appear generous. It is the downward pressure that is most disturbing to the oil companies and the producing countries. They dislike downward pressures on posted prices because every decrease affects their net revenues. The oil companies, in addition, have some reasons of their own to dislike pressures for reduction in the posted prices. First of all because they "guarantee" the latter to the producing countries as a base for calculating royalties and income taxes, and second because they are concerned lest their customers abrogate long-run commitments or refuse to renew them. For all these reasons, it appears unwise for them to relinquish the instrument (that is, transportation) which enables them to preserve the existing delicate structure.

The history of the railroads in the United States during the second half of the nineteenth century provides us with a lesson that has some bearing here. It was not so much for the profits of the railroads that people fought to control them but because they wanted the right to control delivered prices and the markets of the transported commodities. There is one major difference between railroads and tankship transportation, however, which needs to be brought out. No matter how hard the oil companies and the producing countries try to control transportation, they will not succeed in completely eliminating the fluctuations in the spot rates, short of paying dearly for such control.[7] At best they can control the amplitude of the fluctuations through efficient planning and execution of plans. Unlike the railroad beds, which fix railroad investments geographically (like pipelines) and create natural monopolies, tankers are flexible and many. So the competitive nature of the tanker markets should prevail,[8] but for the long run it can be made more efficient.[9]

Note also that the greatest potential control of the fluctuations exists on the high side (upswings) which, as we have pointed out, is the least damaging as far as the oil companies and the producing countries are concerned. But what is the choice that the producers have, one may ask. If they withdraw completely or do not step into the market to acquire ownership control of a substantial part of the necessary transportation capacity, the independents will. This will place the producers at the mercy of the independent tanker owners—the worst possible solution for them, especially if the producers completely abdicate. So logic tells me that we *should* see some changes in the ownership of tanker capacity in the future, with greater representation of the oil companies and the producing countries.[10] The latter ought to be particularly concerned because they do not control refining and distribution.[11]

To summarize this part of our discussion, the C.I.F. approach to selling oil does not appear to be under test or disgrace as far as the producers are concerned. The oil companies, therefore, may attempt, in their effort to exercise tighter control over rate fluctuations, to get more heavily involved in ocean transportation, and so should the exporting countries. My only hope is that any such efforts are accompanied by a thorough study of the dynamics of the tanker markets so that they do not add fuel to future transportation crises that their past actions have already set in motion.

30.3. Financial Requirements for Ocean Transportation

No year passes by without statements from responsible analysts of ocean transportation that the prospects for tankers "are not bright" (Jacobs, 1971, 1972). Even in the face of success, gloominess accentuates at the anticipation of the impending doom that follows so-called "abnormal market behavior." While this diagnosis goes on, the independents plod along becoming wealthy and the oil companies react mostly to crises.

A careful analysis will show that ocean transportation can be a very profitable business, and a growing one. In 1955 the total fleet of oceangoing tankers was less than 40 million deadweight tons (DWT). By December 31, 1972, it grew to over 190 million DWT. The average growth over this period has been 10.2% annually compounded. During the last 10 years the growth rate has been approximately 10.7% annually. The size of the largest new buildings grew during this period (1955 to 1972) from 45,000 DWT to 530,000 DWT.

A recent publication of the Chase Manhattan Bank (1972) estimates that during the period of 1970 through 1985 there will be a need for 247.2 million DWT of new tankers and a total fleet of 450,000,000 DWT. This increase represents approximately 137% of the tonnage as of December 31, 1972, and an annual compounded growth rate of slightly over 6% over the period studied.

No matter how impressive, I do not believe that the preceding forecast is realistic. In the past 10 years, for every 1% growth rate in oil consumption we have required 1.4% increase in transportation to satisfy it. It appears to me that during the 1970s the tanker fleet will increase at an average rate of close to 11% per year. This means a fleet of about 450,000,000 DWT by the end of 1980 and close to 600,000,000 DWT for the period studied by Chase.

The main reasons behind my projection are as follows:

1. Scheduled deliveries of vessels in backlog at this time will add over 100,000,000 DWT over the next 4 years. And even if there is a spillover or stretching, this is not expected to be greater than 6 months. Furthermore, deliveries of vessels not in backlog now are to be expected during these years.
2. The United States, which was importing only 600,000 barrels per day from the Middle East in 1970, is expected to increase this dependence by tenfold by 1980. The Middle East oil is very intensive in transportation.
3. The rate of growth of petroleum consumption in the free world is estimated at 8% annually over this period, and that of the United States at about 5%.
4. The Alaskan North Slope, according to the experts, will not contribute more than 2 million barrels per day. This will most probably flow to the U.S. West Coast, and this not before 1975–1976.
5. Canada will be able to provide only about 2 million barrels per day and South America about 4 million barrels per day.
6. The North Sea finds will not become important before the late 1970s. And even if the output from these fields reaches the impressive figure of 4 million barrels per day, it will not satisfy the increase in the European demand.

In addition to the new buildings of 260,000,000 DWT by the end of 1980, we have the replacement of 52 million DWT that will be over 20 years old by that time. This total of 312 million DWT will require an investment of about $47 billion at current construction costs. If I am correct in my projections that shipbuilding cost should be coming down, the total investment may not exceed $40 billion, inflation included.

The amount of $40 to 47 billion over the next 8-year period is rather insignificant when compared with the present capital expenditures of the oil industry which ran about $24 billion annually in 1972. The problem is that

(a) Ocean transportation never ranked high in terms of the budgetary priorities of the oil industry.[12] Over the period of 1960 to 1970 all expenditures for marine operations in all facets, were less than 10% of the total capital budget.

(b) The oil industry cannot generate enough capital internally to support the anticipated investment for expansion of production capacity during the 70s. According to a long-time oil economist John Winger, vice-president of the Chase Manhattan Bank and head of the Energy Economics Division, the oil industry will need $1 trillion for the period of 1970 to 1985 (*Petroleum Press Service,* October 1972, p. 364).

Unfortunately, we do not have the backup data to analyze Mr. Winger's capital expenditure projections which indicate a compounded growth rate of 13% annually. But if we were to accept them, one may ask where would all this money come from. His answer is, partly from operations and the rest from borrowing. In his estimation operations are not expected to contribute more than $600 billion, leaving a deficit of $400 billion which "is equal to seven times the demand on capital markets by petroleum companies during the past fifteen years." (*Petroleum Press Service,* October 1972, p. 364.)

I am rather pessimistic that such a deficit can be satisfied through the *normal* capital market operations. During the past 7 years the oil industry has been raising 30% of its needs by *borrowing* and the rest through the internal cash flow (Standard and Poor's, 1971, 1972). In order to be able to sustain a Debt-to-Equity ratio of one-half, which today the financial community considers magical for the oil industry, the internal cash flow of the oil industry must increase at a rate of over 10% compounded annually. This is a task of no mean proportions. Of course one may try the equity route. The international part of the oil industry, however, has also been experiencing difficulties in raising equity capital. The opinion of the financial community is partly reflected in the price earnings (P/E) ratios of the stock of international U.S. oil companies versus those that are primarily domestic.[13] As of December 1972 the average P/E ratios stood at 11.2 for the international versus 20.5 for the domestic U.S. corporations. So when it comes to priorities I am led to believe that history will repeat itself.

I am of the opinion that a large part of the needed capital for transportation must come from the petroleum-exporting countries. The royalties and taxes of the Middle Eastern countries alone are now running at about $25 million per day or over $9 billion a year and increasing. As I have already mentioned, it would seem logical for them to be interested in investing in transportation, but they would, for obvious reasons, prefer that others put (fix) money in exploration and production.[14]

To conclude then, I feel that the $40 to 47 billion needed for transportation over the next 8-year period will be found not because of the initiative of the oil companies but rather because of their apathy and lack of appreciation of the role and contribution of transportation. I am not as optimistic, however, about the ability of the integrated oil companies or the independents by themselves to raise $1 trillion over the next 15 years for exploration and production, if indeed this much is needed, especially since $400 billion of the total must be found from outside sources. A new era, therefore, may be dawning, one of direct government participation in the planning for and financing of global energy requirements. A logical consequence of such a development will be the *explicit* introduction of politics into the energy field by both producing and consuming countries, and the end of the oil industry as an economic entity as we now know it.

Notes

1. Spot rates refer to the cost of transporting oil for a given run, and are expressed in monetary terms per ton of oil delivered. They are to be distinguished from time-charter rates that refer to the cost of renting a vessel to carry oil for a certain specified time period. The time-charter rate is usually quoted in monetary terms per deadweight ton (DWT) of carrying capacity per month. We can convert, of course, time-charter rates to spot-rate equivalents. For convenience, spot rates are quoted in terms of percentages of a standard, the latter now being "Worldscale."

2. During prolonged periods of low spot rates, the spot market handles approximately 20% of the total tonnage. During periods of high rates, however, the amount drops to about 6%.

3. We assume here that "enough" users of transportation will rationalize their policies and operations by using the information and thus affect the industry. If not, those few who apply this knowledge will benefit at the expense of the rest.

4. In the past we have been witnessing more of the former rather than the latter. The policies of the international oil companies have resulted in other departments having to absorb costs resulting from inefficient transportation decisions.

5. We have also seen recently another supporting evidence. Iran is demanding that the Consortium increases the output from the present rate of 5 million barrels per day to 8 million. Saudi Arabia is also attempting to increase its output. If we continue on the same consumption course, however, I foresee long-run shortages.

6. Although hard data are not available, it appears that over 95% of all oil is sold by the international oil companies on a delivered basis.

7. One way of gaining such control is for each one to have enough capacity to satisfy 100% of his requirements. This solution, however, will increase the cost of transportation for the industry as a whole (Zannetos, 1972).

8. These arguments are based on some theoretical factors that favor more-or-less perfectly competitive markets in ocean transportation (Zannetos, 1966) and an independent fleet (Zannetos, 1972).

9. In an overall cost sense. Opportunities for speculative profits and arbitrage will then be reduced.

10. In addition to benefiting from greater control of the C.I.F. pricing structure, the oil companies and the producing countries, if they make efficient management decisions, will reap some additional profits from transportation through increased ownership.

11. This *does not necessarily* imply that the producing countries should proceed and invest in refining and distribution facilities in the consuming countries.

12. For some of the reasons behind the behavior of the oil companies, see Zannetos (1972).

13. Whether or not this is due to an overreaction to the political situation in the Middle East is rather immaterial as long as it affects the behavior of capital markets.

14. Nor would it be to the advantage of the producing countries to fix their investments in refining and distribution facilities in foreign countries. In my estimation, there are many other strategy possibilities that merit priority before the latter is attempted.

Bibliography

Adelman, M. A. (1972–1973). "Is the Oil Shortage Real?," *Foreign Policy,* Number 9, Winter.

American Petroleum Institute (1971). *Petroleum Facts and Figures,* Washington, D.C.

Chase Manhattan Bank, National Association (annual). *Annual Financial Analysis of a Group of Petroleum Companies,* The Energy Economics Division, New York.

——— (annual). *Capital Investments of the World Petroleum Industry,* New York.

——— (1972). *The Petroleum Situation,* June.

Clarkson, H. and Company, Limited (weekly reports). *Tanker Freight Market,* London, England.

Conrad Boe Ltd. A/S Shipbrokers (1972). *Estimated Tanker Market Rates Single Voyages 1947–1972,* Oslo, Norway.

DeGolyer and MacNaughton (1972). *Twentieth Century Petroleum Statistics,* Dallas, Texas.

Jacobs, John I. (1971, 1972, semiannual). *World Tanker Fleet Review,* especially 30th June 1971 and 30th June 1972.

Koopmans, Tjalling C. (1939). *Tanker Freight Rates and Tankship Building,* Haarlem, Netherlands.

Middle East Petroleum and Economic Publications (1972). *International Crude Oil and Product Prices,* Beirut, Lebanon, 15 April.

O.E.C.D. (1972). *Statistics of Energy 1956–1970,* Paris, France.

O.P.E.C., *Annual Statistical Bulletin.*

Petroleum and Petrochemical International (1972). Vol. 12, No. 3, London, England, March.

Petroleum Press Service (1971, 1972). London, England, issues of January 1971, March 1971, May 1971, December 1971, April 1972, October 1972, December 1972.

Standard & Poor's (1971). *Industry Surveys: Oil Basic Analyses,* January 3.

——— (1972). *Industry Surveys: Oil Current Analysis,* Vol. 140, No. 48, Sec. 1, November.

Sun Oil Company (annual). Economics Department, *Analysis of World Tank Ship Fleet.*

Tinbergen, Jan (1958). *Selected Papers.* In L. H. Klaasen et al. ed., North-Holland Publishing Company, Amsterdam, Netherlands.

U.S. Department of the Interior, Office of Oil and Gas (1971). *Worldwide Crude Oil Prices,* Quarterly Report, Washington, D.C., Fall.

——— U.S. Office of Oil and Gas (1972). *Worldwide Crude Oil Prices,* Second Report, Washington, D.C., Summer.

Wall Street Journal (1973). Wednesday January 24 and Friday February 9.

Zannetos, Zenon S. (1966). *The Theory of Oil Tankship Rates: An Economic Analysis of Tankship Operations,* MIT Press, Cambridge, Mass.

——— (1967). "Time Charter Rates," Sun Oil Company, *Analysis of World Tank Ship Fleet,* 25th Anniversary Issue, August.

——— (1972). *Market and Cost Structure in Shipping,* presentation made at the International Research Seminar on Shipping Management, Bergen, Norway, August 23–26.

31 Oil Transportation Studies

J. G. HALE* AND R. J. DEAM*

31.1. Introduction

In general, the large crude-oil-producing areas of the world are not the areas of the largest energy consumption. The prices of oil products in the consuming countries of the world, therefore, are substantially influenced by the price paid for transporting oil from the producing to the consuming areas. Western Europe, for example, at present imports over 90% of its oil requirements, 60% of which comes from the Middle East. If one considers a landed cost of crude oil at, say, Rotterdam imported from the Persian Gulf of $20 per ton, something like one-half of this cost can represent the price paid for the transportation of the oil. A similar situation applies in Japan, one of the other large energy-consuming areas of the world, and, with the expected increased dependence of the United States on imported crude oil, the price of oil products in the United States can be expected to become increasingly influenced by the landed cost of imported crude oils.

Despite the increasing use of pipelines in many parts of the world for crude oil transportation, by far the largest quantity of oil is moved by oceangoing tankships, and this situation is likely to persist for many years into the future. It is the price paid for crude oil transportation by tankships, therefore, which is one of the important factors making up the landed cost of crude oil in most countries of high oil consumption. These prices paid, however, are highly variable even when considering one specific route because they depend to a great extent upon the fluctuations that take place in the market for tankship hire.

This paper outlines a study of the oil tankship market, which is being undertaken within a model of world energy being built by the Energy Research Unit at Queen Mary College in London. A hypothesis concerning the relationship between tankship "spot" charter hire and the prices of oil products in consuming countries is described and its likely consequences, if true, for current plans to build and expand ports on the East Coast of the United States.

* Queen Mary College, University of London.

31.2. Vessel Sizes and Port Facilities

Ocean transportation of crude oils is made in tankships of sizes ranging from around 20,000 deadweight tons (DWT) capacity up to the giant crude oil carriers of over 300,000 deadweight tons. Although the cost per day of operating tankships naturally increases with the size of the vessel, the cost per ton of oil moved decreases with increasing size. Similarly, building costs per deadweight ton of carrying capacity decrease with increasing vessel size. The amount of capital, however, tied up in any one of these larger vessels is much larger and therefore might be considered to be at greater risk.

Since World War II, many large refining complexes have been built close to the areas of high oil consumption. This has necessitated moving large quantities of crude oil from the producing areas of the world to feed them. As a result of the belief that the employment of larger vessels would lower the price paid for crude oil transportation, there has been a rapid increase in the average size of vessel employed for moving crude oils over the past 10 years.

The desire to employ the largest vessels possible is constrained by the availability of port facilities capable of handling such vessels. Although there are a large number of vessels of over 150,000 deadweight tons capacity in service in the world at the present time, not all crude oil terminals are capable of receiving vessels of this size and greater. A recent survey stated that at the present time there are some 96 ports in the world capable of handling vessels of greater than 150,000 DWT. Of these, 31 are crude oil loading ports in the Middle East and 65 discharge ports situated in Western Europe and Japan. Within the next year or so these figures are expected to increase to 33 loading and 79 discharging ports. Not all of the 65 discharge ports mentioned earlier, however, are capable of handling the very large vessels of 250,000 DWT and greater, and, in addition, there are many more ports in the world which can handle only vessels of a maximum size of 100,000 DWT. One notable area with ports in this latter category, about which considerable concern has been expressed recently, is the East Coast of the United States. At present it is doubtful whether vessels greater than about 80,000 DWT fully laden can use these ports, but many plans are being considered to change this situation in the near future.

Owing to the large sums of capital investment tied up in these very large tankships, there has naturally been an interest in making certain that there are sufficient port facilities to ensure their full employment. Possibly because of this interest it appears that the employment of large vessels has not been restricted for lack of port facilities.

Transportation of crude oils in the very large vessels to certain small-demand areas in Africa and Asia is not always practicable even if harbor facilities are adequate owing to the fact that tankage or refining capacity is similarly small. Until such times as oil demand in these areas increases substantially to justify larger refining capacity, vessels in the 25,000 to 50,000-DWT size range will continue to be required in order to keep deliveries in manageable parcel sizes. Small vessels are also still likely to be required in the high-demand areas in order to ease refinery scheduling problems and for small quantities of specific crudes required to manufacture certain specialized products, for example, bitumen.

31.3. The Tankship Market

The ownership of the world's oceangoing tankship fleet is divided between the world's oil companies and a number of independent crude oil tankships, and the large international companies generally choose to own around 40% of their requirements in the form of their own subsidiary tankship companies. In addition, these companies try to cover around 90% of their requirements by the combination of their own fleet and vessels on long-term charters from independent owners. The final 10% of requirements is covered by the so-called "spot hire" of vessels for one or two voyages.

Besides the large international oil companies, there are other national oil companies and smaller oil traders who enter the tankship market, and in this open market the rate charged for the hire of vessels is highly influenced by the supply-and-demand position existing at any particular moment in time as in any other open market situation.

The tankship market is a highly complex one, part of the ownership of the tankships being in the hands of the users and contracts being affected for different periods of time and under various different conditions for the vessels' operation. There are also strong indications that expected future values of charter rates have significant effects on the rates actually realized. This latter type of effect suggests considerable apprehension of the unknown future, either through insufficient knowledge or the inability to comprehend adequately a very large, highly interactive, international market.

We in the Energy Research Unit believe that sufficient knowledge of the essential factors is obtainable and that a mathematical model of the supply and demand for tankships in the context of a model of world energy is the only satisfactory way of comprehending the most significant interactions. With this model, sound evaluations of expected future situations can then be made to enable forward planning to be carried out with increased confidence. Such planning might also eventually lead to a diminution of some of the more violent swings at present experienced in "spot" charter rates.

31.4. The Oil Transportation Model

The oil transportation model is contained in a larger linear programming model of World Energy. In this model the world is represented as some twenty areas. Shipping logistics are represented by opportunities for the movement of crude oils from each crude oil exporting terminal to each consumption area by, at present, six size categories of ocean tankship. The opportunities for use of each size of vessel is subject to restrictions limiting the total ton-day world fleet capacity available for each category size. The costs on the opportunities represent the operating cost of the vessel size per ton of oil moved on the particular route, and are computed from vessel operating costs per deadweight ton-day and average round-trip times to the importing area, weighted according to the refining capacity in the area served by each receiving terminal. Additional in-port costs are also included. Limitations on the use of different sizes of tankships in a given importing area are represented as restrictions of two kinds: those which demand that a certain amount of the area's refining capacity must be supplied in vessels of less than a given size, and those which require that only a certain maximum amount of crude oil can be supplied by vessels of over a certain size. Limitations on vessels size at exporting terminals are at present modeled by simply not including opportunities for using vessels of a greater size than can be accommodated at the terminals from routes involving them.

Such a representation of shipping logistics can be employed to study the tankship market in the past, present, and future by changing the input quantity values appropriately. When one is examining a position relating to more than 2 or 3 years in advance, however, the current shipping fleet availabilities and port restrictions will require modification. For time periods further into the future, therefore, opportunities are included for building tankships of different sizes limited by world shipbuilding capacity. These opportunities will add capacity to the various shipping availability restrictions and will have as costs annual capital charges for building each size of vessel discounted over the expected life of the vessel in service. Further opportunities will be added to the matrix representing the building and expansion of port facilities with estimated annual capital charges associated with their completion. These opportunities will relax the restrictions on the use of vessels of certain sizes in the corresponding areas.

31.5. Proposed Work Program

Initially the model will be used historically to examine the relationship between "spot" freight rates and marginal values generated against shipping availability restrictions. For this purpose the transportation model can be isolated from the energy model by feeding to it, as crude oil demands in the areas, the quantities of crude oils actually moved during periods in the past together with the actual carrying capacity of the world tankship fleet by size categories available during these periods.

The marginal values generated against the fleet capacity restrictions will represent the reduction in the cost of meeting the given demand for oil which would ensue from the addition of 1 dead weight ton-day of carrying capacity to each size category of ship for the period considered.

If perfect competition exists in the tankship market and "spot" rates are not substantially influenced by future expected rates, we would expect freight rates calculated from marginal values to correspond closely to actual "spot" rates realized in the period considered. Correlations of model-derived values with actual rates realized will be attempted using techniques similar to those for oil product prices outlined in paper A-1 (abstract in the Appendix of Proceedings).

The real value of a model such as this is, however, for evaluating future planning alternatives. Any work concerned with estimates of future situations will need to be carried out in the context of the whole energy model since they are dependent upon the pattern of crude oil distribution throughout the world. For such studies it will be necessary to include in the model opportunities for building tankships of various sizes limited by the capacity of the world's shipyards. With these additions, together with opportunities for expanding harbor facilities, future equilibrium positions for the market can be evaluated.

The first future position examined would be a worldwide optimum. This would provide a reference point from which cost deviations from optimality for any expected situation imposed upon the model could be measured. It is proposed to examine the effects on the tankship market of factors such as the reopening of the Suez Canal, construction of crude oil pipelines (for example, SUMED, the Egyptian proposed Suez to Mediterranean crude pipeline), increasing crude oil imports to the United States, the building of deepwater "entrepôt" crude oil terminals and such political restrictions that might conceivably be imposed (for example, the limitation of crude oil movements from one country to another).

31.6. The Price Paid for Crude Oil Transportation

As previously noted, the prices of oil products in many of the consuming countries of the world depend to a large degree upon the "spot" freight paid for the transportation of crude oil to the country from the various crude-oil-producing areas of the world.

In view of the open and highly competitive market in tankships that exists and the long time lag in building new vessels, we would expect that the "spot" charter rate for each size of vessel would be determined by the shadow value given in our model to another dead weight ton-day of carrying capacity. (A floor for the "spot" rate, will be set by short-term lay-up charges.) Further, the "spot" charter rates for each class of vessel will be the same unless a particular class or set of classes is in a greater relative surplus than the others.

If there were, for example, a surplus of 200,000-DWT vessels over that number which could be accommodated by ports able to receive such vessels fully loaded, charterers would operate the additional vessels part loaded or use two port discharges to enable them to use harbors with shallower draughts. The result of this type of operation would be that 200,000-DWT vessels would have lower "spot" freight rates (for example, if a vessel was only 80% loaded, the rate charged would be 80% of the "spot" freight price of, say, the 50,000 tonner).

Above 50,000 DWT tons, spot freight rates in the past have not shown real variations with size. Certainly, they do not reflect the large differences in operating costs between vessels of 50,000 DWT and those of 200,000 DWT. Further, if freight rates for vessels of 200,000 DWT were lower than those for 50,000-DWT vessels, one would expect to see posted prices for crude oil F.O.B. to be quoted at large "entrepôt" terminals such as Bantry Bay in Ireland. The fact that this has not happened indicates that refiners with port facilities capable of handling only vessels of less than 50,000 DWT can import crude oil from the Persian Gulf at the same spot freight rate as refiners able to handle the very large vessels.

An oil company needing to supply refineries that can receive only 50,000-DWT vessels and who owned a vessel of 200,000-DWT capacity and a deepwater "entrepôt" port to receive them would in the present circumstances do better to charter out their large vessel and charter in four 50,000-DWT vessels to move the same quantity of oil direct to their refineries. In this way they could avoid being concerned with the additional cost of handling and on-shipping the oil from a large "entrepôt" terminal.

Any further construction of deepwater "entrepôt" terminals or deepening of existing harbors will only maintain and prolong the present situation of indifference to vessel size of "spot" freight rates. (In fact, when a two-tier spot freight structure appears, this is indicative that the world needs harbor deepening.)

We maintain that in calculations used for deciding the location of refinery building and expansions an individual refiner, to obtain his answers, should use the value freight expected (spot) and not the cost of freight as a function of port water depth. We thus expect that our World Energy Model will show that in the medium term (that is, up to 10 years ahead) construction and expansion of refineries in areas where no deepwater harbor facilities are available is desirable and, in addition, that present deepwater refining areas such as Rotterdam are now overbuilt with refining capacity.

A rider to this argument is that at present the economy of scale goes to the shipowner and not, as perhaps is generally assumed, to the refiner or consumer. Similarly, the rent from an investment to deepen a harbor or build a large "entrepôt" terminal goes not to the port authority, the refiner, or consumer but again to the shipowner, until most existing 50,000 tonners have reached the end of their life span.

In order to investigate such highly interactive and complex situations as these, we need the total systems approach that our model of World Energy provides.

The preceding discussion has a particular relevance to the plans that are presently being drawn up for the building and expansion of ports on the East Coast of America. It is suggested that these plans be reviewed in the light of the conclusions of the argument.

32 Demand for Energy by the Transportation Sector and Opportunities for Energy Conservation*

A. C. MALLIARIS† AND R. L. STROMBOTNE‡

The existing automobile and truck populations account for about 76% of the energy consumed within the transportation sector of the economy. Consumption of energy by aviation is relatively small but is growing rapidly. We discuss the structure of demand for transportation services and energy both historically and as projected to the year 2020. In the near term, improvements and modifications to existing automobile and truck types offer an opportunity to reduce relative energy consumption. For the long term, novel fuels and electric energy may provide a way to reduce the dependence of surface transportation upon petroleum. We also discuss some nontechnological actions that potentially offer energy savings within the transportation sector. Estimates of energy savings are provided and several important factors are discussed through numerous examples.

32.1. Introduction and Background

The following general statistics relate the transportation sector, as an energy consumer, to the "energy crisis" and to the "petroleum crisis" (U.S. Department of Transportation, 1972 July, 1972 November; U.S. Congress, House, 1971; Associated Universities, Inc., 1972):

1. Civilian transportation consumes directly about 25% of the total U.S. energy budget and is projected to continue to consume at that same rate for the next several decades. Energy consumption shares of the industrial, residential, and commercial sectors are about 43%, 20%, and 12% of the total, respectively. Direct consumption of energy in the transportation sector includes only propulsion and auxiliary power. It does not include the energy consumed for the production and maintenance of vehicles, facilities, fuels, and other components of the transportation sector. Indirect consumption of energy may amount to as much as 50% of the directly

* The authors wish to acknowledge thankfully several informative and stimulating discussions held with R. A. Husted of the Office of the Secretary, U.S. Department of Transportation. The authors gratefully acknowledge the value of insights gained by their participation in the work of the Transportation Energy R & D Goals Panel. Finally, the authors remind the reader that the views expressed in this paper do not necessarily represent the policies or positions of the U.S. Department of Transportation.

† Transportation Systems Center, U.S. Department of Transportation.
‡ Office of the Secretary, U.S. Department of Transportation.

consumed energy in transportation (Hirst and Herendeen, 1973). Energy consumption by military vehicles and by various vehicles used off the road for agricultural purposes is in the range of 10 to 15% of the civilian transportation energy consumption (U.S. Department of Transportation, 1972 November).

2. Transportation is a major user of petroleum. About 55 to 60% of the petroleum consumed in the United States is used by transportation. This share is projected (National Petroleum Council, 1972) to remain relatively constant in the foreseeable future, on the assumption that current policies and trends continue and that fuels are available. Further, transportation is intensively dependent on petroleum. More than 95% of the transportation energy consumed is from a petroleum source (U.S. Department of Transportation, 1972 November). The consumption of electricity by rapid rail transit, electric locomotives, electric rail cars and pipelines amounts to about 0.3% of the total transportation energy (U.S. Department of Transportation, 1972 November),[1] with an additional 0.3% estimated for the battery powered industrial lift trucks and golf carts. In addition, the engines that power transportation vehicles require specific fuels and have little adaptability to use coal, natural gas, or nuclear energy. Section 32.2 summarizes demand data and projections of transportation service and discusses energy efficiency usage in the transportation sector. Section 32.3 introduces the factors that bear on petroleum consumption as a prelude to the more detailed discussion in Section 32.4 of some ten specific actions that have been proposed to reduce fuel consumption. Section 32.5 provides a preliminary assessment of the usefulness of novel fuels and of electrical power as substitutes for petroleum based fuels.

32.2. Transportation Services Demand and Associated Energy Demand

Total volume and the modal distribution of transportation services are closely related to demographic and economic characteristics of the country. The data of the 1972 National Transportation Report (U.S. Department of Transportation, 1972 July) account for (both historically and in projection) the influence on transportation demand of factors such as population, income, population distribution patterns, quality of transportation, and cost of transportation relative to other costs. Anticipated growth or change in these factors has been used for projections of traffic growth up to 1990. No major changes have been assumed in public policy, technological change, fuel availability, and fuel costs. Tables 32.1 and 32.2 present results from U.S. Department of Transportation (1972 July). Growth rates of passenger-miles and ton-miles are given in 3 time frames: 1965–1970, 1970–1980, 1980–1990.

The energy consumed directly for transportation is closely coupled to passenger-miles (PM) and ton-miles (TM) since the useful work performed by this energy is mainly the sum of the frictional and inertial forces acting on the vehicle over its path. The energy efficiency of a given transport mode is given as passenger-miles per gallon or ton-miles per gallon.

Table 32.3 gives the distribution of passenger and cargo service over six modes for U.S. transportation in 1970, based on data from U.S. Department of Transportation (1972 April, 1972 November). It also reports the distribution of energy consumption, cost, and energy efficiency by mode. The highway mode includes all automobiles, trucks, and buses except the urban transit bus, which is included in Urban Transit. Urban Transit also includes commuter rail and rapid transit rail. The Air mode includes domestic and international passenger and cargo service, as well as private aviation. The Water passenger mode includes commercial and recreational boating, while the Water cargo mode includes both domestic and international activity. (The Water cargo data may be in some error, because of incomplete data in international shipping.) The Rail passenger numbers refer to intercity travel only. The last two columns of Table 32.3 provide a summary of the distribution of costs and energy use for all traffic. Further detail will be presented as required in the following discussion.

Figure 32.1 shows the results of a projection of energy consumption by transport modes through the year 2020. The projection is based on the projected passenger and cargo traffic, with minor modifications, and current values of energy efficiencies. The growth of consumption by the Highway submodes reflects constant consumption per capita (saturation) before 1990, while the Air mode saturates after 2000. The growth rate projected beyond 1990 for all transportation energy consumption is closer to the projected 1.1% population growth than to the projected growth of GNP. This projection is conservative since transportation demand historically has grown with GNP. The total energy used by transportation in the United States is projected to grow from about 15×10^{15}Btu in 1970 to

Table 32.1 Annual Growth Rates of Passenger Travel Expenditures by Mode[a]

Mode	Annual Growth Rate 1965–1970 (Actual)	1970–1980 (Estimate)	1970–1990 (Estimate)
Automobiles			
Personal	4.2	3.8	3.3
Business	3.9	5.1	4.4
Buses			
Intercity	1.3	0.6	0.5
Local transit	−2.2	2.1	2.0
School	1.6	−0.7	0.2
Taxicabs	0.1	0.9	0.7
Airlines			
Domestic	13.7	8.9	8.1
International	15.1	12.1	10.3
Business	19.6	11.9	9.4
Railroads	−9.3	−2.5	0.0
Total, Passenger Transport	4.3	4.8	4.3

[a] Source: U.S. Department of Transportation (1972 July), Table IV-6.

Table 32.2 Annual Growth Rates of Freight Transportation Expenditures by Mode[a]

Mode	Annual Growth Rate 1965–1970 (Actual)	1970–1980 (Estimate)	1970–1990 (Estimate)
Trucking			
Commercial	3.9	5.4	4.4
Private	3.2	4.7	4.1
Government	5.8	4.7	4.3
Railroads	0.3	2.8	2.5
Pipelines	3.3	4.2	3.7
Air			
Domestic	13.7	14.0	11.6
International	12.1	15.5	10.4
Water			
Domestic	3.7	3.6	3.3
Overseas	0.7	5.4	3.9
Total, All Freight	3.0	4.9	4.2

[a] Source: U.S. Department of Transportation (1972 July), Table IV-7.

Table 32.3 Modal Distribution of Passenger and Cargo Traffic, Energy Consumption, Energy Efficiency and Cost for U.S. Transportation in 1970. All data from U.S. Department of Transportation (1972 July, 1972 November). PM: Passenger-miles, TM: Ton-miles, E_i or E_j: Energy consumption in a passenger or cargo mode, $(PM/E)_i$ or $(TM/E)_j$: Energy efficiency in PM/gallon or TM/gallon

	Passenger				Cargo				All Traffic	
Mode	$(PM)_i$ %	E_i %	$(PM/E)_i$ PM/gallon	Cost %	$(TM)_j$ %	E_j %	$(TM/E)_j$ TM/gallon	Cost %	$E_i + E_j$ %	Cost %
Highway	90.0	54.0	31	42.2	21.6	22.0	20	40.6	76.0	82.8
Air	7.6	9.0	16	5.6	0.3	1.0	5	0.7	10.0	6.3
Water	0.2	0.7	5	1.5	26.0	4.2	175	1.4	4.9	2.9
Rail	0.3	0.1	52	0.2	33.3	3.4	200	6.1	3.5	6.3
Pipeline	—	—	—	—	18.8	5.1	75	0.7	5.1	0.7
Urban Transit	1.9	0.5	57	1.0	—	—	—	—	0.5	1.0
All Modes	100.0	64.3	29	50.5	100.0	35.7	60	49.5	100.0	100.0

Passenger traffic in all modes: about 2.2 trillion passenger-miles
Cargo traffic in all modes: about 2.3 trillion ton-miles
Energy consumption by all traffic modes: about 115 billion gallons (about 15.2×10^{15} Btu)
Cost of all traffic in all modes: about 195 billion dollars.

Table 32.4 Projected Percentage Distribution of Energy Consumption within the Transportation Sector for Highway, Air, and All Other Modes in Selected Years

	1970	1990	2020
Highway	76%	60%	55%
Air	10%	30%	33%
All Other	14%	10%	12%

30×10^{15}Btu in 1990 and then to about 50×10^{15}Btu in 2020. Table 32.4 gives the percentage distribution of energy consumption among highway, air, and all other modes for selected years.

The automobile and the airplane dominate the passenger service picture in terms of passenger-miles, costs, and energy consumption both now and in projections. The ubiquitous automobile provides convenient, flexible, door-to-door, personally secure, reliable transportation service with increasing safety. It is less energy efficient than its line-haul competitors on the ground which have had smaller growth rates for the last 25 years.

The airplane has had the highest rate of growth of passenger-miles of

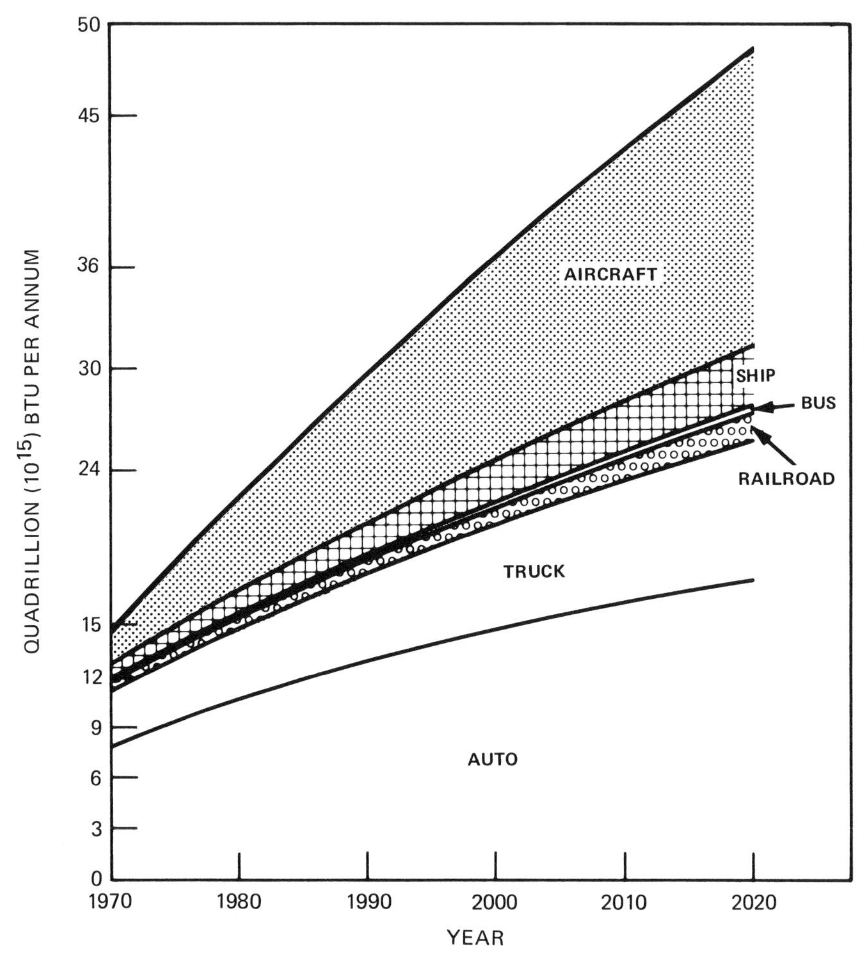

Figure 32.1. Projected transportation energy consumption.

any of the passenger modes and is projected to continue to grow more rapidly in the future, although not so rapidly as in the past. It is the least energy efficient of the major passenger carriers. Its high growth rate, coupled with its low energy-efficiency, combine to make it relatively more important as a consumer of energy in future projections.

32.3. Options for Petroleum Conservation in Transportation

Consider the total passenger and cargo traffic (PM and TM) each as a sum of their modal components:

$$\mathrm{PM} = \sum_i (\mathrm{PM})_i \qquad \mathrm{TM} = \sum_j (\mathrm{TM})_j. \tag{32.1}$$

Next, consider the corresponding expressions for the energy consumed by passenger and cargo traffic:

$$E_p = \sum_i \frac{(\mathrm{PM})_i}{u_i e_i} \qquad E_c = \sum_j \frac{(\mathrm{TM})_j}{u_j e_j}. \tag{32.2}$$

In these expressions u_i and u_j are the vehicle occupancy and payload coefficients in (passengers/vehicle) and (tons/vehicles), while e_i and e_j are the corresponding vehicle efficiencies in traditional units of vehicle-miles/gallon. (The vehicle efficiency of automobiles is commonly known as fuel economy.)

Equations 32.2 display the interrelationship of four types of options that are generally available for petroleum conservation: (a) increase vehicle efficiency; (b) increase the occupancy (or payload) of vehicles; (c) shift demand from less to more energy efficient modes; and (d) reduce overall demand. An additional option is to diversify the sources of transportation energy. In the first two categories one starts with the factors appearing in the denominators of Equations 32.2. The vehicle-efficiency e may be increased if the vehicles, in one or several modes, are made more efficient energy converters and/or if the driving cycles and driving conditions are altered to minimize energy consumption. Demand reduction may be attempted by a variety of actions. Fuel rationing, travel rationing, 4-day week, television links to replace some travel, urban and community design to minimize travel, and walking and bicycling instead of driving have all been suggested as possible ways to reduce demand.

Diversification of the transportation energy supply is a petroleum conservation measure, although it does not necessarily conserve energy. Typical examples include nonpetroleum based fuels and electricity derived from nuclear, geothermal, hydro, and coal sources. In view of the intensive dependence of transportation on petroleum, diversification on a large scale would be a substantial change from the present situation.

Assessment of the options available for transportation energy conservation requires consideration of such matters as the conservation potential of a contemplated action; the impact of such action on the American economy, life-style and on the transportation industry; the time frame and the investment required for implementation; the cost to the transportation user; existing or required government (federal, state, and local) policy; existing trends favorable or unfavorable to the action for conservation; strength and nature of required incentive to accept the action; required administrative or legal enforcement of whatever changes are made. The discussion in Section 32.4 of several action options emphasizes the potential for fuel conservation. Detailed discussion of the other factors is beyond the scope of the paper although some are addressed.

32.4. Discussion of Actions

Ten actions have been chosen to illustrate the conservation potential of several options and to discuss the important factors that enter into the computations. Table 32.5 briefly describes the selected ten actions and summarizes their estimated fuel savings. Actions 1 to 4 refer to different options for increasing the fuel economy of highway vehicles. These vehicles are of primary interest because they now consume 76% of the transportation energy and, according to Table 32.4, may consume more than 50% for the next 50 years. Action 5 is an example of increased vehicle occupancy. Modal shifts are illustrated by actions 6 to 9, while action 10 is an illustration of an attempt to reduce traffic demand. In all cases, a 50% change has been assumed, and the petroleum conservation potential has been computed as a percent of the total transportation energy, under 1970 transportation conditions. The conservation potential of action 1, conversion to small cars, is computed in a straightforward fashion. The present passenger car population is approximated by a two-component mix, namely: 90% family-type cars with a fuel economy of 13.1 mpg and 10% small cars at 22 mpg. The fuel consumption of this mix is equal to the specific consumption of the existing passenger car population, 13.6 mpg.

Table 32.5 Summary of Discussed Actions and the Corresponding Petroleum Conservation Potential as Percent of Total Transportation Energy for 1970

Numbers (#)	Action	% Fuel Conservation
1.	Convert 50% of passenger car population to small cars (22 mpg)	9.0%
2.	Introduce in 50% of highway vehicles a 30% reduction of fuel consumption	11.5%
3.	Eliminate 50% of urban congestion	1.1%
4.	Achieve 50% success in limiting highway speeds to 50 mph	2.9%
5.	Persuade 50% of urban commuters to car-pool	3.1%
6.	Shift 50% of commuters (to and from city centers), to dedicated bus service	1.9%
7.	Shift 50% of intercity auto passengers to intercity bus and rail, evenly	3.0%
8.	Shift 50% of intercity trucking to rail freight	3.4%
9.	Shift 50% of short haul air passengers to intercity bus	0.29%
10.	Persuade 50% of the people to walk or bike up to 5 miles, instead of driving	1.6%

A conversion to a 50-50% mix results in 9% fuel conservation. Conversion to small cars is known as one of the best ways to reduce fuel consumption. This action has the additional benefits of lower initial and maintenance costs to the user. Some disbenefits include inconvenience to those who need larger vehicles, a substantial impact on the auto manufacturers and satellite industries, and greater chance of injury in the event of an accident. A massive conversion of manufacturing facilities taking many years would be required to greatly increase the share of the market taken by small cars. Small cars have been taking a larger share of the market in recent years. The market share of standard-sized cars has decreased from 64% in 1960 to 38% in 1972.

Action 2 in Table 32.5 supposes a 30% reduction in fuel consumption in half of all highway vehicles. The idea here is to introduce fuel conservative aspects in the design of vehicles, possibly at greater cost. A combination of some of the possible technical changes to improve fuel economy, not all of which are immediately practical, are shown in Table 32.6, along with estimates of fuel savings.

These estimates of potential fuel savings are approximate and averaged

Table 32.6 Estimates of Potential Highway Vehicle Fuel Savings for Various Technical Changes (Fuel Economy Technology)

Potential Fuel Savings	Fuel Economy Technology
(a) 5 to 15%	improvements in carburetion, injection, ignition, and air-induction.
(b) 10 to 15%	smaller engine used with a booster to meet peak demand requirements.
(c) 10 to 20%	engine used with an infinitely variable transmission (engine—road load match).
(d) 15 to 20%	present engine replaced by a lean mixture engine.
(e) 3 to 8%	accessories driven at constant speed.
(f) 5 to 10%	radial-ply tires used.
(g) 3 to 5%	modest redesign to reduce the aero-drag at highest speeds.
(h) 3 to 8%	weight reduction by 5 to 10%.

over all highway vehicles and all driving conditions. They represent the views of the authors as modified by expert opinion expressed recently (U.S. Department of Transportation, 1972 September; Anderson, 1972, 1973; Thur et al., 1972). Not all of the improvements are additive. For example, only one of the three improvements (b), (c), and (d), may be selected, since all three aim to eliminate the excessive fuel consumption of the present engines at low loads. The compounded 30% reduction in fuel consumption reflects two considerations: the technological readiness of the industry to implement these improvements within the next few years, given an appropriate incentive, and a subjective balance between the incremental cost of the improved vehicle and the resulting fuel cost savings.

The efficiency of a vehicle may also be improved by a weight reduction, without compromising size. A lighter (not necessarily smaller), car is more efficient because much energy is used for acceleration in stop-and-go driving. However, a substantial effort would be required in redesign and in production engineering so that size, shape, performance, safety, comfort, and durability are not unduly compromised.

Elimination of driving conditions that require high fuel consumption also leads to fuel economy improvements. Elimination of congestion and reduced top speed are evaluated in actions 3 and 4 of Table 32.5. In the case of congestion, the computation accounts for the following facts: congestion occurs in urban areas (except in local streets), mainly during peak hours and in the fraction of highway mileage which is deficient from a traffic flow point of view. Based on available statistics (U.S. Bureau of the Census, 1972; U.S. Congress, House, 1972), the congested fraction of highway travel (vehicle-miles) is given by 42% × 34% × 20%, which equals 2.8% of the total highway travel. In deriving the estimate given in Table 32.5, we have assumed that vehicles in congested driving conditions consume twice as much fuel as cars out of congested areas.

The effect of highway speeds on fuel economy has been estimated from a distribution (Anderson, 1972) of the vehicle-miles traveled by passenger cars and by trucks, as a function of speed. In addition, actual measurements (Highway Research Board, 1971) of fuel economy at uniform speeds for several computations yield 2.9%, the result shown in example 4 of Table 32.5 as an upper limit to potential savings. Moreover, the fuel conservation benefit is only 1.3% if the speed limit is set at the higher level of 60 mph.

Action 5 in Table 32.5 illustrates the merits of increased vehicle occupancy with urban commuting selected as the most promising target. The relevant statistics here are: All highway vehicle-miles are split (U.S. Congress, House, 1972) 52% urban and 48% rural. The urban highway miles are split (French, 1972) 84% passenger and 16% cargo. Finally, the passenger vehicle-miles are split (U.S. Bureau of the Census, 1972) 34% commuting and 66% other. The resulting savings must be discounted by about 10% to account for extra mileage in car pools. Moreover, an increase of about 25% must be applied to account for the fact that fuel is being conserved at about 10 mpg while the highway average is about 12.2 mpg.

Actions 6 to 9 in Table 32.5 refer to modal shifts. The results here show limited potential for energy conservation in spite of the rather optimistic scenarios. In action 6 shifting to buses, we consider the vehicle-miles made by passengers commuting to and from city centers. These are estimated as 52% × 84% × 34% × 40%, or about 6% of all highway vehicle-miles. The first three factors in the above product have been discussed in association with action 5. The fourth factor, 40%, is a high estimate of the fraction of urban commuters who enter the center of cities. In this case, where the passenger destinations are clustered, a dedicated bus service might be attractive. Accordingly a 50% shift is assumed. The rest of the computation is based on about 890 billion total vehicle-miles for passengers, 1.5 passengers and 10 mpg per commuting vehicle, and on 25 passengers per round trip of bus service at 4 mpg.

In action 7, intercity travel is arbitrarily defined as a trip length of 50 miles, one way, or more. Such trips account (U.S. Congress, House, 1972) for about 24.7% of the 890 billion vehicle-miles. Moreover, in such trips a vehicle occupancy of 2.4 passengers/vehicle is observed (U.S. Bureau of the Census, 1972), at about 15 mpg. Arbitrarily, a shift of 50% of the aforementioned passenger-miles is made to the intercity bus (about 100 passenger-miles/gallon) and the intercity train (about 50 passenger-miles/gallon), evenly. The resulting savings are about 2.2% and 1%, respectively. This shift would require a 6.5-fold and a 25-fold increase of the intercity bus and train service, respectively.

Annual mileage has been used as the criterion for separating intercity trucking from local trucking and nonfreight service. Only trucks with more than 10,000 miles/year have been considered in example 8, Table 32.5. Intercity cargo traffic was about 410 billion ton-miles in 1970 (U.S. Department of Transportation, 1972 November; French, 1972) at about 41 ton-miles/gallon. An arbitrary shift of 50% of this traffic to rail freight (about 200 ton-miles/gallon) results in 3.6% savings of the total transportation energy. Since there is a 7.2¢/ton-mile differential in shipping costs, the trucking industry would have to absorb a $15 billion/year impact.

Short-haul air travel refers to trips shorter than 250 miles, one way. Analysis of statistical data (Civil Aeronautics Board, 1971) shows that the short-haul fraction of total air passenger-miles is about 3.8%, although the same fraction for passengers is about 20%. Since the total air passenger traffic accounts for about 9.0% of the total transportation energy, a complete elimination of the short-haul air service would yield 3.8% × 9.1%, or about 0.35% savings of the transportation energy. This amount must be discounted by about 15% to account for fuel consumption by the bus or rail service which would accommodate the shifted passenger traffic. The discount reduces the potential savings to 0.29% of total transportation energy.

The last action in Table 32.5 evaluates one popular approach for reducing the demand for auto passenger-miles. The suggestion is made often and seriously, that substantial fuel savings may result if relatively short automobile trips are replaced by walking or bicycling. The authors have used detailed data (U.S. Bureau of the Census, 1972) to evaluate potential savings involved. Table 32.7 gives the results for various trip lengths. The

Table 32.7 Percentage Trips, Vehicle Miles, and Energy Used in Transportation Sector for Automobile Trip Lengths up to 5 Miles

Car Trip Distance (One-Way Miles)	1/2	1	2	3	4	5
% Trips[a]	8.5	24.0	38.0	48.0	54.4	62.7
% Vehicle Miles[a]	0.3	1.5	3.5	6.0	8.2	11.5
% Transportation Energy[a]	0.2	0.8	2.0	3.4	4.7	6.5
Round Trip, Miles	0.75	1.5	3.0	5.0	7.0	9.0

[a] Cumulative Percentages.

first row specifies a maximum one-way trip distance. All shown percentages are cumulative. For example, the fraction of all automobile vehicle-miles, traveled in one-way trips, shorter than 3 miles, is 6%. The last two rows in this table give the transportation energy savings and the (up to) average round-trip distance that one must walk or bicycle in order to allow these savings to be realized. Table 32.7 shows that no substantial transportation energy savings would be realized unless unrealistically long walks are assumed. Besides, the percentage numbers of trips show that massive public cooperation would be required. It is unrealistic to expect such participation. Many people are not always willing or able to walk because of age, fatigue, weather conditions, loads, schedule, and so on.

32.5. Transportation Energy Diversification[2]

As mentioned in Section 32.3, diversification of the transportation energy supply is a petroleum conservation measure, regardless of its potential for energy conservation. Generally speaking, the diversification of the present transportation energy requires two things: (a) the technological readiness of transportation to use an energy form other than the petroleum-based fuels, and (b) the availability of such an energy form. Two particular cases are of interest: "novel fuels" and electrical energy.

The authors define "novel fuels" as fuels that are not now obtained from petroleum derivatives. In other words, gasoline or diesel fuel derived from shale, coal, or by any other means is not considered a "novel fuel." A simple assessment has been made of the potential of several novel fuels for applications in transportation, especially for the case of the full size passenger auto. Methane and methanol have been chosen as typical examples of relatively simple derivatives from coal. Propane and ethanol are examples of less straightforward coal derivatives. Hydrogen might be produced from nuclear energy (either by electrolysis or by more efficient thermochemical schemes, if successful). Finally, magnesium hydride, ammonia, and hydrazine have been included as examples of chemically stored hydrogen. Other possibilities of chemically stored hydrogen have been tentatively discarded. The alkaline hydrides such as the lithium, sodium, calcium, and strontium hydrides are too expensive to produce, react vigorously with water, and some ignite spontaneously in air. On the other hand, the metallic hydrides are logistically unattractive because they require rare metals, for example, palladium.

Table 32.8 summarizes the assessment of novel fuels for combustion applications without reference to specific heat engines. The entries in the first two columns, namely gallons per Btu and pounds per Btu, are data from standard tables, normalized with respect to gasoline. These numbers are useful in providing a quick rating. However, more informative are the numbers appearing under the columns labeled "Weight" and "Bulk." These are the weight and the volume estimated for the novel fuels and their tankage, necessary to provide a full performance family auto with a range comparable to that provided by 18 gallons of gasoline. A heat engine with the same efficiency is assumed in all cases. The fuel plus tank weight is higher than for gasoline, but nevertheless tolerable in most cases, except perhaps liquid hydrogen plus oxygen and magnesium hydride.

More serious difficulties are evident regarding the bulk of the fuel plus tank. The three hydrogen forms appear to require unacceptable volumes for a passenger car. Large uncertainties in these volumes arise from the many assumptions and design tradeoffs to be made about the cryogenic hardware and hydride fuel tankage.

Next, the question of fire and explosion hazard is examined. A rating is presented in four steps: Poor (P), Fair (F), Good (G), and Excellent (E). The ratings are compounded judgments based on considerations of data regarding flash point, autoignition point, vapor pressures, heat of vaporization, and explosion limits. Toxicity ratings appear in the next column. These are standard ratings as follows: (0: no harm), (1: slight but reversible harm), (2: moderate harm; could be irreversible), and (3: severe; could be fatal). Wherever two numbers appear, the first refers to inhalation and the second to ingestion.

The combustion rating is a rough measure of the thermochemical properties of the fuel, including considerations of combustion (mainly in in-

Table 32.8 Relative Properties of Certain Novel Fuels for Reference Purposes[a]

	Relative Gallons per Btu	Relative Pounds per Btu	Weight (lb)	Bulk (cu ft)	Fire Hazard Rating	Toxicity	Combustion Rating	Distribution Logistics	Tankage Cost
Gasoline	1.0	1.0	125	3	F	1–2	G	E	E
Methane (liquid)	1.6	0.9	210	5	F	0–1	E	F	F
Propane	1.1	1.0	185	4	F	0–1	E	F	G
Methanol	1.8	2.1	250	6	G	1–3	G	F	G
Ethanol	1.4	1.6	180	3	G	1–2	G	G	E
Liquid Hydrogen	3.9	0.4	150	>13	P	0	G	P	P
Liquid Hydrogen/ Liquid Oxygen	5.7	3.6	550	>18	P	0	E	P	P
Magnesium Hydride	4.1	4.9	700	>14	P	0	E	P	P
Ammonia	2.0	2.3	300	7	G	3	P	P	F
Hydrazine	1.6	2.3	265	5	E	3	P	P	F

[a] See text, Section 32.5, for discussion.

ternal combustion engines), corrosion, and exhaust products. Ammonia and hydrazine are rated poor mainly because of their corrosive properties in combustion. The hydrogen/oxygen combination is rated excellent because of its extraordinary low pollution potential. Methane and propane, as lower hydrocarbons, are rated excellent only by reference to gasoline or the alcohols.

The next column in Table 32.8 deals with the logistic problems of central storage, distribution, and local storage of the novel fuels before they are stored in the automobile. These ratings are important in providing a rough idea of the potential deviations from the current status. Thus, gasoline is rated excellent and only ethanol is rated good, since it requires comparable storage space per Btu without any undesirable characteristics on other counts. This means that ethanol could replace gasoline without appreciable modifications in the fuel storage and distribution logistics. By contrast, methanol is rated only fair because it requires almost twice as many gal/Btu as gasoline. This means that fuel stations and distribution trucks would have to double their capacity. The hydrogen cases are rated poor for automotive application, mainly because their distribution represents a major departure from the present fuel distribution system. (Explosion and fire hazard considerations are also relevant.) Ammonia and hydrazine are rated poor mainly because of their poor toxicity rating but also because of volumetric considerations. Finally, liquid methane, although in very limited application today, is rated poor in comparison to propane, which does not require cryogenic storage.

The last column in Table 32.8 summarizes information regarding the cost of the tankage and other handling hardware for storing the novel fuel on board a passenger car. Gasoline, naturally, receives an excellent rating and so does ethanol, because it is not appreciably different. Propane and methanol are relatively good. The first requires a moderately pressurized tank (about 100 pounds per square inch), while the second requires a tank almost twice as large as a gasoline tank, for the same Btu content. Cryogenic hydrogen requires expensive hardware and is rated poor, while cryogenic methane is rated fair, since it requires less sophisticated cryogenics. Ammonia and hydrazine are rated fair, mainly because of special precautionary measures that might be required to deal with their toxicity.

In view of the results of this preliminary assessment, ethanol ranks second behind gasoline or other related petroleum derivatives for automotive application, from a transportation point of view. The first choice is to derive synthetic gasoline from shale, coal, or by any other means, provided that it is economically competitive with gasoline. Second, ethanol should be considered as an alternative fuel, if it can be made economically competitive. Propane, methanol, and liquid methane, in that order, are next in the ranking. Cryogenic hydrogen, hydrogen/oxygen and magnesium hydride appear relatively unattractive. Further, ammonia and hydrazine are less attractive. These last two fuels are basically hydrogen storing compounds. Their unacceptable combustion rating may be avoided if these compounds are treated for hydrogen extraction, before combustion. However, this adds complexity, bulk, weight, and cost to the system. The poor toxicity rating of these fuels is still a serious problem.

As mentioned earlier, diversification of the transportation energy requires both the technological readiness of the transportation system (for a conversion to another form of energy) and the availability of this other form. Regarding electrical energy, the answer to the question of technological readiness depends on how much deviation from the current status the present transportation system can absorb. In the case of the electric train, technology is available now. The electric, full-performance, automobile is a dream for the future. The most important bottleneck here is the storage battery. Batteries presently under development are inadequate, although some promising concepts of high-performance batteries are still in the laboratory. The electric aircraft is of course absurd.

Next, the question is asked regarding the impact of all-electric surface transportation on the present and projected national generation capacity of electric power. An idea of this impact may be obtained from Table 32.9, which presents the projected energy demand of surface transportation as a percentage of projected electrical power generation capacity. The electrical power projections have been taken from Associated Universities, Inc. (1972). The entries for the surface transportation modes have been computed as follows.

Basic oil demand for these modes was obtained from the projections discussed earlier. A 20% efficiency was assumed, roughly, in converting the energy of gasoline to energy at the wheels of the transportation vehicles. This energy requirement, for the electric converted vehicles was traced back to the terminals of a generation plant, assuming a compounded efficiency of about 44%. This accounts for average efficiencies of 85% for

Table 32.9 Energy Demand of Surface Transportation as Percentage of Electric Power Generation Capacity (in Projection)

Mode	Year 1969	1977	1985	2000	2020
Auto	71%	56%	46%	24%	13%
Truck	26%	22%	17%	9.5%	4.9%
Rail	5.1%	4.5%	3.8%	2.9%	2.8%
Auto, Truck, and Rail	102%	82%	65%	37%	21%
Electrical Power (10^9 kWh/yr)	1,553	2,450	3,672	8,000	18,000

traction motors, 90% for power conditioning, 80% for battery charging, 80% for discharge, and 90% for transmission and distribution of electrical power. An inspection of the results in Table 32.9 shows that before the end of the century the impact of all-electric surface transportation on electrical power generation is quite substantial. However, electrical power generation capacity is projected to grow much faster than the transportation energy requirements. The projection shows that all-electric surface transportation could be accommodated not far beyond the year 2000.

Notes

1. Conversions of kWh to Btu are in error, in this reference. All conversions of electric energy for railroads and rapid transit should be multiplied by a factor of 3 to reflect electric power plant efficiency and transmission losses. This is also true for the conversion of pipeline electricity which, in addition, is in error by a factor of 1000. Total energy used for pipelines (in Btu) is approximately correct, but is in the form of natural gas.

2. The authors have drawn upon their contribution to U.S. Department of Transportation (1972 September) for much of this section.

References

Anderson, Wayne S. (1972). "R & D for Fuel Economy in Automotive Propulsion," USATACOM (U.S. Army Tank-Automotive Command), June 19.

——— (1973). "Is a 40 mpg Car Possible?" SAE Paper, January.

Associated Universities, Inc. (1972). *Reference Energy Systems and Resource Data for Use in the Assessment of Energy Technologies,* April.

Civil Aeronautics Board (1971). "Handbook of Airline Statistics," Washington, D.C.

French, A. (1972). Private communication, Federal Highway Administration, U.S. Department of Transportation, Washington, D.C.

Highway Research Board (1971). "Running Costs of Motor Vehicles as Affected by Road Design and Traffic," Report 111.

Hirst, E., and Herendeen, R. (1973). "Total Energy Demand for Automobiles," SAE Paper No. 730065.

National Petroleum Council (1972). "U.S. Energy Outlook, Vol. 1, Summary," Washington, D.C.

Thur, G.; Barber, K.; Murrell, J.; Schulz, R.; and Sebestyen, T. (1972). Private communication, U.S. Environmental Protection Agency, November.

U.S. Bureau of the Census (1972). "National Personal Transportation Study," conducted for the U.S. Department of Transportation, Washington, D.C.

U.S. Congress, House (1971). "Energy, The Ultimate Resource," study submitted to the Task Force on Energy, October.

——— (1972). *1972 Highway Needs Study,* House Document No. 92-266.

U.S. Department of Transportation (1972, July). "1972 National Transportation Report—Present Status—Future Alternatives," Office of the Secretary, Washington, D.C.

——— (1972, September). "Research and Development Opportunities for Improved Transportation Energy Usage," Summary Technical Report of the Transportation Energy R & D Goals Panel, available from National Technical Information Service, Washington, D.C.

——— (1972, November). "Summary of National Transportation Statistics," Assistant Secretary for Policy and International Affairs, Office of Systems Analysis and Information, Washington, D.C.

33 The Impact of Automotive Emission Controls on Future Crude Oil Demand in the United States

N. D. CARTER* AND W. T. TIERNEY†

The growth of the American standard of living can be largely attributed to our abundance of natural resources and the effectiveness with which they have been converted to useful assets or commodities. The development of river and rail transport systems and later the advent and rapid expansion of the use of road vehicles and aircraft have given us the flexibility and mobility to continue our growth and enhance our standard of living.

All of this has not been accomplished without an effect on the environment. As the centers of population move toward the urban areas and at the same time the transportation systems become global, man is having to learn to live with his environment and also to live with the limited resources available to him.

Obviously, through his very existence man is going to have impact on the environment. He cannot fulfill his daily needs or accomplish the things that are important to him in enhancing the quality of his life without changing the environment in some way.

The important thing is that he must seek balance between his needs and the quality of his life on the one hand and the environment and the resources available to him on the other. In attempting to seek these balances, it is inevitable that conflicting needs will arise and that carefully evaluated compromises must be found.

One such conflict that involves automotive engine exhaust pollutants and the energy penalties associated with their reduction to legislated levels is the particular ecology vis-à-vis energy demand relationship we wish to deal with today.

There is no question in our minds that remedial steps must be taken to reduce deleterious exhaust emissions. The first steps were taken in California, where the first standards for automotive emissions were established, as given in Figure 33.1. As you can see, the three most harmful products must be reduced gradually in accordance with the standards that were established. The most drastic of these decreases is imposed by the Clean Air Act of 1970 that mandated a 90% reduction of HC and CO by 1975 as

* Coordinator, Environmental Protection Department, Texaco Inc., Beacon, N.Y.
† Project Manager, Automotive Engine Developments, Environmental Protection Department, Texaco Inc., Beacon, N.Y.

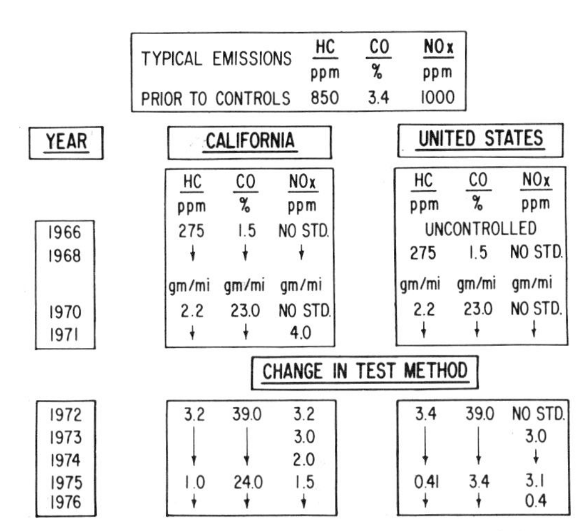

Figure 33.1. Exhaust emission standards for light duty vehicles.

compared to 1970 models and also a 90% reduction of oxides of nitrogen (NO_x) by 1976 as compared to the 1971 models. However, the specific values for the 1970 and 1971 bases were not stated. By 1970 steps had already been taken to reduce exhaust emissions. The selection of this time period as a base case had the net effect of requiring a 97% reduction in hydrocarbon, a 96% reduction in carbon monoxide, and a 93% reduction in nitrogen oxides as compared to the uncontrolled cars of the 1960s.

The most severe automotive air pollution problem occurs in Los Angeles and the few other areas in the United States which are subject to photochemical smog formation (for example, Denver). These limited area conditions have spawned the standards that will have to be supported by residents throughout the United States. Most other areas have no possibility of creating photochemical smog. Even experts dealing with the Los Angeles smog problem believe that the federal standards are more severe than are required for alleviation of the problems in California's South Coast Air Basin. Is it right that the reductions that must be borne in California because of its high traffic density and unique geography and meteorology have to be imposed on the rest of the nation?

Let us look at the impact of the emission reduction levels required by these regulations on our future energy demand, as illustrated in Figure 33.2 (Coppoc, 1972). The base curve here is the crude demand estimate pre-

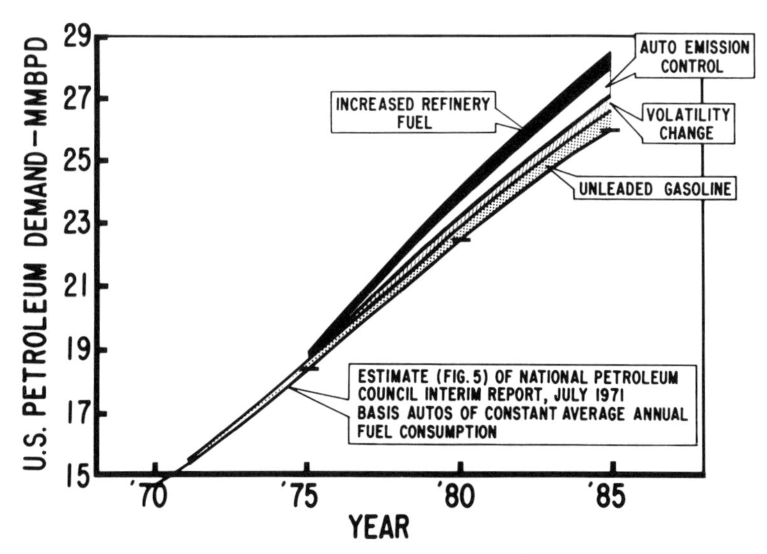

Figure 33.2. U.S. crude demand as affected by pollution controls.

sented in the National Petroleum Council Study Interim report of July 1971. It assumed that cars would have roughly the fuel economy of those in 1970. Car population growth was projected as a continuation of the recent past. This same base curve is believed valid for the case where the current small car fraction of total sales continues and there is no additional energy demand associated with meeting the emission standards of 1975 and beyond.

This base demand curve itself caused concern because refinery construction is not able to keep up with demand projections for many reasons, not the least of which is the resistance of communities to the siting of new refining capacity. Another concern is the inability of domestic crude production to meet our needs. The demand for 1985, for example, will require that 56% of our crude oil be imported. This alone will increase the U.S. trade deficit by 22.5 billion dollars per year as compared to 1970.

Let us now add to this demand the crude supply necessary to satisfy the 1975–1976 emission standards.

The drastic levels imposed on automotive HC and CO emissions in 1975 are such that thermal oxidation of these unwanted gases will not suffice. DuPont, Ethyl, and others have demonstrated the technical feasibility of these leaded-fuel compatible devices, but achieving the last small increment of HC and CO reduction to mass production target levels will undoubtedly require catalytic units. Since most of these catalysts are poisoned by the lead and lead salts, tetraethyl lead, one of our most effective additives for efficient gasoline utilization in an engine, must be eliminated. This results in lowering the available fuel octane number that requires reduced engine compression ratio. Extensive additional refining must be performed to alter the gasoline components to provide the required fuel octane value with little or no lead. This additional processing added to the economy loss from lower compression ratio engine design will require a 3.5% increase in crude runs.

The need to achieve low HC vehicle emissions will require fast engine and catalyst-bed warm-up. One suggested aid would be to change the volatility of the gasoline. Acceptance of one proposed volatility change would require an additional 2.6% of crude to provide an amount of the new gasoline equal to that of the base demand case.

In 1976 the imposition of the low (0.4 gm/mile) NO_x regulation will require the most severe addition to the fuel demand. Both exhaust gas recirculation (EGR) and catalytic reduction have been demonstrated as effective in experimental vehicles. The fuel economy penalty in each case has been shown to vary from about 10% to values as high as 30%. We have assumed a 10% loss to cover the needs of NO_x reduction and the requirements, such as air pumps, and so on, for HC and CO oxidation.

Returning now to the reduction of NO_x, Figure 33.3 (The Aerospace Corporation, 1971) shows that the penalty for excessive reduction of NO_x is severe.

Now return to Figure 33.2 to see the final "add on." The upper segment represents the fuel required to operate the air and water emission control equipment that must be added to refineries in order to meet the federal regulations.

What can be done?

Figure 33.4 (Coppoc, 1972) shows both the base case and the total addition due to emissions control from Figure 33.2. The major fuel savings shown as attributable to "novel engines," such as stratified charge units, is based on demonstrated performance. In this case we have used our experience with our Texaco Controlled-Combustion System (TCCS), a stratified combustion concept. Our engine's fuel economy and its ability to operate on a fuel having no octane or cetane requirement, leading to reduced re-

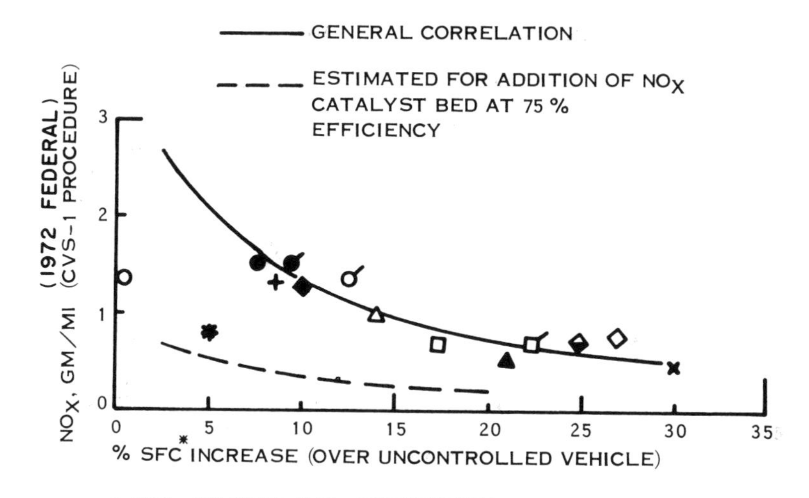

Figure 33.3. Relationship of NO_x reduction to fuel consumption.

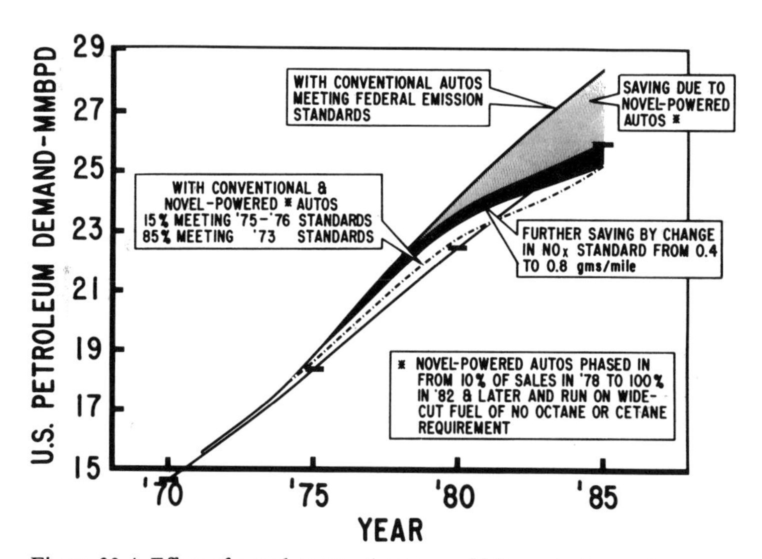

Figure 33.4. Effect of novel-powered autos on U.S. crude demand.

finery processing, has also been taken into account. As you can see, the assumed introduction of such an engine after 1978 results in a return to the base case crude demand in 1985.

Further benefits to the energy crisis are demonstrated under either of two assumptions relative to levels for NO_x emissions. First, the effect of relaxation from 0.4 to 0.8 gram per mile is shown. This is below that passed by the California Air Resources Board and would be more than needed for the remainder of the United States.

The broken line represents the second assumption that the 1976 limits be applied to smog areas and the 1973 level of 3.0 grams per mile be stipulated for all other areas. Car population distribution to these two areas was assumed as 15% and 85%, respectively.

We would be remiss were we to let the discussion rest at this point, for there are many pressures being brought to bear other than auto emissions that are forcing a shortage of crude oil and its products.

Figure 33.5 (adapted from Gaucher, 1971) shows the total energy demand of the United States broken into six source categories:

1. Miscellaneous (wood, batteries, geothermal, and others)
2. Coal and coal gas
3. Petroleum liquids and gas. These are derived from crude oil. A very small amount of liquid products from coal hydrogenation is included here. Oil shale and tar sands synthetics are also included.
4. Nuclear fission
5. Future potential includes improved sources such as fusion reactors, collection and conversion of solar energy, or new concepts.
6. Hydropower

Fuel sulfur regulations have greatly reduced the sources of coal that can be used. As a result, many plants are being converted to burn gas or fuel oil. The change in the central power station field alone is dramatic.

Nuclear power plants, largely because of community resistance to their siting as well as engineering and construction delays, are not coming on stream at anywhere near the estimated rate. The current deficiency amounts to 300 million bbl of crude oil per year (Gambs, 1972).

This means that the petroleum source of energy must expand. Although these supplies can be supplemented by the production of shale oil or tar sands, these sources are not without problems. If, for example, one oper-

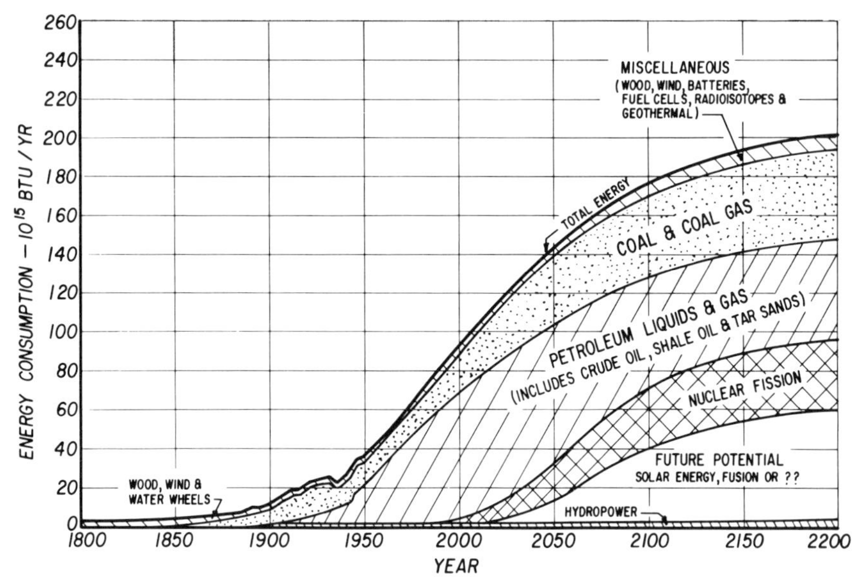

Figure 33.5. Energy consumption in the United States: past, present, and future.

ates a refinery of 100,000 bbl per day on a shale of good quality (25 gallons oil per ton of shale) the waste shale dust, normally a talcum-powder-like substance, would fill a city block to a depth of 1200 feet every month.

The extraction of oil from tar sands requires a tremendous amount of extraction-type processing to which must be added the equipment and energy for sand excavation and handling. The unit heating demands are high because the largest deposits of tar sands are in Northern Canada where the mean temperature is very low.

Recognizing the limitations imposed on the growth of the top four energy sources one can but wonder if technology is capable of bringing to function a growth of the yet unproved "future potential" sources such as solar and fusion power.

In closing then, let me express again our concern over the impact, not of automotive emission control per se but over the establishment of any unrealistically low standards that can only worsen an already critical energy supply picture.

References

The Aerospace Corporation (1971). "Final Report—An Assessment of the Effects of Lead Additives in Gasoline on Emission Control Systems which Might be Used to Meet the 1975–6 Motor Vehicle Emission Standards," Report No. TOR-0172 (2787)-2, November 15.

Coppoc, W. J. (1972). "Challenges in the Utilization of Fossil Fuels with Necessary Environmental Protection," AIChE New York City 65th Annual Meeting, November 26–30.

Gambs, Gerard C. (1972). "An Overall View," presentation at Energy Crisis Forum —Session I, ASME Annual Winter Meeting (93rd), New York, November 26–30.

Gaucher, L. P. (1971). "Energy in Perspective," *Chemical Technology,* Vol. 1, p. 153, March.

VI Conservation

34 The Fuel Shortage and Thermodynamics—The Entropy Crisis*

JOSEPH H. KEENAN,† ELIAS P. GYFTOPOULOS,†
AND GEORGE N. HATSOPOULOS‡

34.1. Introduction

The purpose of this paper is to bring to the attention of the broad spectrum of specialists currently concerned with the energy question in the United States the necessity for using a yardstick other than energy for assessing the effectiveness of fuel usage.

In establishing present patterns of energy consumption, the standard procedure is to find the total amount of energy used in each sector of the economy and, thus, to determine the needs of each sector for different fuels. The term energy in this connection is ambiguous. For example, the heat required in an industrial process may be added to the electrical work required or to the "heating value" of the fuel consumed in producing the electrical work required. None of these quantities represents energy consumed in the process, because it is known from the first law of thermodynamics that energy, rather than being consumed in any process, is always conserved. When opportunities for fuel conservation are to be assessed, it becomes necessary to use a measure other than energy.

For example, every engineer knows that a Btu (British thermal unit) of enthalpy in the circulating water of a power plant is less marketable and less valuable than a Btu of enthalpy in a steam main. He realizes also that a cold battery that is charged is more valuable and useful than a discharged battery having the same energy by virtue of being hot.

Typical conditions of process steam used in industry are 270°F and 30 psig. The heat required to change water from ambient conditions (55°F) into typical process steam conditions is 1150 Btu per lb of steam. Accordingly, since the typical heating value of hydrocarbon fuels is about 20,000 Btu per lb of fuel, it is often concluded that 0.057 lb of fuel per lb of process steam is needed. If this amount of fuel were used then, according to the customary definition, the effectiveness of fuel utilization would be 100%. By virtue of the first and second laws of thermodynamics, on the other

* Invited paper presented by the last author. The text is a revised version of the lecture.

† Massachusetts Institute of Technology, Cambridge, Mass.
‡ Thermo Electron Corporation, Waltham, Mass.

hand, it can be shown that the minimum amount of fuel required to accomplish the task just cited is only 0.015 lb of fuel per lb of steam and, therefore, only when this minimum amount is consumed is the process 100% effective. Conversely, if 0.057 lb of fuel per lb of steam is consumed, the effectiveness of fuel utilization is only $0.015/0.057 = 26\%$.

The preceding simple examples illustrate the necessity for using a yardstick other than energy for assessments of fuel needs and of effectiveness of fuel utilization. The laws of thermodynamics indicate that neither energy, nor heat, nor enthalpy, nor Gibbs free energy are in general satisfactory yardsticks. The relevant quantity is a property called available useful work that is in turn uniquely related to another important property called entropy. Hence, the subtitle is chosen to be "The Entropy Crisis" in suggestive distinction to the fashionable characterization of the problem as "The Energy Crisis."

This paper is concerned with the thermodynamic arguments that lead to available useful work or entropy as the objective measure of fuel needs and of effectiveness of fuel utilization.

34.2. Available Useful Work

It was observed by Lazar Carnot about the year 1800 that the perfect hydraulic engine would produce enough work to return all the working fluid to its elevated source if that work were used in a perfect pump. Another way of saying this is that the minimum requirement for lifting water from one level to a higher one is that an equal weight must be lowered from the higher level to the lower one. Moreover, more than this minimum is required if any part of the mechanism involved is less than perfect.

Such observations are generalized in the laws of thermodynamics from which the following theorem may be proved: for any physical task that is to be performed within an environment that is essentially in a stable equilibrium state, a certain minimum of work is required. The task to be performed may be as simple as the raising of a mass of material from one level to another in the gravity field, or as complex as the conversion of iron ore into steel. The minimum work will be required when the task is performed reversibly, that is, in such a way that all systems involved in the process can be restored to their initial states after the process has occurred, leaving no physical evidence that any process had taken place.

It will be observed from the preceding theorem that the work of the reversible process is fixed by the nature of the system and the initial and final states of that system in the process. Even though the number of such possible reversible processes between the same end states may be infinitely large, the work required is identically the same for all of them.

Associated with each system in each state will be, then, an amount of work which is the minimum required to create that system in that state out of materials from the atmosphere or in mutual stable equilibrium with the atmosphere. This minimum amount is also equal to the maximum useful work that can be done by the system starting from a given state and ending in a state in mutual stable equilibrium with the atmosphere. It is called the available useful work of the system in that state.[1]

Values of available useful work are, of course, associated with the fuels we use. If a fuel such as the molecular species CH_2 were to be formed reversibly from carbon dioxide and water in the atmosphere, the work required would be the available useful work of the fuel. This work could be recovered completely for use on other systems if the fuel were combined with oxygen from the air in a reversible process that restores the carbon dioxide and water as constituents of the gaseous mixture we call air. Clearly, each task that is to be performed on any system within the atmosphere may be carried out by means of the oxidation of fuel without consuming any resource other than the infinite atmosphere, because only work is required and the oxidation of fuel is entirely equivalent in the reversible ideal process to a supply of work.

Indeed in the great majority of all physical tasks undertaken by modern American society, whether they be the lifting of a steel beam to the top of a building or the heating of a house, the process drawn upon is the oxidation of fuel. The proliferation of these tasks in an affluent society has increased the magnitude of the basic process, namely oxidation of fuel, to the point where such questions as where is the fuel to come from? how do we get it? and how do we pay for it? have resulted in what is called the energy crisis but which would be more appropriately called the fuel crisis or fuel shortage.

It is appropriate, therefore, to give thought to the minimum magnitude of the oxidation process that will permit the desired tasks to be performed —that is, the minimum fuel requirement. This minimum would be attained if the oxidation process and all subsequent operations were to be executed reversibly within the terrestrial (air and water) environment. It can be shown that the annual minimum would consume a tiny fraction of the fuel consumed in the year 1972. That tiny fraction would provide all the lifting jobs, all the metal-forming jobs, and all the industrial and domestic heating to which society is now accustomed or is likely to become accustomed in the foreseeable future.

Why then do we consume the massive quantities of fuel that we do? The answer is in part economic; namely, that we use the amount of fuel that minimizes the sum of capital and operating costs. It is also in part technological; namely, that we have not organized our technological skills so as to reduce substantially the fuel requirement.

The economic optimization usually performed by either consumer or industry is faulty. It would be valid only if the prices of capital goods and fuel reflected their total social costs. That is, the price of coal would have to include the cost of rehabilitating the landscape after strip mining, and the price of imported oil would have to reflect in some way the effect of the import on the national balance of payments. On the other hand, the price of iron and steel for power plants and other machinery would have to reflect the cost of cleaning up atmospheric and water pollution caused by production of iron and steel.

The technological argument is not, of course, entirely independent of the economic one. With proper price allocation the demand for competent engineers and engineering to reduce fuel consumption would rise considerably above present levels. As prices and social pressures for fuel conservation increase, however, engineering efforts will be more and more directed toward devising means for realizing in some degree the vast potential for reducing consumption of fuel. It is to this task that the present paper is directed.

The thermodynamic measure of the task-performing value of fuel and air or their products of oxidation is, as stated earlier, the maximum possible work, the available useful work, that could be obtained from them. The amount of this work is dependent on the temperature and other properties of the environment (ambient air and water) *to* which or *from* which the fuel-air system may *deliver* or *receive* energy in any quantity.

In a reversible process the available useful work is conserved. That is, when fuel is oxidized in order to perform a specific task on a specific material, the available useful work is merely transferred from the fuel-air system to the material if the process is executed reversibly. It is informative, therefore, to examine the performance of any particular task so as to trace the losses in available useful work that occur in each step of that task.

The available useful work Φ of a system and atmosphere can be shown to be given by the equation

$$\Phi = E + p_0V - T_0S - \sum_i^n \mu_{i0}N_i, \tag{34.1}$$

where E denotes energy, V volume, and S entropy of the system, N_i for $i = 1, 2, \ldots n$ the number of moles of molecular component i in the system, p_0 the pressure of the atmosphere, T_0 the temperature of the atmosphere, and μ_{i0} the total potential of component i in the atmosphere or in mutual stable equilibrium with it.

The quantity Φ may be evaluated for any system in any state whether stable equilibrium, nonstable equilibrium, or nonequilibrium. In particular its value is zero when system and atmosphere are in mutual stable equilibrium. That is, Φ is zero when the system is in a stable equilibrium state such that its temperature is T_0, its pressure p_0 and its total potentials are μ_{i0} for $i = 1, 2, \ldots n$.

The close relationship between the quantity Φ and the entropy is evident from Equation 34.1. For given energy, volume, and composition of the system, Φ decreases with increase in entropy of the system. For states of small entropy the available useful work is large, and vice versa. A deficiency in available useful work corresponds to a surplus of entropy. The value of Φ may exceed that of the energy of the system. It cannot, of course, exceed the energy of the system and atmosphere taken togther.

In general, the available useful work Φ is different from the Gibbs free energy Z and enthalpy H. Unlike Φ, which has a value for any state of the system, equilibrium or nonequilibrium, Z and H are defined ($\equiv$) for stable equilibrium states only, as follows:

$$Z \equiv E + pV - TS \quad \text{and} \quad H \equiv E + pV, \tag{34.2}$$

where p and T denote, respectively, pressure and temperature which are not in general identifiable for states that are not stable equilibrium states.

The available useful work Φ was devised by Gibbs (1948, p. 77, Equation 54) in 1875. In an earlier paper Gibbs (1948, p. 58) had introduced the function

$$E + p_0V - T_0S \tag{34.3}$$

for application to a system the constituent substance of which did not enter into or mix with the atmosphere. The maximum possible decrease in the value of this function as the system (body) proceeded toward pressure and temperature equilibrium with the atmosphere (medium) he called the "available energy of the body and medium." Gouy (1889) introduced the abbreviated form of Gibbs function

$$E - T_0S, \tag{34.4}$$

the change in which for any change of state is the available work Ω used to introduce the property entropy.

A closely related property

$$H - T_0S, \tag{34.5}$$

which is particularly useful for calculating the available work in steady-flow processes, was used by Darrieus (1930) for processes in turbines and by Keenan (1932) for analysis of steam power plants and for cost accounting when both process steam and power are produced. It appeared in an engineering textbook by Bosnjakovic (1935) and more recently in European literature (Brennstoff-Warme-Kraft, 1961),[2] where it has been given the name exergy.

When applied to a hydrocarbon fuel, the quantity Φ is the minimum useful work required to form the fuel in a given state from the water and carbon dioxide in the atmosphere. Since this minimum will be the useful work of a reversible process, the quantity Φ is also the maximum useful work that could be obtained by oxidation of the fuel and return of the products to the atmosphere. Moreover, any change in state of the fuel-air-atmosphere system will produce an amount of useful work less than, or in a reversible change, equal to the corresponding decrease in Φ.

34.3. Examples

The curves of Figure 34.1 are calculated for 1 pound-mole of a liquid hydrocarbon fuel which may be described as CH_2 and which has a so-called heating value of 280,000 Btu (20,000 Btu per pound). This heating value is the decrease in enthalpy when 1 pound-mole of fuel burns in theoretical air to carbon dioxide, water, and nitrogen:

$$CH_2 + \frac{3}{2} O_2 + 5.65N_2 \rightarrow CO_2 + H_2O + 5.65N_2. \tag{34.6}$$

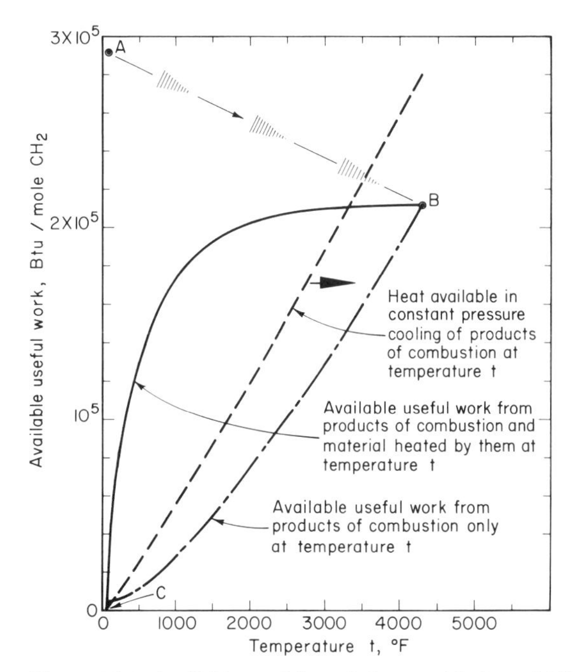

Figure 34.1. Available useful work from oxidation of CH_2 (Equation 34.1).

The available useful work for the reactants of this reaction can be shown to be about 292,000 Btu per pound-mole, namely about 4% greater than the heating value. In order to obtain this work, the following process, as one among many possible ones, could be used: (1) the oxidation is carried out in a reversible fuel cell, at some temperature T_c, which delivers electrical work to the surroundings; (2) the products of the fuel-cell process are cooled from T_c to the temperature of the environment T_0 as they provide heat to Carnot engines that produce further work; (3) each of the products CO_2, H_2O, and N_2 is separated from the mixture reversibly by means of a semipermeable membrane and expanded reversibly and isothermally at T_0 in an engine cylinder until it attains a pressure equal to the partial pressure of that constituent in the atmosphere; (4) each molecular species is introduced reversibly into the atmosphere through a semipermeable membrane. The term semipermeable membrane refers to a device that is impermeable to all molecular species except one. The final state after step (4) corresponds to zero available useful work at atmospheric temperature (55°F). It is the base state C in Figure 34.1.

Such a fuel-air process would produce the maximum possible work, the available useful work. Moreover, the work so obtained could be used in an inverse process to create the original quantity of fuel from the carbon dioxide and water vapor in the air. Any oxidation process that produced less work than the available useful work would be irreversible, and the loss in available useful work would be a measure of the irreversibility of the process.

Work is produced for use in one or more tasks. Because these tasks are almost always performed with a high degree of irreversibility, the available useful work remaining in the material operated on is usually only a small fraction, and more often a tiny fraction, of that available at the beginning of the task. When a steel beam is lifted to the top of a tall building, the available useful work remaining at the end of the task may be of the order of 10% of that in the fuel-air system that was the source of the work. In the manufacture of the steel beam, on the other hand, the available useful work remaining in the beam is a fraction several orders of magnitude less than in the lifting process. The search for potential savings of fuel, therefore, must involve, first, the oxidation process and, second, the task that uses the work produced by the process.

Because fuel cells for efficient oxidation of a hydrocarbon fuel are not presently obtainable, although they are in various stages of development, fuels are almost always burned in a combustion chamber without production of electrical current. For CH_2 and the reaction cited earlier, the temperature at the end of the combustion process is about 4300°F, the adiabatic flame temperature for the stoichiometric mixture. Being irreversible, the combustion process, which is suggested in Figure 34.1 by the broken line AB, results in an increase in entropy and a loss of available useful work. Here the loss is 80,000 Btu per pound-mole of our assumed fuel or about 27% of the original value of 292,000 Btu. The remaining available useful work is 212,000 Btu, which is the maximum amount of work that could be obtained, for example, by transferring heat to Carnot engines and by expansion of the product species to the limit imposed by the environment at C.

The combustion process A to B in Figure 34.1 is a constant-enthalpy process. That is, the capacity for solely transferring heat in steady flow to surrounding systems remains unaltered by the combustion process. By virtue of the irreversibility and the associated increase in entropy in the adiabatic combustion process available, useful work has been lost at constant enthalpy. An analogous process is the flow of a perfect gas through a throttle valve from high to low pressure. Enthalpy remains constant while entropy increases. In the throttling process available useful work is lost while temperature remains constant, whereas in the combustion process available useful work is lost while the temperature rises. Both are adiabatic processes in which entropy increases because of irreversibility.

Beginning with state B, the available useful work can be altered in a number of ways. For example, the energy of the combustion products may be transferred in the form of heat to any material at a temperature t less than 4300°F, and the transferred energy may be used reversibly to produce work. Because the temperature difference between the combustion products and the material is finite, the transfer process is irreversible and, therefore, the available useful work decreases. The solid curve in Figure 34.1 shows the available useful work in the products plus that in a material at temperature t°F which has cooled the products to t°F without itself changing temperature. For this purpose the material is assumed to be of infinite heat capacity because otherwise some heat would be transferred to material at temperatures less than t°F and the loss in Φ would be correspondingly greater. That is, the solid curve shows the maximum available work useful after cooling the products to t°F.

The dash curve of Figure 34.1 shows the heat available from products of combustion in simple steady-flow cooling from temperature t to the temperature of the atmosphere (55°F). It is indeed the variation of enthalpy with temperature. The dot-and-dash curve shows the corresponding available useful work of the products of combustion at temperature t. It reaches atmospheric temperature at a small positive value of the ordinate corresponding to the work obtainable upon reversibly mixing the products with the constituent gases of the atmosphere. The difference in ordinate between solid curve and dot-and-dash curve at any value of t is the available useful work from infinite-heat-capacity material that has cooled the products from 4300°F to t°F.

It is evident from the solid curve that as the temperature of the heat-receiving material is lowered below 2000°F the loss in Φ increases rapidly with decrease in temperature. At a temperature of about 600°F, the value of Φ is about 48% of that for the fuel initially or 140,000 Btu per mole of fuel.

A typical average temperature of the heat-receiving water-steam working fluid in a central steam power plant is 600°F. Accordingly about 25% of the available useful work of the fuel is lost in the irreversible process of transferring energy from the products of combustion to the working fluid across a finite temperature difference. Magnetogasdynamic and thermionic devices have been proposed to bridge these temperature differences and to salvage losses in Φ by reducing irreversibilities.

In an industrial plant, on the other hand, the products of combustion might be used to make process steam at, say, 270°F. The loss of available useful work is then 292,000 minus 90,000 or about 69% of that in the fuel. The difference between 69% and 52% (that is, 100% − 48%) represents the fraction lost because a steam power plant was not interposed between the products of combustion and the process steam. It can be shown by a simple calculation that for the same amount of process steam an interposed steam power plant may produce 60,000 Btu of electrical work for an additional 0.23 mole of fuel consumed. Thus, 60,000 Btu of electrical work would be obtained at the expense of 0.23 × 292,000 or 67,000 Btu of available useful work. A central station power plant would consume about 0.55 mole of CH_2 (as compared with 0.23) and 161,000 Btu of available useful work (as compared with 67,000) to produce the same amount of electrical work.

The preceding paragraphs discuss an example of topping a heating process with a power-producing process in order to reduce the loss of available useful work. Many industrial heating processes, on the other hand, require such high temperatures that a topping process would require either unobtainable or prohibitively expensive materials. Moreover, the saving to be realized per thousand degrees of temperature interval, as shown by the upper portion of the solid curve of Figure 34.1, is small.

In these high-temperature processes, such as the manufacture of steel or cement, emphasis should be placed on salvaging available useful work from the material in process and from the products of combustion leaving the heating process. For example, the billets leaving a heating furnace contain 15 to 50% as much potential to supply heat to other processes as the fuel originally consumed. A similar range of potential may be found in the products of combustion leaving the heating process. The opportunities for reclaiming available useful work through heat transfer to power-producing or lower-temperature heating processes are evident.

The possibilities of arranging processes in series in order to reduce consumption of fuel and available useful work in a simple industrial process may be illustrated by the distillation of seawater to make freshwater. In the best modern plants about 13 pounds of freshwater are made for the amount of fuel that would make 1 pound in a simple distillation. The effectiveness of the process in terms of available useful work produced to that consumed is still very small, but it is thirteen times as great, and the fuel consumption is one-thirteenth as great as in the simple distillation.

It has been estimated that about a quarter of the annual fuel consumption in the United States is attributable to generation of electrical work in central station power plants and another quarter to industrial process heating. In terms of loss of available useful work the central station power plant is far less wasteful than the industrial plant. The product of the power plant, which is pure available useful work, accounts for about one-third of the available useful work consumed, whereas the product of the industrial process accounts for small fractions of 1% of the available useful work consumed. The opportunities for reducing fuel consumption appear, therefore, to be far greater in industrial processing than in power production.

Engineers have been aware of these opportunities during the past century. It is a curious fact that topping of industrial heating with power production was more common 40 years ago than it is today. The reasons for this change are of such vital interest in view of present and future shortage of fuel that they should be thoroughly investigated and understood. They include, no doubt, the improved efficiency of central station power production and the relatively high cost of the licensed personnel required for the operation of a topping power plant. With more realistic pricing and controlled distribution of scarce fuel supplies, the economical practice may prove in the near future to be more nearly that of 40 years ago than that of the recent past.

34.4. Summary

A measure of fuel requirement other than the ambiguous energy or the less ambiguous enthalpy or heating value is needed. The most appropriate one is the function Φ, which was devised by Gibbs. It is the maximum useful work that can be obtained from a fuel by oxidizing it and diffusing the products of combustion into the atmosphere. This maximum will be realized in any reversible process between the initial and final conditions. Similarly, any task to be performed, such as the lifting of a steel beam or the manufacture of steel has a minimum requirement of available useful work which, when compared with the maximum obtainable from the fuel, determines the minimum amount of fuel required. Although this figure taken from current practice usually proves to be discouragingly small, it indicates the great potential in present processes for saving fuel.

Among these processes, that of the central station power plant uses fuel most effectively. Topping of process-steam units with power plants will probably return to fashion, and the corresponding bottoming of high-temperature industrial processes may come into fashion as the price of fuel comes to reflect more accurately the true economic and social cost of producing it. These are but two examples of what must become a reinvigorated engineering attack on irreversibility.

Notes

1. The adjective useful is included here because some work may be done on or by the atmosphere as the system changes volume. This part of the work is excluded from the quantity called here the available useful work. It is not excluded from another availability function which is used in the derivation of the property entropy.

2. This *Fachheft* includes, pp. 506–509, a bibliography which, although it excludes the work of Gibbs, is an excellent list for the years 1889 to 1961.

References

Bosnjakovic, Franjo (1935). *Technische Thermodynamik,* Vol. 1, p. 138, and Vol. 2, p. 2, Steinkopff, Dresden, Leipzig.

Brennstoff-Warme-Kraft (1961). *Fachheft Exergie,* Vol. 13.

Darrieus, G. (1930). *Revue Générale de l'Electricité,* Vol. 27, pp. 963–968, June 21; also *Engineering,* pp. 283–285, September 5.

Gibbs, J. W. (1948). *The Collected Works of J. Willard Gibbs,* Vol. 1, Yale University Press, New Haven, Conn., p. 77, equation (54).

Gouy, M. (1889). *Journal de Physique,* 2nd Series, Vol. VIII, pp. 501–518.

Keenan, J. H. (1932). *Mechanical Engineering,* Vol. 54, pp. 195–204, March.

35 Conservation via Effective Use of Energy at the Point of Consumption

CHARLES A. BERG*

35.1. Introduction

The basic energy problem of the United States is that increased consumption of high-quality nonpolluting fuels is straining the *national* capacity to provide these fuels. With the passing of the era of inexpensive clean forms of primary energy, rising costs will compel everyone to examine the effectiveness with which energy is used. The conservation of energy may be essential to assuring the quality of life, the economic well-being, and even the national security of the United States.

35.2. The Issue of Conservation and Effective Utilization

There are two basic ways to conserve energy: curtailment of fuels and electrical power and improvement of the efficiency of utilization at the point of consumption. Until recently it has not been widely recognized that improvement in the efficiency of energy utilization at the point of consumption could yield significant reductions in the national requirements for high-quality fuels without requiring sacrifices in the comfort, safety, or health of building occupants or in the productivity of industrial processes. Ultimately, effective measures for conservation will probably require both curtailment and improvement of efficiency of energy consumption. But if the potential for energy conservation through efficient utilization at the point of consumption is recognized and developed, programs of conservation can be instituted with far less stringent effect on industry and society than if conservation were to be attempted through curtailment alone.

Any approach toward economic optimization of the national energy system will require careful examination of the effectiveness with which energy is used at the point of consumption; it does not make good sense, neither in economic terms nor in any other, to continue excess consumption of significant quantities of a vital and increasingly costly natural resource, through ineffective practices.

35.3. Areas of Excess Energy Consumption: A Framework for Discussion

To identify areas in which significant energy savings are possible, one must know how much energy is consumed in various practices and how effectively

* Deputy Director, Institute for Applied Technology, National Bureau of Standards, Washington, D.C. 20234.

the energy is consumed. The second question, with what effectiveness is energy consumed, involves technical issues to be discussed later. Basic data have been compiled by Stanford Research Institute (1972). The summary data of the report, given in Table I, provide an itemized breakdown of the amount of energy consumed in various practices throughout the United States. These data represent the most recent and one of the more reliable attempts to account for the use of energy *at the point of consumption* and reveal several significant aspects of energy consumption. Of the three main areas of energy consumption, transportation has been the subject of public scrutiny by government agencies, public interest groups, industry, and academic institutions; it will not be treated further here.

However, the potential for conservation through improved practices of energy consumption in buildings and in industrial processes has received small notice until recently. In both areas ineffective practices allow large quantities of energy to escape utilization. Study of data on energy consumption (Stanford Research Institute, 1972) reveals significant comparisons between areas of energy consumption. For example, industrial electrolytic processing, which includes aluminum refining, consumes approximately 1% of the primary fuels used in the United States. It is widely known that aluminum refining and other industrial electrolytic processes require large amounts of energy. However, Stanford Research Institute (1972) also reveals that hot water heating in residences and commercial buildings accounts for approximately 4% of the consumption of primary fuels in the United States. It would appear that efforts to conserve primary fuels might be as fruitfully applied to hot water heating as to electrolytic processing, even though much greater public attention has been focused on the latter. This observation really illustrates a basic point that is reflected numerous times by the data of Stanford Research Institute (1972). Many of the seemingly mundane practices of everyday life, such as hot water heating, space heating, and cooking consume vast quantities of energy. The effectiveness of practices in these areas is, as a rule, rather low and also rather easily correctable. Improvements in these practices offer excellent opportunities for conservation of primary fuels.

To broach the question of effectiveness of energy consumption, one must define both "consumption" and "effectiveness." Consumption of energy means the ultimate use of energy to operate a process or to provide a service.

Effectiveness, on the other hand, implies some measure of the minimum energy required to operate the process. In some instances, it is possible to determine a physically irreducible minimum energy requirement for a process, as in the case of drying fabrics, thermal conversion of limestone to portland cement, or refrigeration of food. In other instances, it is not clear whether physically irreducible minimum energy requirements can be identified.[1] And while it is useful to know this, it is often more useful to know the amount of energy that might be required to operate a process if full application of available technology were made, up to an economically justifiable limit. It is this amount of energy that will be used here to establish a measure of the effectiveness of a thermal process. The criterion for economic justification proposed here is minimum combined initial cost and operating cost of equipment, including, in particular, fuel cost.[2]

35.4. The Technological Potential for Improved Energy Utilization at the Point of Consumption

A few examples of opportunities for improving effectiveness of energy consumption in buildings and industrial processes are offered here.[3]

35.4.1. Energy Use in Buildings

The effectiveness of energy use in buildings is determined by three items:

1. *Design* (including insulation, fenestration, selection of heating and ventilating equipment, and so on),
2. *Construction Practices in Implementing Design,*
3. *Occupant Practices in Using Buildings.*

Moyers (1971, p. 28) has studied the economic aspects of insulation and fenestration of residences. He determined the savings from increased wall insulation and calculated the life cycle costs of insulation and storm windows by combining initial costs (with suitable interest and taxes applied) with 1970 fuel costs. Figure 35.1 shows some of Moyers' data, which indicate that by using 3½ inches of wall insulation, 6 inches of ceiling insulation, and applying storm windows one can cut the heat losses through the walls of a typical residence in New York or Minneapolis by somewhat more than 40%, as compared with the heat losses that would obtain with 1⅞-inch ceiling insulation, and no wall insulation or storm windows (points A on the figure).

The residences built prior to 1970 were designed for a level of thermal performance typical of the point A. With the issuance of the 1971 FHA

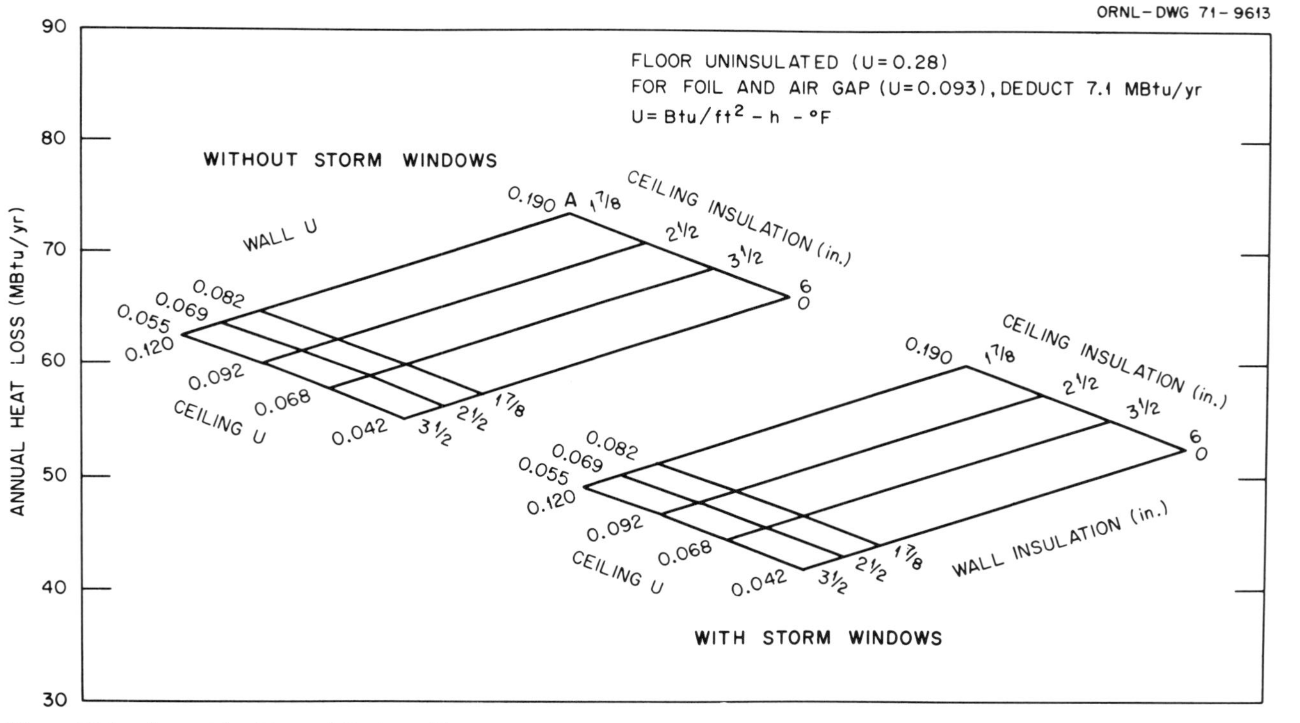

Figure 35.1a. Annual heat loss: Atlanta residence.

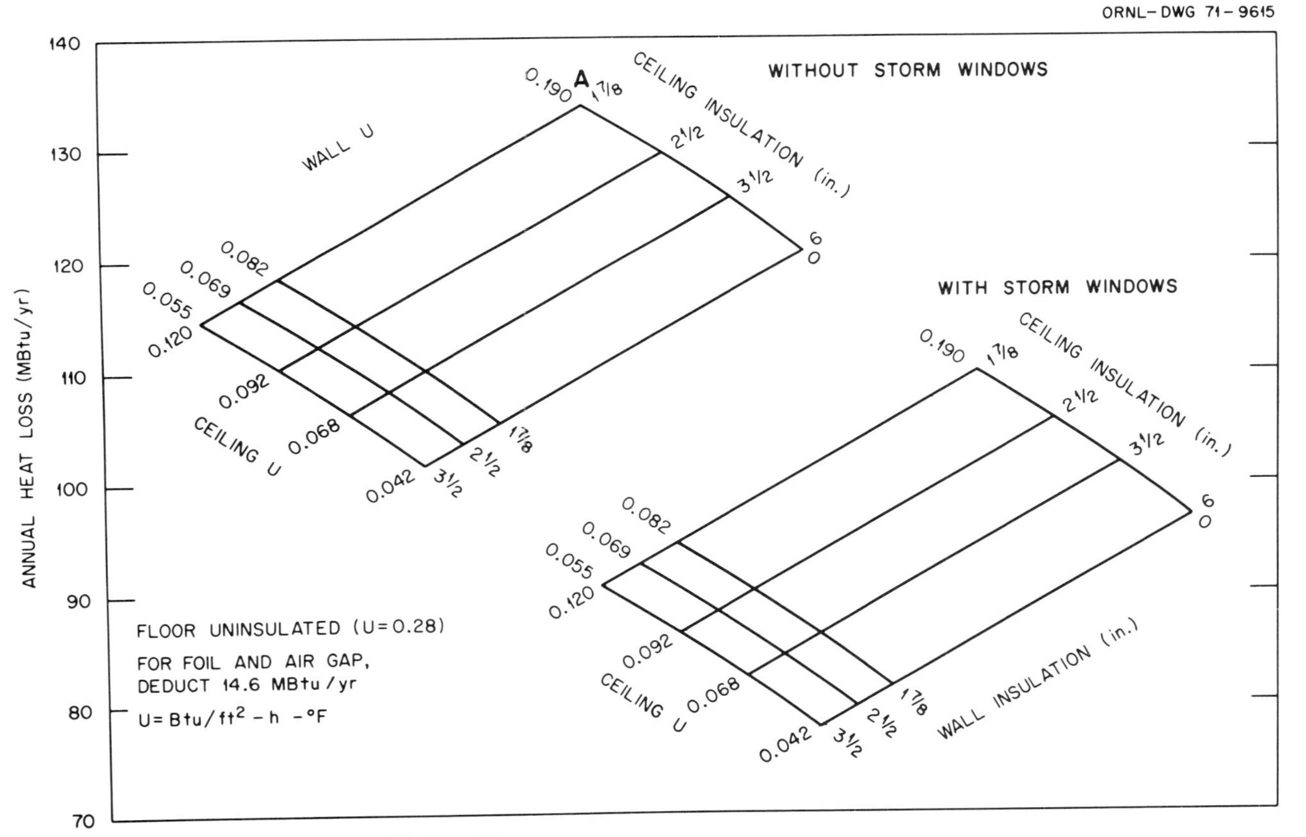

Figure 35.1b. Annual heat loss: New York residence.

Minimum Property Standards (no storm window requirements), the insulation levels of residences affected by these standards was substantially improved. It should be noted that federal standards have a direct effect on approximately 35% of new construction; the indirect effect of federal standards on residential construction is much larger than 35%. Thus, for the most of the extant residences in the colder climates of the United States and for a great number of those yet to be built, application of storm windows and insulation could reduce direct heat loss through walls by approximately 40%.

Figure 35.2 shows Moyers' calculation of net savings to the consumer, realizable through insulation and storm windows. There are several observations on these data which bear mention. First, not only is it economically attractive to install insulation and storm windows, but the economically optimal level of insulation for residences in cold climates is found to be the extreme of the range of insulation which Moyers considered (3½ inches of wall insulation, 6 inches of ceiling insulation plus storm windows). Second, the data show that not only the net savings at the economic optimal condition but the character of the optimal point is extremely sensitive to the price of energy; in Figure 35.2 the only difference between gas and electric heat is price. As the price of energy goes up, small departures from the optimal condition, as represented in Figure 35.2, can cause large decreases in net annual savings. In fact, the true economic optimal design for either electric heating or heating with higher priced gas, may actually be found at some greater levels of insulation and control of fenestration than the extreme of the range shown in the figure. Of course, to install more than 3½ inches of wall insulation would require either modifying the wall cavity or devising some insulation system which could be mounted on the wall rather than in it. Either of these steps might be costly; but, with rising energy prices being a certainty in the future, the character the data in Figure 35.2 suggests that it might not be too soon to start considering these steps.

During the past few decades energy prices have increased at a slower rate than prices of construction and other prices; relative to other items energy has been an increasingly good bargain. Thus, if it is now economically justifiable to install insulation and storm windows, it was even more easily justified in the past. The fact that insulation was not widely used in the past illustrates that the rational economic criteria of life cycle cost have not, as

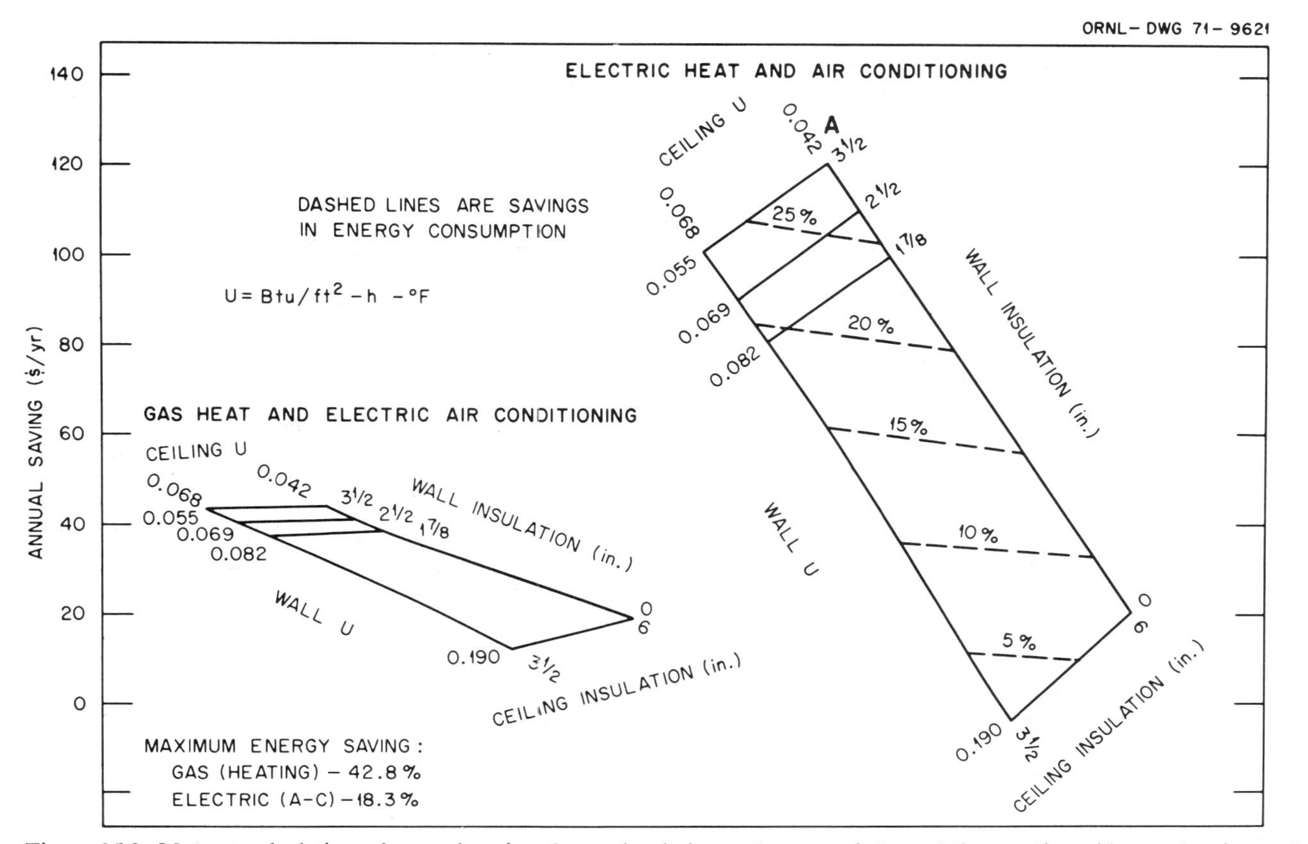

Figure 35.2. Moyers' calculation of annual saving due to insulation and storm windows, Minneapolis residence—heating and air conditioning.

a rule, been effectively applied in housing purchases. We will return to this point later.

The selection of building equipment is another important aspect of thermal design of buildings; Table 35.1 gives performance data for air conditioners of various rated capacity and price (Dubin, 1973). The power consumption rate is the ratio of cooling delivered to electrical energy consumed by the device. Air conditioners of 4000 Btu/hour capacity have electrical energy requirements that decrease by approximately a factor of 2 as the price increases by 35%. On a life cycle cost basis, the least expensive unit to own is the one with the greatest initial price. The same general pattern, with only minor exceptions, can be found in the units of higher rated capacity. These data prompt several observations. First, as with insulated housing, the units with the lower initial price tend to sell very well even though they may be, in the long run, more expensive to own; rational economic criteria are not commonly applied to building equipment. Second, since the market is and has been much more sensitive to initial price than to life cycle costs, the equipment manufacturer is prone to design in high consumption in order to reduce initial price. One interpretation of Table 35.1 is that by approximately doubling the energy consumption of a 4000 Btu/hour air conditioner one can reduce its price by approximately 35%. Thus, a great deal of the building equipment in service today, which was bought because of low initial price, makes rather ineffective use of energy.

Construction practices in implementing design have a very important effect on energy use in buildings. Thermographic studies, for example, indicate that even small construction flaws at the edge of studs are sufficient to reduce the effective insulation of a wall significantly below its design level. As energy becomes more costly, it will probably become worthwhile to make the extra investment required to attain more careful practices in construction or to find designs for wall insulation, and so forth, which are not so sensitive to minor construction flaws.

Occupant practices in buildings are very important in attaining effective use of energy (Sampson, 1972). Also, careful management of large refrigeration plants has been shown to conserve as much as 12% of the electrical energy required by the plant, without sacrifice of plant performance (Rathbone, 1972).

The maintenance of home heating equipment is also an item of great importance. The small-scale combustion equipment used for home furnaces

Table 35.1 Performance Data for Air-Conditioning Units

Rated Cooling Capacity Btu	Rated Current Amps	Retail Value $	Power Consumption Rate Btu/watt	10-Year Total* Cost $/1000 Btu
4,000	8.8	100	3.96	84
	7.5	110	4.65	77.70
	7.5	125	4.65	81.45
	5.0	135	6.96	67.25
5,000	9.5	120	4.58	74.90
	7.5	140	5.80	68.20
	7.5	150	5.80	70.20
	5.0	165	8.70	59.80
6,000	9.1	160	5.34	67.30
	9.1	170	5.24	68.90
	7.5	170	6.96	61.80
	7.5	180	6.96	63.50
8,000	12	200	5.80	67.30
	12	220	5.80	67.80
24,000 (Central)	13.1	—	8.25	—
	15.4	—	7.10	—
	17.0	—	5.85	—

* Based on 886 operating hours per year.

and hot water heaters is designed so that approximately 70 to 75% of the heat of combustion will be transferred to the hot water or air stream (as the case may be) when the equipment is clean, in proper adjustment, and operated in steady state. However, transient operation can produce soot formation that can greatly reduce the effectiveness of this equipment. Sample field observations and theoretical estimates of the effects of minor unattended items of maintenance indicate that the actual effectiveness of these small combustion units in field service may be in the range from 50 to 35% (Hottel and Howard, 1971).

35.4.2. Energy Use in Industry

The use of energy at points of consumption in industry has been much less well studied than energy use in buildings.[4] Until quite recently it has been assumed that industry must make the most effective possible use of energy, because to do otherwise would not be profitable. This assumption has now been reexamined, and found to be inaccurate by industry itself. For example, merely by plugging leaks in air and steam lines, by providing steam at the pressures and temperatures required, and by instituting other straightforward energy management practices, consultants have been able to reduce the fuel requirements of large industrial plants by 7 to 15% (*Business Week*, 1972). Present efforts for energy conservation through application of waste heat management have hardly scratched the surface of this field. In addition, redesign of process equipment offers a great potential for energy savings. Experts in metal processing have estimated that if full application of presently known economically justifiable technology were to be made in furnace design, heat soaking pit design, and thermal management of processes, the overall fuel requirements of steelmaking could be reduced as much as 20% (Nesbitt, 1972). Recent redesign of vacuum furnaces, entailing improved vacuum insulation and the use of direct combustion with a heat pipe to provide heat, in place of electrical heating, has reduced the fuel requirements of vacuum furnace operations by 75% (Shefsiek and Lazaridis, 1972).

Newly emerging developments in thermal process design are also noteworthy. For example, papermaking accounts for approximately 2% of the total primary fuel consumption in the United States (Stanford Research Institute, 1972). The largest part of this energy is consumed in the paper-forming process, which is essentially a drying operation. Industrialists in Europe (Waihren, Norman, and Grundstorm, in *Paper Trade Journal*) and in the United States (Kalmas, private communication) are developing techniques for "high consistency" paper forming. Using these techniques, one begins the papermaking process with a thicker slurry; there is less water to dry out of the paper and thus less energy is required. For example, for certain types of paper it is possible to change from a slurry having 0.5% solids at the outset to one having 3%; thus only one-sixth as much water need be removed from a given unit of paper. Engineers who have studied these processes have estimated that as much as 55% of the energy requirements of paper forming could be conserved through implementation of high-consistency forming technology.

Cement production also accounts for approximately 2% of the primary fuel consumption in the United States. The largest part of this energy is used as direct heat in the cement kiln. Kilns now being introduced on the Western European market are capable of cutting the fuel requirements by approximately 30%. More advanced prototype kilns are under development in the United States. These employ fluidized bed technology to achieve rapid heat transfer and intimate mixing. They promise even greater reductions in fuel requirements (Lazaridis, private communication).

The data cited above support the belief that the practices and equipment used at the points of energy consumption in industry can be substantially improved, and that in many cases the improvements can be economically justified. In the case of industry, economic justification of improvements may be even easier to show than for buildings because reduced fuel consumption often carries with it reduced costs of pollution control. While data are too sparse to support statistical estimates of potential energy savings in industry at high confidence levels, an economically justifiable 30% reduction in primary fuel requirements for energy-consuming processes of industry would seem to be a realistic goal to set for application of effective technology to these processes.

35.5. Information

All who take part in the marketing of energy and in consuming energy need to be apprised of the technical options for more effective use of energy and the economic implications thereof. Marketplace mechanisms do not work effectively unless the participants in the market are informed. In particular, there is need for data on incremental cost benefit ratios of various options for improved effectiveness in utilization of energy; the relative merits of insulation as opposed to control of infiltration in buildings need to be determined. Also, realistic formulas by which trade-offs between initial investment and operating costs might be included in criteria for financing construction or new industrial plant equipment are required; these should take into account estimates of energy price increases.

35.6. Effective Technology

At the base of all efforts to improve use of energy at the point of consumption is application of effective technology. Some of the examples cited earlier have shown that technology can be devised for more effective use of energy. In addition, in those instances where life term costs of energy utilization at the point of consumption have been determined, the minimum life term cost has been found to be attainable by investing initially in equipment that is at or very near the top of the price- and quality-range available on today's market. This suggests that the technical potential for designing effective performance into the equipment used at the point of energy consumption has not been fully exploited. Had it been fully exploited, there should be some very high performance equipment on the market which would be suitable for very long-life operations (say 40 years) but which would be economically unattractive for shorter life applications (say 10 years). But this does not appear to be the case. As a specific example, one may consider electrically driven heat pumps. Comparing the measured performance of heat pumps with the limiting Carnot efficiency of the device, one finds that existing heat pumps, operating under mild climatic conditions where they are most effective, attain only approximately 5% of their limiting efficiency. Under more severe climatic conditions, in which their actual measured performance is rather poor, existing heat pumps attain approximately 20% of their limiting efficiency. Power generation devices, on the other hand, attain 70% and more of their limiting efficiency. The basic technology used to approach the limiting efficiency in a power generation device is not at all unsimilar to that required for design of a heat pump, and by appropriate design, one could produce a more effective heat pump. Of course, until the value of such an improvement is appropriately recognized in the marketplace, there will be little incentive to design one.

Two major causes of ineffective operation of devices and processes at the point of energy consumption are faulty heat transfer and mixing. Many special techniques exist to overcome these problems, but up until now they have not found many applications outside of energy conversion operations. Among these techniques are fluidized beds, heat pipes, induced vorticity for improving mixing processes or combustion, jet impingement, and heat recovery apparatus. These techniques can be most useful in improving effectiveness of energy consumption. The application of these techniques in an economically justifiable way should be encouraged.

Another technical flaw often found at the point of energy consumption is deterioration of equipment through lax maintenance. This problem could be attacked in at least two ways. The householder should be apprised of the fuel costs he will ultimately pay if he permits deterioration of the heat transfer surfaces of his furnace or hot water heater; he should be informed as to when and how to apply maintenance procedures. In addition, designers of equipment should devote some serious efforts to design for maintainability. The general principles of informing the user and designing for maintainability obviously also apply to most other items used at the point of energy consumption.

35.7. Measurement Capability

The capability to measure the thermal performance of buildings in the field and to determine that the potential of the thermal design has actually been realized is a basic requirement for the support of any market system to facilitate investment in effective thermal performance of buildings. Also, the capability to measure the performance of building equipment, major appliances, and items of industrial equipment, in such a way as to reflect actual conditions of field service, is needed to provide those who wish to invest in higher quality performance with the information required to make rational decisions. The present standard test methods that are applied to most major appliances and items of building and industrial equipment are not constructed so as to reflect the actual duty cycle to which an appliance is subject in the field. The maintainability of these items is, usually, not covered by standards, and the susceptibility of equipment to unattended maintenance is not usually determined. In fact, the efficiency of equipment is seldom measured in a standard test method. Reliable and realistic information pertaining to equipment performance is a basic requirement of those who must decide which equipment to purchase. Our present capabilities to provide this information require substantial improvement to support effective efforts for conservation.

35.8. Practices of Assessment and Predictive Modeling

Although a great deal of work has begun to assess energy use and to construct predictive energy flow models, much of this work relies upon trend extrapolation, a technique which is based upon the assumption that the future must resemble the past. In particular, the majority of energy assessment studies and predictive modeling efforts initiated to date have contained the assumption that the technical potential for improvement in practices at the point of energy consumption have been exhausted. As has been shown, this is not the case. Moreover, in a very significant respect the future will not resemble the past; primary energy sources will no longer be inexpensive. With conservation now emerging as an inevitable issue for the future, almost all predictive models are used to study options for conservation. But, a predictive model based upon the assumption of "frozen" technology at the point of energy consumption can yield but one conclusion on conservation, which is that conservation can be achieved only through curtailment.

It would be very useful to have predictive models that admit technological flexibility at the point of energy consumption. Such models would be especially helpful in evaluating the balance of investment in development of new energy sources, effective technology at the point of energy consumption, and systems of energy curtailment. It is at this point that one can apply technical improvements. Studies of energy consumption need to be carried out so as to reflect the quality as well as the quantity of energy consumed and to compare these with the quality and quantity of energy actually required at the point of consumption. Since high (thermodynamic) quality fuels are at the center of energy problems, the measurement of quality is an extremely important, but largely underestimated, aspect of assessing energy-consuming practices. An important illustration of this point is the heat pipe vacuum furnace described earlier. This furnace is more effective in its use of energy because it wastes less of the thermodynamic availability[5] of the energy source than did its predecessors.

35.9. Conclusion

The technical and economic opportunities for energy conservation through improved effective use of energy at the point of consumption are only now beginning to be recognized. If appropriate steps are taken now to apply technology effectively to energy-consuming processes and to institute appropriate market mechanisms to further these applications, measures for energy conservation, which appear to be eventually necessary, can be comprised of an appropriate balance between more effective use of energy and curtailment.

Notes

1. For example, in space heating, heat is required to replace losses, but if perfect insulation and sealing were available, no losses would occur and no heat would be required. However, perfect insulation and draft sealing are not available, and even if they were they would be neither economically justifiable nor desirable. Perfectly sealed buildings would be unlivable.

2. The term life-cycle costing is often used to describe this combination, and will be employed here. Appropriate discounting is, of course, to be applied in calculations of operating costs.

3. Detailed studies of the potential for improved energy use in buildings have been made by a number of investigators, for example, U.S. National Bureau of Standards (1973), U.S. Department of Housing and Urban Development (1972), Moyers (1971), Dubin (1973), and National Mineral World Insulation Association (1972).

In a few instances, specific possibilities for obtaining more effective consumption of energy in industrial processes have been studied in detail, for example, *Business Week* (1972), Shefsiek and Lazaridis (1972), Rosenberg (1972), and Nesbitt (1972).

4. We are not concerned here with electric power generation.

5. See Chapter 34, this volume, for a discussion of this concept.

References

Business Week (1972). "Saving Energy by Cutting Needless Loss," p. 30j, November 18.

Dubin, Fred S. (1973). "A Wiser Use of Electricity and Energy Conservation Through Building Design," *Building Systems Design, 70*(1), pp. 8–14, January.

Hottel, H. C., and Howard, J. B. (1971). *New Energy Technology—Some Facts and Assessments,* The MIT Press, Cambridge, Mass.

Kalmas, O. J. (private communication). Lynn, Mass.

Lazaridis, L. (private communication). Thermo Electron Corporation, Waltham, Mass.

Moyers, John C. (1971). "The Value of Thermal Insulation in Residential Construction: Economics and the Conservation of Energy," Oak Ridge National Laboratory, Oak Ridge, Tenn., December.

National Mineral World Insulation Association, Inc. (1972). "Impact of Improved Thermal Performance in Conserving Energy," p. 35, April.

Nesbitt, J. D. (1972). "Improving the Utilization of Natural Gas in Major Steel Mill Application," Institute of Gas Technology Publication, Chicago, April.

Rathbone, DeForest Z. (1972). "Reducing Energy Consumption in Refrigeration Production," *Building Systems Design.*

Rosenberg, R. B. (1972). "The Future of Industrial Sales," Institute of Gas Technology Publication, American Gas Association, Chicago, May.

Sampson, Arthur F. (1972). "Instrumental Office Building Will Test Energy-Saving Design and Equipment," *Engineering News Record, 188*(22), p. 14, June 1.

Shefsiek, Paul K., and Lazaridis, Lazaros J. (1972). "Development of a Natural Gas Fired Heat Pipe Vacuum Furnace," *Natural Gas Research and Technology.*

Stanford Research Institute (1972). "Patterns of Energy Consumption in the United States," report to Office of Science and Technology, Washington, D.C., January.

U.S. Department of Housing and Urban Development (1972). "Subpanel Reports on Total Energy Systems, Urban Energy Systems, Residential Energy Consumption," submitted to the Committee on Energy Research and Development Goals, U.S. Federal Council on Science and Technology, Washington, D.C., July.

U.S. National Bureau of Standards (1973). "Energy Conservation Through Effective Utilization," National Bureau of Standards Report No. NBSIR 73-102, February.

Waihren, D.; Norman, B.; and Grundstorm, K. "High Consistency Paper Forming," *Paper Trade Journal.*

36 Energy Utilization in a Residential Community*

RICHARD A. GROT† AND ROBERT H. SOCOLOW†

36.1. Introduction

The decision to study how energy is ordinarily used in housing is motivated by the perception that there is a critical need to improve the data and methods commonly employed in the determination of energy requirements in the residential sector. The existing state of knowledge of energy utilization in a residential community is relatively primitive. The data describing the consumption of energy are generally either based on the detailed study of the performance of a dwelling in hypothetical or controlled situations or are based on the requirements of some average household living in some standard unit. The quantitative modeling of the performance of individual structures, as represented by standard heat-load calculations, ASHRAE procedures, or various computer codes such as the so called "Post Office Program" or GATE, has never been verified in detail in a real-life situation. The numerous studies of utilities' aggregated data on sales to residential customers, though giving useful information about the range of situations encountered and trends over time, do not explicate the role of technical factors, such as house construction, in determining actual energy use.

The research program that has been initiated in the Center for Environmental Studies at Princeton University is designed to fill some of the voids in the data on residential energy usage. It was decided to study an actual community, Twin Rivers, New Jersey, to determine how energy is used there, what variables affect energy consumption, and how energy consumption can be reduced. The research has proceeded on three fronts: (1) A statistical study, based on all the utilities' data (monthly meter readings) since the town began, in conjunction with the architects' drawings, sales data on options selected, and available weather data (high and low daily temperatures). (2) An instrumentation program, designed to develop detailed information about the behavior of selected dwelling units. The first

* This research is supported by Contract No. GI34994 from NSF/RANN. The data discussed in this report could not have been gathered without the willing cooperation of many people in Jersey Central Power and Light Company, Public Service Electric and Gas Company, the Twin Rivers Holding Corporation, the Township of East Windsor, and the architectural firm of J. Robert Hillier. We are also grateful to Mr. Edward Pickering, who contributed his personal weather records. The inventiveness, care, and teamwork of John Fox, Harrison Fraker, Jr., David Harrje, Norman Kurtz, Elizabeth Schorske, and Douglas Zaeh are everywhere in evidence in this report.

† Princeton University, Center for Environmental Studies, School of Engineering.

effort has been to establish a weather station at Twin Rivers. (3) A program of interviews, intended to establish how the technological reality came into being. The builders, architects, consulting engineers, public utility officials, and municipal authorities have explained which decisions they participated in, the rationale for their actions, and the factors that would have to be altered for them to choose differently. Later we present a preliminary account of our activities in these areas.

36.2. Twin Rivers, New Jersey

Twin Rivers is a Planned Unit Development (PUD) that occupies a little more than one square mile of what only a few years ago were potato fields, half a mile from exit 8 of the New Jersey Turnpike. As New Jersey's first PUD, it is being watched closely by professional planners and sociologists for whatever special lessons it may hold. From a technical standpoint it is representative of much conventional residential housing being built at this time. It possesses an example of almost every generic type of housing: there are single-family detached homes (168 units), town houses (1626 units), and garden apartments (942 units), and there are plans for 2 high-rise apartment complexes. The current price range of the dwelling units is $32,000 to $41,000. The town houses and condominium and garden apartments are of masonry bearing-wall construction with wood framing floors and roof. The single-family dwellings are conventional wood frame. The construction nominally meets the various applicable codes, ordinances, and FHA standards. The construction of Twin Rivers began in June of 1968 and has proceeded in four stages—called quads. At present, Quads I, II, and III are completed and the construction of Quad IV is scheduled to begin in the spring of 1973. With the exception of 43 town-center apartments, all constructed units have gas space heating. There is electric central air conditioning throughout. The dwellings of Quads I and II have electric hot water heaters, electric ranges, and electric dryers, while the corresponding appliances of Quad III are gas. The town is particularly suited to be a "laboratory" because there are relatively few basic units, each repeated many times, all containing identical major appliances, except for standardized options. Thermopane windows and patio doors are also optional and these have been adopted by approximately half of the residents.

Twin Rivers is a relatively homogeneous community; its residents are drawn mainly from the mobile professional middle class.[1] The average income is $16,180; 65% of the heads of family have at least a bachelor's degree; the mean age of the head of family is 29 years; 96% are white; the average number of children is approximately 1.0 for the town houses, 0.3 for the garden apartments, and 1.4 for the single-family houses. It is interesting to note that 36% of the residents of Twin Rivers come from New York and approximately 51% of the heads of family work in New York. Each weekday morning, 19 buses carrying approximately 850 commuters leave the Twin Rivers Shopping Center for the 55-minute trip to New York City on the New Jersey Turnpike. The land-use plan of Twin Rivers sets aside 35% of the space for residential use, 30% for industrial use, 17% for open space, 10% for commercial use, and 8% for roads.

36.3. Preliminary Analysis of the Monthly Gas and Electric Consumption

Both the gas and the electric utility servicing Twin Rivers have provided us with complete records of their monthly meter readings since the first units were occupied. They have also arranged for us to receive each new set of readings on a regular basis. The results presented here are preliminary. We have not yet analyzed all of our data, and the data for the current winter will be extremely helpful in establishing some of the uncertain features of the data we have looked at most closely.

This summary will emphasize the data on gas consumption during the 1971–1972 heating season, the winter immediately preceding the inception of this project. We have defined a six-month winter season, from October 29, 1971, to April 28, 1972,[2] during which (based on a reference temperature of 65°F) there were 4567 degree-days, 92% of the annual total.

We have further restricted our first investigation to the town houses in the second quad. Among our reasons for this choice are

1. Gas in Quad II is used only for heating, not for appliances.
2. In contrast to Quad I, the meters are located out-of-doors, so utility data include few estimated readings.
3. There are over 400 town houses, and many fewer of the other types.
4. The single-family detached homes have not yet passed through a complete heating season.

Of the 401 Quad II town houses, 153 are split-level town houses ("splits") and 248 are two-floor town houses. In both cases, the town

Table 36.1 Average Winter Gas Consumption (therms[a])

	Split-Level Town Houses	Sample Size	Two-Floor Town Houses	Sample Size
18 ft, 2-bedroom	728	(34)	626	(71)
22 ft, 3-bedroom	958	(98)	836	(138)
24 ft, 4-bedroom	903	(20)	901	(39)
All units	900	(152)	782	(248)

[a] One therm = 10^5 Btu.

houses have either 2, 3, or 4 bedrooms and are 18, 22, and 24 feet wide, respectively. A single structure either contains splits or two-floor town houses, not both, but it contains a mix of the three sizes.

The town houses can be further identified by whether they are end or interior units, by whether the Thermopane option was exercised or not, and by their compass orientation.

The average winter gas consumption by number of bedrooms is shown in Table 36.1. It is apparent that the average winter gas usage is quite closely proportional to the town-house width in the two-floor town houses but not in the splits; in fact, the three-bedroom splits use slightly more gas than their four-bedroom counterparts. Our tentative explanation (put forward by Mr. John Fox, a graduate student working with us) is that the additional consumption is primarily due to an architectural variation in the three-bedroom units, a projecting rectangular element on the second floor. To isolate such a cause will require further study; the four-bedroom units have larger furnaces, for example (in both the split-level *and* the two-floor town houses, however). Improved statistics, for this and other investigations, accrue automatically as time passes and additional years of data unfold.

Similarly any detailed accounting for the increased gas consumption in splits relative to two-floor town houses will require a careful comparison of architectural features. Some of the features contributing to the gas consumption are given in Table 36.2, which presents the results of a standard heat-loss calculation for a three-bedroom interior two-floor town house. It is worth emphasizing a result of this calculation that is well known to professionals but apparently not elsewhere: heat loss through windows and heat loss via air infiltration are both at least as large as heat loss via all other means, once walls, ceiling, and basement have even average insulation (2

Table 36.2 Summary of Heat Loss Calculation for Three-Bedroom Interior Two-Level Town House (70°F design temperature difference)

Component	Area (ft²)	U-factor (Btu/hour/ft²/°F)	Heat Load (Btu/hour)
Walls	514	0.093	3,400
Doors	20	0.48	700
Ceiling	723	0.063	3,200
Basement			4,000[a]
Windows	192	1.13[b]	17,300
Air infiltration			11,600[c]
Total			40,200

[a] Includes heat losses above and below ground, through two small windows and floor.
[b] Single glass. If Thermopane (U = 0.72), heat load reduced by 6300 Btu/hour, or 16% given the other assumptions above.
[c] Assumes 0.75 air exchange per hour, a typical handbook value. We have not yet measured air infiltration rates at the site.

inches in walls, 4 inches in ceiling). The attribution of 43% of the heat load to windows and 28% to infiltration justifies our continuing emphasis on these characteristics in our future research.

A naïve way to predict the winter gas demand for three-bedroom town houses would be to take the heat load, which is calculated for a 70°F design temperature difference, and multiply by the 4567 degree-days in the heating season. Choosing a 75% furnace efficiency gives an average gas demand of 835 therms.[3] The close correspondence with the value in Table 36.1 is fortuitous for several reasons: (1) the average value in Table 36.1 includes end units and units with Thermopane, (2) the heat load calculation omits any consideration of solar heating and internal heat sources (it is intended for sizing a furnace that has to operate under the worst conditions—no sun and an empty house), and (3) neither the furnace efficiency nor the air infiltration rate is yet a directly measured number. An improvement on the methodology of calculating fuel consumption that takes these factors into account is one of the goals of our study.

The number of degree-days per month correlates closely with the average gas consumption, confirming that a reference temperature of 65°F for calculation of degree-days is a good choice.[4] The correlation coefficient between the average consumption of gas in the splits and the monthly degree-days (adjusted for actual dates of meter reading) is 0.997, calculated from 12 months of data.

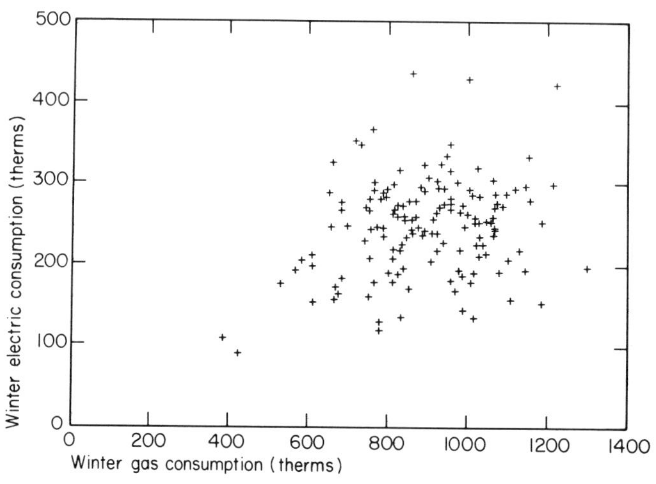

Figure 36.1. Winter gas and electric consumption, in therms, for 152 splits. (The 153rd split was not occupied until midwinter.) The coefficient of correlation for this population is 0.24.

Averaged over all of the splits, the electric consumption for almost the same 6-month interval (the meter reading date is slightly different) is 253 therms (7409 kWh), 28% of the gas consumption. The coefficient of variation (the ratio of the standard deviation to the mean) is somewhat larger for the electric consumption (60.9 therm standard deviation, coefficient of variation of 0.24) than for the gas consumption (162 therm standard deviation, coefficient of variation of 0.18), as one would expect, given the larger number of ways in which individual behavior can influence electric consumption. What came as a surprise was the total lack of correlation between gas and electric consumption in the same unit. A scatter plot of gas and electric consumption for 152 units is shown in Figure 36.1; the coefficient of correlation is 0.24. Apparently, of the two contradictory "explanations"—(1) those who are profligate with one energy source will be profligate with another, and (2) more heat generated by electricity means less heat generated by gas—both are operating.

From the fact that the thermal content of the electricity consumed during the winter period is 28% of the thermal content of the gas consumed in the same period, it is apparent (assuming about 30% efficiency in electricity generation and delivery) that roughly equal quantities of fossil fuels are involved in the production of the gas and the electric services for these homes in winter. The gas consumption in summer (April 29–October 28) is about 9% of the consumption in winter, but the electric consumption in summer is about 125% of the electric consumption in winter, a result, of course, in substantial measure of the air conditioning. This large summer electric load will become an increasingly important focus of our further research.

One of our major objectives is to identify the factors that account for the variations in gas and electric consumption in similar units. When previous studies based on utility sales data have confronted large variations, they have sometimes ascribed these variations to differences in "life-style"; it is occasionally intimated that with people so unpredictable and individualistic, alterations in building practices, appliance design, control systems, and other aspects of the technological reality in which people function cannot be expected to affect aggregated energy consumption substantially (as much, for example, as campaigns to alter each resident's consciousness of how his own decisions are affecting his energy consumption). In our study, where many of the technological factors are standardized, we were prepared to discover that nearly all of the "life-style" effect had vanished, in which case we would have been in a position to emphasize the role of technology and to deemphasize the role of individual behavior.

It is already clear that the truth lies somewhere in the middle. As each technological variable is separated out, the observed variation in gas and electric consumption is reduced, but when many of the technological variables of which we are currently aware are separated out, considerable variation remains. In Figure 36.2, a histogram of gas consumption in all split-level town houses is shown, and, within it, a histogram of that subset of the large sample having no Thermopane, having windows facing east and west, and not occupying end positions. In this subset, the ratio of standard deviation to mean has dropped somewhat, but it is still substantial, as seen in Table 36.3, which also gives the corresponding data for electric consumption.

We have attempted to extract the energy cost associated with being an end unit and with not having Thermopane windows and patio door. We have done a multiple regression analysis on a sample of 98 three-bedroom splits and on a sample of 138 three-bedroom two-floor town houses, using the expression Winter Gas = Constant + A × (Cost of Being an End Unit) + B × (Cost of No Thermopane) where A and B each take on the

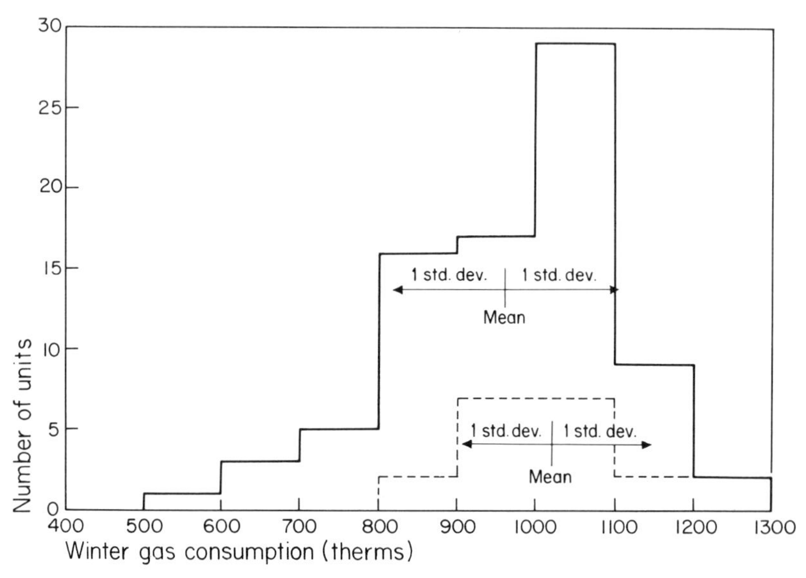

Figure 36.2. Distribution of winter gas consumption for 82 three-bedroom splits and for a subpopulation of 20 three-bedroom interior splits without Thermopane and with windows facing east and west.

values of 0 or 1, depending on which unit is being considered. Our results are shown in Table 36.4.

When we attempt to add the compass orientation to the multiple regression, using the simple form of a weight of 0 for windows facing north and south and 1 for windows facing east and west, the improvement in fit is not significant for the town houses and is barely significant for the splits. (Except for the constants, the values in Table 36.4 change by less than 5 therms.) On further study, we find that the modeling of the solar load term needs to be done quite carefully—we estimate that the shading of the windows by projections and overhangs can easily make a difference of 50 therms in the winter gas consumption. With the help of our weather station, which records direct and indirect solar flux, we hope to study the solar effect on winter heating load in considerable detail.

In our study of solar load, we have come upon one result that appears not to be widely known: a vertical surface facing south receives *more* incident solar radiation over a clear *winter* day than an identical vertical surface facing east or west, and it receives *less* incident solar radiation over a clear

Table 36.3 The Slowly Dwindling Variability of Gas and Electric Consumption as More Nearly Identical Subsets Are Chosen

Sample	Sample Size	Mean Winter Gas (therms)	Standard Deviation in Winter Gas (therms)	Coefficient of Variation for Gas	Mean Winter Electric (therms)	Standard Deviation in Winter Electric (therms)	Coefficient of Variation for Electric
All splits	152	900	162	0.180	253	61	0.241
All three-bedroom splits	98	958	141	0.147	251	58	0.231
All three-bedroom interior, non-Thermo-pane, east- or west-facing splits	20	1023	119	0.116	255	58	0.227

Table 36.4 Multiple Regression for Effect of Being on End, Having No Thermopane, on Winter Gas Consumption[a]

	Splits	Two-floor Town Houses
Sample size	98 units	134 units
Constant (base gas consumption)	879 ± 263 therms	787 ± 247 therms
Added cost for being an end unit	60 ± 54 therms	110 ± 44 therms
Added cost for not having Thermopane	98 ± 57 therms	37 ± 50 therms

[a] Two-standard-deviation uncertainties are quoted.

summer day, for 40° latitude. The difference lies in the fact that the sun is enough higher in the sky in summer than in winter to reduce the heating of a south-facing surface from a value above to a value below the value for the morning heating of an east wall or evening heating of a west wall. Hence, glass on the south walls is advantageous relative to glass on east or west walls, both from the point of view of winter heating and from the point of view of summer cooling.

The cost in Table 36.4 for being an end unit should have been roughly the same for the splits and the two-floor town houses, because the end wall is virtually identical. An elementary calculation like the one for the total gas consumption discussed earlier (see Table 36.2) gives an estimated winter energy cost of 108 therms for the end wall;[5] it is not clear to us why the "cost" of an end wall on a split is so much smaller. The cost for extra outside walls is real, however, and confirms the desirability, from an energy standpoint, of multiple-family relative to single-family structures.

The cost in Table 36.4 for not having Thermopane should have turned out (and *did* turn out) larger for the splits than for the two-floor town houses because a larger area of the splits is glass. (For the same reason, the base gas consumption for the splits is larger.) A direct calculation of the difference in the conduction losses through Thermopane is cited in Table 36.2: a 16% reduction in the heat losses is predicted, versus a 10% reduction in average winter gas consumption in the splits and a 5% reduction in average winter gas consumption in the two-floor town houses. Thus, our tentative conclusion is that the Thermopane is less effective in reducing winter fuel consumption than naïve calculations would predict.

In dollar terms, a 100-therm reduction in winter gas consumption corresponds to a saving of $12.80 per year, which is about 3% of the cost of the Thermopane option for the Twin Rivers town houses (between $400 and $500, depending on window area). To this return on investment one needs to add the dollar saving for reductions in summer air conditioning (not yet estimated) and the advantages in comfort from not having cold interior surfaces.

Once we have factored out the effects of architectural style, size, orientation, Thermopane, and end wall, considerable variation remains. (Recall Figure 36.2.) Of course, this variation, too, must have an explanation in terms of technological variables. A critical distinction is between variables over which the consumer has little control and variables that the consumer dominates. We will be extending our research in the next months, hoping in the process to understand what part of the residual variation is attributable to variables of the first kind that we have not yet studied (like variations of tightness of fit of windows and variations in performance characteristics of "identical" furnaces) and what part is attributable to variations of the second kind (like variations in thermostat settings and in frequency of door openings).[6]

Among the factors that are not proving helpful in analyzing variations in energy consumption are the family income and the family size. The family income and the family size are inversely correlated, in part because there are often two incomes when there are no children, and in part because wealthier large families would have bought larger homes. But even among families with children there is no perceptible increase in gas consumption and only a small increase in electric consumption as the number of children increases, holding number of bedrooms constant. Our (by now familiar) population of 20 three-bedroom interior splits without Thermopane and with windows facing east and west has annual family incomes ranging from $15,000 to $34,000 and family size ranging from two to five, yet other than the inverse correlation of family size and income, no correlations of either winter gas or winter electric consumption with either family income or family size are statistically significant. A correlation matrix for these four variables for this 20-unit population is given in Table 36.5.

36.4. Future Plans for Data Acquisition

Early in this project it became apparent that there was little hard data on the usage of energy in residential communities and that readily available data sources such as gas and electric bills would probably serve only a limited objective. There is a need for data that could (1) describe how energy

Table 36.5 Correlation Matrix for Special Population[a] of 20 Identical Splits

	Winter Gas Consumption	Winter Electric Consumption	Number of Occupants	Family Income
Winter Gas Consumption	1.00	0.14	0.13	−0.24
Winter Electric Consumption	0.14	1.00	0.31	−0.20
Number of Occupants	0.13	0.31	1.00	−0.57
Family Income	−0.24	−0.20	−0.57	1.00

[a] Three-bedroom, interior, non-Thermopane windows facing east or west.
Note that for a 20-member population correlations as large as 0.44 occur randomly 5% of the time.

is used in urban communities, (2) be utilized to evaluate various proposed technological and socioeconomical innovations that would affect energy consumption, and (3) be used to generate data statistically for other urban mixes. Though the components of the Twin Rivers project that will be described briefly here are still mainly in the planning or development stage, a knowledge of the methodology employed and the data being generated may be useful for those planning similar projects on other aspects of urban communities and for those attempting to evaluate the effects of specific modifications in energy technology. For example, the data that will be collected should provide a base for evaluating modular integrated utility systems and will be compatible, hopefully, with data that will be collected by the National Bureau of Standards in its evaluation of the total energy plant at the Jersey City Operation Breakthrough site.

By type, the data being collected at Twin Rivers can be classified as describing energy consumption, local meteorology, physical characteristics of the town and of individual residences, the internal environment of the dwelling, and household characteristics. In each instance, we are beginning with the readily available data (when it exists), then refining it by limited sampling and field studies, and finally, where it seems warranted, preparing a large-scale monitoring effort. The various stages and components of this scheme are depicted in Table 36.6.

We have just finished installing an automatic weather station atop the bank at Twin Rivers. The station has the capability of measuring wind velocity, total wind run, wind direction, air temperature, ground temperature,

Table 36.6 Data Sources

1969–1973 Available Data	1973–1974 Limited Sampling	1974–1975 Large-Scale Monitoring
Consumption Gas and electric utility bills	**Consumption** Temporal record from electrical primaries and gas mains, temporal record by use in 2 to 8 units	**Consumption** Expand number of units monitored
Meteorological Data Daily maximum and minimum temperature (Hightstown)	**Meteorological Data** Hourly air temperature, humidity, wind velocity, wind direction, rainfall, barometric pressure, total solar flux, diffused solar flux, ground temperature	**Meteorological Data** Possible spatial resolution, add net radiation, measure of turbulence
Physical Characteristics of Units VA-FHA "Material Description," architectural drawings, sales records, observation of construction	**Physical Characteristics of Units** Field study of air infiltration, heat flow, air flow, laboratory evaluation of furnaces and appliances	**Physical Characteristics of Units** Modification of several units, monitoring of furnaces and appliances
Internal Environment No data	**Internal Environment** Field study of two to eight instrumented units	**Internal Environment** Expand number of units monitored
Household Characteristics Size, age, income, occupation for some units	**Household Characteristics** Usage of appliances	**Household Characteristics** Usage of appliances, door and window openings

barometric pressure, dew point, rainfall, total solar flux, and diffuse solar flux. Normally, these data are gathered and digitized hourly and transmitted from Twin Rivers to the Engineering Quadrangle at Princeton University where they are recorded on paper tape to be processed later on a computer. Several field studies are currently being initiated, primarily on air infiltration measurements, heat flow losses, and temperature variations within dwellings. Three single-family homes have been instrumented to give a continuous record of gas consumption and electricity used for hot water. This work is being done in conjunction with Hittman Associates, Columbia, Maryland. Our data will be used to evaluate a computer program that they have been using for residential energy consumption. The data also provide us with a preliminary indication of the details of energy use, which allows a more rational means of planning future instrumentation.

A furnace and some of the appliances used at Twin Rivers have been installed in a laboratory in the Engineering Quadrangle, for the dual purpose of checking instrumentation and evaluating performance. One or two prototype data acquisition systems for monitoring the internal environment, energy consumption, and user habits are being assembled for installation at Twin Rivers in a few months. The quantities that will be measured are listed in Table 36.7. The idea is to instrument a few homes in detail in order to determine what will be required in a large-scale effort.[7] Finally, with the electric and gas utilities we are planning instrumentation of the mains and the primary distribution system.

36.5. The Program of Interviews

In an effort to untangle the web of ideas, decisions, and actions that resulted in the creation of Twin Rivers, we are carrying out a series of interviews that are producing a mass of material, an oral history of the decision making. In the space available here, it is impossible to do justice to the tapestry of relationships that we have found. These were presented in Fraker and Schorske (1973).

Table 36.7 Proposed Measurements in Individual Units

1. Temperature
a. Air temperature in each room
b. Air supply temperature in each room
c. Exterior wall and window temperatures (inside, outside, and difference)
d. Basement temperatures
e. Garage temperature (if applicable)
f. Attic
g. Main return duct
h. Supply duct at furnace
i. Hot water temperature
j. Cold water temperature
k. Furnace flue gas temperature (also gas hot water heater flue if applicable)

2. Humidity
a. Kitchen
b. Bathrooms
c. Main return duct
d. Living room
e. Upstairs
f. Main supply duct

3. Electric Consumption
a. Range (if applicable)
b. Refrigerator
c. Hot water heater (if applicable)
d. Clothes dryer (if applicable)
e. Total living area
f. Furnace fan
g. Air-conditioner compressor
h. Air-conditioner condenser fan

4. Gas Consumption (pulse meters and counters)
a. Furnace
b. Hot water (if applicable)
c. Range (if applicable)
d. Clothes dryer (if applicable)

5. Water Consumption
a. Cold
b. Hot

6. Air Flow
a. Supply ducts
b. Return duct
c. Exhaust fans

7. Time of Usage
a. Various appliances
b. Door open time
c. Windows
d. Fans

8. Pressure across Exterior Surfaces

9. Heat Flow
a. Exterior walls
b. Glass
c. Basement walls
d. Ceiling
e. Through flue

10. Continuous Monitor of Air Infiltration

Notes

1. The data in this paragraph are drawn from Burchell (1972).

2. The meters are read in all Quad II town houses on each of these days.

3. $$\frac{40{,}200 \text{ Btu/hour}}{70°\text{F}} \times \frac{24 \times 4567 \text{ hour-°F}}{0.75} \times \frac{1}{10^5 \text{Btu/therm}} = 835 \text{ therms.}$$

4. Let T be the average temperature on a given day; the degree-days in a time interval are found by summing $(65 - T)$ for all days where this is a positive number.

5. Assumptions: 4567 degree-days; 556 sq ft of wall of U-value 0.093; 9 sq ft of glass of net U-value 1.24; 12 cu ft per minute infiltration through cracks; 75% furnace efficiency.

6. Further statistical analysis of the monthly gas and electric consumption data is to be found in Fox (1973). The quantitative conclusions in this chapter have been confirmed with only small modifications. Final numerical results will be published shortly.

7. This work is being done in collaboration with the National Bureau of Standards, which has a supporting grant from the National Science Foundation.

References

Burchell, Robert W. (1972). *Planned Unit Development: New Communities American Style,* Center for Urban Policv Research, Rutgers University, New Brunswick, N.J.

Fox, John (1973). "Energy Consumption for Residential Space Heating: A Case Study," M.S.E. dissertation, Princeton University, Princeton, N.J.

Fraker, Harrison, Jr., and Schorske, Elizabeth (1973). "An Analysis of the Development Process at Twin Rivers," Center for Environmental Studies, Princeton University, Princeton, N.J., September.

37 A Thermodynamic Valuation of Resource Use: Making Automobiles and Other Processes

R. S. BERRY,* M. F. FELS,* AND H. MAKINO*

37.1. General Considerations

Problems of the use of energy and other resources bring one quickly to the realization that quantifiable and unambiguous *objective* general indices could be very useful guides for analyzing and responding to these problems. The conventional economic indices act as measures of current preferences and may reflect perceptions of the implications and impacts of alternative policies. However, there need be no correspondence between these perceptions and any real impacts, particularly any consequences near or beyond the time horizon of the participants in the market. It is therefore desirable to try to develop indices of resource use that separate physical or objective characteristics from normative or perceptive characteristics. The purpose of our efforts is to explore and develop at least one such index, to relate it to economic indices, and to integrate the use of such indices into institutional decision making.

To establish a physically based quantitative index of resource use, one must ask "What is it that we really consume?" In the somewhat more precise language of economics, we ask "What resources are actually in shortage, in the sense that there is any limitation on their allocation?" Some of the negative parts of the answer are easy to recognize: there is no real shortage of any *substance,* in the sense that the earth's crust contains far more than enough atoms of all the natural elements to make all the materials we could possibly want, in the foreseeable future. The problems associated with materials are not really problems of their amounts but of their flows. Analyses of their flows, such as those of Kneese, Ayres, and D'Arge (1970) will clearly be necessary components for understanding and adapting to changing patterns of resource use. One can say that goods or services of specific kinds are in shortage; this is a truth that can be stated in a more precise and more useful way. The one real shortage in the world of goods and services is the capacity to transform things. Or, in the language of the scientist, the item in shortage is the capacity to do work.[1]

The stored capacity to do work and its rate of consumption form the central topic for a large part of the science of thermodynamics. In particular, certain variables, the *thermodynamic potentials,* are defined precisely

* The University of Chicago.

to be equal to this stored capacity under well-defined circumstances.

Under some rare conditions, energy itself takes the role of a true thermodynamic potential; nearly always, energy is a very important component of the thermodynamic potential. The other major component of thermodynamic potential comes from entropy. High-grade ore is better than low-grade ore because of its lower entropy content, not because of its higher energy content. We do less work, or draw less from our storehouse of thermodynamic potential when we can select a low-entropy starting point than one of high entropy, because it is costly in thermodynamic potential to reduce entropy. Moreover, pure materials and (deliberately) intricate structures are low-entropy forms, while random aggregates, solutions, and mixtures are high-entropy forms. Only if we really must do we draw upon materials in very high entropy forms; there are very few materials that we recover, for example, by sweeping them up from seawater, although iodine, gold, and magnesium have been examples.

In principle, we could evaluate our available supply of thermodynamic potential. The problems in making such an evaluation are similar to the problems of estimating fuel reserves; the evaluations can be done just as easily and no doubt will be. They will be important for long-range modeling and planning, and for selecting policies of resource use. At this time, however, we are much less concerned with absolute amounts of thermodynamic potential than with the amounts of this potential that we use, and with possible changes in these amounts that could be considered for the host of well-known reasons.

For the present context, we would like to focus on the analysis of the consumption of thermodynamic potential in some very specific real processes. We do this to show how one can identify particular activities in which it appears possible to make more efficient use of thermodynamic potential —to save, as it were. We are concerned with distinguishing short-term savings that require little change from existing practices, medium-term savings that require larger changes but for which the means are more or less well known, and long-term savings that may require basic scientific, technological or attitudinal changes for their accomplishment, if indeed we opt to work toward them.

Underlying the analyses at this level is an assumption that it is desirable to be thrifty in our use of thermodynamic potential. The grounds for the assumption are clear: energy husbandry, as Socolow has called it, is valuable both as a way to buy time to solve problems of energy supply and of short-term impacts, and also to adapt our life-style toward a time when we face the limited carrying capacity of our habitat. At another level, which we do not consider here, one meets the problem of relating the valuation based on the consumption of thermodynamic potential (or on any other "objective" or physical index) and the valuation that occurs in the marketplace, the valuation of traditional economics.

Analyses of the consumption of thermodynamic potential differ very little in many cases from analyses restricted to energy use. In fact, the energy analyses are an important component of any evaluations of *free energy*—free energy being essentially a more common and less cumbersome synonym for thermodynamic potential. Most commonly, energy and free energy figures differ by some 5 or 10%. In a few cases, however, the differences are larger, enough to have qualitatively different implications. These situations (which do *not* correspond to the examples we shall discuss) generally involve transformation of useful materials into gases or the dispersion of rare materials; a rather obtuse example is the escape of helium from the earth's atmosphere. Apart from these special situations, we can generally expect to use analyses based on energy and on thermodynamic potential almost interchangeably. It is also becoming increasingly clear that we can usually make the numerical conversions with ease, if the analyses are themselves specific enough about the materials and processes.

37.2. Specific Analyses: Methodology and Method

To carry out analyses of energy and free energy use, one must select a clearly defined system (as distinguished from the surroundings, whose contributions are neglected), establish a data base, and create validity checks. It has been our experience that overcasual treatment of any of these obvious aspects of the procedure has led to confusion and error, sometimes large enough to affect the implications rather seriously. For example, apparent discrepancies between our own analysis of primary metal production and that of Stanford Research Institute (1972) were sometimes as large as a factor of 4. However, these discrepancies were traced to differences in our definitions of system: our own definition includes the collection and preparation of *all* the materials used in the recovery and treatment processes, whereas the Stanford Research Institute (SRI) system omits all the secondary contributions, such as those of the limestone, coal, and coke preparation for iron smelting. In another case, a preliminary analysis of automobile manufacture and discard by one of us (Berry, 1972) had figures that turned out to be too erroneous to be useful for policy evaluation because an apparently self-consistent data base was simply much too small to display its own errors and inadequacies.

We therefore want to emphasize here the importance of careful treatment of procedures and to pass on a few useful points that we have learned. It has been our experience that two principal sorts of errors are likely: errors of omission and errors of redundancy. Errors of omission tend to be the more common ones, and naturally lead to evaluations that are on the low side of the truth. Errors of redundancy, on the other hand, tend to generate positive errors, making calculated results higher than the true values. Omission errors, in our experience, were best avoided by site visits and direct consultation with people in the specific fields. Redundancy errors tended to result from overlapping categories in secondary data sources such as the U.S. Census of Manufacturers (U.S. Bureau of the Census, 1971a). Only by direct consultation with people involved in preparing these secondary sources could we avoid this sort of redundancy error. Errors could sometimes be picked up by making checks for consistency; whenever two independent data sources are available, they should both be used to check the energy or free energy consumption in a specific process. We have occasionally found discrepancies in this way that sometimes amounted to a factor of 10. In most cases, fortunately, the independent data sets give fairly similar results, which can then be used to estimate the uncertainty of the calculation.

In our experience, the largest single difficulty in doing energy and free energy analyses at the present time is the unavailability of some of the data. In particular, proprietary considerations have prevented free disclosure of energy consumption figures from certain industries, even when it is public knowledge that such figures have been compiled. This problem will have to be resolved, clearly, if any kind of energy-use analysis is to be incorporated into a national energy program.

The specific studies we have been doing are an analysis of the manufacture, discard, and recovery of automobiles (Berry and Fels, 1972), and a study of consumer services, meaning packaging, storage, transport, and merchandising of consumer goods. The automobile study, now complete, involved the analyses of mining and smelting, of metal treatment and fabrication, of parts manufacture, and of scrap recovery. Figure 37.1 shows how the system is defined and broken into manageable parts. We selected the automobile as a prototype of a manufactured article, and thus as a means to develop methods and procedures by studying a well-defined and reasonably well documented collection of manufacturing (and related) industries. Our own study did not deal with the automobile as a part of the transportation sector; we were aware of other groups studying transportation (Hirst, 1972; Leach, 1972), and found that, apart from the allocation of spare parts for maintenance, the manufacture and use of automobiles could be separated quite unambiguously.

The system for the automobile study in the thermodynamic sense, consists of the constituent materials used in the automobile itself and also all the other materials required in the preparation and discard of the automobile. Hence we include the limestone required for the blast furnace and the fuel used to transport the materials as well as the steel of the auto body. We do not include the resources needed to feed, clothe, or shelter the workers. We made a deliberate assumption, partly for convenience in limiting the scale of the project and partly because of an assumption about the way our society operates, that the workers would subsist at roughly the same level whether they were engaged in making automobiles or in some other activity. This can be taken as a normative social judgment; however, as more and more activities are analyzed by various groups, the data will accumulate that tell us the thermodynamic potential associated with human subsistence at various levels. When that information is available, the problems can be carried to the level of studying labor displacements in terms of thermodynamic potential.

The study of consumer services, still in progress, presents a very different sort of problem. In automobile manufacturing it is not difficult to define the vertical system of materials recovery, preparation, and fabrication, and to evaluate the contributions from the industries at each step to the manufacture of the automobile. In consumer services, some components of the picture, such as paperboard container manufacture, can be analyzed relatively easily. However, packaging made of plastics has required considerably more care and effort to analyze, partly because of the variety of materials involved, partly because of problems of proprietary information, and, at least as important, because of the large variety of uses other than packaging to which plastics are put. The packaging aspect is, naturally, the

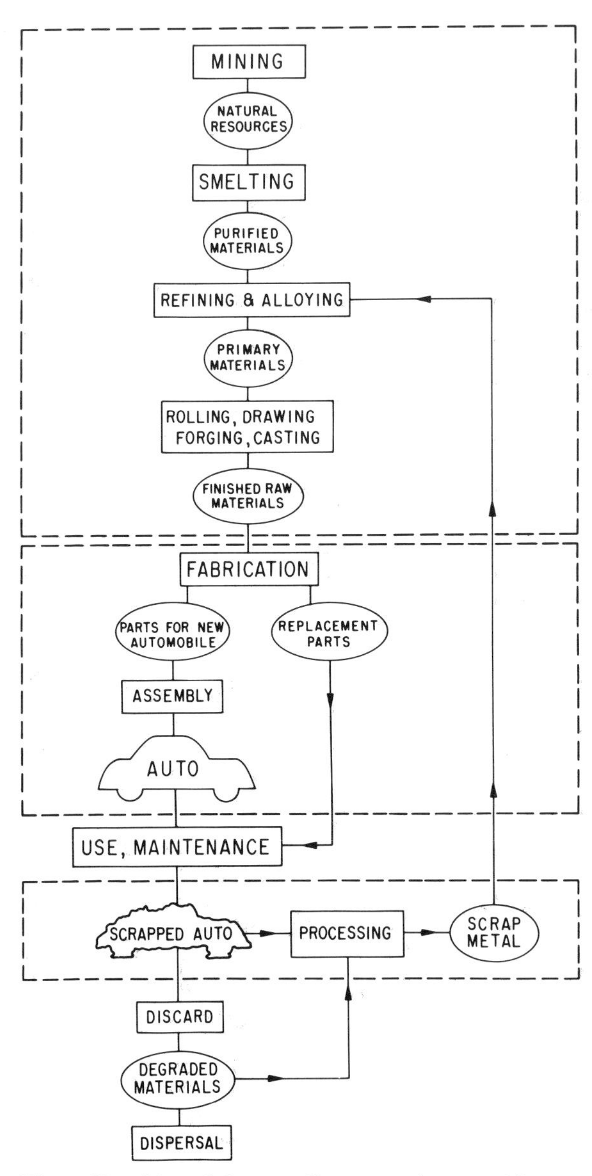

Figure 37.1. Map of the overall process of automobile manufacture, discard, and reuse.

most difficult part of the analysis of consumer services because of its diversity.

One very interesting point is particularly cogent in the present context. In the analysis of consumer services, we have found that there are now enough other studies, usually of specific detailed products, to be very helpful in terms of supplying data, and also to give us points of verification of our own data base. For example, Hannon's well-known study of beverage containers (Hannon, 1971, 1972), and the Midwest Research Institute study of plastic and cardboard meat trays (Franklin and Hunt, 1972) have been of considerable use to us.

In general, the largest contributions to the data base for our studies have come from the well-known sources such as the U.S. Census of Manufactures (U.S. Bureau of the Census, 1971a), the Census of Mineral Industries (U.S. Bureau of the Census, 1971b), the Census of Transportation (U.S. Bureau of the Census, 1971c), the *Minerals Yearbook* of the Bureau of Mines (Smith, 1973), and the industrial yearbook and standard industrial monographs. However, invaluable and sometimes key pieces of information have come from direct contact with individuals. We cannot emphasize too much the importance that these interactions have had; we are, quite frankly, left with an attitude that any energy analysis must be treated with considerable skepticism and care if it does not have a component of direct contact with individuals involved in the system under analysis.

A word is in order about the manner in which the results have been analyzed. The analyses can be conceived in terms of the time scale to which they apply, or to whether they deal with preparation, use, or discard. In the case of automobiles, it turns out that the short-term savings fall primarily into the area of discard and reuse, while at least some of the potential longer-term savings fall in the areas of use and manufacture. The most obvious approach has been the one we used to identify means to achieve short-term savings. Here, we compared alternative *existing* procedures, and considered the savings that could be achieved by altering procedures. For example, we compared various balances among different ways to deal with junked automobiles, to determine how much thermodynamic potential would be saved if various policies were followed. A second level of analysis involved consideration of changing patterns of *use,* and of the changing patterns of energy consumption that would accompany a change in usage. The third and presumably longest-term level of analysis was the identification of areas in which one might try to make basic technical changes in the industrial processes themselves. Two criteria appeared to be useful guides for identifying the processes ripe for improvements. One is the difference between the actual thermodynamic potential consumed in a process, per unit of product, and the ideal thermodynamic limit for this figure. We refer to this difference as the *free energy waste.* The other criterion is the ratio of the free energy waste to the actual consumption, according to present practice; this quantity is called the *waste factor.* Typically, current values for waste factors are about 0.90 for efficient processes (90% of the consumed thermodynamic potential is wasted) and inefficient processes have waste factors greater than unity—meaning that we supply potential to systems that should be returning potential to us.

The economic concept of *added value* has a direct parallel in the concept of *added potential* that occurs with each step of a process. In a thermodynamic analysis, we can compute both the increment of thermodynamic potential that an item gains in a particular process, and the expenditure of thermodynamic potential associated with that process, per unit of product. The difference between these two is precisely the free energy waste, that we have just introduced.

37.3. Specific Systems: Some Results

We shall not try to present the full set of results of the automobile analysis here; these are available elsewhere (U.S. Bureau of the Census, 1971a). However some of the more generally useful results merit discussion in this report. First, the most important basic component in auto manufacture is the preparation of iron and steel. Figure 37.2 shows the schematic representation of this system. We have found that tabular representation is far less helpful than diagrammatic representation. To make the diagrams as useful as possible, we have developed a systematic form of presentation, which we have now made standard for our analyses. We use *rectangles* to label *processes, ovals* to represent *products, triangles* (point down) to represent *free energy consumption,* and *carts* to represent *transportation.* Figures in ovals represent amounts of materials (tons, unless otherwise indicated) for a particular sequence; figures in triangles and carts are the consumed amounts of free energy for the process and for the associated transportation, respectively. Our units for energy have been kilowatt-hours. Lines with arrows indicate flow patterns, with nonintersecting flows indicated by the old-

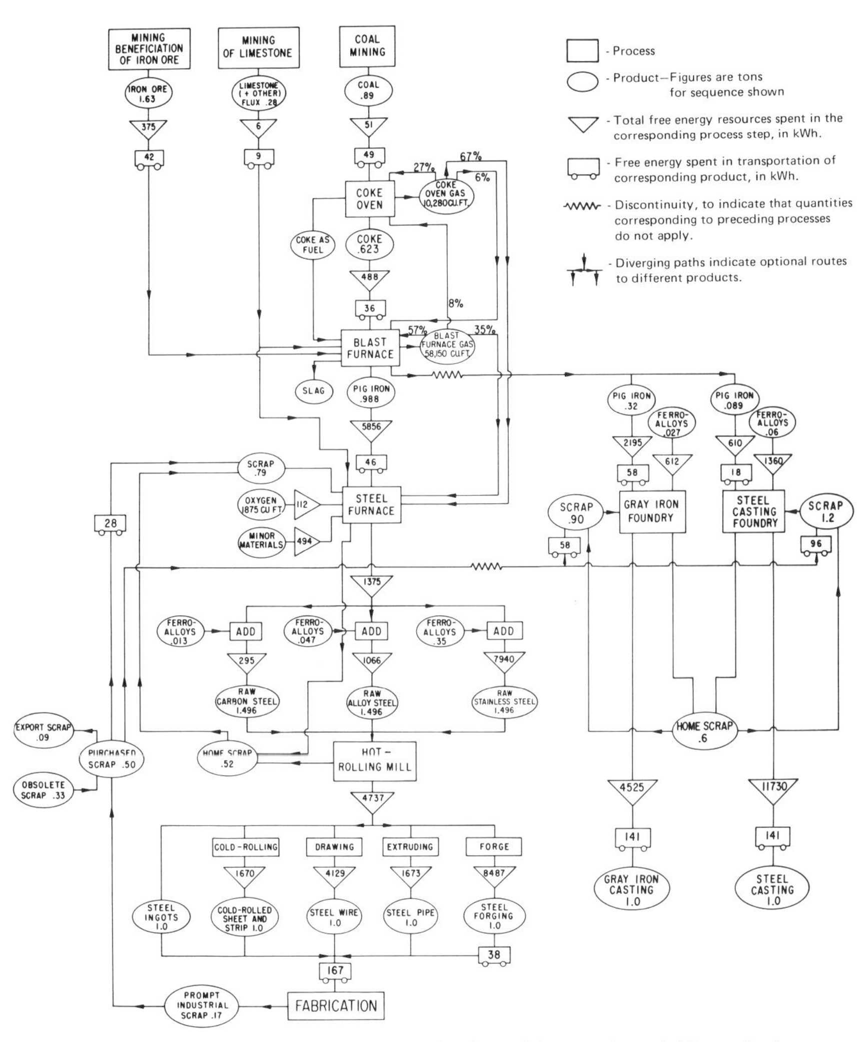

Figure 37.2. Free energy analysis of the production of iron and steel. The units in ovals are tons, and in triangles and carts, kilowatt-hours.

Table 37.1 Total Free Energy (ΔF) per Ton of Finished Metal

			ΔF (kWh per ton)
Carbon steel:	cold-rolled		15,455
	wire		17,915
	pipe		15,460
	forging		22,275
Alloy[a]		(add)	770
Stainless		(add)	7,645
Automotive sheet[b]			14,870
Iron casting			7,330
Steel casting			13,680
Aluminum:	rolled		73,440
	forging		73,345
	casting		62,770
Copper:	rolled		37,500
	wire		31,130
	casting		36,485
Zinc:	rolled		23,225
	casting		25,705

[a] The alloy and stainless counterparts of the four types of finished carbon steel are obtained by adding the increment indicated to the corresponding figure for carbon steel.
[b] 65% of the automotive sheet is cold-rolled (AISI).

fashioned wiring diagram for unconnected wires. Zigzag breaks (like resistors in wiring diagrams) indicate breaks in the scale of quantities of materials. The net cost in free energy for each of the basic metals is given in Table 37.1; this table is based on current (1967) practices. These figures are approximately 5 to 10% higher than the values for energy alone. They include the primary hydrocarbon fuel costs required to generate electricity used in processing.

The free energy analysis of the production of automobiles is summarized in Figure 37.3. Table 37.2 gives a breakdown of the principal contributions to the overall process. The total cost is just over 37,000 kilowatt-hours (kWh), of which 26,000 are used in manufacture of metallic materials, and 9300 are used in fabrication.

The analyses of minor components and scrapping and discard are given elsewhere (U.S. Bureau of the Census, 1971a). The recoverable materials, approximately 1.3 tons of steel, 0.25 ton of cast iron, 0.04 ton of aluminum, 0.03 ton of copper, and 0.02 ton of zinc, can be compressed into "No.

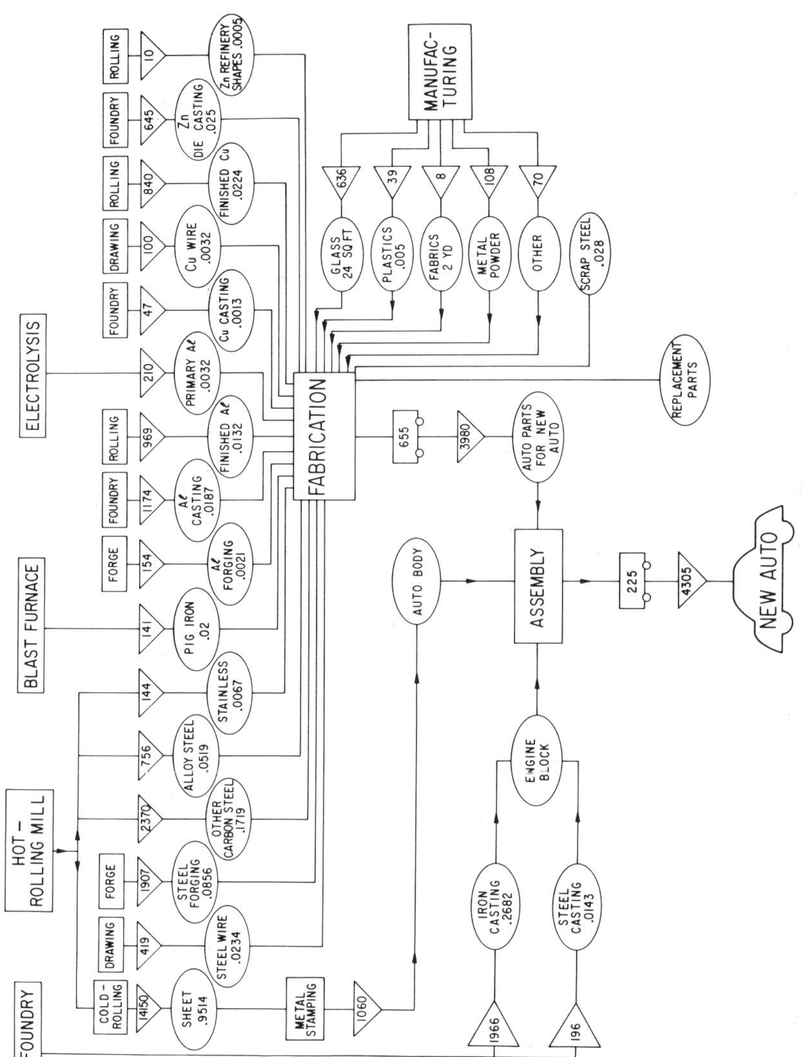

Figure 37.3. Free energy analysis of automobile manufacture. Units are the same as for Figure 37.2.

Table 37.2 Contributions to the Total Free-Energy Change Due to the Manufacture of One Automobile

	ΔF (kWh)
Manufacture of Metallic Materials	26,185
Manufacture of Other Materials	865
Fabrication of Parts and Assembling of Automobile	9,345
Transportation of Materials	655
Transportation of Assembled Automobile	225
Total (ΔF_{total})	37,275

2 bundles" or shredded and segregated. (Some segregation is done by removal of radiators, engine blocks, and other parts in almost all scrapping operations.) Because of the impurities retained in a No. 2 bundle, scrap in this form can only be used in limited amounts; only 15% of a steel furnace's charge can be No. 2 bundles if the product is to be steel suitable for making automobiles.

The free energy savings from scrap recovery are 7040 kWh per ton of iron or steel (replacing a ton of pig iron), 63,500 kWh per ton of aluminum, 22,100 kWh per ton of copper, and 17,800 kWh per ton of zinc. On this basis, the scrap in an automobile has a **gross value** of 14,500 kWh. When we subtract the transportation costs and costs of scrap processing, we find a **net free energy value** of 12,600 kWh per auto for shredded, segregated scrap and 10,600 kWh for No. 2 bundles (because no nonferrous metals are recovered). However, the scrap from No. 2 bundles cannot be used without dilution, as we pointed out. In fact, only 1585 kWh per *new* automobile can be saved if the maximum permissible amount of No. 2 bundle scrap is used, and all the rest of the ferrous materials come from newly smelted materials.

We can now analyze the scrap recovery system to find what savings could be made by various choices of policy. We find (U.S. Bureau of the Census, 1971a) that if S million scrapped autos are processed to make P million new autos, and a number Y from S goes through shredding (so $S - Y$ is the number made into No. 2 bundles), then

$$\text{Savings per new automobile, in kWh} = (10{,}560S + 2080Y)/P \qquad (37.1)$$

Note that the requirement of dilution limits $S - Y$ to be no more than $0.15P$. Practice in 1967, our data base year, corresponded very closely to

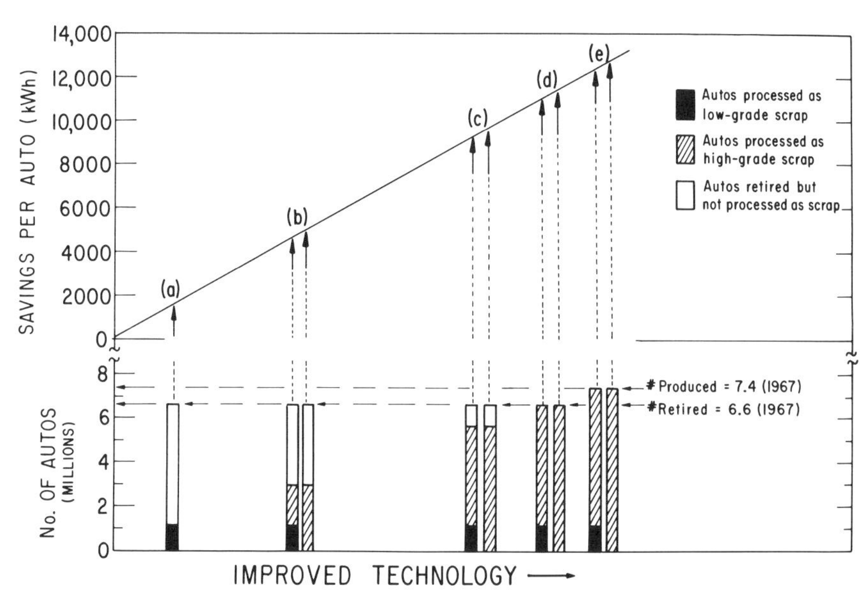

Figure 37.4. The free energy analysis of policies for scrapping automobiles. The lower graph (bar graph) shows numbers of scrap automobiles handled by simple discard without recovery (white), by compression into No. 2 bundles (black), and by shredding and segregation (shaded). The horizontal axis represents increased use of scrap recovery, and especially of shredding. The shorter bars refer to the number of hulks retired in 1967; the longer bars at right refer to the total number of autos *produced* in 1967. The upper graph (straight line) shows the amount of free energy saving per new automobile associated with the allocation indicated on the lower graph at the point directly below. Thus (a) represents the free energy saving per new car if the policy indicated by the leftmost bar were followed.

$Y = 0$ and $S = 0.15P$, corresponding to maximum use of compressing, and no shredding. Figure 37.4 displays this practice as the leftmost bar in the lower graph and as the point (a) on the energy scale in the upper graph. Other policies for scrap recovery correspond to the other bars and points on the slope: if all the recovered automotive scrap (in 1967) were used for automobile manufacture, this would correspond to point (b), with the left and right bars indicating the use of the maximum amount of bundled scraps, and of shredding all the scrap, respectively. Point (c) corresponds to a policy of reusing all the *collected* hulks; point (d) corresponds to processing all the hulks discarded in 1967, and point (e), the point of maximum savings (given present practices), corresponds to recovery of all the metallics for the next year's cars from scrap, which would entail mining junkyards. The maximum achievable savings, 12,600 kWh per new car, is realized this way and corresponds to about one-third of the present cost of the car. This would, in turn, reduce the national power needs by about 11,400 megawatts, or the energy needs by about 10^{11} kWh per year. (It is worthwhile noting that the energy cost of manufacturing a car is roughly equal to the cost of operating the vehicle for one year.)

Longer-term policies can be based on more fundamental changes than modifying scrap recovery. Extending vehicle life, for example, would require closer tolerances and much better inspection procedures, which would require virtually no extra thermodynamic potential, and more replacement parts, which do require free energy. We estimated a cost of about 900 or perhaps 1000 kWh for the parts required to double the present lifetime of an automobile, about 3400 kWh for parts to triple the lifetime and about 1500 kWh for an upper limit to the added initial manufacturing costs. Hence the savings resulting from tripling the lifetime would be about 23,000 kWh per present vehicle lifetime.

Finally, we examine the processes currently in use to identify where technology might be improved. Table 37.3 lists ΔF_{ideal}, the ideal free energy expenditures for presently used processes if they ran at their thermodynamically perfect limits, ΔF_{real}, the actual expenditures according to present practice, $\Delta F_{\text{work}} = \Delta F_{\text{real}} - \Delta F_{\text{ideal}}$, the free energy waste, and the waste factor $w = \Delta F_{\text{waste}}/\Delta F_{\text{real}}$. This table lets us begin to apply Sutton's rule, in order to select processes to improve. Clearly, the processes with large ΔF_{waste} are processes with low waste factors; somebody else has gone this route before us, which is not surprising. In fact the announcement was made

Table 37.3 Comparison of ΔF_{ideal} and ΔF_{real} for Various Processes and for the Total Automobile

Process	ΔF_{ideal}[a] (kWh)	ΔF_{real} (kWh)	ΔF_{waste} (kWh)	w	To Produce
Coke Oven	−105	785	890	1.13	1 ton coke
Blast Furnace	565	5,925	5,360	0.90	1 ton Fe
Steel Furnace	−260	1,375	1,635	1.19	1 ton steel
Refining of Bauxite	0	3,220	3,220	1.00	alumina for 1 ton Al
Smelting—Aluminum	4,610	60,970	56,360	0.92	1 ton Al
Smelting—Copper	−425	11,730	12,155	1.04	1 ton Cu
Smelting—Zinc	−390	16,115	16,505	1.02	1 ton Zn
Total Process	1,035	37,275	36,240	0.97	1 automobile

[a] Based on the reactions of the dominant technology now in use.

recently (Smith, 1973) that aluminum recovery will soon be achieved in a still more efficient way. The important point is that we can examine data such as those of Table 37.3 to help decide our priorities for new technological development.

Our analysis of consumer services is still in progress. This area, in contrast to the automobile cycle, is highly fragmented and contains contribution from a wide variety of processing industries. Consequently the problems of conflicting data are greater than in simpler industries. We also find occasional large variations in total energy costs for a given product, depending on the starting materials; this is analogous to the situation with new steel and reprocessed scrap. Paper pulp made from groundwood, for example, requires over 5000 kWh/ton, while the average energy requirement to make wood pulp is only about 370 kWh/ton, according to our best figures thus far.

Steel (tinplate) cans cost between 1.0 and 1.3 kWh/lb or about 1.0 kWh/12-oz can, but figures currently available on aluminum cans span a range from 4.2 kWh/lb to 17.0 kWh/lb or between 2.1 and 8.5 kWh/12-oz can. Tentatively, we give more credence to the higher figures, based on our experience with other conflicting data. We present these numbers here to illustrate the discrepancies that one faces in analyzing the data, rather than energy costs of containers. The comparison between plastic and paper containers offers an interesting little point on which we might close. Paper containers cost between 2.5 and 5 kWh/lb, according to the type of material; paper bags are among the most costly. Estimates of polyethylene energy costs, like aluminum, show wide variations; our own favored value is about 4 or 4.5 kWh/lb, perhaps as high as 5. These figures have the amusing implication that a polyethylene bag, which weighs about two-thirds as much as a paper bag with the same volume, has more capacity for reuse, *and costs less energy to make,* than the paper bag! Reconciling the implications of energy-use analysis with aesthetic attitudes may be part of the next stage of dialogue in this field.

Note

1. Strictly, the shortage is in the capacity to do work at a desired rate. Anticipating the ensuing discussion, we must point out that the qualification "at a desired rate" extends some aspects of the discussion considerably. To consider rates, one must

examine not only the absolute availability and consumption of thermodynamic potential but the available *flux* of thermodynamic potential and the behavior of a dynamic system rather than a static system. At this level, one confronts the problems of flexibilities and elasticities, the cost of changing processes and other "transaction costs," all in a quantitative way. A new process may be more economical than an old one (in either the monetary or energetic sense), but the net cost may be higher to exchange an old process for a new one and operate the new one. This is perhaps the lowest level at which the dynamic aspects enter into an analysis of the sort we are studying. At a higher level, one must examine the costs of supplying thermodynamic potential at different rates. This question brings in not only the means of transporting thermodynamic potential but also the problems of the direct impact of its use and of its secondary impacts, *as functions of the rate at which it is used.* The present discussion is limited to aspects of the subject that do not depend sensitively on rate considerations.

References

Berry, R. S. (1972). "Recycling, Thermodynamics and Environmental Thrift," *Bulletin of the Atomic Scientists* (Science and Public Affairs), p. 8, May.

———, and Fels, M. F. (1972). "The Production and Consumption of Automobiles. An Energy Analysis of the Manufacture, Discard and Reuse of the Automobile and its Component Materials," report to the Illinois Institute for Environmental Quality, Chicago, July.

Franklin, W. E., and Hunt, R. G. (1972). *Environmental Impacts of Polystyrene Foam and Molded Pulp Meat Trays,* Final Report, MRI Project No. 3554-D, Midwest Research Institute, April.

Hannon, B. M. (1971). "System Energy and Recycling, A Study of the Soft Drink Industry," report to the Illinois Institute for Environmental Quality, Chicago, August.

——— (1972). *Environment, 14,* p. 11.

Hirst, E. (1972). "Energy Consumption for Transportation in the U.S.," ORNL-NSF-EP-15, Oak Ridge National Laboratory, Oak Ridge, Tenn., March.

Kneese, A. U.; Ayres, R. U.; and D'Arge, R. C. (1970). *Economics and the Environment, A Materials Balance Approach,* Resources for the Future, Inc., Washington, D.C.

Leach, G. (1972). "The Motor Car and Natural Resources," report to the Organization for Economic Cooperation and Development, Division of Urban Affairs/Environmental Directorate, U/CKO/72.800, Paris, October.

Smith, W. D. (1973). *New York Times,* p. 43, January 12.

Stanford Research Institute (1972). *Patterns of Energy Consumption in the United States,* report to the Office of Science and Technology, Executive Office of the President, Washington, D.C., January.

U.S. Bureau of the Census (1971a). *1967 Census of Manufactures,* U.S. Department of Commerce, Washington, D.C.

——— (1971b). *1967 Census of Mineral Industries,* U.S. Department of Commerce, Washington, D.C.

——— (1971c). *1967 Census of Transportation,* U.S. Department of Commerce, Washington, D.C.

38 Energy Conservation in Perspective of International Energy Requirements

J. D. ADAMS, R. L. FOLEY, AND R. L. NIELSEN*

The purpose of this paper is to take a broad look at world energy demand to the year 2000 and to describe in general terms the problems of supply. To many it is clear that supply of oil is the major constraint. It is felt that reasonable conservation steps could and should be taken to "buy" time for the development of alternative resources. Reductions in energy demand result in large decreases in capital requirements. Capital requirements for the "Greater Efficiencies" Scenario are discussed in Appendix 38.A.

38.1. The "Trends Continue" Scenario—Energy Consumption Approaches the Worldwide Oil Resource Base

Figure 38.1 is a picture of the world energy demand growth from 1950 to 2000 under a "trends continue" situation in which uses of energy and growth rates of GNP and population follow currently established patterns. Of course, this situation can only continue to the point of supply limitation consistent with end-use requirements (for example, coal and nuclear power for electricity, oil for transportation). As can be seen, industrial energy demand is projected to be the major energy consumer but energy losses in the conversion process for electricity will become almost one-third of the total.

Overlaid on Figure 38.1 is an indication of the amount of total energy that would be expected to be filled by oil in the absence of supply constraints. However, if one considers the amount of oil discovered to date and also that which has already been produced, the difficulties looming on the horizon are obvious. Figure 38.2 shows the relationship between current and projected use of oil compared to a frequently cited estimated ultimate recoverable quantity (EUR) of approximately 2000 billion barrels (B bbl) mentioned by the National Academy of Sciences in their book *Resources and Man* and attributable to W. P. Ryman (1969). The data result from a calculation of cumulative production plus 10-year reserve required to supply the demand projection for oil shown earlier. (It should be remembered that these EUR estimates are imprecise at best and our experience has been that independent estimates vary considerably.)

The major energy-consuming areas are now drawing much of their growth in supplies from overseas oil, as shown in Table 38.1. Thus, all

* All authors with Exxon Corporation, New York.

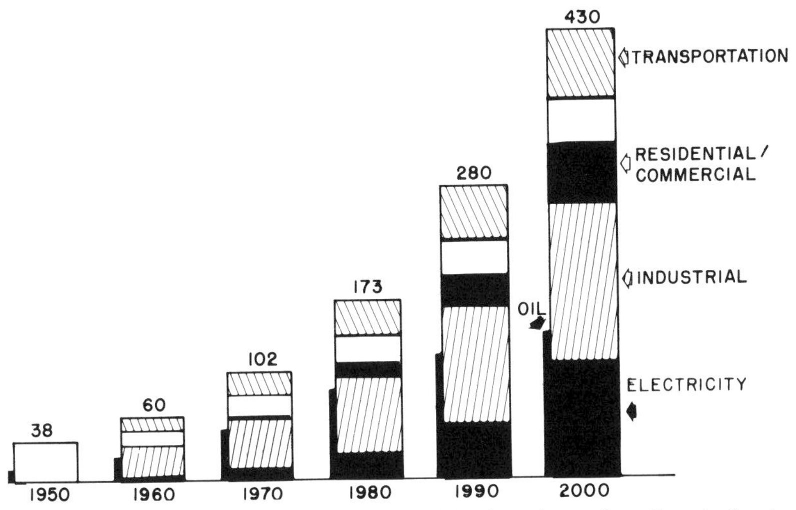

Figure 38.1. Total world energy demand (million barrels per day oil equivalent).

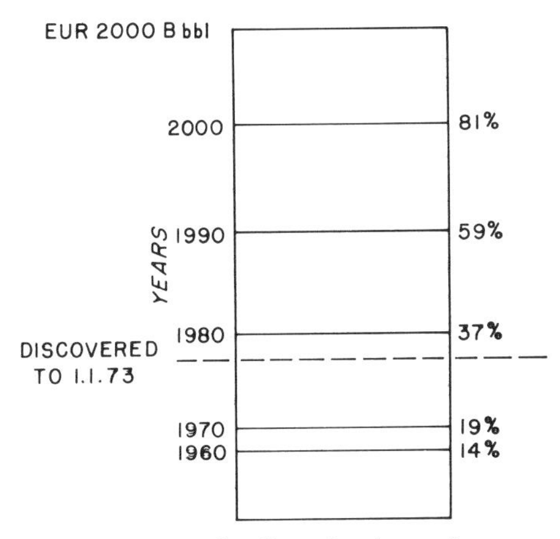

Figure 38.2. World oil production and resources.

Table 38.1 Growth in Imports of Overseas Oil as Percentage of 1970–1980 Energy Growth

United States	35–50
Western Europe	45–50
Japan	Approx 60

should have an interest in eliminating waste in the use of energy while maintaining economic growth and improving environmental quality.

38.2. Other Energy Resources Can Fill the Growth Needs—At Higher Costs and Given Enough Time

There are of course abundant energy resources on earth to last all mankind for a very long time to come. However, we can act only on the basis of reality. Time, money, and scientific breakthroughs are needed to unlock all these resources. There is about 500 million times as much energy in sea-water as there is conventional oil, but the feasibility of controlling the fusion reaction has not yet taken place. There is about 50 times as much oil in shale beds around the world as there is conventional oil, but most of this is in concentrations of only 10 gal/ton of shale (about 4%), so huge mountains of rubble would result from efforts to extract this (for every unit volume of shale lifted two volumes of sand or "spent" shale results).

Table 38.2 contrasts the current resource availability with this energy in place and it is fairly obvious why nuclear power, especially from breeders, is being encouraged in many quarters and why coal also holds great promise.

However, until technology develops so that coal can be mined and used under ecologically satisfactory conditions, it cannot play a significantly larger role in supplying the world's growing energy needs. Starting in the late 1970s, ecologically acceptable modes of coal use such as stack gas desulfurization and manufacture of synthetic methane are expected to be introduced. Another 10 to 15 years will be needed for such uses of coal to reach significant proportions in terms of worldwide energy demand. At present, nuclear reactors are essentially confined to electricity generation, although high-temperature reactors are being considered for producing usable process heat (*Nucleonics Week,* 1973). Breeders offer, perhaps, a hundredfold increase in the utilization of uranium but will not have any significant commercial impact in the next 15 years.

Table 38.2 Energy Resources of the World

	In-Place Trillion Barrels Oil Equivalent	Recoverable with Present Technology
Coal[a]	80	40
Uranium Oxide in Breeders	300	20[2]
Oil[b]	6	2
Gas[1]	2	1.7[3]
Oil Shale[b]	300	0.2
Tar Sands[c]	2	0.1

[a] Ryman (1969).
[b] U.S. Bureau of Mines (1970).
[c] National Petroleum Council (1971).

38.3. The "Greater Efficiencies" Scenario—Achieving Energy Savings While Maintaining Equivalent Functional Service

The problem to be solved is one of overcoming short-term dislocations in energy supply and adjusting to a changing structure with greater reliance on currently nonconventional forms of energy. Conservation measures, taken seriously but imaginatively, will help to diminish the problems arising out of this period of adjustment and can have sufficient impact so that rationing or other drastic solutions need not be resorted to.

In considering this problem, we have found it most useful to employ a systems approach, in which all the elements required to provide energy for a given consumer need are treated as a whole. With this approach, overall system efficiency, investment, and environmental impact can be assessed in a way that allows meaningful comparisons of alternatives. In addition, it provides a means for relating consumer energy requirements with changing technology, social attitudes, and ecological constraints. Thus, it is a useful tool for constructing alternatives to the "trends continue" scenario.

One alternative we have considered is the "greater efficiencies" scenario, in which we assume a high level of emphasis on conservation of energy resources while continuing to meet consumer needs, with the same level of economic growth as in the "trends continue" scenario. Emphasis on environmental conservation is also assumed to continue at its current high level.

Examples of the analysis of energy systems applicable to the "greater efficiencies" scenario are given in the following section, for selected major energy-consuming sectors in the United States.

38.4. Residential Energy System Conservation Potential

We have analyzed today's typical residential energy system. Electricity is produced in a conventional coal-fired plant and transmitted and distributed via overhead lines. Natural gas is transmitted and distributed by underground pipes. For an average home in the Northeast, annual delivered energy requirement in a single-family residence is 200 million Btu. This needs 257 million Btu of energy at the minehead and wellhead. The investment required (*ex* return) to produce, convert, and deliver the electricity and gas is $1440 in constant 1970 dollars. Atmospheric emissions of SO_x and NO_x would be 140 lb/yr and 40 lb/yr, SO_2 and NO_2 equivalent, respectively. Our studies of possible future systems indicate that the highest efficiency system for providing energy needs and meeting environmental standards would still be gas for heating, and electricity for lighting and appliances, both derived from coal when natural gas is no longer available. The energy requirement of the possible future house is only half that of today's, without reducing forecast use of convenience and recreation appliances and without reducing the comfort of the house. The added consumer investment could be less than 5% of average housing costs. System investment per unit of energy delivered is much higher, and therefore costs of energy per unit would have to be much higher, but the total cost, *ex* inflation, would increase less than 15%. Note that through use of advanced technology in coal gasification and combined-cycle power generation, system emissions would also be dramatically reduced to 0.5 lb SO_2 equivalent/year and 2 lb NO_2 equivalent (technically possible, not EPA requirements). Also, investment was provided for change to underground distribution of electricity, in line with the trend in consumer preferences.

Based on this analysis of what is possible in the future, we have estimated the potential savings in the U.S. residential energy consumption. Assuming that a little over half the pre-1970 homes, two-thirds of the units built in the period 1970 to 1985, and all the post 1985 units could have improved insulation and tighter construction by the year 2000, the weighted average annual per unit demand for energy for space heating would decline to about 90 million Btu from 130 million Btu for single-family dwellings.

In the commercial sector, space heating and cooling account for about 60% of energy consumption. It would appear that existing structures are not so straightforward to improve as private homes. However, on a new building, a chromium coated dual-wall insulating glass has been shown to be so effective that the heating system was 53% smaller and the air conditioning system was 65% smaller than it would have been.[4] In other words, savings of the same order of magnitude as in private dwellings can be obtained, at an economic cost, on new structures.

38.5. Transportation Energy System and Conservation Potential

Turning to personal transportation, our analysis gives some interesting insights. The present system considered is for a family in suburbia who might typically have two cars, one standard and one intermediate. The average annual mileage of the intermediate-sized family car is taken to be 10,000 miles at 13.2 mpg. The father daily commutes 25 miles each way to the office in his standard-sized car at 11.8 mpg. The annual energy demand is 234 million Btu requiring 262 million Btu of oil resource due to losses in producing, transporting, and refining the oil. If by 1990 a standard-sized car could be developed (using, say, a gas turbine or diesel) with a consumption of 20 mpg and father commuted in a bus to work, the energy requirement for the same annual mileage drops to 70 million Btu, or on an input basis 78 million Btu. This is a 70% "saving." We have *not* considered such savings possible for all families and cannot imagine that it would be feasible except under extreme authoritarian control. Incidentally, if the emissions regulations already enacted for future cars are met, the weight of pollutants would drop to 2.5% of the 1970 level for the same mileage.

In the United States specifically, we have considered gradual changes in the composition of the aircraft fleet, favoring jumbo jets; we have considered gradual replacement by the year 2000 of 25% of current car commuting by bus or train which would take a doubling of our estimate of 140,000 commercial buses in the United States by the year 2000; we have considered a gradual shift in the mix of larger versus small cars so that sports-compact or smaller types account for 50% of the car population compared to 30% today and, finally, a gradual improvement in efficiency of about 15% in each engine size (over 1970 levels) by 2000.

The potential for efficiency improvement in personal transportation is greatest in the United States but would also be significant in other industrialized nations of the world. How might such changes be quantified in an alternate scenario for the future from trends continue? Appendix 38.B shows such a calculation for the United States in the year 2000. Assuming that average adult annual mileage reaches a plateau of 8500 miles, 140,000 *extra* buses operating at 1970 levels of occupancy and efficiency would reduce the average annual adult mileage traveled by car to 7500. (There are about 90,000 buses now, and this number is expected to grow to 140,000 by the year 2000.) This change can be directly used in estimating the energy demand for the "greater efficiencies" scenarios.

38.6. Industrial Energy System Conservation

The industrial sector is significantly more heterogeneous than either the transportation, residential, or power generation sectors. The U.S. Census of Manufactures gives great detail on materials used in each Standard Industrial Classification (SIC) and energy consumed by fuel type. Further breakdown into energy use by function (process heat, mechanical drives, lighting, and others) would be useful for a systems approach. Some spotty data are available, but for this study more general comments on the subsectors of industry will have to suffice in order to give perspective while indicating potential for change and the relative impact this might have.

In 1969, six SIC groups accounted for two-thirds of the U.S. industrial consumption of energy, as shown in Table 38.3.

Growth rates of each subsector vary widely so, with improvements in technology which are taking place continually, one must caution those who would compare forecasts to ensure complete understanding of the "base case." For example, with the demise of open-hearth steelmaking and the concurrent growth of the Basic Oxygen Furnace and Electric Furnace, the

Table 38.3 U.S. Industrial Energy Demand by SIC Group, 1969

	Trillion Btu[a]	Growth Rate 1962–1967[b] % p.a.
20 Food	1,189	2.5
26 Paper	1,184	4.6
28 Chemicals	4,248	5.5
29 Petroleum Refining	2,994	4.1
32 Stone, Clay, Glass, and Concrete	1,099	3.0
33 Primary Metals	4,610	3.7
All Others	7,583	
Total	22,907	

[a] U.S. Bureau of Mines (1969).
[b] Rand Corporation (1971).

per ton energy requirement for steel production will decline from 26 million Btu to about 18 million Btu.

Of major interest to many people is the great potential of usefully using the waste heat from generating electricity.

Electricity is generated principally from fossil fuels and uranium, and 65 to 70% of the energy is dissipated to the elements (that is the thermal efficiency is 30 to 35%). A factory buying electricity also burns fossil fuels to produce steam. It would seem logical to seek ways to interface these two systems. The factory could generate its electricity on-site and use the waste heat directly, or the power plant could be situated near or in an industrial complex. Let us just take a look at the impact this might have on overall energy demand. Appendix 38.C shows a forecast of industrial electricity demand in the United States. If 50% of electricity demand growth were generated on-site and one-half of the waste heat from this were utilized to replace the steam from boilers, 3 million barrels per day (oil equivalent) of energy would be saved by the year 2000 in the United States alone. However, this leads us into an apparent dilemma. The electricity generated on-site would probably use scarce oil or gas because there are environmental constraints on the use of coal. This would only aggravate the shortage of oil and gas. The priority we have is conservation of oil, then gas, then electricity, and not the other way around. Advocates of "nuplexes"—a nuclear power plant situation near an industrial complex providing both electricity and process heat—appear to have, at least theoretically, a reasonable compromise.

38.7. Greater Efficiencies Should Be Sought to Assist in Continuing Economical Growth During Transition to Other Energy Resources

Table 38.4 summarizes the scope of reductions in energy demand, without rationing or other "quick fixes," which may be possible in the world.

The degree of conservation possible in countries other than the United States will not necessarily parallel the estimates discussed in this paper. In particular, the potential for savings in transportation in Japan is probably much smaller than in the United States, as illustrated in Table 38.5.

However, after allowing for such differences among countries, the potential for the "greater efficiencies" scenario is still quite significant, as shown in Figure 38.3. This compares the "trends continue" scenario shown earlier

Table 38.4 Energy Conservation Estimates (million barrels per day oil equivalent)

	1980	1990	2000
Transportation			
Increased average efficiency of engines	—	6.0	8.8
Increased use of public transport for commuting	1.0	1.4	2.5
Accelerated trend toward jumbo jets	0.3	1.0	1.5
Residential/Commercial			
Improved insulation, construction, and airtightness	2.7	5.3	9.1
Industrial			
Waste heat utilization	2.2	4.0	13.9
Recycle of steel and aluminum	0.8	1.2	1.8
Reuse of glass, incineration of paper	0.3	0.7	0.9
Power Generation			
Decreased average heat rate	—	4.5	15.1
Total	7.3	24.1	53.6

Table 38.5 Relative Use of Various Modes of Intercity Transport (percentage)

	United States[a]	Japan[b]
Private Cars	86.6	26.9
Buses	2.2	19.0
Trains	1.1	52.1
Inland Waterways	0.3	0.7
Aircraft	9.8	1.3
Total	100	100

[a] U.S. Bureau of the Census (1971).[5]
[b] Statistics Department of the Bank of Japan (1971).

with the aggregate "greater efficiencies" scenario. Cumulatively about 140 billion bbl of oil would be conserved by the year 2000.

The conservation measures described here could result in significant extra time for adapting to the necessary change in the world's sources of energy. They will require the positive actions of governments, industries, and consumers. We believe that the need for energy conservation will be recognized by consumers widely and soon, and that the resources needed to achieve significant effects will be brought into action by both government and industry.

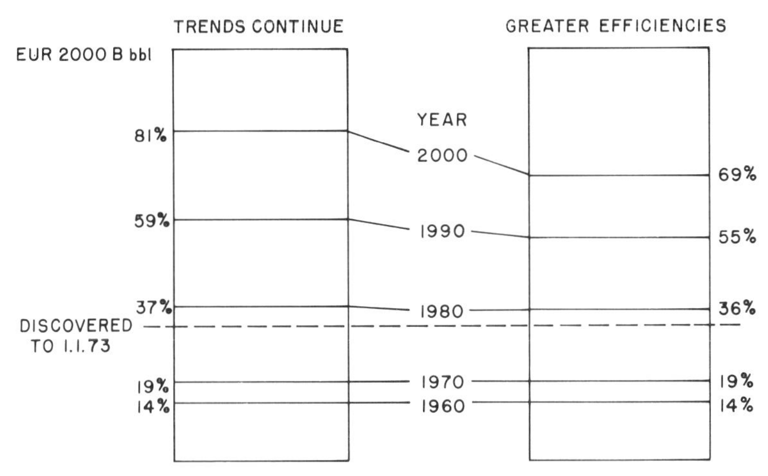

Figure 38.3. Oil production and resources.

Appendix 38.A

Capital Investments for Energy Supply of Large Developed Non-Communist Countries

This section summarizes our projections of capital investments for supplying the combined energy needs of the United States, Canada, Western Europe, and Japan over the three remaining decades of this century (Foley and Nielsen, 1973). These capital investments (constant 1970 dollars) are based on the "greater efficiencies" scenario described in previous sections of this paper. In the 1960s, capital investments to supply these developed countries with energy totaled $340 billion. This amount represented 2.0% of their cumulative GNP, 11% of their total domestic investments, and 24% of all capital invested for machinery and equipment during this decade.

The total energy demand in these developed countries is projected to grow at 4.3% per year, to a total of 224 million barrels per day by 2000. Fossil fuel supplies are projected to grow in absolute terms during the 1971 to 2000 period, but the shares of total energy demand to be supplied by oil, gas, and coal are each projected to decline. Table 38.A.1 shows the com-

Table 38.A.1 Primary Energy Supply for "Greater Efficiencies" Scenario in United States, Canada, Western Europe, and Japan

	Percent of Total Primary Energy Demand				
Primary Energy	1960	1970	1980	1990	2000
Oil	40	52	54	45	31
Gas	18	20	19	18	16
Coal	34	21	15	15	17
Hydro	8	7	5	4	3
Nuclear Fuel	—	[a]	7	17	31
Geothermal	—	—	[a]	1	2
	100	100	100	100	100
Total (million barrels/day)	38	63	102	158	224
% to Power Generation	21	25	28	36	43

[a] About 0.5%.

ponents of the total supplies by the various forms of primary energy. In 1970 oil supplied 52% of total demand as opposed to 31% by 2000. During this same interval, that fraction supplied by gas declines from 20% to 16%, and coal from 21% to 17%. Most of the decline in fossil fuels is supplanted by nuclear energy. Nuclear's share increases from essentially nil in 1970 to 31% by 2000.

An increasingly large portion of the total primary energy input is being consumed by electric power generation. In 1970, 16 million barrels per day input for power generation represented 25% of total energy demand. By 2000, the input to power generation consumes 43% of total energy input, as also shown in Table 38.A.1.

Capital investments by decade, major function, and energy form are given in Table 38.A.2. Cumulatively over the three decades, some 3.4 trillion dollars of investment capital are indicated to be needed to provide energy for the United States, Canada, Western Europe, and Japan. From the data shown in Table 38.A.2 the following features merit emphasis:

1. The growing share of the total devoted to electricity supply reflects the high capital intensity and faster-than-average growth for this energy form; growth in new investments for electricity essentially parallels demand growth at 6.1% per year.
2. There is a shifting in share away from oil, consistent with volumetric supply patterns as oil production approaches its eventual resource limit.
3. Although nuclear's volumetric supply share is projected to grow to 31%

Table 38.A.2 Capital Investments for Energy Supplies in United States, Canada, Western Europe, and Japan (Billions 1970 $)

	1960s	1970s	1980s	1990s
Raw Materials				
Oil	62	85	119	103
Gas	19	22	36	40
Coal	4	5	10	22
Oil shale/tar sands	—	4	20	31
Uranium	—	3	9	22
Total for Exploration and Extraction	85	119	194	218
Conversion				
Oil refining	22	59	62	55
Synthetic fuel manufacture	—	8	34	77
Nuclear fuel cycle	—	7	23	53
Electricity generation	84	164	377	630
Total for Processing	106	238	496	815
Distribution				
Oil (conventional and synthetic)	42	67	75	53
Gas (conventional and synthetic)	18	35	76	96
Coal	4	5	9	18
Electricity	85	141	271	465
Total for Transportation and Marketing	149	248	431	632
Total Supply System	340	605	1121	1665

by the year 2000, investments in nuclear fuel rise to only 13% of the primary energy total. This disparity simply reflects the low capital intensity, or conversely, high energy release per ton of uranium.

Capital investments shown in Table 38.A.2 represent our estimates of capital needed to build facilities that link the ultimate consumer to the resources. The general methodology of determining capital investments are

1. From demand for specific energy form obtain growth component for each 5-year period.
2. Obtain sustaining components for each 5-year period based on production decline or "wear-out" rate for facilities.
3. Sum of (1) plus (2) to yield to capacity that must be provided for.
4. Multiplication of unit investments by total capacity to obtain required capital investment.

Table 38.A.3 gives these energy capital investments in the perspective of the total economies of these developed non-Communist countries. When

Table 38.A.3 Capital Investments for Energy Supply as a % of GNP in United States, Canada, Western Europe, and Japan

	Percent of Cumulative GNP Each Decade			
Function	1960s	1970s	1980s	1990s
Raw Materials	0.87	0.88	0.94	0.86
Conversion	0.62	0.85	1.09	1.10
Distribution	0.50	0.43	0.42	0.29
Total	1.99	2.16	2.45	2.25

viewed thusly, energy investments are projected to remain at less than 2.5% of combined economies. Although this is a relatively small percentage, the question of capital availability to individual segments of these industries will continue to depend on their financial performance, investors' criteria, and governmental policies.

Appendix 38.B

Calculation for the Reduction in Miles/Adult Traveled by Car Due to Increased Number of Buses (Year 2000)

Assumed adult population = 188.4 million.
Desired annual mileage = 8500 (This is the assumed long-term saturation level).
Therefore, yearly passenger miles = 188.4 million × 8500 = 1601.4 billion miles.
We are interested in the reduction of these miles, probably in commuting due to increased use of public transport.
Say there will be 140,000 *more* buses in the year 2000 than forecast.
Average annual bus mileage = 30,000 miles.
Average occupancy = 44 persons.
Therefore, reduction in adult miles traveled by car would be
140,000 × 30,000 × 44 = 184.8 billion miles.
Therefore, average annual mileage *by car* per adult is

$$\frac{(1601.4 - 184.8)\text{ billion}}{188.4\text{ million}} \approx 7500.$$

If this additional number of buses were introduced by the year 2000 in the

United States, the saving in total mileage of private cars, used for commuting in the year 2000, would be approximately 25%.

Appendix 38.C

Waste Heat Recovery in the U.S. Industrial Sector (million barrels per day oil equivalent)

	1980	1990	2000
Electricity Purchased (1970 was 1.1)	2.1	4.4	9.3
Reduction in Electricity Purchased	(0.5)	(1.1)	(2.5)
Increased Fossil Fuel to Generate Electricity	1.5	3.3	5.5
Reduction in Fossil Fuel Required Due to Heat Recovery from Self-generated Electricity	(0.5)	(1.1)	(1.5)
Net Change in Industrial Fuel Requirement			
Fossil	1.0	2.2	4.0
Purchased Electricity	(0.5)	(1.1)	(2.5)
	0.5	1.1	1.5
Saving on Input Energy in Power Generation Sector	1.4	3.2	7.2
Overall Energy Saving Due to Waste Heat Recovery	0.4	1.0	3.2

( ) = Negative

Notes

1. See Ryman (1969, p. 194). $12{,}000 \times 10^{12}$ cu ft $\times$ 1035 Btu/cu ft $\div$ 5.85×10^{6} Btu/bbl.

2. "Nuclear Energy Its Growth and Impact." 11th Annual International Conference of the Canadian Nuclear Associates, June 20–23, 1971. Paper 71-CNA-301. This gives cumulative U_3O_8 demand to 1990 for 1945 thousand tons. U.S. Bureau of Mines (1970) gives total resource at <$30 of 4887 thousand tons. Shortt (1971) gives 475×10^{9} Btu energy per ton of normal U_3O_8. Breeders give 50 times that energy (*Technology Review*, 1971 October).

3. Eighty-five percent recovery.

4. The Toledo Edison Building. Architects Sawborn, Steketee, Otis, and Evans.

5. Volume of domestic intercity-passenger traffic, see U.S. Bureau of the Census (1971).

References

Foley, R. L., and Nielsen, R. L. (1973). "Capital Requirements for Energy Supply of Developed Non-Communist Countries," Exxon Corporation, New York, February.

National Petroleum Council (1971). *U.S. Energy Outlook 1971–85,* Vol. II, N.P.C., Washington, D.C., March.

Nucleonics Week (1973). p. 4, January 18.

Oil and Gas Journal (1972). December 25.

Rand Corporation (1971). "Interim Report: The Growing Demand for Energy," Rand Corporation, Santa Monica, Calif., p. 44, April.

Ryman, W. P. (1969). *Resources and Man,* National Academy of Sciences, W. H. Freeman & Co., San Francisco, p. 194.

Shortt (1971). AMR Energy Crisis Seminar, New York, November 10.

Statistics Department of Bank of Japan (1971). "Passenger Traffic," in *Japan Statistical Yearbook,* 20th Edition, International Publications Service, New York, p. 267.

U.S. Bureau of Mines (1969). *1969 Minerals Yearbook,* U.S. Department of the Interior, Table 10, p. 26.

——— (1970). *Mineral Facts and Problems,* Bulletin 650, U.S. Department of the Interior, Washington, D.C., p. 190.

U.S. Bureau of the Census (1971). *1971 U.S. Statistical Abstract,* Department of Commerce, Washington, D.C., Table 832, p. 525.

Appendix

A-1

Energy Economics

F. M. O'CARROLL, Queen Mary College, University of London

A Model of World Energy, representing resources and constraints, has to be supplemented by information on the behavioral dynamics of the system if it is to be used for planning in mixed economies. This means that the link between supply and demand via prices has to be quantitatively understood. Historical series of product prices are being studied for this purpose in conjunction with data on volume of demand and supply costs. In so far as the regulating mechanism is based on free competition, prices will tend towards marginal values as determined by the LP optimization of the system. Distributed lag relationships based on this hypothesis are being sought. An immediate obstacle is the lack of coherent series of market prices. The application of the profit method to market data is being developed in an attempt to remedy this. An alternative approach, using profits rather than prices, is also being pursued.

A-2

A Linear Programming Model of the Nation's Energy System*

KENNETH C. HOFFMAN, Brookhaven National Laboratory

A linear programming model of the U.S. energy system was developed for the purpose of technology assessment and planning. The model focuses on interfuel substitutability and the mix of resources and technologies that may be used to satisfy energy demands. A variety of electric and non-electric supply categories are included along with specific energy intensive end-use activities. The solutions indicate the optimal supply-demand configuration of the energy system. The load-duration structure of electrical demands may be expressed and environmental effects are included as constraints. The optimization may be performed with respect to cost or some combination of cost and effects. The model is applied to an investigation of alternate technologies in the year 2000. The quantities of major air pollutants and waste heat produced by the energy system are considered as environmental effects and the impact of several new utilizing technologies is determined in terms of resource consumption, environmental effects, and cost. Further applications and extensions of the model are discussed.

* Editor's note: See *Energy Modeling—Art, Science, Practice,* Milton F. Searl, Editor, Resources for the Future, Inc., 1755 Massachusetts Avenue N.W., Washington, D.C. 20036, March 1973.

A-3
The Problem of Development Decisions for an Advanced Energy Technology: An Illustration in MHD Power Generation

DAVID A. OLIVER, M.I.T.

A new idea for the generation of power holds forth the *promise* of improved efficiency of energy conversion with reduced cost and minimal detrimental environmental impact. The promise exists in the face of *uncertainty* because the ultimate performance of the system embodying the idea rests on the development of new materials, processes, and physical conditions which are as yet unproved. The uncertainty is accompanied by *risk* since the development of the idea may require a heavy cost both in direct funding and in the diversion of resources from the development of alternative concepts. In this paper a systematic way of dealing with the three elements of promise, uncertainty, and risk in decision making for the development of a new energy conversion concept is presented. The decision analysis is then illustrated in detail for the case of open cycle MHD power generation.

A-4
The Assessment of Research and Development Options in Energy

KENNETH C. HOFFMAN, PHILIP F. PALMEDO, and V. L. SAILOR, Associated Universities, Inc.

An important decision area in the development of national energy policy is the determination of research and development priorities. If unified analysis rather than fragmented decision is to guide the choice of research priorities, that analysis can best be carried out as part of a hierarchical assessment process. "Below" the level of the comparative assessment of all competing potential technologies there must be carefully structured evaluations of each technology. Likewise, "above" the assessment of technologies there exists (at least in principal) a level of goal formulation and overall policy evaluation which imposes definite requirements on the output from the technology assessment. The assessment of technologies inevitably involves subjective judgment but can be structured by means of a reference energy system–perturbation approach. The input data required in such an approach—cost data, energy demand projections, and environmental effects—have quite individualistic requirements. Relative to other uses, for example, high accuracy is not required in demand projections whereas a high degree of disaggregation is desirable.

A-5

Modeling of Electric Power Demand Growth

JAMES B. WOODARD, MARTIN L. BAUGHMAN, and FRED C. SCHWEPPE, M.I.T.

This paper describes a modeling approach, presently under development, directed at the growth in demand for electric power. The emphasis of this effort is to develop a model which can be used for analysis of detailed questions such as: How will changes in air-conditioning power demand, electric rate structures, population, and so on, affect the daily load shapes (kW vs. time) as well as the overall electric energy consumption (kWh)? Detailed answers to these questions are needed for generation planning of capacity and plant mix (nuclear, fossil, pumped-hydro) as well as for the evaluation of resulting environmental and economic impacts. These issues require detailed models combining economic models with the engineering considerations affecting the dynamics of load behavior on a daily, weekly, and seasonal basis. Modeling of this kind can be limited by the data availability, and an important aspect of this effort is to identify the data required to increase our understanding of load behavior and growth. The approach is to combine state space dynamic structures driven by stochastic processes with econometric models into a hierarchical description of the stochastic load behavior. System identification techniques will be used to determine the parameters.

A-6

Determination of the Total Energy Costs of Rapid Transit Systems

TIMOTHY J. HEALY, University of Santa Clara

The total energy costs of a rapid transit system include both direct costs, for propulsion of vehicles, and indirect costs, for construction, operation, and maintenance of the system. A general approach to obtaining these energy costs for any rapid transit system is presented, with some preliminary results obtained for the Bay Area Rapid Transit (BART) system. For many systems direct propulsion energy is in the form of electric energy. It is obtained from power consumption data available from the system operator or the local electric utility. Energy in this form must be converted back into energy into the power generation plant if it is to be compared in a useful way with alternative nonelectric transportation modes. Similar approaches are used to obtain operation and maintenance costs. Construction costs are obtained either by vertical analysis or by the use of energy input/output matrices. Both approaches are discussed with emphasis on the matrix approach. Practical problems associated with this approach are considered. These include the effect of regional and yearly cost differences, the accuracy of the matrix, and the accuracy of the system cost data.

A-7

Energy Use Patterns and Conservation Potential for Transportation

ERIC HIRST, Oak Ridge National Laboratory

Historical, present, and possible future energy use patterns for transportation in the United States are examined. Between 1950 and 1970, annual energy consumption for transportation increased 89% to 16,500 trillion Btu, a per capita growth of 40%. During this period, energy use for intercity freight fell slightly in spite of an increase in freight traffic. This reflects the large increase in railroad energy efficiency. However, energy consumption for passenger transport grew more rapidly than did passenger traffic levels. This reflects the shift to less energy-efficient modes and a decline in efficiency for most modes. Various strategies for increasing energy efficiency of transportation are briefly discussed. These include: shifts from energy intensive transport modes (airplanes, automobiles, trucks) to more efficient modes (mass transit, trains, buses); increased utilization of existing equipment; reduced speeds; and use of advanced transport technologies. Finally, possible ways to implement these energy conservation schemes are described.

A-8

Oil Imports, the Wellhead Price of Natural Gas, National Energy Policy, and Joint Costs in Oil and Gas Exploration*

R. M. SPANN and EDWARD W. ERICKSON, Virginia Polytechnic Institute and North Carolina State University

A significant component of the energy crisis is a regulation induced short-

* Editor's note: See *Energy Modeling—Art, Science, Practice,* Milton F. Searl, Editor, Resources for the Future, Inc., 1755 Massachusetts Avenue N.W., Washington, D.C. 20036, March 1973.

age of natural gas. U.S. oil producing capacity is becoming a progressively smaller fraction of U.S. oil demand. Oil and gas exploration are carried on under conditions of joint cost. It has been argued (see the minority report of the Cabinet Task Force on Oil Import Control) that new natural gas discoveries are sensitive to real wellhead prices of crude oil in such a way that liberalizing oil import policies to limit increases (or cause reductions) in the real price of domestic crude oil will have a serious adverse effect on new natural gas discoveries and aggravate the long-run natural gas shortage. This view presupposes a positive cross-elasticity of supply between oil and gas discoveries with scale effects dominating substitution effects and supply effects in turn dominating any effects from positive cross-demand relations. Constraints developed from joint cost theory are used in an econometric test of the cross-supply relations. The results are then used to discuss the relation between deregulation of the wellhead price of natural gas and oil imports in the context of a national energy policy.

A-9
Residential Demand for Natural Gas

WADE P. SEWELL, Federal Power Commission

Market data on residential consumption of gas reflect different demand characteristics by connected consumers than by those demanding connection, different demands for the various services households derive from the use of gas, and different stages of market penetration in various parts of the country. The econometric study presented attempts to encompass these considerations, as well as the more usual ones, and to assess their effect on the usefulness of predictions.

A-10
The Potential Impact of North Slope Gas on the Natural Gas Industry of North America

LEONARD WAVERMAN, University of Toronto

Econometric models measure aggregate demand and supply elasticities. Little attention, however, is placed on the transportation element of the final cost of energy to the consumer. For natural gas, transportation costs vary widely across North America. At city gates, transportation costs range from 15 to 25% of the price in markets near fields to 40 to 60% in markets far away from fields. The proposed shipments of North Slope gas to eastern markets would involve transportation costs of some 60 to 75% of city gate prices. The modeling of the geographic nature of the natural gas industry from locations of fields to locations of markets is important. Any policy set by the government will have differing impacts on fields and markets. For example, a uniform tariff will lead to differing degrees of protection.

This paper has two major aims. The first is to simulate in a simple way the actual gas network in North America for 1971. The simulation involves a static single year linear programming model. The other is to answer questions such as: What impact will North Slope gas have on consumers and producers?

Given the large cost differential between northern gas and supplies from existing wells, what prices are expected to prevail in fields and markets when these new supplies are shipped?

A-11

A Dynamic Programming Approach to Estimating Household Demand for Electricity

MARJORIE B. MC ELROY, Duke University

Demand projections are typically made on the dubious assumption that some parameters, say price and income elasticities, are constant over the relevant range of income and prices. As the range of income and prices expands, this is an increasingly untenable assumption. A technique that avoids this severe restriction and at the same time is more informative than results obtained by stratification of data into income and price classes is "Curve Fitting by Segmented Straight Lines." * This is a dynamic programming technique where the desired function is approximated by a continuous sequence of straight lines. Information from the entire range of income and prices is used to fit each segment. This paper discusses this technique as applied to the household demand for electricity.

A-12

An Analysis of World Energy Supplies

H. R. LINDEN and J. D. PARENT, Institute of Gas Technology

The rapid increase in demand for energy has created great concern for the

* R. Bellman and R. Roth, *Journal of the American Statistical Association*, Vol. 64, 1969.

adequacy of long-range supply, not only in the United States, but in the world as a whole. It is the purpose of this work to examine the worldwide supply of energy in the form of fossil and nuclear fuels combined. Estimates have been made of future rates of production, maximum future rate of production, and the time at which the maximum can be expected. Although desirable to examine the effect of many diverse variables, it was decided to apply the simple empirical model developed by Elliott and Linden which successfully accounts for the effects of the various technological, economic, and geological factors which have affected United States natural gas and crude oil supplies in the last quarter century. It was not known a priori if this model could be adapted to worldwide energy supplies. However, projections for the future obtained with this model in the present work seem reasonable.

A-13

Electrical Energy Demand Projection Methodology*

W. E. MOOZ, The Rand Corporation

A summary of the electrical energy demand projection methodology is produced for the state of California. The methodology combines both traditional methods of projecting demands and more detailed methodologies which allow greater analysis by policy makers. Features previously unavailable to the state include the ability to examine the industrial sector by size, composition, and the electrical energy intensiveness of individual manufacturing and mining industries, the ability to consider explicitly changing energy prices, and a somewhat more detailed method of examining the commercial sector. The projection methodology was developed for consistent economic and demographic inputs on a statewide basis, while preserving the effects of regional and climatic differences. The methodology is intended for use in making projections of future demands for electrical energy, for analyzing these demands as to their origins and growth rates, and for testing the effects of alternative futures upon the demands for electrical energy.

* Editor's note: See *Energy Modeling—Art, Science, Practice,* Milton F. Searl, Editor, Resources for the Future, Inc., 1755 Massachusetts Avenue N.W., Washington, D.C. 20036, March 1973.

A-14

An Assessment of Solar Energy as a National Energy Resource*

FREDERICK H. MORSE, National Science Foundation

In response to the President's 1971 energy message to the U.S. Congress, the Solar Energy Panel, along with ten others, was established within the Committee for Energy R&D Goals, under the Federal Council of Science & Technology in the Office of the White House. The Panel was charged with assessing the potential of solar energy as a national energy resource and the state of the technology in the various solar energy application areas, and with recommending necessary research and development programs to develop the potential in those areas considered important. Three areas evolved where solar energy could supply significant amounts of the nation's future energy needs: (1) energy for heating and cooling of buildings, (2) the production of fuels, and (3) the generation of electrical power. It was concluded that, with adequate R&D support over the next 30 years, solar energy could provide at least 35% of the heating and cooling of future buildings, greater than 30% of the methane and hydrogen needed in the United States for gaseous fuels, and greater than 20% of the electrical power needs of the United States. All of this could be done with a minimal effect on the environment and a substantial savings of nonrenewable fuels.

A-15

Direct Conversion of Solar Energy, on Earth, Now

J. A. ECKERT, R. W. WILLIS, and E. BERMAN, Exxon Enterprises Inc.

Photovoltaic energy systems are discussed which compete economically with other power sources for terrestrial applications. A silicon solar cell module specifically designed to meet the environmental and cost requirements for various earth uses is described. Systems design considerations including weather and seasonal effects on insolation, optimum sizing of converter and energy storage, and power conditioning are examined for specific solar-powered equipment. The results of field tests are also discussed.

* Editor's note: See *Solar Energy as a National Resource NSF/NASA* Solar Energy Panel, December 1972. Document available from Department of Mechanical Engineering, University of Maryland, College Park, Md. 20742.

A-16

Schottky Barriers for Terrestrial Solar Energy Conversion*

WAYNE A. ANDERSON and A. E. DELAHOY, Rutgers University

Natural energy sources (sun, wind, tide) are new areas of interest for solving the growing demand for energy and the depletion of conventional sources. The sun is most abundant and may be used as a local source (household, industrial) or a regional source (city, county) of energy. Fifty percent efficient solar cells on a 40 ft × 40 ft roof would supply 85,000 watts, assuming 100 mW/cm^2 sunlight. Conventional p-n silicon solar cells have a 22% theoretical maximum efficiency with a 14% actual efficiency. Schottky barrier diodes have recently been fabricated by evaporation of elemental metals and sputtering of alloys on p-type silicon. Current devices exhibit an efficiency close to 6% which would approach 12% with proper antireflection coatings. Computer calculations show that the use of new thin-film design techniques would theoretically boost the efficiency to beyond the 22% limit for silicon p-n devices. This would require suitable antireflection coatings on alloy barrier metals using thin-film semiconductors. These techniques could lead to a flexible and inexpensive solar cell for terrestrial solar energy conversion.

A-17

Technical and Economic Factors in the Implementation of Solar Water Heaters

E. DAVIS, R. CAPUTO, and G. SPIVAK, Jet Propulsion Laboratory, Pasadena, California

The Environmental Quality Laboratory (EQL) of Caltech has embarked on a program to facilitate the introduction of solar water heaters into multiple units being built in Southern California. Working in partnership with energy utilities and Builder/Developer and HVAC equipment manufacturing industries, the EQL is exploring scenarios for establishing beachheads in the U.S. energy marketplace for solar-related technology for water heating and space conditioning. A methodical process of partnership development—business potential evaluation—implementation-oriented engineering development—and prototype implementation leads to an "adequate model" which allows the established institutions to decide to proceed with beachhead implementation on a significant scale. Successful beachhead

* This work is supported by the National Science Foundation, Grant GI-32726.

implementation will validate the use of solar energy for those institutions outside the partnership and diffusion of the technology can proceed. In this paper we will discuss the impact of both user requirements and various implementation scenarios.

A-18

A Solar House System Providing Supplemental Energy for Consumers and Peak Shaving with Power-on-Demand Capability for Utilities

K. W. BÖER, University of Delaware

A solar harvesting system is described using flat-plate solar collectors covered with CdS/Cu_2S solar cells for combined thermal and electrical solar conversion. A power processing system including approximately one-day storage is outlined for space comfort conditioning and electricity supply. Solar energy will be used supplementing conventional means in conjunction with electricity supply from utilities. An economic analysis of a typical single-family dwelling is given indicating feasibility to supply electricity and heat for prices compatible to estimated values in the late 1970s. Means are described to use part of the storage as standby for peak power shaving with power release on demand through power utilities. Such solar harvesting system incorporated into an experimental solar house is currently in construction in Delaware with first test results of some of the units (CdS/Cu_2S solar panels and heat storage/heat exchanges) to be reported.

A-19

The Potential Impact of Solar Energy on the Energy Household of the United States of America

MARTIN WOLF, University of Pennsylvania

Technological development is in progress for the utilization of solar energy, to provide a complement to other energy resources in the future. The performance and present status, problems to be solved, and expected achievements for, and cost in the various approaches will be reviewed. These include: heating and cooling of buildings, obtaining storable chemical fuels through photosynthesis, generation of electric power via thermal or photovoltaic conversion in distributed or central station systems, including satellite in geostationary orbit, and indirect utilization via wind power or ocean thermal gradients. The expected times of commercial readiness, introduction rates, and energy delivery impacts will be discussed, as will be, time permitting, other economic impacts, land use, and environmental effects.

A-20

Solar-to-Thermal Energy Concepts Applied to Large-Scale Electrical Power Generating Systems

P. B. BOS, R. A. FARRAN, H. J. KILLIAN, J. R. SMITH, J. B. SCHROEDER, and W. H. WETMORE, Aerospace Corporation

This paper will present the results of systems analyses* regarding design and development of large scale ($\sim$1000 MWe) electrical power generation plants, using solar energy. The paper will first discuss the rationale for the United States' interest in using solar energy for terrestrial power plants. This discussion will be followed by a definition of the required system characteristics, leading to an overall systems analysis. Emphasis will be placed upon the various subsystems, that is, solar collectors, thermal transport fluids, thermal storage, heat exchangers (unique to solar systems) and turbogeneration units and waste heat management. The system economics, both investment and operating costs, will be discussed. The final portion of this systems analysis will discuss the major steps necessary for successful demonstration and implementation.

A-21

Solar Sea Power

C. ZENER and A. LAVI, Carnegie-Mellon University

Scientists and engineers commonly assume that nuclear power plants will furnish the dominant part of our energy by the end of this century. To avoid thermal pollution they visualize enormous offshore plants. To solve energy transportation costs, and to solve the problem of energy storage, they further visualize that the electric power will be used on site to electrolyze water into hydrogen and oxygen, thereby establishing the *hydrogen economy*. Because of the rising costs of nuclear power plants, we recognize the high probability that hydrogen may be supplied to the coastal regions of the United States more economically from solar sea power plants located in the tropics. Accordingly, our research is specifically directed toward the

* The study reported in this paper would cover work now being performed by Aerospace Corporation under contract from the National Science Foundation (NSF-C716).

energy system in which solar energy is absorbed by the surface of the tropical oceans, converted to electric power by solar sea power plants, then converted by electrolysis into chemical energy, transported by ship to the United States, and then distributed to heat our homes, power our transportation, and form a basic ingredient in materials processing.

A-22

Oil and Gas: A Case Study of Institutional Irrationality

C. CICCHETTI and O. S. GOLDSMITH, The University of Wisconsin

Crude oil and natural gas are two commodities that are usually jointly discovered (note, natural gas may be found without crude oil but not vice versa) and often jointly produced. In many uses they or their direct products are close substitutes in consumption. Economists have filled many scholarly treatises describing how the prices and quantities of such jointly produced and close substitute goods should be related in a smoothly functioning competitive world. No such world exists for oil and gas. A combination of federal and state legislation, taxes, and regulation makes the markets for these goods far from competitive norms. Furthermore, there is an asymmetry in the incidence of these market interventions by the public sector on each of these private goods. The most notable form this takes is the price difference between equivalent energy units of oil and gas. In this paper we will untangle some of these market interventions.

A-23

Institutional Capacity to Implement Energy Conservation Proposals*

EDWARD BERLIN, Berlin, Roisman and Kessler

It is appropriate that we intensify the dialogue addressed to the energy demand crisis and that we strive for innovations directed at dampening consumption. In so doing, however, a major focal point must be the institutional process that now governs decision making in the energy sector which itself may exacerbate growth and lack the capability to respond to new direction. For example, is it possible, given present regulatory constraints, to internalize the externalities of electric production; can regulatory commissions, consistent with their industry promotional responsibilities, implement inverted rate structures assuming that demand is price elastic; what less

* Editor's note: This has been one of the four invited papers. We regret that the author did not send his manuscript for inclusion in the proceedings.

promotional options can be implemented in place of rate base regulation? On a broader basis, is the present institutional fragmentation on the federal level and the bifurcation of federal-state responsibilities, conducive to the realization of conservation objectives?

A-24

Establishing Regulations for Allocating Energy: Sociological Considerations

SAMUEL Z. KLAUSNER, Center for Energy Management and Power, University of Pennsylvania

A rule governing rationing specifies the locus of the authority to allocate and criteria for allocative decisions. Criteria establish consumption priorities among persons (including corporate persons) and among functions. Allocation rests on how strategic an activity is for the social system and how strategic a particular form of energy is for the conduct of that activity. For establishing priorities, social activities may be divided into four classes: activities relating a population to its resources, activities relating a community to other communities, activities concerned with the establishment of social order, and activities concerned with maintaining the values which direct a society and the motivation of its members to fulfill its roles. The paper concludes by comparing political, economic, administrative, and judicial processes for setting priorities. Given our current society, priorities are most likely to be set through political processes with legislative mechanisms at the forefront and the technical system lurking in the background.

A-25

Natural Gas Stimulation by Underground Nuclear Explosion

C. STERN and E. VERDIECK, University of Connecticut*

The Bureau of Mines estimates that about 317 trillion cubic feet of natural gas are present in the Rocky Mountain area but, because of the low permeability of the geologic formations, cannot be recovered economically by present techniques. The AEC has experimented with nuclear explosives to fracture these tight formations and has estimated that the flow rates could be increased to 10 to 30 times that of a conventionally drilled well. However, there are major disadvantages associated with nuclear stimulation—

* Assistant Professor of Environmental Economics and Research Associate, respectively. Research supported in part by EPA.

principally, radioactivity. The contaminants of prime concern are tritium, present as HTO, HT, and CH_3T, and krypton 85. Because most of the radioactivity is contained in the first year's gas production the AEC has proposed either (a) to dilute the radioactive gas with uncontaminated supplies or (b) to dedicate the first year's gas production from each well to electric generating stations with high stacks. The paper discusses the potential for the nuclear stimulation technology in the light of the above problems.

A-26

Energy Requirements for Future Transportation

RICHARD D. THORNTON, M.I.T.

Today most of the energy required for moving people is consumed by internal combustion powered highway vehicles and jet powered aircraft. However, important research is now directed at electrically propelled alternatives. For low-speed personalized travel, both the electric car and electrically propelled dual mode systems are possible, and for high-speed ground transportation the magnetically suspended and propelled vehicle could be competitive with aircraft. This paper examines the probable changes in energy requirements that these developments would produce. It is concluded that not only do these new systems make it possible to use alternate energy sources, but they make possible reduction in energy consumption per passenger-kilometer of travel.

A-27

Atmospheric Pollution

J. R. ISAAC, Queen Mary College, University of London

The main themes are sulfur in fuel oil and TEL in motor gasoline. The full cost implications of desulfurization and of reductions in TEL dosages, leading to complete lead elimination, will be determined so as to indicate how they will affect the U.K.'s national fuel bill and the balance of payments. Economic repercussions in other countries and the effects on competitive conditions in the general energy market will also be examined.

A-28

Electric Energy Requirements for Environmental Protection

ERIC HIRST, Oak Ridge National Laboratory, and TIMOTHY HEALY, University of Santa Clara

The amount of electricity needed for (or saved by) operation of several environmental quality strategies is examined. These strategies include: electric mass transit, waste water treatment, solid waste management, air quality control, waste heat dissipation, and energy conservation. Energy requirements of existing electric mass transit systems are compared with the new BART system, buses, and autos. Electric energy costs, as a function of plant size, are examined for primary/secondary sewage plants. Electricity costs and savings are computed for solid waste disposal, recycle, and use as fuel. Electricity needs for air pollution control at stationary sources and from motor vehicles are evaluated. Electricity needs for use of cooling towers at power plants are reviewed. Finally, potential energy savings which reduce air and thermal pollution levels are examined. The electricity required in 1980 to meet the needs discussed here—based on the assumptions in this study—are small relative to total projected kWh consumption.

Index